When Cooperation Fails

When Cooperation Fails

The International Law and Politics of Genetically Modified Foods

Mark A. Pollack
Gregory C. Shaffer

OXFORD

UNIVERSITY PRESS

OXFORD
UNIVERSITY PRESS

Great Clarendon Street, Oxford OX2 6DP

Oxford University Press is a department of the University of Oxford.
It furthers the University's objective of excellence in research, scholarship,
and education by publishing worldwide in

Oxford New York

Auckland Cape Town Dar es Salaam Hong Kong Karachi
Kuala Lumpur Madrid Melbourne Mexico City Nairobi
New Delhi Shanghai Taipei Toronto

With offices in

Argentina Austria Brazil Chile Czech Republic France Greece
Guatemala Hungary Italy Japan Poland Portugal Singapore
South Korea Switzerland Thailand Turkey Ukraine Vietnam

Oxford is a registered trade mark of Oxford University Press
in the UK and in certain other countries

Published in the United States
by Oxford University Press Inc., New York

© Mark A. Pollack and Gregory C. Shaffer 2009

The moral rights of the authors have been asserted
Database right Oxford University Press (marker)

First edition published 2009

British Library Cataloguing in Publication Data

Data available

Library of Congress Cataloging in Publication Data

Pollack, Mark A., 1966
    When cooperation fails: the international law and politics of genetically modified foods /
Mark A. Pollack and Gregory C. Shaffer. —1st ed.

    p. cm

    Includes bibliographical references and index.
    ISBN 978-0-19-923728-9 (hardaback)—ISBN 978-0-19-956705-8 (pbk.) 1.Genetically modified foods—
Law and legislation 2. Genetically modified foods—Law and Legislation—European Union countries.
3. Genetically modified foods—lwa and legislation—United states. I. Shaffer, Gregory c.. 1958-II. Title.

    K3927. P65 2009
    344.04'232—dc22

Includes bibliographical references
Typeset by SPI Publisher Services, Pondicherry, India
Printed in Great Britain
on acid-free paper by Clays Ltd., St Ives plc.

ISBN 978-0-19-923728-9 (Hbk.)
      978-0-19-956705-8 (Pbk.)

For Rita, Cameron and Fiona
Mark Pollack

For Michele, Brook and Sage
Gregory Shaffer

Contents

List of Tables

Acronyms

APHIS	Animal and Plant Health Inspection Service, US Department of Agriculture
BIO	Biotechnology Industry Association
BRS	Biotechnology Regulatory Service (of APHIS)
CBD	Convention on Biodiversity
Coreper	Committee of Permanent Representatives (EU)
DNA	deoxyribonucleic acid
EC	European Community
EFSA	European Food Safety Authority (EU)
EIS	environmental impact statement
EPA	Environmental Protection Agency (US)
EU	European Union
FAO	Food and Agricultural Organization (UN)
FDA	Food and Drug Administration (US)
FDCA	Food, Drug and Cosmetics Act (US)
FIFRA	Federal Insecticide, Fungicide and Rodenticide Act
GATT	General Agreement on Tariffs and Trade
GM	genetically modified
GMO	genetically modified organism
GRAS	generally recognized as safe
IPPC	International Plant Protection Commission
LMO	living modified organism
NIH	National Institutes of Health
NTA	New Transatlantic Agenda
OECD	Organization for Economic Cooperation and Development
OIE	International Office of Epizootics (Office International des Epizooties), now also called World Organization for Animal Health
OLF	other legitimate factors
OSTP	Office of Science and Technology Policy (US)
PIP	Plant-Incorporated Protectant
r-BST	recombinant bovine somatotropin

Acronyms

rDNA	recombinant DNA
SPS	Sanitary and Phytosanitary Agreement (WTO)
TABD	Transatlantic Business Dialogue
TACD	Transatlantic Consumer Dialogue
TBT	Technical Barriers to Trade Agreement (WTO)
TEP	Transatlantic Economic Partnership
USDA	United States Department of Agriculture
UN	United Nations
US	United States
WHO	World Health Organization
WTO	World Trade Organization

Acknowledgements

The transatlantic dispute over the regulation of genetically modified (GM) foods and crops has been troubling us almost as long as it has troubled the United States and the European Union. Back in 1999, we—a lawyer (Shaffer) and a political scientist (Pollack) at the University of Wisconsin-Madison—began a joint project on transatlantic governance, which we saw as an ambitious, if flawed, effort to facilitate and oversee the operation of a transatlantic market-place in the absence of formal transatlantic institutions. More concretely, we saw the United States and the European Union building increasingly close links and engaging in joint governance at three levels: the intergovernmental level of high-level contacts between Washington and Brussels; the transgov-ernmental level of direct agency-to-agency links among lower-level officials in day-to-day domestic regulation; and the transnational level of direct civil-society cooperation among American and European businesses, labor unions, environmentalists, and consumer advocates. We were impressed, back then, by the depth of day-to-day cooperation in areas like competition policy, yet we also recognized that some areas were proving more resistant to joint, techno-cratic governance, and we undertook a preliminary case study of GMO regu-lation to understand why some issue-areas seemed more difficult to govern jointly than others. Put simply, we knew back then that GMOs posed a tough issue for transatlantic cooperation, one in which the US and EU regulators were protective of their respective regulations and regulatory frameworks, and in which interest groups and civil society could be easily mobilized to set limits on any cooperative efforts upon which regulators might agree.

Almost a decade later, we see the transatlantic dispute over GMO regulation not simply as a tough nut to crack, but as a remarkable case study of the failure of cooperation. The United States and the European Union, two long-time democratic allies, have continued and in some cases deepened their economic integration and efforts at regulatory cooperation during the past decade across a range of issue-areas (despite US-European political differences, such as over the invasion of Iraq). Yet in the case of GMOs we have seen ongoing stark dif-ferences in regulatory standards and regulatory frameworks; stillborn efforts at bilateral regulatory cooperation; persistent battles and logjams in multilateral regimes like the World Trade Organization (WTO) and the Convention on Biodiversity; and finally a so-far unsuccessful effort to litigate a resolution of the problem through the WTO dispute settlement process. Why, we wanted to know, had cooperation repeatedly failed? Why were the US and the EU so implacably opposed? Why couldn't US and EU regulators come to a meeting of the minds in their frequent joint meetings? Why didn't multilateral regimes facilitate cooperation, as regime theorists would have predicted? Could litigation resolve the issue by deciding clearly in favor of one side or the other, or would it backfire, prompting a backlash against both GMOs and the WTO itself? Is there any sign of convergence between the two sides, or is the now decade-old conflict likely to continue and spread to the rest of the world—and if so, how will this affect the dynamics of the conflict? What law and policy lessons, in sum, can we draw?

These are the questions that we address in this book, and our core arguments are spelled out briefly in Chapter 1, and elaborated in detail in the chapters that follow. We shall come to these arguments presently, but, as one might suspect in a project that is a decade in the making, we have many thanks to offer to many people. This project began in Madison, Wisconsin, where our efforts were funded and encouraged by the University of Wisconsin's European Union Center of Excellence, funded by a generous grant from the European Commission, where we held a conference on GMO regulation organized by Gregory Shaffer who was then the Center's Director.

We have both since left Madison, and we have received additional encouragement, and good advice, from a number of friends and colleagues—many of whom, we fear, we are forgetting. We are particularly grateful to Tim Büthe, Sungjoon Cho, Jeffrey Dunoff, Emilie Hafner-Burton, Ronald Herring, Christian Joerges, Neil Komesar, Ambassador Richard Morningstar, Daniel Naurin, Kal Raustiala, Adam Sheingate, David Vogel, Helen Wallace, Alasdair Young, four anonymous reviewers, as well as participants at conferences and workshops at the Council of European Studies (2005), Princeton University (2006), the University of Wisconsin (2006), the European Union Studies Association (2007), Hebrew University (2007), London School of Economics (2007), the Global Administrative Law conference in Viterbo, Italy (2007), the American Branch of the International Law Association (2007), Northwestern Law School (2007), University of Georgia Law School (2008), and the Law and Society annual meetings in Las Vegas (2005) and Berlin (2007), for their comments on earlier versions of portions of the manuscript. We also thank Mario Bifano, Erin Chalmers, Rebecca Estelle, Matt Fortin, Geoff Seufert, Timo Weishaupt, and Anna Woodworth for their excellent research assistance. Mark Pollack would like to thank the College of Liberal Arts at Temple University and the BP Chair in Transatlantic Relations at the European University Institute for funds to undertake research on the project in Washington and Brussels. Gregory Shaffer would like to thank the University of Wisconsin, Loyola University Chicago School of Law and the University of Minnesota Law School for their research and research travel support. He would also like to thank the Fulbright European Union Scholar-in-Residence Program for its semester of funding during his stay in Rome, as well as the Legal Office of the Food and Agricultural Organization and Judge Sabino Cassese of the Italian Constitutional Court for providing him with wonderful working space during that stay.

We would also like to thank the dozens of government officials, international civil servants, and representatives of business and non-governmental organizations, who have met and shared their views with us in Washington, Brussels, Paris, Rome, Geneva and elsewhere. Many of these practitioners, we note in Chapter 1, have requested anonymity, and we have honored that request in each case, identifying those individuals only by broad institutional affiliation, city and date of interview, although some officials have been willing to be identified by name, and in those cases we do so. As social scientists we have tried to rely as far as possible on replicable, publicly available sources—yet given the sensitivity of GMO regulation as an issue, it was inevitable that we would rely on interviews with practitioners for some facts of the case, for context, and, quite often, as a detailed corrective to the big-picture views and predictions offered by theories of international law and politics. If we have come close to getting the details of our story right, the credit goes largely to the many participants who shared with us their time and insights, and, whether named in these pages or not, they have our heartfelt thanks.

Oxford University Press, and in particular our editor, Dominic Byatt, have been supportive of our efforts, patient with our occasional delays and constant updates in covering a moving target, and efficient as ever in moving the manuscript from review to revisions to publication. We also wish to thank Louise Sprake and Lizzy Suffling at OUP and Kay Clement for their care and diligence with the editing and production.

If we had been worried about the dangers of interdisciplinary cooperation when the project began, any such doubts have long been dispelled. We have, throughout this project, learned from each other, addressed issues that we would otherwise have ignored, and indeed discovered issues that our two disciplines both considered important, but often while using different terminology and involving scholarly literatures that often took little or no notice of the other. We hope to have built bridges across these two disciplines, and we ourselves would have great difficulty in establishing paternity over any given idea in this volume, on which we worked equally and collaboratively (with the exception of Chapter 5, which is based primarily on Shaffer's legal analysis). The order of names on the cover, we hasten to add, is solely alphabetical, and the intellectual partnership of the volume is entirely equal.

Finally, it goes without saying that the burden of traveling to foreign capitals for research, and of writing and editing on nights and weekends, has fallen not just on the two of us but on our families, who have been endlessly patient and supportive of our efforts. We dedicate this book, with love and thanks, to them.

1

Introduction and Overview: Biotechnology, Risk Regulation, and the Failure of Cooperation

In 1992, the United States (US) Food and Drug Administration (FDA) approved the first genetically engineered food—Calgene's Flavr Savr Tomato—for sale and marketing in the US. Encouraged by a favorable US regulatory system and the lack of serious domestic political challenge, US scientists have subsequently created, farmers have grown, and companies have marketed a wide range of genetically modified (GM) foods and crops. By the end of the 1990s, in "the most rapid adoption of a new technology in the history of agriculture," some 60 per cent of the processed foods available in US grocery stores were derived from transgenic varieties, more popularly referred to as genetically modified organisms (GMOs).[1] By the end of 2003, the estimate had risen to "between 70 and 75 per cent of all processed foods available in US grocery stores."[2] By 2007, approximately 89 per cent of soybeans, 83 per cent of cotton, and 61 per cent of corn grown in the US consisted of genetically modified varieties, and these figures have been rising annually.[3] US farmers also grow genetically engineered canola, potatoes, tomatoes, papaya, squash, and sunflowers among other foods, although to much lesser degrees.

By contrast with the US embracing agricultural biotechnology, European Union (EU)[4] regulators and the public have taken a far more cautious approach to GMOs, treating genetically modified foods and crops as different from their conventional counterparts, and adopting increasingly strict and complex regulatory procedures for their approval and marketing. Unlike in the US, GM foods and crops face considerable regulatory hurdles in the EU, including requirements for mandatory pre-approval of all GM products, as well as provisions on the mandatory labeling and traceability of all GM products, which have made it difficult and sometimes impossible for US farmers to export GM foods to markets in Europe. They also face greater social resistance, with activists campaigning against GM foods and ripping up GM crops from fields, and public opinion far more mobilized over GM foods than in the US.

In an age of increasing international trade and economic interdependence, these sharp and persistent regulatory differences have resulted in ongoing transatlantic and now global disputes where economic interests and social values clash, in what some political scientists have called "system friction."[5] By the late 1990s, stricter European regulations and slower European regulatory approval processes for new GM varieties raised potentially serious obstacles to the export of agricultural products, first from the US and then from other agricultural producers. A potential international trade war loomed. More broadly, the regulatory systems of the two largest and most powerful markets on earth came into conflict.

Throughout the past decade, US and EU representatives have concurrently dueled and attempted to find some common ground, or at least manage the conflict over their respective approaches to biotechnology regulation. They have formed numerous bilateral networks of government officials, scientists and civil society representatives to engage in joint consultation and efforts to coordinate, where possible, their distinct approaches.[6] They have also discussed and negotiated the issues in multiple multilateral contexts, such as before the Organization for Economic Cooperation and Development (OECD); the international food standard setting body, the Codex Alimentarius Commission; the international trade body, the World Trade Organization (WTO); and an international environmental body, the Conference of the Parties to the Convention on Biodiversity and its Biosafety Protocol. Despite these efforts, the two transatlantic partners have failed, either bilaterally or in multilateral regimes, to reach any fundamental agreement on the regulation of agricultural biotechnology. After considerable internal debate and delay, the Bush administration finally filed a legal complaint before the WTO in May 2003, joined by Canada and Argentina, maintaining that the EU's regulatory decisions over GM crops and foods violated the EU's international trading commitments, which finally resulted in a panel decision adopted by the WTO Dispute Settlement Body in November 2006. Even after the WTO ruling, however, the fundamental differences between the US and EU regulatory systems remained deeply entrenched and resistant to change and the fate of GM foods and crops remained deeply contested and uncertain.

The transatlantic GMO dispute has brought into conflict two longtime allies, economically interdependent democracies with a long record of bilateral and multilateral cooperation in both economics and security. Yet the dispute has developed into one of the most bitter and intractable transatlantic and global conflicts, resisting efforts at resolution in bilateral networks and multilateral regimes alike, and resulting in a bitterly contested legal battle before the WTO. Indeed, our account, with its emphasis on the intractability of the GMO dispute and the strikingly limited contribution of bilateral networks and multilateral regimes, contrasts sharply with most other books and articles about international cooperation, which are essentially optimistic accounts of how cooperation under anarchy is possible and can be facilitated by international regimes. The general form of such studies falls into what might be called a Home Depot theory of international cooperation: "You can do it, we can help."[7]

Our account of agricultural biotech regulation, by contrast, emphasizes the difficulties, the limits, and in many instances the outright failure of international cooperation in regulating GMOs. How can we explain the origins of this dispute, its intractability, and the repeated failures of cooperation? What happens to international law and to international regulatory institutions when the two economic powers clash? Just as importantly, what role, if any, can international regimes play when cooperation fails? What domestic legal, administrative and commercial responses take place in the face of prolonged diplomatic stalemate? To answer these questions, we argue, analysts need to focus on multiple levels: on the domestic sources of the dispute, on the international efforts at bilateral and multilateral cooperation, and on the interaction between the domestic and international levels. This book specifically addresses the following questions:

- *What are the domestic sources of international disputes?* Starting at the domestic and EU levels, we ask what explains the different policies on agricultural biotechnology in the US and EU. What are the respective roles of interests, institutions, and ideas in explaining these differences? Do these differences reflect deep philosophical divisions between Europeans and Americans about the regulation of risk to their societies? How deeply entrenched or path-dependent have their policies become?

- *What are the obstacles to cooperation and to "deliberation" at the bilateral level?* In light of the considerable externalities that each side's policies had on the other in a global economy, and the prospects of a potential trade war, many scholars and practitioners placed their hope in the promise of informal transatlantic networks to deliberate jointly and identify the best policies for biotech regulation. What evidence is there of deliberation in these encounters, and why has joint deliberation failed to resolve the dispute?

- *What are the obstacles to cooperation at the multilateral level?* What happens within multilateral regimes when bilateral cooperation between powerful states fails? What strategies do powerful states use to attempt to export their policies internationally, such as forum shopping among international institutions in a fragmented international law context? How does distributive conflict affect the functioning of institutions, including the interaction among them? What happens to the interaction of hard (binding) and soft (voluntary) international law regimes?

- *What role can the WTO play when cooperation fails?* In light of the failures of both bilateral and multilateral cooperation, the US and the EU

3

became embroiled in WTO litigation. How does the WTO judicial process attempt to exercise influence in managing such a conflict, in the face of entrenched differences between the WTO's two most powerful members? Can the WTO facilitate a resolution of such a dispute, or will it only exacerbate the conflict and create a public-opinion backlash to its own legitimacy?

- *Finally, what influence do international political, legal, and market pressures have on the domestic laws and policies of each side?* Do we find evidence that the US is "trading up" to the EU's more precautionary standards, or has WTO pressure led the EU to liberalize or "trade down" its restrictive policies? Or, alternatively, are each side's policies too entrenched to be changed significantly even in the face of intense international pressure?

In this book, we investigate these challenges—the obstacles to reconciling regulatory differences through international cooperation, and what happens when cooperation fails—through the prism of the US–European dispute over the regulation of agricultural biotechnology or GMOs. The book addresses the dynamic and reciprocal interactions of domestic law and politics, transgovernmental and transnational networks, international regimes, and global markets, through a theoretically grounded and empirically comprehensive analysis of the governance of GM foods and crops.

Crucially, while the primary focus of our analysis alternates between the domestic and international levels in the various chapters of the book, we do not analyze these two levels in isolation, but examine the interactions between them. Thus, for example, we demonstrate that the deeply politicized, entrenched and path-dependent nature of GMO regulation in the US and the EU has fundamentally shaped negotiations and decision-making at the international level, limiting the prospects for deliberation and providing incentives for both sides to engage in hard bargaining and to "shop" for favorable international forums. In addition to this bottom-up perspective, however, we also take a top-down approach, examining the impacts and the limits of transnational and international pressures on domestic laws and politics. Both foreign market pressures and international regimes, we argue, have exerted pressures for change in the US and the EU, empowering some actors and weakening others within domestic political and legal processes; yet the impact of these international pressures has been blunted by deeply entrenched, path-dependent patterns of interests and institutions on both sides of the Atlantic.

In this introduction and in the subsequent chapters of the book, we offer five inter-related arguments, five sets of answers to the above questions about the origins of the transatlantic and global disputes, the obstacles to resolving these disputes both bilaterally and multilaterally, the risks and potential rewards of legal recourse to the WTO, and the impact of international and transnational developments on domestic regulatory systems in the US, the EU and other countries. In analyzing the ongoing struggle over the regulation of agricultural biotechnology, we draw upon and seek to contribute to rich literatures on politics and law, at both the domestic and international levels.

First, at the domestic level, we ask (in Chapter 2) why the US and EU systems for the regulation of GM foods and crops look so different, and we survey theories of comparative law and politics that attribute differences in domestic regulation to differences in organized interests, political institutions, culture and ideas, and contingent events, respectively. Although multiple factors contributed to the polarization of US and EU regulatory approaches, we show why the best explanation for the differences lies neither in innate or "essentialist" forms of culture (such as US and European attitudes toward food, risk or technology) nor in institutions alone (such as US specialized agencies compared to European political processes), but in the ability of interest groups to capitalize on preexisting cultural and institutional differences, with an important role played by contingent events such as the European food-safety scandals of the 1990s. We ask, in particular, about the development of the two regulatory systems over time, drawing on the historical institutionalist literature to understand the conditions under which different regulatory systems are subject to inertia or path-dependence, resisting pressures for change or displaying the change only at the margins. The stark differences in the US and EU regulatory systems were not preordained, we argue, by the interest-group, institutional or cultural configurations of the two sides; but the differences are real and strongly resistant to change. The friction between them has led to increasing trade conflicts and legal disputes, spurring calls for greater transatlantic and international coordination to avoid, in particular, the threat of a transatlantic and global trade war.

Second, at the international level, we draw upon a growing body of international relations and international legal scholarship that focuses on the promise of regulation through transnational networks, with a particular emphasis on the prospect of "deliberation" as a form of decision-making in which governmental and nongovernmental actors put aside fixed positions and negotiating tactics in favor of a collective search for better understanding and better policy. We find, however, that the record of transatlantic cooperation on GMOs has largely been one of failure, despite hopes for a new type of bilateral collaboration through flexible and deliberative networks of government regulators and civil-society groups.[8] Although we find a wealth of transatlantic governmental, scientific, business, and civil society networks arising to address the regulation of GMOs (examined in Chapter 3), the record of US–EU regulatory cooperation on agricultural biotechnology has shown only limited evidence of genuine deliberation, particularly deliberation with concrete policy consequences. Deliberation, we argue, is a hothouse flower that flourishes only under restrictive conditions, and the sharp disagreements, intense politicization, and distributive conflicts that characterize agricultural biotechnology have all prevented US and EU policymakers from engaging in a joint deliberative search for the best policy in this area.

Third, we argue that the record of multilateral cooperation (undertaken within overlapping regimes such as the WTO, the Convention on Biodiversity, the OECD, and the Codex Alimentarius Commission) has been similarly limited, characterized largely by strategic maneuvering by both sides to "export" their own standards and their own principles for risk regulation, and to "forum shop" among the regimes most likely to produce each side's favored outcomes, imposing most of the costs of adaptation to new global norms on others. We find (in Chapter 4) that cooperation has been frustrated in practice by the existence of severe distributive conflicts between the two sides (involving both economic and political costs), which has given rise to overlapping and (sometimes purposefully) inconsistent regimes for trade, the environment, and food safety. Although a growing amount of scholarship has addressed the roles of "soft" law (which is formally non-binding) and "hard" law (which is formally binding and enforceable) as complementary and mutually reinforcing means for international problem solving, we find that hard and soft law regimes can constrain each other's operations, at least as they were initially intended to function. Our study shows how "soft" law regimes such as the Codex Alimentarius can become "hardened" because of the implications of their decisions in "hard" law regimes, while "hard" law regimes such as the WTO can become "softened" and less certain when they seek means to avoid substantively deciding contentious issues. The interaction of "hard" and "soft" law regimes, rather than progressively moving toward a new consensus, may instead perpetuate the substantive deadlock in member-state preferences, undermining the effectiveness of both types of regimes in the process.

Fourth, we suggest that despite considerable risks, the US's complaint before the WTO Dispute Settlement Body has offered the prospect of some clarification and mutual accommodation that had hitherto eluded the two sides in other bilateral and multilateral fora. In Chapter 5, we apply a comparative institutional analytic framework to examine the radically different institutional implications of the interpretive choices that the WTO judicial panel faced in the *EU-Biotech* case. We demonstrate how interpretive choices by a WTO judicial body can attempt to allocate decision-making to different institutional processes in which constituencies of different countries, with varying priorities, perceptions, and abilities to be heard, participate to varying and always imperfect degrees. At the same time, we show how these choices are themselves constrained by challenges to powerful members' acceptance of WTO judicial decisions, which, in turn, can inform those decisions. While we pay close attention to the WTO legal texts, the optic here is to see the WTO judicial process through the lens of *governance* and not through a judio-centric perspective focused on judicial interpretation and review. In particular, we show how the WTO has constrained the conflict by channeling it into a "legal" process and thereby deflecting pressure within the US to retaliate aggressively and unilaterally against Europe, which might have occurred had there been no WTO. We find that the WTO panel took a procedural approach in its decision, refusing to articulate a single substantive standard on GMO regulation, but instead insisting on certain procedural requirements that all states must observe in adopting their own domestic regulations. In the process, we contend, the WTO has empowered domestic political actors (such as the European Commission) with an interest in complying with WTO law, and as a result, has encouraged regulators on both sides of the Atlantic to operate more transparently, taking into greater account the effects of their actions on third parties.

Finally, we return to the domestic level to assess whether several decades of discussion, negotiation, and litigation have resulted in significant reform and/or convergence of the two regulatory systems. We demonstrate that, despite some domestic changes on each side, the US and EU regulatory systems for agricultural biotechnology show few, if any, signs of convergence toward a common regulatory model, but continue to differ fundamentally in their respective approaches to the regulation of GM foods and crops. There has, to be sure, been some domestic change on both sides of the Atlantic, due at least in part to external pressures from international markets and international regimes. In the EU, the Commission and biotech companies have been somewhat empowered by international developments to resume approvals of new GM varieties after a long moratorium and to challenge member-state bans against those already formally approved. On the US side, meanwhile, regulators have increased the requirements for trials before the commercial release of many GM seeds, so that these varieties, in fact, are treated distinctly from more conventional ones, despite official US proclamations to the contrary. Even in the absence of tightened regulation, moreover, US farmers have demonstrated a reluctance to adopt new GM foods and crops which, they fear, will be rejected in the EU and other large export markets. In both the US and the EU, however, these developments have been profoundly path-dependent, taking place at the margins of regulatory systems that remain unchanged in their fundamental approaches. These developments reflect long-standing adaptations to existing regulations by actors on either side of the Atlantic who have managed to impede significant reforms even in the face of considerable pressures for change. In some ways, in Europe we have seen much *regulatory reform without fundamental change* in outcomes, and in the US, some mostly market-driven *change without regulatory reform*.

In sum, the story of the transatlantic GMO conflict is largely one of failed attempts at bilateral and multilateral cooperation. The GMO story is, therefore, particularly interesting as a case study of the many obstacles—both domestic and international—to successful cooperation. These obstacles include the politicization and path-dependence of domestic policies which stymie deliberation and mutual accommodation, as well as the existence of transnational distributive conflicts and the strategic opportunities provided by overlapping international regimes. The result is frequent stalemate, fragmentation of international law, and conflict between soft and hard law regimes. Examining those obstacles and showing them in operation is one of the major contributions of the book. Yet our account is not a counsel of despair, for in addition to examining how and why cooperation fails, we also address ways in which states and regimes can facilitate the ongoing management of regulatory conflict, and, over time, together with transnational market forces, influence national regulatory and commercial practices in a (somewhat) more accommodating manner.

The final chapter of the book offers a summary of our findings, examining the implications of our study for the future of agricultural biotechnology, in particular for developing countries, as well as the study's broader implications for the prospects and limits of international cooperation in other areas. We maintain that in a world of rapid technological change, new conflicts over divergent regulations will continue to arise. New and existing transnational networks and multilateral regimes will become sites where underlying conflicts manifest themselves. We show the severe limits of international political and legal regimes where domestic policies are strongly path-dependent, and powerful members have conflicting distributive stakes. In these contexts, international hard and soft law regimes do not progressively lead to harmonized mutually beneficial outcomes, but rather can constrain and even undermine each other's operations. Yet, even here, we find roles for international law and international institutions. While international institutions have been demonized by some and dubbed irrelevant by others, we show how, even with highly politicized issues over which state representatives engage in strategic maneuvering and little deliberation, they can channel political conflict into legal processes, and in this way diffuse conflict and bide time. We show, as well, how they can empower domestic actors in domestic playing fields, leading (potentially over time) to some accommodation of difference and some convergence of practice. We conclude the book by offering our prediction that, in the end, the technology is likely to be gradually accepted in most regulatory jurisdictions, but within significant market and regulatory constraints for GM foods. With the rise of China, India, and Brazil as players in the world economy, and as growers of GM products in particular, we address the importance of these countries' responses to agricultural biotechnology for its future.

The rest of this introductory chapter lays out the essential elements of our arguments, chapter by chapter, regarding the reasons for US and EU regulatory difference, the resulting difficulties of bilateral and multilateral cooperation, the interactions of fragmented international legal orders, the deeply disputed role of the WTO in helping to resolve the dispute, the lack (thus far) of fundamental US/EU convergence, and the implications of the US–EU conflict for developing countries and the future of agriculture.

1.1. The regulatory context: Agricultural biotechnology and risk regulation

Genetic engineering, the process used to create GM seeds, crops, and the foods produced from them, is a technology used to isolate genes from one organism, manipulate them in the laboratory, and inject them into another organism. In Chapter 2, we introduce the technology of genetic engineering and address the arguments of both its advocates and its skeptics. Supporters of agricultural biotechnology consider genetic manipulation to be merely the latest step in an ongoing process of human intervention in nature to stabilize and increase crop yields and improve the nutritional quality, appearance, and taste of plant varieties. Biotechnology's skeptics, by contrast, raise concerns over food safety, environmental harm, agribusiness power, ethics, and broader cultural concerns.

Advocates and opponents of the technology, as in many debates over risk regulation, argue over the applicability of the key concepts of "risk" and "uncertainty." Advocates of the technology, who have been more successful in the US, focus on cost-benefit approaches to risk, weighing the probability and magnitude of harm against prospective benefits, in this case in relation to conventional plant varieties. Opponents and skeptics, who have been more successful in the EU, in contrast, focus on the concept of uncertainty which belies the measurability of risks, in particular as regards long-term effects. They thus call for the exercise of greater precaution before approving GM varieties, pursuant to application of a rather open-ended precautionary principle. We review these broader concepts and their use in the specific arguments raised about the benefits and opportunities offered, and the risks and uncertainties posed, by the technology.

We also examine the key concepts of "risk assessment" and "risk management" and their application in regulatory systems. Risk assessment refers to the technical assessment of the risks in question. Risk management, in contrast, concerns the policy responses to these assessments. These two concepts are ideal types, as completely segregating risk assessment and risk management determinations is difficult (and some contend impossible) in practice. We will see how deliberation over risk assessment principles and approaches has been more fruitful in various international fora, while bilateral and multilateral consultations and negotiations over the appropriate risk management approach to the technology have resulted in hard bargaining, strategic maneuvering, and frequent stalemate.

Regulators have applied a range of risk management policies over time, including product bans, setting standards to minimize risks "to the extent feasible," enacting requirements to eliminate "significant risk," and engaging in cost-benefit analysis. A number of scholars contend that, while the US took a lead in risk regulation from the 1960s to the mid-1980s, there has since been a switch, with Europe becoming the more stringent regulator since the 1990s.[9] Although there is countervailing evidence that the level of precaution exercised in the US and Europe varies with the regulatory sector (from nuclear energy to chemicals and novel foods), we focus on the evidence as to how and why Europe and the US have taken starkly different approaches to the regulation of agricultural biotechnology, with the EU taking a much more stringent regulatory approach.

1.2. The domestic sources of the conflict

In the US, the basic regulatory framework for biotechnology was set in 1986 by executive action, when the Reagan administration's Office of Science and Technology Policy (OSTP) issued a "Coordinated Framework for the Regulation of Biotechnology" that continues to shape US biotech regulation to the present day. Simplifying slightly, the US regulatory framework was based on the premise that the techniques of biotechnology are not inherently risky and that biotechnology can, therefore, be adequately regulated by existing federal agencies under existing statutes, obviating the need for new legislation dedicated to GMOs. The Coordinated Framework established a division of responsibility among three primary US regulators, with the FDA serving as the primary regulator of GM foods, the United States Department of Agriculture (USDA) charged with oversight of the planting of GM crops, and the Environmental Protection Agency (EPA) responsible for overseeing the environmental and food safety impact of GM crops that have pesticidal characteristics. Crucially, these three agencies have generally regulated GM foods and crops in terms of the characteristics of the product, rather than in terms of the process by which they are produced.[10]

By comparison with the US, EU regulation of biotechnology was far more decentralized, and the EU adopted a decision-making process in which the key decisions were taken not by a specialized regulatory agency such as the FDA, but by political bodies such as the Council of Ministers, Commission, and European Parliament, in an uneasy cooperation with "competent authorities" in each of the member states. The EU has also taken a far more precautionary and stricter approach to biotech regulation, adopting by legislation a specific and increasingly demanding regulatory procedure for the environmental release and marketing of GM foods and crops. According to the terms of a 1990 directive (as amended and elaborated by subsequent legislation), all GM foods and crops are subject to a special authorization procedure, requiring scientific risk assessment by national and/or European regulators, and featuring much greater involvement by political officials in the Union's regulatory committees and in the Council of Ministers. In practice, this procedure has led to a much slower approval of new GM varieties in Europe, and to a six-year (1998–2004) moratorium on the approval of GM varieties in the Council. Even officially approved varieties, moreover, have had to meet additional hurdles, including EU provisions for the labeling and traceability of GM crops, the prospect of national-level bans on specific GM foods and crops approved at the EU level, and boycotts of GM products by consumers and retailers.

The causes of these starkly different approaches to the technology are examined in Chapter 2. Surveying the various accounts of GMO regulation in the US and the EU, we identify four classes of explanation for the observed differences. As we shall see, some analysts stress *cultural differences* in European and American attitudes toward food or toward risk, with the US purportedly more risk-acceptant than the EU; some point to *institutional differences* in US and European assessments and management of risk, including the existence of specialized regulatory agencies in the US, and the larger number of institutional actors or "veto players" in the EU; some highlight differences in *interest group configurations*, with the US being characterized by a larger and more politically influential biotech sector; and some note the differential impact of *contingent events* such as the European food-safety crises of the 1990s.

Against this theoretical backdrop, we argue that the very different approaches to GMO regulation that we observe between the US and the EU, were not determined in any straightforward way by either the institutions, the political culture or the interest-group configurations present on either side of the Atlantic. It was *not* inevitable that US regulators would adopt a product-based approach to GMO regulation, nor was it obvious from the outset that the EU would adopt the strict, politicized, and highly precautionary system that emerged over the course of the 1990s. The best explanation for the observed transatlantic differences, we believe, is multi-causal, lying in the ability of interest groups to capitalize on preexisting cultural and institutional differences, with an important role played by contingent events such as the European food-safety scandals of the 1990s.

In the US case, powerful interest groups, including the biotech industry and farmers' associations, sought a regulatory framework that would treat new GM varieties as substantially equivalent to their conventional counterparts. In doing so, they were able to draw upon a supportive institutional and cultural context, including a regulatory system featuring strong, specialized government agencies, a diverse consumer protection movement that was divided on GMO regulation, and a cultural tradition of accepting the use of new technologies in food production. Yet, contingent events also played a role, including the preferences of a Reagan administration that shaped a Coordinated Framework giving primary responsibilities to the USDA and FDA at the expense of the more precautionary EPA.

11

In Europe, by contrast, pro-GMO interests were weaker, with a smaller biotech sector and an agricultural community that did not take up GM crops in light of consumer responses to US imports and thus, never emerged as a champion of the new technology. Pro-GMO interests in Europe encountered an institutional and cultural framework that provided multiple veto points (and in particular at the member-state level where national governments exercised considerably more regulatory authority than states in the US federal context) and multiple sources of opposition (on environmental, food-safety, and ethical grounds) to GMOs, resulting in the more demanding, politicized, and process-oriented, multi-level EU regulatory system. The subsequent evolution of the EU regulatory process in the direction of ever-greater precaution, however, is in large part a direct result of contingent events, namely the food-safety scandals of the 1990s which undermined public support for the technology and trust in regulators at a crucial time for the introduction of GM foods and crops.

Neither the US nor the EU, then, was preordained by its interest-group, institutional or cultural characteristics to adopt the precise regulatory framework that it did. By the same token, however, once the respective US and EU regulatory frameworks were adopted, they proved remarkably resilient in their essential characteristics. The explanation for this resilience, we argue, can be found in historical institutionalist theory, which examines the effects of institutions on politics *over time*, maintaining that institutional choices taken at critical junctures can persist or become "locked in," thereby shaping and constraining the actors later in time. Political institutions and public policies, in this view, are subject to "positive feedback," insofar as those institutions and policies generate incentives for actors to stick with and not abandon the existing laws and institutions, adapting them only incrementally to changing political environments. These positive feedbacks may reflect constraints *from above*, in the form of legally binding rules that are difficult or costly for political actors to change, or *from below*, as societal actors adapt to and develop a vested interest in the continuation of specific public policies.[11]

Insofar as political institutions and public policies are, in fact, characterized by positive feedbacks, regulatory politics will be characterized by *inertia* or *lock-ins*, whereby existing institutions may remain in equilibrium for extended periods despite considerable political change; *a critical role for timing and sequencing*, in which relatively small and contingent events that occur at *critical junctures* early in a sequence shape events that occur later; and *path-dependence*, in which early decisions provide incentives for actors to perpetuate institutional and policy choices inherited from the past, even when the resulting outcomes are manifestly inefficient. At the extreme, institutions and policies can become *self-reinforcing*, such that the operation of an institution or a policy bolsters its societal support base, making it more difficult to change and more stable in the face of external shocks over time.[12]

In this context, we argue that the US and EU regulatory frameworks for agricultural biotechnology, while not themselves determined by preexisting institutional constraints, have since generated significant positive feedback, lending each system considerable resistance to change, and indeed making each system self-reinforcing. In each case, timing and sequencing have proven vital, as the initial regulatory frameworks were adopted at critical junctures that shaped subsequent developments. In the case of the US, the critical juncture occurred in the mid-1980s, when the introduction of agricultural biotechnology and of GM foods and crops presented policymakers with a crucial set of choices. In this context, the Reagan administration laid down a comprehensive regulatory framework within existing statutory authority, with results that came close to the preferences of the biotech industry.

The critical juncture in the EU came a few years later, in the context of a EU with a relatively weakly organized biotech industry, diverse preferences among EU governments, and a decision-making system with a large number of veto points (involving key roles for member states demanding increasingly strict EU-level requirements), with the result that the EU's initial regulatory framework laid down a more demanding regulatory procedure closer to the preferences of GM opponents. In addition, a further critical juncture arguably came during the second half of the 1990s in the EU, when various food-safety scandals strengthened the position of those actors who sought to make the EU's regulatory framework even more restrictive, resulting in the post-1998 moratorium and the subsequent strengthening of the regulatory framework early in the following decade.

Just as importantly, in each case, the regulatory frameworks adopted have generated positive feedback, both by creating institutional rules that could be changed only with difficulty and by generating adaptations among interest groups and public opinion that contributed to the stability of the two respective frameworks. In the US, the early adoption of a relatively welcoming regulatory framework contributed to significant investment, the rapid growth of the biotech industry, and the equally rapid adoption of GM crops by American farmers, who have represented the bulwark of political support for the existing framework.

In the EU, by contrast, the early adoption of a relatively restrictive regulatory framework, together with the turn against GMOs in public opinion and the subsequent declaration of a *de facto* moratorium, discouraged farmers from planting GM crops, prompted retailers to resist GM foods, and led to the flight of agricultural biotech investment from Europe, further undermining interest-group and societal support for GM foods and crops. At the same time, the EU's convoluted legislative process, requiring qualified majorities among the member states in the Council of Ministers and an absolute majority of a European Parliament that has turned largely against GMOs, has created a huge institutional hurdle to the fundamental reform of the EU regulatory framework. Hence, the politics of GMOs, which were arguably fluid during the early years of the technology, have become increasingly rigid in both polities, with strong resistance and high thresholds to fundamental change on either side.

Whatever their causes, the stark and persistent differences between the two systems have led to serious transatlantic tensions, as US biotech producers and farmers have found themselves increasingly unable to export GM foods and crops that have been found to be safe by US federal regulators to Europe, or to countries following Europe's example. In response, the US has brought increasing pressure on the EU to facilitate the approval of new GM varieties, culminating in the bringing of a WTO complaint against the EU in May of 2003. The stark contrast between the US and EU regulatory systems, therefore, is not simply a compelling case of comparative public-policy analysis. It has become the source of serious transatlantic and international trade and regulatory disputes, reflective of conflicts over divergent regulatory approaches that we will continue to see in the future.

We conclude Chapter 2 by adapting a two-level (and in the EU case, three-level) game from international relations theory to assess how national and international politics and law interact. According to Robert Putnam's "two-level games" model, all international negotiations take place simultaneously at two levels.[13] At the international level (Level 1), negotiators (also known as statesmen, chiefs-of-government or COGs) bargain with their foreign counterparts in an effort to reach mutually beneficial agreements, while at the domestic level (Level 2), the same negotiator engages in bargaining with domestic constituencies, who must ultimately accept the contents of any agreement struck at Level 1.

In such two-level games, we demonstrate, the bargaining power and the ultimate outcomes of political cooperation depend not only on the preferences of COGs—in our case, the US federal government and the European Commission negotiating on behalf of the EU—but also on the nature and intensity of preferences among their respective political constituencies. As we shall see, the intense, highly mobilized, and increasingly entrenched preferences of interest groups and publics on both sides of the Atlantic act as a source of bargaining power for both COGs, who can claim that their "hands are tied" by intense domestic pressures over the issue of GMO regulation. Those same domestic constraints, however, also decrease the "win-sets" of each side and hence the likelihood of successful cooperation, by making it difficult for either COG to engage in genuine deliberation or in significant concessions to the other side, for fear of having the resulting agreement rejected "back home." As we shall also see, much of the past decade has witnessed an effort to overcome these obstacles, through deliberation, negotiation, and, ultimately, litigation.

1.3. The promise and failure of transatlantic deliberation

The relationship between the US and the EU is not, of course, purely conflictual, despite the real and significant differences between them over the regulation of GM foods and crops. While the trade impact of different regulations presents a clear potential for conflict, the US and the EU remain each other's largest trading partner and source of direct foreign investment, as well as political and military allies, and these common interests provide a strong incentive for both sides to cooperate to achieve a harmonized approach to the regulation of GM food and crops—or, failing that, to prevent the GM issue escalating into a full-scale transatlantic trade war.

Toward this end, the US and the EU have engaged in efforts at both bilateral and multilateral cooperation on GM issues since the 1990s, seeking common understandings, if not common standards, for the regulation of agricultural biotechnology. These efforts reflected scholarly claims about the promise of transgovernmental networks of regulators and about the ability of international regimes to encourage international cooperation, without the need for a hierarchical political or legal order at the international level. In practice, however, both of these routes—bilateral cooperation between US and EU regulators, as well as multilateral cooperation in various international regimes— have proven disappointing, providing as yet no clear accommodation of the fundamental differences between the two sides' approaches.

International relations and international legal scholars have pointed to the prospect of international governance through the so-called "transgovernmental networks" of lower-level government officials cooperating directly on a day-to-day basis with their counterparts in other jurisdictions.[14] By contrast with the image of two-level games, in which high-level international negotiators (COGs) monopolize the representation of domestic interests abroad, transgovernmental networks raise the promise of direct interactions between technocratically oriented domestic regulators, who might be expected to work pragmatically and deliberatively to solve common problems.

Transgovernmental networks are now commonplace in the European Union, where national regulators have established formal and informal EU-wide networks in most areas of policy-making, from competition policy, financial services, and environmental policy to utilities regulation. By the turn of the twenty-first century, Anne-Marie Slaughter has argued, national regulators had emerged as "the new diplomats," bypassing traditional foreign-ministry channels to cooperate in a "fast, flexible, and efficient" manner with their counterparts.[15] Significantly, we and other scholars have pointed to the transatlantic relationship and in particular, to the 1995 New Transatlantic Agenda and the 1998 Transatlantic Economic Partnership (TEP), as an emerging arena for such regulatory networks, with US and EU regulators interacting directly and fruitfully, despite their differences, in areas such as competition policy and data privacy protection.[16]

Some scholars have gone even further, maintaining that these emerging transgovernmental networks could provide the setting for a sort of international deliberative democracy, in which national experts would meet, set aside their preconceived notions about the national interest, and deliberate together in search of the best available policy in a given issue-area. This emphasis on deliberation derives largely from the work of Jürgen Habermas, whose theory of communicative action has been adapted to the study of international relations and to the study of EU governance.[17] In Habermasian communicative action, or what Thomas Risse calls the "logic of arguing," political actors do not simply bargain based on fixed preferences and relative power, they may also "argue," questioning their own beliefs and preferences, and being open to persuasion and the power of the better argument.[18]

Habermas and his followers concede that genuine communicative action or argumentative rationality is likely to flourish only under a fairly restrictive set of conditions. In international politics, Risse argues, deliberation, or a logic of arguing, are most likely under certain specific conditions, including a "common lifeworld" among the participants, provided by "a high degree of international institutionalization in the respective issue-area"; "uncertainty of interests and/ or lack of knowledge about the situation among the actors"; and "international institutions based on nonhierarchical relations enabling dense interactions in informal, network-like settings."[19] These conditions are by no means satisfied everywhere in international politics, but where they are present, Habermasian scholars predict that international actors will engage in arguing rather than bargaining, presenting their arguments in a common language, such as those of law or science, and proceeding to decisions on the basis of "the better argument" rather than the bargaining power of the respective actors.

Faced with a situation of growing economic interdependence, US and EU policymakers in the 1990s onward engaged in extensive efforts at bilateral cooperation, enlisting networks of scientists, civil-society groups, business representatives, and especially government regulators from both sides to exchange views in the hope of fostering better understanding of each other's regulatory approaches. In this context, they identified biotechnology as an area in which structured dialogues might build mutual understanding and trust, provide early warning of disputes, and perhaps contribute to a gradual convergence of regulatory approaches to GM foods and crops. Starting in the 1990s, the US and the EU established a series of working groups on GM foods and crops, bringing together government regulators, scientists, and representatives from business and civil-society groups, in order to deliberate, separately and together, and possibly find a common ground.

As we shall see in Chapter 3, however, these groups generally did not produce the level of deliberation desired, or at least any deliberation that has so far had any significant impact on the ongoing transatlantic conflict. US and EU regulators did meet regularly and exchange information and views during the 1990s, but they also brought to the table and sought to defend starkly different regulatory approaches, and none of these groups was able to reach an agreement on practical cooperation in the approval of GM foods and crops, much less on harmonized regulations. Just as importantly, even if regulators from the two sides had been able to bridge their differences and move toward a common approach, both sides found themselves operating in a highly politicized issue-area characterized by strongly mobilized interest groups and by a volatile public opinion that made it difficult, if not impossible, to engage in any substantive compromise. Because of this, US and EU regulators declined to invest significant resources in bilateral regulatory cooperation, but rather brought their clashing perspectives and demands to multilateral fora.

1.4. The move to multilateral regimes

The regulation of agricultural biotechnology did not remain simply a bilateral issue. By the late 1990s, other countries were adopting their own regulatory approaches to GM foods and crops. The choices made by those countries and by the various international regimes whose competences touched in one way or another on the issue of agricultural biotechnology, could bolster or undermine the US and European positions. Both sides, therefore, sought to advance their interests through a variety of multilateral regimes such as the WTO and its Agreement on Sanitary and Phytosanitary Measures (SPS Agreement), which address trade-related aspects of GM foods and crops; the Cartagena Biosafety Protocol signed in 2000, as an amendment to the UN Convention on Biodiversity (CBD), which deals with the environmental implications of GMOs; the OECD, which examines cross-cutting trade, regulatory, and technological issues among advanced industrialized countries; and the Codex Alimentarius Commission, which sets "voluntary" food-safety standards for conventional as well as GM foods.

For decades, regime theorists have argued that multilateral regimes could help states to cooperate, by reducing the transaction costs of negotiations, by facilitating deliberative decision-making among states, and by monitoring and facilitating state compliance and implementation of agreed rules, standards, and principles. Consistent with the basic tenets of regime theory, the US, the EU, and other countries have undertaken negotiations on many aspects of agricultural biotechnology regulation. Once again, however, successful cooperation—and in particular, a resolution of the fundamental transatlantic dispute—has proven elusive, for two interrelated reasons.

First, many issue-areas in international politics are characterized by stark disputes about the *distribution* of costs and benefits from cooperation. While all parties to a regime may agree about the *desirability* of cooperation, they may and often do disagree about the *terms* of cooperation, which may result in unequal benefits and costs for the parties.[20] International regimes can contribute to cooperative outcomes in these situations by facilitating negotiations and establishing common rules, but the presence of distributive conflicts is likely to impede deliberation and foster hardball bargaining, in which each side maneuvers to press for an agreement closest to its own preferences.

In the case of agricultural biotechnology, we argue, the US, the EU, and other countries share a common interest in avoiding a global trade war, but they differ sharply in their preferred solutions. Each side has sought to promote international standards and international cooperation *on its own terms*. For this reason, within each of the various multilateral fora that we examine in Chapter 4, the US has sought to promote what it terms a "science-based" approach to biotechnology regulation, while the EU has sought to secure international recognition for its more "precautionary" approach. We do find some evidence of deliberative decision-making on agricultural biotechnology within multilateral regimes, most notably within the OECD and the Codex Alimentarius Commission. Such deliberation is most likely, we find, where negotiations focus on issues of scientific risk assessment, where the negotiators are themselves scientific and technical experts, and where network deliberations are insulated from domestic public opinion. By contrast, however, where the subject turns from risk assessment to risk management, where international trade negotiators and trade concerns play a key role in negotiations, and where negotiations take place in the shadow of domestic public opinion, negotiators typically revert to a logic of consequentiality, bargaining from essentially fixed positions and seeking international codification and support for their established domestic positions. Deliberative decision-making, therefore, is by no means impossible, but it is challenging and relatively rare in the deeply politicized sphere of GMO regulation.

A second, related impediment to cooperation has to do with the inherently cross-sectoral nature of agricultural biotechnology, which implicates numerous ministries and agencies within government, which in turn represent governments in different multilateral regimes in diverse areas such as trade (WTO), the environment (CBD), food-safety standards (Codex), and those of a cross-cutting nature (OECD). Within such a "regime complex" in which different institutions offer different opportunities for strategic actors, the states frequently engage in "forum shopping," favoring the specific regime or forum most likely to produce their preferred outcomes.[21] The existence of multiple overlapping regimes with no clear hierarchy in a fragmented international legal system, moreover, tends to produce legal inconsistencies among regimes reflective of underlying differences among powerful actors, further clouding the prospects for successful cooperation. This phenomenon of regime complexes, moreover, is related to the problem of distribution, insofar as distributive conflict provides the states with incentives to forum-shop among different regimes within a regime complex, or to create new regimes deliberately to support their own positions and undermine those of the other side.

This is indeed the pattern we find in the case of agricultural biotechnology. Both the US and EU sought to promulgate global standards for agricultural biotechnology that reflected their own domestic standards, protecting their own carefully negotiated regulatory standards and exporting the costs of adjustment to the other side.[22] In the process, both sides also shopped actively for the international forum most likely to support their respective efforts. We find that the US demonstrates a clear preference for the WTO forum, with its emphasis on trade and its disciplines on the use of non-tariff barriers, while the EU has shown a preference for the CBD and the Biosafety Protocol, with its greater emphasis on environmental impacts and on the importance of precaution.

We also find substantial and as yet unresolved inconsistencies among the various fragmented regimes, which place variable emphasis on the importance of free trade and of environmental protection, with no overarching hierarchy to resolve conflicts among them. For these reasons, none of the various multilateral regimes has yet resolved the fundamental differences between the US and the EU over the regulation of GMOs, which remains fundamentally contested after more than a decade of multilateral negotiations.

Indeed, one of our novel findings is that the inconsistencies and conflicts among regimes have influenced the nature of the regimes themselves, including the long-standing distinction between "hard" or formally binding law on the one hand, and "soft" or non-binding law on the other. As we shall see, the reputedly "hard" rules of the WTO have been "softened," made more flexible and less predictable, as the WTO's judicial process has sought to accommodate environmental and health concerns, such as those reflected in the Biosafety Protocol and debates within Codex. The so-called "soft" law mechanisms within Codex, by contrast, have been "hardened" because of concerns over the possible implications of Codex decisions in WTO dispute settlement, as large states have sent trade delegates along with food-safety technical experts to discuss the adoption of new "voluntary" standards and principles. For this reason, we argue, Codex, which earlier was more deliberative in its standard-setting processes, is more frequently characterized by the same disputes, coalitions, and hardball bargaining that characterize negotiation and litigation before the WTO. Hence, we find that hard and soft law regimes do interact, as previous scholars have argued, but they do so in ways that can also undermine, rather than take advantage of, their respective strengths.

Yet, we also maintain that these tensions among international law regimes should not necessarily be lamented. The conflicts reflect underlying differences among states and state constituencies (and in particular, powerful ones) in a diverse, pluralist world. Overlapping regimes, involving "hard" and "soft" law mechanisms, can also provide a service to each other, signaling the states and international decision-makers to tread softly in applying their particular rules by taking account of related international developments.

1.5. The peril and promise of WTO adjudication

Transatlantic tensions over the regulation of GM foods and crops built steadily over the course of the late 1990s and into the following decade, yet for much of this period the US chose not to avail itself of options within its preferred international forum, the WTO. During this period, the US government came under increasing pressure from agricultural producers to bring a case before the WTO's Dispute Settlement Body. US biotech firms joined agricultural associations in arguing that the EU's strict regulation of GMOs, and in particular, its unofficial moratorium on the approval of new GM varieties, damaged US interests and violated the provisions of WTO law. Despite these pressures, the Clinton administration, and for a time the George W. Bush administration, resisted the temptation to bring a legal complaint against the EU at the WTO. In May 2003, however, the Bush administration's forbearance gave way, and the US, joined by Canada and Argentina, brought a WTO complaint against the EU, alleging that the Union's *de facto* moratorium on new approvals, as well as the national bans on approved varieties, constituted a violation of the SPS Agreement. In Chapter 5, we examine the reasons for the US decision to bring a complaint before the WTO, analyze the legal issues before the WTO panel in terms of their broader institutional implications, and weigh the possibility that the case, despite its obvious risks, might have a beneficial impact by clarifying the parties' obligations under WTO law, stabilizing the conflict, and encouraging greater transparency and accommodation in GMO regulation on both sides of the Atlantic.

By 2003, we argue, the Bush administration had come to believe that the costs of bringing a WTO case (further backlash against GMOs in Europe, spread of the anti-GMO movement to the US) had partially abated, while the global stakes of the debate, and thus the potential benefits of a WTO case, had substantially increased. During this period, a number of advanced industrialized countries had followed the EU's lead in requiring special approval and labeling procedures for GM foods and crops, and instituted moratoria on GM plantings, while some less developed countries in Africa had gone so far as to reject the provision of GM corn offered as food aid. If the US failed to act promptly, Bush administration officials feared, these policies could become entrenched beyond Europe and difficult to change later. At the same time, however, the number of countries growing significant acreage of GM crops had increased to include major agricultural producers such as Argentina, Brazil, Canada, China, and India, and though to a relatively minor extent, even Spain and

Germany within the EU.[23] In this context, the US had a strong incentive to try to arrest the spread of the EU's precautionary approach to a growing number of countries that might share Europe's views.

Significantly, the US and other complainants did not challenge the EU's legislative framework for GM approvals as such, nor did they challenge (despite loud complaints from producers) the EU's more recently adopted labeling and traceability provisions. Instead, the complaints focused on the EU's implementation of its regulatory framework, challenging three specific EU actions: (1) the EU's *de facto* "general moratorium" on new approvals; (2) "product-specific moratoria," or failure to approve particular GM varieties found to be safe by the relevant EU scientific bodies; and (3) the persistent use of "safeguard provisions" by individual EU member states to ban GM varieties that had been approved as safe by the Union's own scientific experts. In all three cases, the complainants argued, the Union had failed to base its regulatory decisions on scientific risk assessments as required under Article 5.1 of the SPS Agreement, and those decisions were, therefore, inconsistent with EU obligations under WTO law.

The EU, by contrast, denied the existence of any moratorium, maintaining that new approvals needed to wait until after the completion of the EU's regulatory framework. It further argued that the SPS Agreement did not apply (or applied only in part) to the regulation of GMOs, since the EU was concerned with the protection of its environment which fell outside of the SPS Agreement's scope. The EU contended that the legal claims should thus be assessed under other WTO and non-WTO agreements, including that of the EU's preferred forum, the Cartagena Biosafety Protocol.

In September 2006, the WTO dispute-settlement panel issued its decision, which was over one thousand pages of text. The panel expressly avoided examining many crucial issues and most particularly the questions "whether biotech products in general are safe or not" and "whether the biotech products at issue in this dispute are 'like' their conventional counterparts." The panel ruled primarily in favor of the US, but largely on procedural and not substantive grounds. It first found that the SPS Agreement applied to all of the legal claims. It then held that the EU had engaged in "undue delay" in its approval process in violation of Article 8 and Annex C of the SPS Agreement, and thus had yet to take an "SPS measure." In this way, the panel avoided determining whether the EU had based a decision on a risk assessment or whether the assessments showed actual risks or greater risks than for conventional plant varieties, because these SPS Agreement obligations only applied to "SPS measures." Ironically, by deciding that the EU had engaged in undue delay before taking an "SPS measure," the panel which took almost three years to issue a decision (in a legal process that was not to exceed nine months), itself avoided making any substantive evaluation.

Regarding safeguards enacted by EU member states, in contrast, the panel found that all of them were "SPS measures" that violated the EU's substantive obligations under Article 5.1 of the SPS Agreement because they were "not based on a risk assessment." It noted, in particular that the EU's "relevant scientific committees had evaluated the potential risks...and had provided a positive opinion." Thus, while the panel refrained from substantively evaluating decisions at the EU level, it expressly found that the member-state bans were inconsistent with the EU's substantive WTO commitments. Since the European Commission was already opposed to such member-state safeguards and member-state delay in the approval process, the WTO panel decision effectively reconfirmed the Commission's position in intra-EU politics, and thereby could potentially lead to greater transatlantic accommodation.

In Chapter 5, we apply a framework of comparative institutional analysis to assess the WTO panel's interpretive choices in terms of their institutional implications. This framework allows us to see how the WTO judicial process can attempt to allocate power structurally from one institution and one level to another, thus affecting who participates and how they participate in deciding which substantive policies to pursue. We examine and evaluate five radically different institutional alternatives available to the panel through its interpretation of the relevant WTO texts, ranging from deference to national law-making, to "judicial balancing," to allocation of decision-making to international political processes or to international markets. By shifting authority among institutional alternatives, the WTO judicial process alters relations between the decision-makers and the affected public. The optic here is to see the WTO judicial process through a broader lens of governance and not through a judiciocentric perspective focused solely on judicial interpretation and review. We show how, in the end, the panel chose to focus on the *procedures* of the EU approval process, and in this way effectively deferred to substantive decision-making in the EU, although subject to internationally imposed procedural constraints. We further show the similarities and differences of our comparative institutional approach with analytic frames currently used in the legal academy, such as global constitutionalism, conflict of laws/legal pluralism and global administrative law.

We then examine the WTO panel decision in light of the sociological legitimacy constraints confronting the WTO. In this way, we assess how national legal contexts reciprocally affect WTO legal decisions in an interactive process. The WTO judicial system, while striving toward objectivity in its rulings and deploying highly legalistic analysis, is necessarily concerned with compliance by the parties to the dispute and the general acceptance of its decisions by WTO members, especially those on whom the WTO's effectiveness depends. Even before the panel's GMO decision, the jurisprudence of the Appellate Body indicated a willingness to provide significant discretion to domestic regulators in determining the appropriate level of risk for the members of their society. In this case, we argue, the WTO dispute-settlement panel again left discretion to the EU in determining the level of acceptable risk, while spelling out the procedural decision-making requirements that it must meet before implementing trade-restrictive measures on GM foods and crops.

More generally, we contend, the WTO judicial process, while undertaken by unelected officials operating far from the purview of ordinary citizens, may nevertheless play a positive role, although subject to real constraints that we address. To start, it can help to manage trade conflicts by channeling them into a legal process. In our case, for example, it has helped US officials counter pressures to retaliate unilaterally against the EU's measures, which might well have occurred in the absence of WTO dispute resolution. In addition, the WTO judicial process can help clarify members' procedural obligations to justify their decisions to other members who may be adversely affected, including by reference to a scientific risk assessment. In this way, the WTO judicial process can help to correct the parochialism of national decision-making. WTO decisions, while frequently controversial and contested, have arguably pressed national decision-makers to make decisions that better take account of the impacts on foreigners while still meeting their regulatory objectives, and to justify those decisions in a transparent manner, knowing that they are potentially subject to scrutiny and review before the WTO dispute-settlement system.

Put differently, WTO rulings have not required *substantive convergence* of US and EU approaches to risk regulation, but they have spelled out *procedural obligations* that members have toward each other when seeking to protect their own societies from the various risks of modern life. In doing so, they can lead to greater accommodation of divergent regulatory systems in at least two ways. First, procedural obligations based on the use of public reason, building from scientific risk assessments, can empower certain actors in domestic processes (such as the European Food Safety Authority, the European Commission and the European Court of Justice, in our case) that lead to different substantive decisions on particular matters. Second, in cases where regulation decisions are challenged and are found to have met the procedural requirements, the WTO legal process may lead to greater acceptance of the foreign decision by the other side. In both ways, the WTO dispute-settlement system can facilitate accommodation of divergent regulatory decisions. Nonetheless, as we will see, where social and regulatory approaches to technology and its risks are deeply engrained, the impact of the WTO or any other international or transnational body on substantive regulatory convergence, at least in the shorter term, is significantly constrained.

1.6. US and EU regulatory developments since 2000: Continuity, change, and (lack of) convergence

Many analysts have expressed hopes that the US and EU regulatory systems might converge, relieving the friction between them, whether through joint deliberation or as a result of pressures exerted by international organizations such as the WTO, by national and transnational interest groups and public opinion, or by market forces. Many American observers, for example, hoped that the EU, under pressure from the WTO, might move toward what the US calls a more "science-based" and less "politicized" system of regulation which would, in turn, facilitate the approval of new GM varieties. Many European observers, by contrast, hoped that either public opinion or market pressures would prompt a process of "trading up" in the US, which might become more precautionary in its own regulations and thus more accommodating of European regulatory choices.

In Chapter 6, we review the impact of transnational political, legal, social, and market pressures on EU and US policies toward agricultural biotechnology. We identify the direction and the sources of change, and assess the evidence for convergence between the two systems. In both cases, we find, pressures from international markets and international institutions have provided some incentives for domestic change, but the impact of these pressures has been limited and channeled by the entrenched and path-dependent nature of interests and institutions on both sides of the Atlantic.

In the EU, the period since 2000 has witnessed a root-and-branch reform of the regulatory system for agricultural biotechnology, as the European Commission has sought to reassure the public about the completeness and the rigor of the EU regulatory system, and thereafter resume the stalled approval process for new GM foods and crops. Beginning with the publication of a White Paper on Food Safety in 2000, the Commission has proposed, and the Council of Ministers and European Parliament have adopted, a raft of new legislation regulating every aspect of GM food production "from farm to fork." Among other measures, the EU has strengthened the original 1990 directive on the release of GMOs into the environment, extending the scope of the regulation to include GM feed as well as products derived from, but no longer containing, GMOs. In addition, the Union has adopted binding legislation providing for mandatory labeling and traceability of all GM foods and crops, as well as a recommendation on the coexistence of GM and conventional crops, and new rules on the approval and cataloging of GM seeds. The Union's new legislative framework incorporates some elements of US practice and of WTO jurisprudence. Most notably, it requires scientific risk assessment of each GM variety by a newly created independent agency—the European Food Safety Authority (EFSA). The adoption of this strict and comprehensive regulatory framework was aimed, in part, at reassuring the European public about the adequacy of regulatory controls and hence at ending a six-year moratorium on the approval of new GM varieties in the Council of Ministers, which was one of the targets of the US legal complaint before the WTO. The *de facto* moratorium did indeed end in May 2004, in the middle of the WTO case, with the approval by the Commission of a new variety of GM maize.

However, the controversy over GM foods and crops shows no signs of abating in the EU. Public opposition to GMOs remains high throughout Europe (including in the twelve new member states that joined the Union since 2004). This opposition has been reflected in the Council of Ministers, which has consistently deadlocked on the approval of new GM foods, leaving the final decision to the unelected European Commission. Although the Commission, for its part, has wanted since 2004 to overturn a series of national bans on specific GM varieties, which the EU's own scientists have found are not supported by scientific evidence, the Commission's efforts were repeatedly rebuffed by the Council, where a large number of member states opposed the initiation of a legal challenge by the Commission.[24] Moreover, other aspects of the revised EU legislative framework, such as its labeling and traceability provisions, represent a further move *away* from the more accommodating US model, so that the EU system has actually gotten worse for many US growers. Even though the EU has developed a complex framework for the approval of GM crops and foods, whether they will be approved and, in light of the new labeling and traceability requirements, actually marketed, remains in doubt.

For these reasons, we conclude that the EU regulatory system, despite its many modifications over the past decade, remains, as it has been, a strict and highly precautionary system. It continues to regulate GM foods and crops stringently in terms of the process of their production (their use of genetic engineering), rather than the characteristics of the product. More importantly, in practice it continues to impede the commercialization of GM foods and crops in Europe as well as around the world because of the importance of the EU market for foreign farmers and the overall normative influence of the EU in global politics.[25] In sum, we argue that while the EU regulatory system has been overhauled, EU policies and practices remain similar in their effects. In this sense, the EU's increasingly complex, Byzantine system for authorizations and marketing of GM varieties, incorporating multiple governmental actors and non-government stakeholders, can be viewed as something of a Potemkin village. The *de jure* regulatory system, with its many procedures for consultation, looks impressive on the surface. But for many non-EU constituencies the system appears to be largely a sham to meet formal WTO requirements, as key member-state veto players ensure that no GM crops or foods are marketed, regardless of scientific risk assessments. The result, in their view, is lots of costly show, but with a predetermined outcome. We call this *reform without change*.

In the US, meanwhile, national regulators had adopted a more flexible, product-oriented regulatory system, while biotechnology companies and farmers had embraced GM foods and crops far more readily than in Europe. By the end of the 1990s, the US faced some pressures for change, leading some scholars to speculate that the US might "trade up" to the precautionary and process-based European approach.[26] These pressures took the form of three inter-related phenomena: (1) commercial adaptation, which occurs when US firms or farmers voluntarily comply with EU standards, in order to gain access to the EU market (e.g., growing only EU-approved GM varieties); (2) political mobilization, which occurs when domestic US interest groups, spurred (at least in part) by events in Europe, mobilize for stricter GM regulations; and (3) policy change, when US authorities adopt stricter domestic regulatory practices, whether to protect Americans from risks or to reassure foreign markets and foreign governments of the safety and content of US products.[27] A careful analysis of recent US events provides some evidence of commercial adaptation and political mobilization, as well as some modest policy change. However, these policy changes largely reflect an incremental elaboration of the traditional US system rather than any regulatory overhaul in the direction of the EU's approach.

With regard to commercial adaptation, US farmers' and growers' associations have based their decisions on which crops to plant at least in part on the regulatory standards of the EU and other important markets such as Japan and Korea. Many farmers, for example, have concentrated production of corn and soybeans in those GM varieties that have been approved for marketing in the EU, while resisting the introduction of new GM wheat or rice varieties not approved for use in the EU. We also find some evidence of US farmers avoiding the use of US-approved and even EU-approved GM crops, in order to appeal to the EU market for GM-free foods and avoid having to comply with the EU's increasingly strict labeling and traceability requirements. The commercial prospects for new GM foods and crops in the US, therefore, remain unclear. On the one hand, US farmers have showed little inclination to abandon established GM varieties, such as soybeans, cotton, canola, and corn, the use of which continues to grow in the US. On the other hand, GM production in the US has increasingly concentrated on these four crops, while notification of new varieties and commercial acceptance of other GM crops have decreased from the rapid pace of the late 1990s.

With regard to political mobilization, the evidence suggests that media coverage of the US/EU dispute, together with certain domestic scandals such as the 2000 Starlink controversy (in which a GM corn approved only for animal feed was found in corn chips and other food products), provided opportunities for US consumer and environmental groups to mobilize in opposition to GM foods and crops. This mobilization has so far been unsuccessful in the US (unlike in Europe), and there is little evidence that US public opinion shares the deep distrust toward GMOs felt by the European public. Polls show relatively high levels of trust in federal regulators such as the FDA, and much less support (and even less intensive political pressure) for stricter regulation of GM foods in the US.

At the level of federal regulation, finally, there have been debates among US legislators and regulators about possible reforms of the US regulatory process, but the US Congress has not produced any significant changes to the statutory basis for US biotechnology regulation. In the absence of legislative action, the most important regulatory developments have come from government regulators such as the FDA and the USDA, which conducted various hearings and studies to consider administrative changes to the existing regulatory system, including the possibility of introducing mandatory labeling or premarket approvals of new GM varieties. These hearings led the FDA to make some changes to its procedures, including the issuing of guidelines for companies to undertake voluntary notification to the FDA of new GM foods, as well as guidelines for companies wishing to voluntarily label their products as being organic and thus produced without the use of GM varieties. Nevertheless, the agency declined to follow the EU practice of requiring mandatory prior approval of all GM foods and crops, nor did it endorse mandatory provisions for the labeling and traceability of GMOs. In addition, some farm groups have asked the USDA to take account of administrative authorizations of GM varieties in key export markets before permitting their planting in the US because of the difficulty of segregating grains to ensure they meet that market's requirements, but so far it has not been done. Reform of the US regulatory system thus remains on the US agenda, with the USDA also undertaking reviews of its regulatory procedures, but such reforms are likely to be piecemeal and relatively modest in comparison with Europe's regulatory requirements. Hence, by comparison with the EU, where we found much reform with little or no fundamental change, in the US we find some, primarily market-oriented, *change without reform* of the regulatory framework.

In both cases, moreover, we find striking evidence of positive feedback, self-reinforcing systems, and path-dependent development. In the US case, the early adoption of a welcoming regulatory framework in the 1980s contributed to the growth of a strong biotech industry and the widespread acceptance of GMOs among farmers and (to a lesser extent) public opinion, creating a powerful constituency for the new technology from below. At the same time, US institutional rules privilege the status quo, in which GMOs continue to be regulated under the two-decades-old Coordinated Framework, which has changed only at the margins in the absence of new Congressional legislation. In the EU case, by contrast, the early adoption of a highly restrictive regulatory framework, together with the food-safety crises of the mid-1990s, discouraged farmers from planting GM crops, prompted retailers to resist selling GM foods, and led to a flight of biotech investment from Europe, all of which undermined political support for GM foods and crops. Furthermore, the EU's supermajoritarian legislative rules, requiring a qualified majority among disparate states in the Council of Ministers, as well as a majority in the European Parliament, have created a huge institutional hurdle to any fundamental reform of the EU regulatory framework.

The story of GM crops and foods is an ongoing one, and there could be more convergence in the future in response to increasing understanding by regulators of the risks or benefits of GM foods and crops, or to exogenous shocks such as a future food-safety or environmental crisis, or to a change in the framing of the issues in light of developments in large developing countries such as Brazil, China, and India. We conclude that indeed, the period since 2000 has seen some changes in US and European regulatory procedures and market behavior. Some elements of these changes can be interpreted as responses to external pressures, and as modest steps by each side toward some movement that accommodates the other. Yet despite these changes, we find at best limited evidence of fundamental convergence between the two systems. Notwithstanding more than a decade of negotiation, deliberation, and dispute, the differences between the US and EU regulatory systems have proven to be robust and enduring.

1.7. Conclusions

In the conclusion (Chapter 7), we draw out a series of five policy lessons from our study. We find in particular that the transatlantic dispute over transgenic foods and crops is unlikely to be *resolved* in the near future, whether by bilateral deliberation, multilateral negotiation, gradual market-driven convergence, or international litigation. Yet, we argue, the dispute can and should be *managed* through a combination of these and other methods. Bilateral consultations have thus far failed to live up to the high expectations of deliberative decision-making. Yet, we do find evidence of deliberation among scientific experts over risk assessment, and this effort to establish scientific (if not political) consensus is worth pursuing, preferably within multilateral regimes like the OECD and Codex, which have the additional advantage of including third parties who engage in the regulation and trade of GM products. Multilateral regimes, while subject to distributive conflicts and to forum-shopping and inconsistency across regimes, also have a potentially positive role to play, by lowering the transaction costs of negotiations, providing a common vocabulary for discussing GMO regulation, clarifying at least some of the mutual obligations of the parties, and contributing to regulatory knowledge and capacity-building. We find that the WTO, while treading relatively lightly in a deeply politicized area, can have—and has already had—a positive impact on the conflict, by channeling the dispute into legal processes, clarifying the procedural obligations of both sides, and catalyzing the pressure to increase the transparency of domestic regulatory processes.

During our study, an important new factor arose—the emerging role of large developing countries in the struggle over genetically modified crops. China and India have adopted GM cotton for textiles, and Brazil and Argentina have adopted GM soy. With the rise of these countries as players in the world economy and in its governance, other developing countries may look to them when making their own choices about GM varieties. We therefore also conclude the book by examining how the US–EU dispute has affected the developing countries, and the role that these countries may play in shaping the future of agricultural biotechnology in a dispute that is no longer purely transatlantic but increasingly global in scope.

Not surprisingly, in light of the many domestic and international pressures on developing-country policy, and in the absence of a generally accepted global standard, developing countries have adopted a wide range of regulatory stances toward agricultural biotechnology, running the gamut from US-style acceptance and widespread commercialization to strict rejection of the new technology. Looking across the developing world, we nonetheless see two broad and somewhat contradictory trends in the regulation and commercial adoption of GM foods and crops. On the one hand, cross-national surveys of GMO regulation indicate that with a few exceptions, the general trend in regulation has been toward greater stringency. A growing number of countries have enacted new regulatory approval and monitoring requirements, as well as more restrictive rules for the importation and labeling of GM foods and crops, reflecting in part transnational capacity-building efforts catalyzed by the Cartagena Biosafety Protocol. At the same time, however, studies on GMO cultivation and commercialization reveal impressive increases in the adoption of GM crops, with the greatest growth rates in the developing world. Indeed, it seems likely that larger developing countries such as Brazil, China, and India will continue to invest in GM technology, cultivating a select group of GM crops, and most others will agree to import at least some GM products.

In sum, if we extrapolate from current trends, it seems likely that agricultural biotechnology will be increasingly accepted over time, albeit within significant market and regulatory constraints for GM foods directly consumed by humans. Nonetheless, the development of GM foods and crops, their regulation and commercialization, are also likely to continue to be shaped, for good or ill, by future contingent events. A major food-safety or environmental crisis involving GM foods and crops is, in our view, unlikely in light of over a decade of experience—yet such a crisis, if it occurred, would likely have a profoundly negative impact on the commercial acceptance and future regulation of GM foods and crops around the world. In contrast, the development and adoption in developing countries of second-generation GM crops that significantly benefit the world's poor could also help swing public opinion, markets, and regulators in favor of agricultural biotechnology around the world. To the extent that a resolution of the conflict is possible, it is most likely to be catalyzed neither in the courtroom nor at a negotiating table, but in the laboratory and in the fields.

1.8. A note on methods and theoretical implications

Finally, before proceeding in the next chapter to the challenges of risk regulation and genetic engineering, we close this chapter with a brief discussion about the nature of our chosen topic, the generalizability of our findings, and the research methods we have employed. We have chosen to delve deeply into a particular policy area in which political and legal institutions at multiple levels interact within a global market context, rather than address multiple issue-areas, as we have done in earlier work.[28] The dangers of generalizing broader conclusions from a single case study are well known, and we take care throughout the book to acknowledge the aspects of the dispute that are distinctive to the subject of agricultural biotechnology. We, nonetheless, decided to focus on this case for three reasons.

First, the regulation of GM foods and crops is an area of major public policy import involving revolutionary technologies that could bring significant benefits, yet are also considered by many to pose considerable risks. The transatlantic dispute over agricultural biotechnology matters not only for farmers and biotech companies that produce GM foods and crops, but also to each of us, for whom it will affect the food we eat, the clothes we wear, and the environment we inhabit. In a world where agricultural trade continues to grow faster than agricultural production, and where the global food distribution system cannot guarantee the segregation of seed varieties, the resolution of the regulatory conflicts between the US and EU will facilitate or impede the adoption of agricultural biotechnology, and in this way, will shape the future of agriculture.[29]

Second, we believe that the GMO conflict is emblematic of issues that will arise in the future, in an economically globalized world characterized by rapid technological changes having uncertain effects. Current disputes over Europe's new legislation for chemicals, and potential future ones over nanotechnology are two examples. Future technological developments, including agricultural biotechnology, will affect a broad spectrum of concerns, ranging from international competitiveness, trade and investment, research and development, environmental risk, human and animal welfare, consumer protection, poverty eradication, human rights such as food security, the ethics of new research, the relative roles of scientific and political oversight of regulatory approvals, and the impact of foreign and international law and of global markets on national decision-making and local social orders. Understanding how domestic polities have governed the new technology of GM foods and crops, and how international networks and regimes have succeeded or failed in coordinating domestic regulations in the face of risk, is therefore, a crucial step in understanding the regulatory challenges to be faced in a rapidly changing world in the years to come.

Third, and most generally, the GMO case is one in which two large, market-oriented and democratic actors have attempted repeatedly to cooperate in an area that would traditionally have been characterized as "low politics" in international relations, and in which a plethora of bilateral networks and multilateral regimes stood ready to facilitate cooperation. Yet, such efforts at cooperation have failed repeatedly, leading us to believe that a careful understanding of the GMO case would reveal both the impediments to successful cooperation and the contributions that international institutions and international law might nevertheless make when cooperation fails.

Throughout the book, therefore, our aim is not only to investigate and offer a definitive empirical account of the transatlantic GMO dispute—although we hope we have done just that—but also to draw on a series of "mid-range" theories that address the various aspects and stages of the dispute, and to engage in the testing and further elaboration of those theories. Hence, in Chapter 2, we elaborate the theories of comparative public policy-making, path-dependence, and two-level games; in Chapter 3, theories of deliberation and persuasion; in Chapter 4, theories of distributive bargaining, regime complexes, and forum-shopping; in Chapter 5, legal theories of international judicial decision-making; and in Chapter 6, theories of international–domestic law and policy interaction. In one sense, therefore, this book represents "applied theory," drawing on theories of politics and law to analyze and think critically about one of the vital public-policy issues of our day.[30] At the same time, however, we believe that our analysis of the GMO case has implications for broader theoretical inquiry, providing an important test case for theories of public policy-making, network governance, and deliberation, and generating new insights and hypotheses about the workings of international "regime complexes," the interaction of hard and soft laws, and the role of international courts and tribunals when cooperation fails. We, thus, make no attempt to develop any "grand theory" of the GMO conflict, but rather seek, in this book, to borrow from theoretical literatures on each of these topics, while also aiming to make contributions back to those literatures, identifying the implications and lessons of our findings for the study of international relations and international law more generally.

In methodological terms, this study is based on more than eight years of joint empirical investigation, drawing on a wealth of primary and secondary sources, including official documents from the US and the EU, as well as non-governmental organizations and the many multilateral regimes that have dealt with one or another aspect of the conflict. Crucially, moreover, we have cross-checked our findings from these written sources against interviews with a wide range of European and US government regulators, international civil servants, scientists, representatives of business, farmers associations, anti-GMO activists, and other stakeholders holding different ideological and policy orientations about the dispute. In many cases, these individuals have requested anonymity in making their comments, and in those cases we have identified these subjects by their institutional affiliations and the date of the interview; where subjects have agreed to allow it, we have identified them by name. In addition, we have attempted, wherever possible, to corroborate information gleaned from interviews in other sources, to minimize our own and the reader's reliance on non-replicable interview data. Throughout the book we have attempted to process-trace[31] our hypotheses and claims at both the national and international levels, triangulating on our subject using all available sources, whether primary or secondary, quantitative or qualitative, written or interview-based, and we have documented our sources as fully as possible in the footnotes and references to the book.

Although the empirical story of GM foods and crops is a moving target, we hope that this study provides an accurate and insightful analysis of the domestic sources of international regulatory conflict; the attempts at and impediments to cooperation through deliberation and bargaining in transnational networks and multilateral regimes; the roles that international law, international regimes and international markets can nonetheless play, albeit within severe limits that we identify, in domestic systems; and ultimately what happens to international law, international regimes and markets, when cooperation fails.

2

The Domestic Sources of the Conflict: Why the US and EU Biotech Regulatory Regimes Differ

The US and the EU represent the two largest economies on earth and, at least at first glance, possess striking political similarities: both polities are part of the advanced industrialized world, both are democratic, and both feature federal or quasi-federal political systems in which power is shared between the state and central governments in Washington D.C. and Brussels, respectively.[1] Despite these similarities, the US and the EU have taken sharply different approaches to the regulation of agricultural biotechnology, adopting not only different regulatory *standards* but also different regulatory *systems* for the approval and marketing of GM foods and crops. These different approaches, which have aptly been called "regulatory polarization,"[2] have in turn led to bilateral trade disputes and to a contest in which each of the two parties has sought to export its own approach to the rest of the world.

Analysts have identified a number of causes that could explain these differences. Some stress *cultural differences* in European and American attitudes toward food or toward risk; some point to *institutional differences* in US and European assessments and management of risk; some highlight differences in *interest group configurations*; and some note the differential impact of *contingent events*. In this chapter, we present the respective regulatory systems put in place during the 1980s and 1990s by the US and the EU to govern GM foods and crops, offer an explanation for the differences that arose, and trace the dynamics that led to the transatlantic dispute. As we shall see, the US opted to treat GM foods and crops as essentially equivalent to their conventional counterparts, regulating them under existing laws administered by three US regulatory agencies—the Food and Drug Administration (FDA), US Department of Agriculture (USDA), and Environmental Protection Agency (EPA). In contrast, EU regulators, operating in a multi-leveled regulatory system and in the context of widespread public distrust of both regulators and GMOs, adopted a stricter, more precautionary approach, regulating GMOs in terms of the process whereby they were produced, rather than the characteristics of individual products. This approach, which we analyze in detail, by the mid-1990s led to tensions with the US, whose farmers had adopted GM varieties and now saw their exports to Europe increasingly affected by the strictness and slowness of the EU regulatory process.

The chapter is organized in five parts. In part 1, we introduce the new technology of genetic modification, its purported benefits and risks, and the central concepts of risk regulation as well as their application in the US and EU in the context of rapid technological change. After presenting the US regulatory system for agricultural biotechnology in part 2 and the EU system in part 3, we analyze and seek to explain the causes of the regulatory polarization in part 4. We then conclude in part 5 with a brief discussion of two-level games analysis, which helps us to model how domestic constituencies within the US and EU both empower and constrain US and EU officials when they attempt to engage in regulatory cooperation over agricultural biotechnology, which we address in Chapter 3.

Our arguments regarding why the US and EU have taken such different approaches to this technology, and thus the domestic sources of the conflict, are two-fold. First, we argue that the very different approaches to GMO regulation that we observe between the US and the EU were not determined in any straightforward way by the institutions, the risk culture, or the interest-group configurations present on each side of the Atlantic. It was *not* inevitable that the US regulators would adopt a more product-based approach to GMO regulation, nor was it obvious from the outset that the EU would adopt the strict, politicized, and highly precautionary system that emerged over the course of the 1990s. *The best explanation for the observed transatlantic differences, we contend, is multi-causal, lying in the ability of interest groups to capitalize on preexisting cultural and institutional differences, with an important role played by contingent events such as the European food safety scandals of the 1990s.*

Second, however, we argue that once initial choices were made on each side of the Atlantic, these differences became entrenched through statute, through social practices and through public opinion in ways that have made each system highly resistant to change. To the extent that change has proven possible (a subject we address in Chapter 6), it has demonstrated a clearly path-dependent character, introducing incremental changes at the margins of regulatory systems that have remained remarkably stable in their essentials over time. For this reason, US/EU regulatory polarization, although not inevitable from the outset, has proven robust and enduring, as have the trade disputes that it has generated.

2.1. Agricultural biotechnology and risk regulation

Genetic engineering, the process used to create GM seeds, crops, and the foods produced from them, is a technology used to isolate genes from one organism, manipulate them in the laboratory, and inject them into another organism. Supporters of agricultural biotechnology consider such genetic manipulation to be merely the latest step in an ongoing process of human intervention in nature to stabilize and increase crop yields and to improve the nutritional quality, appearance, and taste of plant varieties, from the farmer's "old-fashioned" selection of seeds and Mendelian cross-breeding to the use of radiation to scramble genetic material in crops to the mapping of plant and animal genetic codes.[3] In other words, they argue that the definition of a "conventional" crop is ambiguous in light of the ongoing development of scientific techniques of plant breeding over time. Opponents, in contrast, stress its novelty on account of its mix of genes from different species.

Supporters maintain that the characteristics of these new plant varieties offer significant benefits to both producers and consumers. The benefits to producers have been most evident in the "first generation" of GM crops, as new GM varieties can provide greater efficiency and lower costs in agricultural production. As an extensive report of the Pew Initiative on Food and Biotechnology states, "For the most part, this first generation of agricultural biotechnology products consists of single-gene, single-trait modifications made for agronomic purposes, primarily to make crops pest resistant or herbicide tolerant."[4] The predominant GM trait for pest resistance, known as Bt, genetically incorporates resistance to pests such as the corn borer.[5] The predominant trait for herbicide resistance genetically incorporates resistance to the commercial Roundup herbicide, such as in Monsanto Corporation's Roundup-Ready soybeans.[6]

Direct consumer benefits, in contrast, have been less immediately evident, as the most common GM crops require less maintenance by farmers without enhancing the quality or necessarily reducing the price of the product to the consumer.[7] However, the second generation of GM crops could more significantly benefit consumers, including the poor in developing countries, by increasing crop yields and nutritional attributes.[8] Some GM traits, for example, could provide greater resistance to drought and salinity, critical for food-scarce regions. Greater yields could reduce prices, critical for the urban poor, while helping to preserve forests and biodiversity because of the reduced agricultural clearing. GM crops could also provide important health benefits by adding vitamins and nutrients to conventional crops, resulting in products such as vitamin A-enhanced rice, "heart-friendly" oil, and iron-enriched wheat.[9] For advocates of GM foods and crops, it is these long-term benefits, together with the short-term cost advantages to farmers, which constitute the promise of agricultural biotechnology. Blocking advances in this technology entails its own risks, they contend, to human life and health, especially in poorer countries.[10] A major study by the Food and Agriculture Organization (FAO) concludes,

Thus far, in those countries where transgenic crops have been grown, there have been no verifiable reports of them causing any significant health or environmental harm... On the contrary, some important environmental and social benefits are emerging. Farmers are using less pesticide and are replacing toxic chemicals with less harmful ones. As a result farm workers and water supplies are protected from poisons, and beneficial insects and birds are returning to farmers' fields.[11]

GM supporters maintain, moreover, that the health risks from eating organic foods are much greater than for GM ones.[12] In fact, the US Centers for Disease Control and Prevention "estimate that each year in the US, 76 million people get sick, 325,000 are hospitalized, and 5,000 die from food-related illnesses," so that assuring food safety is rightly a major concern.[13] For example, *E. coli*-tainted bagged spinach from Natural Selection Foods killed at least three people and sickened more than 200 others in the US during the fall of 2006.[14] All the attention given to GM foods, supporters argue, is diversionary, stripping resources from other areas, and thus arguably increasing food safety risks.

Finally, GM crops can provide for more than food and clothing. As governments and private businesses search for alternatives to carbon fuels, GM technology could be important for the creation of more efficient biofuels, which are now being heavily subsidized in the US and Europe for a "greener future," although there is again contention about its prospects.[15] Genetic modification of crops for energy production can improve the chemical processes for converting crops into fuel so that their conversion consumes less energy.[16] Biotechnology products could also be introduced into the environment to clean it. For example, biotechnology could be deployed to clean toxic spills and other wastes more effectively, including through the introduction of a highly radioactive-resistant bacterium that expresses enzymes that degrade toxic materials.[17] Transgenic plants can also be used to produce therapeutic proteins that are essential to the research and production of various pharmaceutical products.[18] Biotechnology has been used to produce biopharmaceuticals such as insulin, as well as drugs used in the treatment of cancer and AIDS.[19] In short, many call for further development of this technology. However, until the benefits of agricultural biotechnology become more manifest, there will remain a "legitimacy" gap between the claims of agricultural biotech advocates and the characteristics of the GM plant varieties in the market.[20]

Skeptics, in contrast, have raised concerns over food safety, environmental harm, agribusiness power, and ethics, pointing to longer-term uncertainties. Many critics question the safety of GM foods, maintaining that they could encourage perverse selection for antibiotic resistance (through the consumption of foods with antibiotic marker genes) or trigger allergenic reactions (through the ingestion of genes introduced from foreign species, such as peanuts). They question whether GM foods with antibiotic resistance markers might not be fully broken down in the digestive tract but enter the human stomach lining, increasing human resistance to antibiotics.[21]

Environmental critics raise fears that the technology could lead to monocultures, impairing biodiversity, and potentially wreaking unintended consequences on other species in the food chain. They stress how GM varieties could give rise to "super weeds" and "super bugs" through cross-pollination and pest adaptation. To the extent that pollen from GM plants travels, it could also cross-breed with organic varieties, undermining the prospect of alternative GM-free organic agriculture. Because GMOs can be "self-replicating," critics fear that adverse consequences could be irreversible.[22]

Some skeptics are less concerned with the technology itself (or in fact may even favor its adoption), but rather are critical of its method of ownership through patents. Defenders of small-scale agriculture maintain that patented GM seed varieties favor agribusinesses, and threaten the livelihoods of small-scale farmers throughout the world who would no longer be able to save harvested seed for the next year's crop.[23] Similarly, those defending the interests of agriculture in developing countries are concerned with the fact that farmers in developing countries could become dependent on highly concentrated intellectual property holders in the US and Europe who could extract monopoly rents for patented seeds and thus control their access to important crop varieties.[24]

These concerns were the driving force behind the uproar surrounding Monsanto's plan to develop "Terminator technology" that would prevent farmers from saving second-generation seeds for replanting, because the seeds would be rendered infertile. For a patent holder such as Monsanto, terminator technology protects its investment in intellectual property through engineering. Critics, on the other hand, see this technology as a scheme to undermine the independence and food security of farmers in developing countries.[25] Terminator technology became a public relations disaster for Monsanto and other biotech seed companies. Just over a year after Monsanto's announcement in 1998 of its proposed merger with the company that owned this technology, Monsanto publicly disavowed any attempt to commercialize it.[26]

A dispute between the Indian arm of Monsanto and the state government of Andhra Pradesh exemplifies critics' concerns, at least in part. Andhra Pradesh believed that Monsanto was charging local farmers exorbitant prices for the company's GM cotton variety.[27] In January 2006, Andhra Pradesh petitioned India's Monopolies and Restrictive Trade Practices Commission, claiming that Indian farmers were being charged royalty fees twelve times higher than paid by farmers in the US, and thirty-six times higher than paid by farmers in China.[28] Monsanto eventually agreed to reduce the royalty fee it charged Indian farmers by 30 per cent.[29]

Many opponents raise further ethical concerns that complement the ones based on risk, uncertainty, and corporate control of seeds. Some question the morality of mankind's manipulation of genes, characterized by a statement of Britain's Prince Charles that the production of GM foods "takes mankind into realms that belong to God and to God alone."[30] The genetic engineering of animals provides a clear example of how ethical issues complement environmental ones. For example, faster-growing GM salmon could potentially cross-breed with wild species, eventually eliminating them and degrading larger ecosystems.[31] Those who maintain that humans owe ethical obligations to other species would find that such genetic engineering violates these obligations.[32] Now that applications regarding genetically engineered animals and products derived from them are proceeding through the US regulatory process, with one GM ornamental fish already being commercialized, debates over ethical concerns could intensify.[33] At the extreme, the development of genetic-engineering technology could lead to the deliberate modification of human genes, which some critics argue could "pose a threat to human dignity" and "diminish our humanity."[34]

Finally, cultural theorists of risk maintain that the management of risk is fundamentally a reflection of cultural "values," and thus raise fundamental issues of democratic control of science. They contend that if "risk disputes are really disputes over the good life, then the challenge that risk regulation poses for democracy is less how to reconcile public sensibilities with science than how to accommodate diverse visions of the good within a popular system of regulation."[35] In the case of multiple jurisdictions, the challenge is greater still, namely how to accommodate multiple, conflicting, and overlapping sets of regulations, each reflecting a distinctive conception of "the good life." In that context, the spread of GM varieties patented by US-based companies is often seen as one more reflection of US cultural hegemony, and an attack on alternative ways of living.

Many of these debates over risks and benefits have largely been framed in the US and Europe, without taking into account the perspectives and priorities of those in other parts of the world. Proponents and opponents in the US and Europe may refer to the needs of developing countries, but they do not represent them. Proponents stress the potential of GM foods to reduce malnutrition and disease, while opponents point to the biodiversity challenges posed by GM crop monocultures. They also point to the adverse social impacts on farmers in developing countries caused by the widespread use of seeds owned and controlled by US and European multinational companies. Many opponents of this new "gene revolution" remain critical of the earlier "green revolution," despite the increased food yields for farmers and lower prices for consumers that it produced. As hundreds of non-governmental groups wrote to the FAO's Director General following the FAO's 2003–04 report: "If we have learned anything from the failures of the Green Revolution, it is that technological 'advances' in crop genetics for seeds that respond to external inputs go hand in hand with increased socio-economic polarization, rural and urban impoverishment, and greater food insecurity."[36] In the context of these struggles over defining principles (e.g., science and precaution) and their enactment in formal law, developing countries must balance their desire for access to European commercial markets and their competition with US and other agricultural exporters, along with local concerns. Although the FAO has noted case studies showing "how biotech can be deployed to help the poor and hungry,"[37] the evaluation of the prospects and risks of agricultural biotechnology for these countries and their constituencies have so far taken a backseat to US–European commercial, regulatory, and cognitive framings. We examine the potential for a shift in such framings in Chapter 7.

2.1.1. Assessing and managing risks under uncertainty

Commentators generally agree that biotechnology regulation concerns the *regulation of technological risk under uncertainty*. Since we live in an economically and environmentally interdependent world, such regulation pits specific regulatory standards and broader regulatory systems of powerful states against each other. However, from here, supporters and opponents of the US and EU approaches quickly diverge. Supporters of the new technology tend to focus on the scientific assessment of its *risks* that can be measured and managed, while skeptics tend to focus on the management of *uncertainty* that belies the possibility of any meaningful measurement of long-term effects, calling for greater precaution and the recognition of different values underlying risk perceptions.

In the world of risk regulation, *risk assessment* refers to the technical assessment of the risks in question, *risk management* to the policy responses to these assessments, and *risk communication* to the way in which government regulators convey information about these risks to the general public.[38] These concepts are ideal types, since completely isolating risk assessment and risk management determinations can be difficult (and some argue impossible) in practice. The concept of risk itself refers to "the combination of the likelihood (*probability*) and the harm (*adverse outcome*, e.g., mortality, morbidity, ecological damage, or impaired quality of life), resulting from exposure to an activity (*hazard*)."[39] In principle, regulators faced with a novel product or process—such as the genetic modification of foods and crops—need to ascertain the potential harm caused by such activities, the probability of such harm, and the costs and benefits of feasible alternatives in order to take a decision on the legality, illegality, or regulatory conditions for that product or process. To do so, they often start with a technical assessment of the likelihood of harm—that is, with a risk assessment.

Skeptics of the technology tend to focus on the *uncertainty* of its effects, insisting that uncertainty is a different concept than risk, and entails not just differences in degree but of kind. They point to the classic Knightian distinction that risk is something that one can calculate, while uncertainty is something that one cannot.[40] Borrowing from sociologists such as Ulrich Beck, some argue that, with such modern industrially produced risks, "the *actual* consequences ultimately become more and more incalculable."[41] They maintain that what one cannot control with GMOs, in particular, are the long-term ecological effects of their adoption, effects that cannot be estimated, modeled or predicted. Skeptics also raise the complementary concept of *ignorance* in which "not only the probabilities, but also some possibilities may be unknown."[42] That is, the very nature of the possible harm and its magnitude are unknown. They thus focus on the need for considerable precaution in risk management decisions regarding the adoption of such new agricultural production processes. Finally, they argue that risk assessments under uncertainty are inherently value-laden, since the perception of risk itself reflects cultural predispositions, including among scientists, so that risk assessments are not purely technical.[43] We will see in this book how these different approaches to risk assessment and risk management have played out domestically in the US and EU (in this chapter) and internationally in different international regimes (Chapter 4).

In many ways, such decision-making over technological risk is similar to what Max Weber identified as "calculation" on the part of the capitalist entrepreneur.[44] The entrepreneur is uncertain of the state of the world, but engages in ventures which involve risk. In making an investment, the entrepreneur must calculate risk under uncertainty. And so it is with scientists and regulators when they assess and manage technological risk. They are uncertain what they know, and there is risk in what they venture. Yet there is a huge difference between the regulator and the entrepreneur. The risk for the entrepreneur is a private one, but for the regulator, it is a public risk—one that, in some cases, could be catastrophic.

Supporters of the technology nonetheless maintain that it is an error to focus on uncertainty, because science admits for little certainty. Rather, science focuses on degrees of risk that can be tested and progressively reduced. They contend that there are not only potential costs, but also potential benefits from the technology that need to be assessed and compared. In fact, Knight developed the very concept of uncertainty in an attempt to explain "profit," as reflected in the title of his famous book *Risk, Uncertainty and Profit*.[45] Were society to attempt to eliminate uncertainty, it would also eliminate great and beneficial technological advances, from electricity to the airplane.[46] As regards agricultural biotechnology, proponents maintain that the GM seeds and foods at issue pose no greater risk than conventional varieties, and offer real (and potential future) benefits. The US, they argue, has provided a free laboratory experiment for the world in which GM-derived crops and foods have been grown and consumed for over ten years without any proven harm. Had there been harm to humans, animals or the environment, proponents argue, surely it would have been uncovered in the US which has the strongest tort-liability system in the world.

Leading sociologists such as Ulrich Beck and Anthony Giddens contend that questions of risk regulation have become defining traits of modern society.

They theorize "modernity" in terms of the emergence of what they call a "risk society" in which risks are increasingly "manufactured," as opposed to being "natural" or "external" to human activity, and in which the management of risks becomes a defining element of societal conflict and social understanding.[47] As Giddens writes, "Risk is the mobilizing dynamic of a society bent on change, that wants to determine its own future rather than leaving it to religion, tradition, or the vagaries of nature."[48] As Beck puts it, "in advanced modernity, the social production of *wealth* is systematically accompanied by the social production of *risks*. Accordingly, the problems and conflicts related to distribution in a society of scarcity overlap with the problems and conflicts that arise from the production, definition and distribution of techno-scientifically produced risks."[49] Since the risks are often global in their potential consequences, the resulting societal conflicts also become transnational and global ones.

Whether or not one accepts the claim that we live in a fundamentally different form of society, in which the regulation of manufactured risks is a defining trait, regulators must act in the face of uncertainty regarding the nature and extent of the risks posed by new products and production processes. They sometimes take *precautionary* measures, restricting or banning certain products or activities in the absence of complete information about the risks posed. Dynamic technological change raises fundamental questions about the governance of technology and risk.

2.1.2. Comparative risk regulation

In a review of comparative risk regulation, Giandomenico Majone categorizes four distinct ways in which government regulators in the US and other jurisdictions have responded to risks, listing them in order of decreasing regulatory severity or increasing regulatory rationality, depending on one's perspective. Regulators have responded to risks by (1) imposing product bans; (2) setting standards that minimize risk "to the extent feasible"; (3) enacting requirements to eliminate "significant risks," typically following a risk assessment procedure; and/or (4) engaging in cost-benefit analysis, prohibiting a product or process only to the extent that one calculates that its risks outweigh its benefits, possibly allowing for a margin of error. Majone maintains that there has been a general trend over time from (what is in his view) the first and the least sophisticated to the fourth and the most sophisticated approach.[50] In particular, Majone concludes that American policymakers, regulators, and courts progressed, in the space of some three decades, to a highly sophisticated approach to risk regulation, relying on scientific assessments of risk and economic assessments of costs and benefits—"an outstanding, and in many respects unique, case of policy learning."[51]

While a useful heuristic device to understand the range of possible approaches to regulating risk under uncertainty, Majone's classification scheme simplifies a complex US response to risk that today combines elements of all four approaches under different laws and in different areas. Even more importantly for our purposes, this ideal–typical progression fails to capture parallel developments in Europe, where risk regulation took place largely within national contexts until the 1980s, when EU institutions began to play an increasing role in harmonizing risk regulation across the EU's various member states.

In the EU context, David Vogel and others have argued that Europe's approach to risk regulation has evolved quite differently than in the US.[52] Whereas the former began with highly precautionary legislation in areas like the environment, consumer protection, and worker health and safety, only to adopt scientific risk assessment and cost–benefit analysis more recently, regulators in Europe have arguably become more precautionary and more risk-averse over time. Vogel writes:

> From the 1960s through the mid 1980s American regulatory standards tended to be more stringent, comprehensive and innovative than in either individual European countries or in the European Union (EU). The period between the mid 1980s and 1990 was a transitional period: some important regulations were more stringent and innovative in the EU, while others were more stringent and innovative in the United States. The pattern since 1990 is the obverse of the quarter-century between 1960 and the mid 1980s: recent EU consumer and environmental regulations have typically been more stringent, comprehensive and innovative than those of the United States...Regulatory issues were formerly more politically salient and civic interests more influential in the United States than in most individual European countries or the EU. More recently, this pattern has been reversed. Consequently, over the last fifteen years, the locus of policy innovation with respect to many areas of consumer and environmental regulation has passed from the United States to Europe.[53]

In effect, Vogel maintains, US and EU risk regulations resemble "ships passing in the night," with the EU becoming more precautionary and the US less so over time. A central cause for this increasingly precautionary approach, Vogel argues, has been the long series of European regulatory failures and crises over the past several decades, including most notably the BSE or "mad cow" crisis discussed below. As we shall see, these crises have weakened public trust in EU regulators and scientific risk assessments, increased support for more precautionary regulations, and called into question European publics' acceptance of the *legitimacy* of EU law and institutions.[54] Responding to this crisis of legitimacy, EU institutions have moved aggressively to overhaul EU risk regulation across a range of areas, adopting strict new regulations for products and processes like GM foods and crops and elevating the "precautionary principle" to play a defining role in EU regulation, as examined in Chapter 6.[55] In contrast, the high cost of compliance with some US federal environmental regulation, such as for the cleaning of toxic wastes, arguably has spurred a backlash within the US against further federal regulation.[56]

Other scholars dispute Vogel's "ships passing in the night" characterization of US and EU risk regulation, noting that the purported "flip-flop" in US and EU approaches to risk regulation draws disproportionately from a few controversial issues. In a wide-ranging survey of US and European risk regulation, Jonathan Wiener and Michael Rogers find a more complex set of outcomes, in which the US is more precautionary in some areas (such as nuclear energy and particulate air pollution), while the EU demonstrates greater precaution in others (such as agricultural biotechnology and hormone-treated beef). They contend that "[t]his broader analysis indicates that neither the US nor the EU is a more precautionary actor across the board, today or in the past. Relative precaution appears to depend more on the particular risk than on the country or the era."[57]

For this reason, we resist characterizing either the US or the EU as the more risk-averse beyond the context of agricultural biotechnology. We focus, rather, on the transnational governance challenges exemplified by biotechnology regulation, which raise central questions about transnational regulatory cooperation and conflict over the application of environmental, food safety and trade laws at the domestic, regional, and international levels. As we now show, the US and the EU have taken starkly different approaches to biotechnology regulation that have become embedded over time.

2.2. Regulating agricultural biotechnology in the US

The regulation of food and environmental safety in the US were traditionally matters for state and local governments. They took primary responsibility, for example, for the inspection of slaughterhouses in the nineteenth century. By the beginning of the twentieth century, however, the growth of interstate trade in the US meant that, in order to be effective, food safety regulation would also have to reach across state lines. The US Congress responded to this challenge in 1906 and 1907 by using its powers under the Interstate Commerce clause of the Constitution to adopt the first comprehensive federal food safety legislation, namely the Pure Food and Drugs Act and the Federal Meat Inspection Act.[58] These acts, and subsequent Congressional legislation, created a federal system for the establishment and enforcement of regulatory standards for food safety and specified the roles of the most important federal agencies in that system. Federal environmental regulation would come much later, with the EPA not being established until 1970.[59] In that year, the US passed the National Environmental Policy Act (NEPA) which requires federal agencies to take environmental concerns into account as part of their decision-making processes.[60] As we will see, NEPA has been used to challenge certain agency decisions regarding GM plant varieties.

The Pure Food and Drugs Act (later expanded to the Federal Food, Drug and Cosmetic Act) delegates primary responsibility for food safety regulation to a federal agency, the FDA, which is authorized to inspect, test, approve, and set safety standards for foods, drugs, chemicals, cosmetics, and household and medical devices. The FDA does not itself test and inspect each individual food product marketed in the US. Rather, the agency establishes safety standards for various foods, while individual food producers bear primary responsibility for ensuring that new products conform to FDA standards. When violations are encountered as a result of FDA inspections, complaints, or the outbreak of illness, the FDA may halt the sale of unsafe products and prosecute the persons or firms responsible for the violation.[61] As we shall see below, the FDA has played the leading role in approving the sale and marketing of GM foods in the US.[62]

Parallel to the FDA regime, a second and much more labor-intensive system has been established for the regulation of agriculture, including the inspection of meat and poultry. The lead agency in this second system is the USDA, founded in 1862, which not only establishes food safety standards, but also employs over 7,000 inspectors who carry out "continuous inspection" of meat and poultry plants and certify meat products as safe and compliant with USDA standards.[63] In the 1990s, the USDA system came under some criticism for its failure to detect and prevent outbreaks of *E. coli* and other harmful bacteria in meat, resulting in the establishment of new safety procedures for meat and poultry in 1996. Nevertheless, the USDA has been generally successful in securing consumer confidence in the safety of US meat and poultry products.[64] As we shall see, the USDA also acted, together with the FDA, as the lead agency in the approval of growth hormones in US beef, which were prohibited in the EU and led to an earlier trade dispute between the two sides.[65]

Third, the US EPA has primary responsibility for environmental issues implicated by agricultural practices, as well as some authority over food safety matters. The use of pesticides, for example, has long been considered among the most severe threats to the environment in the US, and their registration is regulated by EPA under the Federal Insecticide, Fungicide, and Rodenticide Act (FIFRA).[66] Under the Food, Drug and Cosmetics Act (FDCA), EPA also regulates the amount of pesticide residue that can safely be left in food, giving particular attention to carcinogenic risks.

Finally, state and local officials complement the work of the federal agencies, including in respect of enforcement. For example, state and local governments continue to play a dominant role in food safety inspections of grocery stores and restaurants, and for specific categories of foods such as milk and shellfish.[67] The US food safety system is therefore complex and multi-leveled, requiring the cooperation of multiple federal agencies, state and local governments, and the food producers who bear ultimate legal responsibility for food safety.

2.2.1. Regulation of GM products

Genetic modification and GMOs first became a concern for US and other regulators in the 1970s, as biologists began making fundamental advances in recombinant DNA (rDNA) research (see chronology of key events, Table 2.1).[68] The debate over the regulation of such research is often dated to the international meeting of scientists at Asilomar, California, in 1975, which not only pointed to the promise of biotechnology but also called on the scientific community to exercise caution and restraint in the creation of genetically engineered organisms that might prove hazardous. In the US, the Asilomar conference triggered a national debate over the regulation of biotechnology, with a number of Congressional representatives introducing legislation that would ban or regulate rDNA research. At the same time, the National Institutes of Health (NIH) created a Recombinant DNA Advisory Committee, which in June 1976 put forward a set of guidelines for rDNA research in the US. By the late 1970s, initial public fears about the biohazards of biotech laboratory research had abated somewhat and the US had emerged as a world leader in biotechnology research. These developments led to strong support in the US Congress and the executive branch for a regulatory system capable of ensuring the safety of biotechnology research while encouraging the development of a new high-tech sector.

As the technology advanced toward commercialization in the 1980s, however, it was unclear which way the US would go—whether toward greater precautionary regulation of the technology's use or toward its promotion.[69] The EPA in particular supported a "process-based" approach to regulate GMOs, noting that "the most appropriate way to distinguish between 'new' and 'naturally occurring' microorganisms is by the methods or processes by which they are produced."[70] Some EPA officials maintained that the agency held existing authority under the Toxic Substances Control Act to regulate GMOs as "new" chemical substances. Since the criterion for EPA authority was based on "newness," it was in EPA's interest to find that GM varieties were fundamentally novel based on the genetic-engineering *process*. A widely circulated internal 1983 EPA memo to this effect allegedly "created a firestorm" within the pro-industry Reagan White House.[71] In Congress, representative Al Gore chaired a House subcommittee which held hearings on the "Environmental Implications of Genetic Engineering." It issued a report in 1984 suggesting that the EPA take the lead in addressing the environmental risks posed.[72] At the same time, anti-GM activist Jeremy Rifkin sought and obtained an injunction from the federal district court in Washington against the NIH's approval of any applications for the deliberate release of GM varieties into the environment without the court's approval. This aspect of the injunction, however, was soon overturned.[73] During the first half of the 1980s, therefore, it appeared as if the US might take a highly precautionary, process-based approach to

Table 2.1 Key events in US biotech regulation, 1975–99

1975	Asilomar conference on biohazards posed by GMOs
1976	NIH creates Recombinant DNA Advisory Committee
1983	EPA proposes process-based regulation of GMOs
1984	House subcommittee calls for EPA as lead regulator of GMOs
1985	Reagan Administration creates Biotechnology Science Coordinating Committee to devise federal standards for GMOs
1986	Coordinated Framework for the Regulation of Biotechnology adopted by OSTP, calling for product-based regulation, key roles for USDA and FDA
1992	FDA issues guidance on approval of GM foods and crops, to be regulated in terms of product, not process
	First approval of a GM crop, the Flavr Savr tomato
1996	First approval of a GM soy crop, Roundup-Ready soybeans
1999	FDA begins extensive public hearings into adequacy of US regulatory framework

GMO regulation, with a leading role for the EPA and possible Congressional regulation.

The Reagan White House, however, responded quickly to these developments, attempting to forestall new legislation in Congress as well as EPA activism under existing statutes. In November 1985 it formed a Biotechnology Science Coordinating Committee "to help resolve questions of jurisdiction, settle upon common regulatory principles and establish a uniformly product-based regulatory approach."[74] The result was a curtailment of EPA's role and an elevation of those of the USDA and FDA. In 1986, after public notice and comment, the Office of Science and Technology Policy (OSTP) in the Reagan Administration issued a "Coordinated Framework for the Regulation of Biotechnology" that continues to shape US biotech regulation today.[75] Crucially, the OSTP concluded that the techniques of biotechnology are not inherently risky and that biotechnology could therefore be adequately regulated by existing federal agencies under existing statutes, obviating the need for new legislation dedicated specifically to regulating GMOs. The Coordinated Framework established a division of responsibility among the three US regulators, with the FDA serving as the primary regulator of GM foods, the USDA charged with oversight of the planting of GM crops, and the EPA limited to overseeing the environmental and food safety impact of GM crops that have pesticidal characteristics.[76] Each of the agencies has since generally found that GM varieties introduced to date are not fundamentally different from their non-GM (conventional) counterparts, so that the agencies' oversight is much more limited than in Europe. As the Pew Initiative on Food and Biotechnology's extensive 2004 report states, "The central premise of the Coordinated Framework was that the process of biotechnology itself poses no unique risks and that products engineered by biotechnology should therefore be regulated under the same laws as conventionally produced products with similar compositions and intended uses."[77]

This regulatory choice in 1986 has had ongoing effects, as Congressional committee interest in agricultural biotechnology regulation generally waned, and those hearings that were held focused more on the benefits of the technology than its risks.[78] In an excellent study, Adam Sheingate shows how the results of the political contention in the 1980s have since framed US policy debates over biotechnology in terms of its "commercial opportunities" and its importance in "international economic competition."[79] The Reagan Administration was able to move primary regulatory responsibility to the USDA, whose primary constituency is agricultural trade associations, and in the process, shift primary legislative oversight to the agricultural committees of the House and Senate. In the 1990s in the White House, the US Council on Competitiveness would take the lead on biotech policy formation.[80] These choices have made it more difficult for GM skeptics to use the existing regulatory and political framework to impede approval of GM crops and foods in the US.[81]

The USDA is now the leading US agency for regulating the introduction of GM plant varieties into the environment, whether for field trials or for commercial production. In practice, the USDA, operating through its Animal and Plant Health Inspection Service (APHIS), has been responsible for the most regulatory actions involving agricultural biotechnology of any US agency.[82] Under the authority provided by the federal Plant Protection Act, which was enacted in 2000, APHIS categorizes GM plant varieties as *potential plant pests*.[83] If APHIS finds that the GM variety is a plant pest, then the variety will be highly regulated. If APHIS finds that the GM variety is not a plant pest, then it will be *"deregulated"* and "no longer subject to APHIS' legal authority."[84] APHIS first requires the developers of GM crops to notify the agency about field trials or, for higher risk plant material, to obtain a permit to conduct them. Notification to APHIS is a streamlined process for low-risk plant material, pursuant to which the agency will simply acknowledge receipt of the notification, which constitutes authorization for the trial. Notifications (as opposed to permits) are used for around 90 per cent of all the field trials, and they have generally been used for GM varieties which are not for pharmaceutical or industrial use.[85] Companies may engage in these field trials for years, gathering data before seeking to commercialize the variety. In total, by the end of March 2008, APHIS had "acknowledged" 12,145 notifications and issued 1,188 release permits.[86]

On the basis of these field trials, a developer may then petition APHIS to find that the GM variety is not, in fact, a "plant pest," so that the variety may be deregulated and the GM crops may be planted without restriction. Although APHIS' regulations provide that it will make a decision within six months, in practice, it first needs to verify the completeness of the file and may request further information, so that the approval process can take a year or more.[87] By the end of March 2008, APHIS had agreed to deregulate seventy-three GM varieties.[88] These APHIS decisions to deregulate a variety must be accompanied by either an Environmental Assessment or an Environmental Impact Statement under the NEPA, and they can be challenged before federal courts. APHIS decisions have been generally based on an Environmental Assessment which may take just a few weeks to prepare, pursuant to which the agency finds that no Environmental Impact Assessment is necessary.[89] Because of successful court challenges to APHIS decisions in 2007, APHIS has begun to prepare some Environmental Impact Statements which are much more labor intensive and can take up to a year to complete. USDA has also proposed to expand APHIS' regulatory oversight to include the oversight of GE organisms that have the potential to be "noxious weeds," again under existing authority pursuant to the Plant Protection Act.[90]

Some analysts criticize the central role of USDA in agricultural biotech regulation, because USDA's regulatory authority focuses on harm to plants from pests (and, in particular, to commercially valuable plants) and not on environmental protection in general.[91] Moreover, a primary role of USDA is to serve and protect US agricultural interests, so that USDA may not prioritize environmental protection.[92] Large agricultural producers have become responsible for an increasing amount of US farm production,[93] so that the agribusiness constituency has become relatively more important for USDA over time.

USDA's close relationship with the agricultural biotech industry has allegedly been facilitated by "revolving door" careers of personnel moving between key government agencies and biotech industry associations, such as the Biotechnology Industry Organization (BIO). For example "Val Giddings went from being responsible for biotechnology regulation within the US Department of Agriculture and part of the US negotiating team in the biosafety negotiations to become Vice President for Food and Agriculture of the Biotechnology Industry Organization." In the other direction, after spending seven years as Monsanto's lawyer, Michael Taylor became "the FDA's deputy commissioner for policy and responsible for drafting the guidelines on GE cattle drug rBGH."[94]

Overall, the USDA has been a strong supporter of agricultural biotechnology. Its Agricultural Research Service not only undertakes biotech research, but also owns some biotech patents.[95] Biotech skeptics thus question the reliability of what the US calls its "science-based" approach, claiming that the use of the term "sound science" can become "little more than Orwellian code for bending scientific data to meet political ends."[96]

The EPA's responsibility for the regulation of genetically engineered crops has been limited, particularly in contrast with the central role of the EU's Directorate General for Environment (see below). Today, the EPA only exercises regulatory authority over GM varieties that have *pesticidal characteristics*. It does so pursuant to the FIFRA and the FDCA.[97] These long-standing statutes respectively grant to the EPA the power to regulate the use of pesticides to oversee their environmental safety and their safety for consumption. Technically, the EPA does not regulate the GM plant itself, but only the pesticidal substance that the GM plant expresses.[98] In this sense, it is regulating GM crops on the basis of the varieties' particular *characteristics*, and not on account of the process used. GM varieties without pesticidal traits are thus *not* subject to EPA oversight or control. In 2001, the agency created a new regulatory category for GM varieties that incorporate resistance to pests through genetic modification, which it calls "plant-incorporated protectants" (or PIPs).[99] Examples of PIPs that have been approved and commercialized include varieties that contain the Bt toxin, such as Bt corn and Bt cotton.

The EPA maintains that it holds this regulatory authority over these GM varieties under existing statutes. However, even this authority could be subject to challenge under the Administrative Procedures Act on the grounds that these GM varieties are not pesticides. Agricultural biotech companies have so far not challenged EPA's assertion of authority before the courts, possibly because EPA regulators have so far not restricted these GM varieties' sale or use to any significant extent. Biotech companies may fear that such a regulatory challenge of EPA authority could trigger greater pressure for legislation that could result in stricter EPA regulatory oversight over all GM varieties. Moreover, biotech companies can benefit from EPA oversight which can reassure traders in US and foreign markets.

The EPA regulates GM crops with pesticidal traits in three primary ways. First, prior to commercialization, the EPA regulates them (as with any pesticide) through the grant of "experimental use permits." These permits are not required for field tests that do not exceed ten acres, an exemption that has raised some controversy as applied to GM varieties which could reproduce and spread.[100] Second, the EPA is empowered to impose conditions on the commercial use of a pesticide so as to ensure environmental safety. It has exercised this authority to require growers of approved Bt crop varieties to plant a portion of their fields with conventional counterparts in order to prevent pests from developing Bt resistance, although EPA relies on state enforcement in this area. Some analysts are concerned that farmers are not complying with these conditions.[101] Third, under the FDCA, the EPA establishes tolerance levels at which pesticide residues are deemed safe in food. If it finds that the pesticide is of such a low risk that no tolerance level is required, it can grant a "tolerance exemption."[102] In each case, the GM varieties have been granted a "tolerance exemption" for use in food and animal feed so that they may be freely marketed, although the exemption has sometimes been limited only to food or feed.[103]

As we will see in Chapter 6, GM varieties with pesticidal characteristics have been the subject of some of the most significant public controversies in the US. For example, the "Starlink" controversy involved a GM corn variety with pesticidal characteristics that was approved under an EPA "tolerance exemption" only for animal consumption, but the harvested corn found its way into food products.[104] Similarly, the monarch butterfly controversy raised concerns that GM plants expressing pesticides could threaten non-targeted species such as butterflies. In addition, skeptics maintain that plants expressing pesticides could facilitate the development of insect and weed resistance which could be more difficult to control when the pesticide is genetically engineered into plants that reproduce themselves.

The US Federal Food, Drugs, and Cosmetic Act delegates the primary responsibility for food safety regulation to the Food and Drug Administration. Faced with the first applications from producers for the licensing of GM foods and crops, the FDA issued a Statement of Policy in 1992 (reaffirmed in 2001) that GM foods were *not* substantially different than regular foods, and that it would therefore approve foods based on the health risks of the individual product, and *not* on the process by which it is produced.[105] In 1992, the FDA also ruled that neither any pre-market approval process nor any specific labeling would be required for GM foods that were "substantially equivalent" to conventional foods and thus *"generally recognized as safe"* (GRAS).[106] Pre-market approval is only required by the FDA for products where the genetic manipulation has altered the substance and safety of the product (for example, by introducing new allergenic properties or changing the nutritional content of the food in question), in which case it is regulated as a "food additive."[107] As a result, GM crop developers have significant incentives "to characterize GM products as GRAS," so as to avoid triggering regulatory requirements for "food additives," where possible.[108] In short, FDA does not review "novel" (such as GM) foods for safety, but rather only takes action against foods that constitute "food additives" or have otherwise been "adulterated."[109] The FDA, like the other two agencies, thus decided to regulate GM foods only if a specific GM food presents adverse product *characteristics*, and not on the basis that the food was derived from genetic engineering.

Since the FDA made its determination that there would be no pre-market approval process for GM products that are "substantially equivalent" to conventional counterparts, it has approved all GM varieties *without any labeling requirement*. While the FDA later prepared guidelines for manufacturers wishing to *voluntarily* label their foods as either containing or not containing bioengineered ingredients, there is no US requirement for consumers to be informed that foods may contain such ingredients.[110] In fact, mandatory labeling that products contain GMOs may be subject to constitutional challenge in the US on "free speech" grounds. In a related matter, a US federal appeals court held that state laws mandating the labeling of milk products derived from cows treated with the hormone r-BST (recombinant bovine somatotropin) were likely unconstitutional on First Amendment grounds on account of a person's right "not to speak."[111] As a result, in practice, only those selling milk produced without the hormone typically label their products that the milk is "r-BST free" or "produced with no artificial hormones."

FDA is also the primary US regulator to set rules regarding food product labeling. Product labeling can be mandatory or voluntary, and it can be positive or negative in its focus. Mandatory labeling is that required by government, while voluntary labeling is that permitted to be done at the private sector's initiative. Voluntary labeling, however, can be subject to restrictions—such that labels cannot be misleading or deceptive to the consumer. Positive labeling refers to that designed to show the alleged positive attributes of a product, such as labeling a product as GMO-free. Negative labeling refers to that designed to inform consumers of the negative (or potentially negative) risks of a product, such as labeling that a product contains GMOs. While the latter label could be deemed to be neutral on its face (as consumers could find a GM trait to be desirable), in the current context, it is maintained by GM technology supporters that such labeling, especially if required by government, could misleadingly stigmatize the product as potentially dangerous.[112]

Those labeling their products as GM-free are thus subject to *labeling restrictions*. Producers of r-BST-free milk, for example, must add a disclaimer to the effect that "the FDA has found the hormone to be safe."[113] If they do not, they could be subject to legal challenge for false or misleading statements, whether by the FDA or a private actor, such as a biotech trade association. As the FDA states in its labeling guidelines,"Under section 403(a) of the act, a food is misbranded if statements on its label or in its labeling are false or misleading in any particular. Under section 201(n), both the presence and the absence of information are relevant to whether labeling is misleading."[114] As a likely result, products have been more frequently labeled as "organic" in the US than as containing "no GMOs," pursuant to 2001 FDA regulations described in Chapter 6. As we will see, Europe, in contrast, has mandated negative labeling of all products that contain or may contain GMOs and thus clearly permits a product to be labeled as GM-free. In fact, many retail establishments in the EU boldly advertise that all of the products that they sell are GM-free, a practice that could be subject to legal challenge in the US.

In sum, no US statute specifically regulates genetic modified crops or foods *per se*. Rather, three US agencies regulate them under the agencies' existing regulatory authority which applies to all crops and foods—namely, authority for the regulation of plant pests, noxious weeds, pesticides and food additives. The result has been a strange fit between the definition of regulatory authority and the agencies' risk assessments of GM products. We summarize this situation in Table 2.2.

US regulation of food safety by specialized agencies is sometimes contrasted with European regulation of GMOs by politicians. It is contended that decision-making by these US agencies is more neutral and technocratic.[115] However, none of the three US agencies (USDA, EPA, or FDA) are technically "independent agencies" in the sense used in the US political context in which independent agencies refer to agencies that Congress has created to be independent of the executive branch, such as the Federal Reserve Bank, Federal Trade Commission, and Federal Communications Commission. The USDA is

Table 2.2 Regulatory authority of US agencies

Agency	Statute	Regulatory Authority relating to GM products
USDA	Plant Protection Act	Over plant pests and noxious weeds
EPA	FIFRA; FDCA	Over pesticides
FDA	FDCA	Over food additives and labelling

a cabinet-level department within the executive; the FDA operates within the US Department of Health and Human Services; and the EPA, although not a "department," operates under executive branch control. The executive branch is periodically accused of using these agencies for political purposes, as when a senior USDA economist came under fire "for suggesting that the Bush administration could maximize votes in key dairy states by keeping milk prices high through the election,"[116] or when EPA political appointees are accused of suppressing scientific studies showing global warming trends, arsenic levels in water, or particulate concentrations in air.[117]

Moreover, all federal agencies, whether "independent" or not of the executive branch, are subject to various legislative-branch control devices, such as Congress's ability to pass new legislation, to allocate or withhold funds, or to object to key appointments, which limit their autonomy.[118] US regulatory agencies' actions are likewise subject to extensive administrative law requirements under the US Administrative Procedure Act, requiring prior notice and comment of all proposed regulations, backed up by judicial review before the federal courts.[119] Interest groups can use these procedures to constrain agencies' ability to operate, especially when coupled with constraints on these agencies' enforcement budgets.[120]

Nonetheless, unlike in Europe, *the US system for biotechnology regulation has been determined almost exclusively by regulators operating under existing federal statutory authority, while the legislature (Congress) has played a relatively passive oversight role.* Bills have been introduced at the state level regarding GM foods, as we will see in Chapter 6. Yet if enacted, they could be subject to lawsuits on the grounds that federal law and policy, expressly and implicitly, preempt them.[121] Also, unlike in Europe, the federal government has played the dominant role in GMO regulation, with a relatively minor supporting role for individual state governments.[122]

Because agricultural biotech regulation has not attracted much attention in the US Congress, federal regulatory agencies have operated with significantly greater *de facto* independence from political pressure. As a result, overall the US system has been characterized by strong federal institutions, significant independence of regulators from political pressures, extensive reliance on scientific risk assessment in regulatory decisions, and industry self-regulation, all of which stand in contrast to the historically more decentralized and increasingly politicized food safety system of the EU. This remains the case despite some degree of administrative fragmentation among the three lead agencies, and some concern about gaps in the regulatory framework under existing US legislation (examined in Chapter 6).

Before moving to regulation within the EU, we stress one last factor that could affect regulation of GM products in the US marketplace compared with that in Europe. The US legal system imposes *tort liability* on producers whose products cause harm, which can lead to substantial damage awards. The US legal system couples these liability rules with procedures that facilitate the bringing of lawsuits. "Class action" procedures initiated by entrepreneurial attorneys and individual claims financed on a "contingency fee" basis, facilitate the ability of parties with less income to bring legal claims. These liability rules and legal procedures create market incentives for sellers of GM products in the US to take precautions. Market actors falling within the potential liability chain can (informally) oversee and enforce precautionary procedures.[123] European legal systems, in contrast, generally neither provide for significant tort damage awards, nor for analogous procedures that facilitate individual and class action law suits. It is thus not surprising that European regulators take a more stringent regulatory approach on account of their greater responsibility for any potential harm that could occur. Were the FDA forced to decide whether to approve a GM food, and in the process explicitly or implicitly certify its safety, GM food developers could have an affirmative defense in the US that could exculpate them when faced with a tort liability claim, unless a statute or regulation expressly provided otherwise.[124]

As a result, some will argue that the incentives of private actors to ensure product safety would actually be reduced in the US were the FDA to assume certification responsibility. This difference in the operation of US and European legal systems is often forgotten in political and sociological assessments of their different approaches to the risks posed by genetically engineered varieties. While the US administrative law system may be more *politically centralized* than in Europe, enforcement is more *legally decentralized* in the US. The threat of a liability claim can exercise pressures in the US marketplace in a "bottom-up" manner. This potential, in turn, has led to pressures on state legislators in the US to limit liability by statute, in contrast to pressures in Europe to create a specific EU liability regime for GM products.

2.3. Regulating agricultural biotechnology in the EU

Today's EU, as well as its policies on GM foods and crops, can be traced to the creation, in 1957, of a European Economic Community (EEC). In theory, the EEC was to be a common market characterized by the "four freedoms"—the free movement of goods, services, labor, and capital—supplemented by a common external tariff, a collective trade policy vis-à-vis third parties and a Common Agricultural Policy (CAP).[125] The EEC Treaty itself made no explicit mention of an EU policy for biotechnology or even for the closely related areas of environmental and food-safety policy, which remained primarily a national responsibility within each of the Community's member states.

The EU nevertheless developed a *de facto* policy on food products, including eventually over products produced from agricultural biotechnology, as the EU's policies on agriculture and the establishment of an internal market for food and agricultural products "spilled over" into the regulation of the approval, content, and labeling of European foods, including of GM foods. The term "regulatory spillover" refers to a process in which the goal to assure the free movement of goods in Europe triggers new regulation at the European level to remove non-tariff barriers to trade caused by member-state regulations. More specifically, the Treaty provided for the possibility that the Council of Ministers, acting on a proposal from the executive Commission, could adopt "approximated" or "harmonized" regulations, typically in the form of a Directive which would establish minimum EU standards, leaving member states to transpose and implement those standards in their national legal systems.[126] These harmonized directives were supplemented by the principle of "mutual recognition" pursuant to which member states must accept products produced by each other, subject to certain conditions.[127]

By and large, the EU food safety system achieved its aim of establishing a single market for food products, increasing consumer choice, and removing non-tariff barriers to trade among the member states. In comparison with the US, however, the EU food safety system, and the EU regulation of biotechnology in particular, remained a much less centralized and incomplete regulatory patchwork, with a decision-making process in which the key decisions were taken not by a specialized regulatory agency like the FDA, but by political bodies such as the Council of Ministers, Commission, and European Parliament, in an uneasy cooperation with competent authorities in each of the member states.

2.3.1. The EU as a political system

The contemporary EU began life as an international organization of sovereign member states, but it has gradually evolved into something akin to a domestic political system, characterized by both a vertical or "federal" division of powers between the EU and member-state levels, and the horizontal "separation-of-powers" among the legislative, executive, and judicial branches of the Union.[128]

In terms of the vertical separation of powers between the Brussels-based EU institutions and the member states, the EU is increasingly characterized as a federal or quasi-federal system. This is not to say, of course, that the EU's central institutions are as powerful as those of the US. Indeed, many scholars question whether the EU can or should be accurately described as a federal state, noting the Union's secondary role in many important areas of policy, as well as the relatively weak role of the Commission as the EU's primary executive body.[129]

Nevertheless, if the EU's "federal" or supranational institutions are weaker than those of the US, the EU as a political system can be described as federal. The term federalism has been the subject of numerous overlapping definitions, but most of these formulations rely on three elements emphasized by R. Daniel Kelemen, who has described federalism as "an institutional arrangement in which (a) public authority is divided between state governments and a central government, (b) each level of government has some issues on which it makes final decisions, and (c) a federal high court adjudicates disputes concerning federalism."[130] In most federal systems, moreover, the structure of representation is two-fold, with popular or functional interests represented directly through a directly elected lower house, while territorial units are typically represented in an upper house whose members may be either directly elected (as in the US Senate) or appointed by state governments (as in the German *Bundesrat*). In both of these senses, the EU *already* constitutes a federal system—with a constitutionally guaranteed separation of powers between the EU and member-state levels, and a dual system of representation through the European Parliament and the Council of Ministers. In this regard, we can compare the EU to the US federal system, while remaining cognizant of the greater legal competence and budgetary resources of the US federal government.[131]

Unlike the parliamentary states of Western Europe, but like the US, the EU can also be characterized by a horizontal separation of powers in which three distinct branches of government take the leading role in the legislative, executive, and judicial functions of government, respectively. This does not mean, of course, that any one institution enjoys sole control of any of these three functions; indeed, as Amie Kreppel points out, the Madisonian conception of separation of powers "requires to a certain extent a co-mingling of powers in all three arenas (executive, legislative, and judicial)."[132] In the case of the EU, for example, the legislative function is shared today by the Council of Ministers and the European Parliament, with an agenda-setting role for the Commission; the executive function is shared by the Commission, the member states, and (in some areas) independent European regulatory agencies; and the judicial function is shared by the European Court of Justice (ECJ), the Court of First Instance, and a wide array of national courts bound directly to the ECJ through a preliminary reference procedure.

The legislative function in the EU—the adoption of legally binding Regulations and Directives governing the behavior of the EU's member states and their citizens—begins with the European Commission, which in most issue-areas (including those of interest to us here) enjoys the "sole right of initiative" to propose legislation for consideration by the EU's primary legislative bodies, the Council of Ministers and the European Parliament. The Commission therefore, although normally considered to be the "executive" branch of the EU system, plays a vital role in shaping draft legislation in its early phases. The Commission itself is led by a "College" of Commissioners, who vote by simple majority, although in practice EU legislation is drafted in the first instance by the Commission's services, including its various Directorates-General (DGs) devoted to individual portfolios such as the internal market, agriculture, and the environment.[133] (In practice, as we shall see, DG Environment has played a leading role in the drafting of legislation governing GM foods and crops, with a secondary role for other relevant DGs.)

The precise legislative procedure governing any particular piece of draft legislation is spelled out in the Treaty that specifies the rules applicable to a particular issue-area. For much of the history of the EU, the dominant procedure was the so-called consultation procedure, according to which a relatively weak European Parliament would be "consulted" about the contents of the proposed legislation, but the final decision would rest with the Council of Ministers, a meeting of domestic ministers from the EU's various member states. Although nominally a single body, the Council meets in various configurations, with, for example, the Environment Council being composed of national environmental ministers, the Agriculture Council of agriculture ministers, and so on. (The Environment Council has played the leading role in the regulation of GM foods and crops, as we shall see below.) Political negotiations in the Council are most often taken by consensus, and indeed the Treaties initially specified the use of unanimous voting for most issue-areas. Over time, however, the Treaties have been amended to provide for increasing use of qualified majority voting (with more powerful member states enjoying greater voting weight than smaller ones), and the Council increasingly resorts to formal votes on draft legislation—particularly in especially controversial issue-areas such as GM foods and crops.[134]

Alongside the Council, the European Parliament (EP) has grown from being a weak and primarily symbolic body to being, in many issue-areas, an equal partner with the Council in a bicameral legislative process. Over the course of the 1980s and the 1990s, the legislative powers of the EP were increased sequentially in a series of Treaty amendments, from the relatively modest and non-binding "consultation" procedure of the EEC Treaty through the creation of the "cooperation" procedure in the 1980s to the creation and reform of a "co-decision procedure" in the 1990s. Under the contemporary co-decision procedure, spelled out in the 1997 Treaty of Amsterdam, the legislative process begins with a Commission proposal, which is sent both to the Council and the European Parliament. The Council and the EP deliberate and vote in a series of readings, and any discrepancies between the two bodies are negotiated in a "conciliation committee," roughly analogous to a US House–Senate conference committee, the results of which must be approved by the requisite majority in both chambers. As a result of this procedure, "the Council and the Parliament are now co-equal legislators and the EU's legislative regime is truly bicameral,"[135] and the EP has, alongside the Council, played a vital role in the adoption of the EU's legislative framework for GM foods and crops.

Just as the legislative process is fragmented among multiple actors, so too is the EU's executive power distributed among several bodies, four of which deserve mention here. First, the Commission has been delegated the task, under many EU Regulations and Directives, of applying broad EU legislative frameworks to specific cases, as for instance in the approval of specific GM foods and crops. Second, however, the Commission, as an agent of the member governments that created it, is not free to act as it likes, but in most instances must submit its draft decisions to a so-called "comitology" committee of member-state representatives, who examine the Commission's decisions and in some cases may block its implementation.[136] In the case of GM foods and crops, for example, the Commission is supervised by a "regulatory committee," which may block a draft Commission decision by a qualified majority vote of its members, or transmit a draft decision to the Council of Ministers for reconsideration. Both the regulatory committee and the Council of Ministers have, in fact, been the scene of bitter disputes over the Commission's proposed decisions, thus blurring the line between legislative decision-making and executive implementation. The Commission is also assisted in the biotech sector by a third set of executive actors, who conduct scientific risk assessments of individual GM foods and crops. This task was originally played by a committee of member-state experts under the terms of the EU's 1990 Directive, although the task of risk assessment was turned over in 2002 to a new European Food Safety Authority (analyzed in Chapter 6). Fourth, unlike the US with its large federal bureaucracy, the relatively small Commission relies almost entirely on the member governments to implement EU laws "on the ground"—a reliance that has proven problematic in many instances, as we shall see below.

The judicial function in the EU, finally, is entrusted primarily to the Luxembourg-based ECJ, which has over the years interpreted EU laws broadly, helping to move along the process of European integration.[137] Under the terms of the EU's founding Treaties, the Commission can, and often does, bring "direct actions" against member states for failing to transpose or implement EU legislation into national law, and the Commission has made limited use of this power to bring so-called "infringement proceedings" against member states for failure to transpose EU Directives on GM foods and crops in a timely fashion. In addition, the ECJ can also decide on "indirect actions," in which individual litigants in the various member states raise questions of EU law, claim rights under EU law, or challenge the actions of their own governments for contravening EU law; in these instances, national courts typically request a preliminary ruling from the ECJ, which can and often does overturn the national law in question in favor of EU law. Thus far, however, the ECJ has been a relatively minor actor in the regulation of GM foods and crops in the EU, although the Court has in some instances been called upon to rule on the legality of certain member-state measures, and the standards for member-state courts to review them (see Chapter 6).

2.3.2. The EU regulates GMOs: Directives 90/219 and 90/220

Agricultural biotechnology is regulated at both the EU and the member-state levels. We focus here on the EU level since the US and other parties deal primarily with the EU regarding conflicts over GMO policy. As we will see, however, member-state regulation remains central to managing conflicts between European, US, and other national agricultural biotech regulations.

The first comprehensive legislation for the regulation of biotechnology in the EU came in 1990, with the adoption of two directives by the EU Council of Ministers: Directive 90/219 on the Contained Use of Genetically Modified Microorganisms, which regulates the use of GMOs in laboratory settings; and Directive 90/220 on the Deliberate Release into the Environment of Genetically Modified Organisms, which governed for over a decade the approval, planting, and marketing of GM foods and crops within the EU and is therefore particularly important for our purposes here. As in the US, however, consideration and debate over an EU policy for GMOs began with the creation of the new technology itself, in the 1970s (see chronology of events, Table 2.3).

From the 1970s through the mid-1980s, the development and marketing of GM foods and crops lay in the distant future, and regulators in Europe as

Table 2.3 Key events in EU biotech regulation, 1978–99

1978	Commission proposes directive requiring prior notification and authorization of GM research (withdrawn 1980)
1983	Biotechnology included in EU Framework R&D Programme
1984	Commission forms Biotech Steering Committee
1986	Commission report, *A Community Framework for the Regulation of Biotechnology*
1988	Commission proposes twin directives on Contained Use and Deliberate Release of GMOs
1990	Council adopts Directives 90/219 and 90/220, establishing legal framework for the approval and marketing of GM foods and crops
1996 (March)	Start of BSE crisis, questioning of EU food safety regulation
1997 (Jan.)	Council adopts Novel Foods Regulation
1997 (Jan.)	Commission approves sale of GM maize; three member states invoke safeguard clause
1998 (Oct.)	Start of *de facto* moratorium on approval of new GM varieties
1999 (June)	Declaration of moratorium on GM approvals by five member states

elsewhere therefore focused primarily on issues relating to laboratory research and to the development of a competitive biotechnology industry. In 1978, the Commission Directorate General for Science, Research and Development (then DG XII) proposed a Community R&D programme in molecular biology, together with a draft Directive requiring notification and prior authorization by national authorities for all biotechnology research.[138] The Commission, following the advice of DG XII, withdrew the latter proposal in 1980, wishing to avoid legislation on account of reduced concerns over the risks posed.[139] Instead it supported the adoption of a non-binding recommendation from the Council of Europe, calling for notification of rDNA research to national authorities.[140] At the same time, the Commission remained concerned about the competitiveness of the EU's biotechnology industry, and in 1983 the Commission incorporated biotechnology into its multi-annual Framework Programme for research and technological development, and in 1985 adopted a Biosafety Action Plan highlighting the need for "research" and "concertation" to develop the new technology.[141]

By the mid-1980s, with the rapid development of genetic engineering and the early efforts to regulate biotechnology among the member governments, the Commission began more actively to explore the development of a Community framework for biotechnology regulation. Such proposals created major challenges for the Commission due to the multi-sectoral nature of biotechnology, which would implicate internal market policy, industrial policy, research and technological development, environmental policy, food safety, agriculture, and international trade. As a multi-sectoral issue, the regulation of agricultural biotechnology raises the challenge of coordinating policymaking horizontally among a large number of public actors with diverse perspectives about biotechnology. The Commission responded by creating several inter-departmental coordinating bodies, most notably the Biotechnology Steering Committee (1984) and the Biotechnology Regulation Inter-service Committee (1985).[142] Through the mid-1980s, the Science, Research and Development DG (DG XII) played the predominant role, for, "the other DGs...saw the mysteries of biotechnology as still playthings of [it] and the scientific community..." Hence, "[t]he Commission communications on biotechnology, although presenting a strategic approach, were in practice largely drafted by the research service, DG XII, with marginal additions by the other services."[143] Over time, however, policy leadership within the Commission shifted gradually away from the research service, and towards other DGs, as biotechnology moved out of the laboratory to planting in crop trials and the marketing of GM seeds and foods.

The two most logical places for the drafting of legislation on agricultural biotechnology were DG Environment or DG Agriculture, but DG Environment took the lead, maintaining that GM crops needed to be assessed in environmental terms. DG Agriculture was seemingly more focused on the Common

Agricultural Policy and the challenges of managing the EU agricultural surpluses that the CAP fostered at this time.[144] Increasingly, according to observers such as Cantley, DG Environment cut out DG Science, Research and Development from policy influence, "prefer[ing] to consult its own experts," and reluctant to accept other experts' advice, especially those wary of the advisability of new legislation.[145] DG Environment knew it lacked allies in DG Science, Research and Development for its desired regulatory role, which DG Science found to be duplicative and without scientific justification. Cantley notes how, after 1986, "the inter-ministerial conflicts (and within the European Commission, the inter-DG conflicts) were continual, bitter, and angry."[146]

With DG Environment in the lead, the Commission released a new Communication in 1986, "A Community Framework for the Regulation of Biotechnology," which laid out the Commission's rationale for a European regulatory regime and its plans for specific EU regulations.[147] The Commission noted that the European Parliament had already called for Community-level regulation, and that a number of member states had already moved to adopt a range of national measures, threatening to disrupt the EC's single market: "The internal market arguments for Community-wide regulation of biotech-nology are clear," it argued. "Microorganisms are no respecters of national frontiers, and nothing short of Community-wide regulation can offer the necessary consumer and environmental protection."[148] The Commission therefore indicated its intention to come forward with concrete regulatory proposals covering both laboratory use and deliberate release into the envir-onment of GM organisms. As many commentators have noted, the biotech industry was not as well organized in Europe and was unable to mobilize political resources to prevent the enactment of process-based GM regulation that was framed in environmental terms.[149] Thus by 1986, the year of both the US Coordinated Framework and the EC Community Framework, Europe and the US had started down different paths.

The Commission's 1988 proposal for a "deliberate release" Directive began by noting the extraordinary diversity of existing national regulations across the various member states, including: (a) a ban on deliberate release (sub-ject to exceptions) in Denmark and Germany; (b) a case-by-case approach to the release of individual GMOs in a number of member states (UK, France, Belgium, Netherlands, and Luxembourg); and (c) an absence of legislation in other member states (Ireland, Greece, Italy, Spain, and Portugal).[150] The Commission's proposal emphasized the scientific uncertainty associated with genetic engineering and therefore proposed an EU regulatory scheme that would provide for case-by-case assessment and authorization of the release of all new GM varieties into the environment. In contrast with US agencies, which elected to regulate GM foods only in terms of their final characteristics as *products*, the European Commission elected to apply distinctive regulations to GM foods as a function of the *process* through which they were developed.

More specifically, the Commission's proposal would require any individual wishing to release GMOs into the environment to notify and provide a detailed risk assessment to the competent regulatory authority of the EU member state in which the release was proposed. That particular member state would then be charged with evaluating the application in line with the provisions of the directive. If the member state rejected the proposal, the procedure would end, but if the member state accepted the proposal, the dossier would then be forwarded to the Commission and to the other member governments, which would have a limited period to object to the authorization. If no objections were put forward, the product would be authorized for release and/or placement on the market throughout the EU. By contrast, if one or more member governments or the Commission objected, the Commission would then undertake its own assessment and formulate a decision to approve or deny the application. The Commission's draft decision would be circulated to an advisory committee of member-state representatives, of whose opinion the Commission would have to take "utmost account"; the final decision, however, would remain with the Commission. In a final acknowledgement of member-state prerogatives, the Commission proposed a "safeguard procedure" whereby a member state could, if it had evidence of a serious risk to people or the environment from a previously approved GMO, "provisionally restrict or prohibit the use or sale of that product on its territory." Once again, however, the member state in question would have to inform the Commission of its actions and give reasons for its decision, and the Commission would retain the power to approve or reject the measures in question.

The European Parliament—which has emerged as a consistent champion of strict regulation of biotechnology over the past two decades—criticized the European Commission's proposal as being too lax on a number of points and proposed a number of amendments that would have substantially tightened regulatory restrictions on the approval of new GMOs. The US government, in contrast, criticized both the Commission's proposal and the Parliament's proposed amendments as unnecessarily strict and as arbitrary, particularly insofar as they proposed to regulate all GMOs regardless of the characteristics of the products to which they gave rise. The official US position went as follows:

> By basing the Directive on the technique by which the organism is modified, the EC is regulating organisms produced by a given process. This is not a functional category directly related with the characteristics of the organism. As expressed in the US coordinated framework for the regulation of biotechnology, the US generally regulates products rather than the process by which they are obtained. We are concerned whether differences in approaches and their implementation *may lead to difficulties in our attempts to achieve international harmonization*. It is important to understand that whether an organism is "unmodified" or "genetically modified" is, in itself, not a useful determinant of safety or risk."[151]

61

In other words, once the US internally determined how it would approach the regulation of genetically engineered agricultural products in 1986, it wished to export this approach through "international harmonization." It saw the EU's proposed directive as a clear threat to these efforts.

The Council of Ministers followed the broad lines of the original Commission proposal, thus rebuffing the core US objections, while at the same time rejecting the Parliament's most far-reaching amendments. The Council did, however, modify the procedure under which the Commission could issue approvals for new GM varieties: Whereas the original text provided for the Commission decision to be subject only to an *advisory committee* of member-state representatives, the final text featured a more constraining *"regulatory committee,"* which could approve a draft Commission decision by a qualified majority vote. If the regulatory committee did not approve the decision, however, it was to be sent to the Council of Ministers, which could approve the Commission decision by qualified majority or reject it by a *unanimous vote.* If the Council failed to act within three months, the directive provided that "the proposed measures shall be adopted by the Commission" (Article 21). By switching oversight of Commission implementation from an advisory to a regulatory committee, the member states increased their leverage by subjecting Commission decisions to a potential vote in the Council of Ministers (the rough analogue to the US Senate). Yet by requiring a unanimous vote to overturn a Commission decision, the Council's powers remained extremely constrained, at least formally, in terms of the approval or rejection of individual GM varieties.

Finally—and significantly, in light of later developments—the Council retained a slightly modified version of the Commission's safeguard clause, whereby a member state could, on the basis of new evidence about risks to human health or the environment, "provisionally restrict or prohibit the use and/or sale of that product on its territory" (Article 16). The member state in question would be required to inform the Commission, which would approve or reject the member-state measures in cooperation with the regulatory committee mentioned above.[152] As we will see, the regulatory committee would later reject the Commission's efforts to challenge member-state "safeguard" bans imposed on approved GM varieties.

Thus, in contrast to the US system, politicians were involved at three levels in the EU decision-making process. First, it was politicians who enacted a new regulatory framework for the growing and marketing of GM foods and crops. Second, it was the Commission, consisting of political officials designated by member-state governments, which would make the decision whether to approve individual GM varieties. Third, the Commission's decisions are subject to review by committees of member-state representatives, and ultimately by national politicians in the Council of Ministers. This EU system of a one-house legislative veto (by the Council) of the executive branch's administrative decisions regarding individual GM varieties would be unconstitutional in the US.[153]

In 1997, the regulatory structure of Directive 90/220 was supplemented by Regulation 258/97, the so-called Novel Foods Regulation.[154] The regulation defined "novel foods" as all foods and food ingredients that had "not hitherto been used for human consumption to a significant degree within the Community." The definition included both GM foods as well as foods produced from, but not containing, GMOs (for example, oils processed from GM crops but no longer containing any traces of GM material).[155] The regulation established an authorization procedure similar to that of Directive 90/220, as well as labeling requirements for all approved GMOs used in food and foodstuffs. Significantly, however, the regulation created a simplified regulatory procedure for foods derived from, but no longer containing, GMOs, provided that those foods remained "substantially equivalent" to existing foods in terms of "their composition, nutritional value, metabolism, intended use and the level of undesirable substances contained therein." Such a determination would be made by the competent authority in the member state receiving the application and would be notified to the Commission, which would in turn notify the other member states. Although the procedure still involved pre-market approval (unlike for foods "generally recognized as safe," by the US FDA), the simplified procedure significantly eased EU approvals for processed foods that no longer contain GMOs. In practice, the procedure would prove to be significant in the coming years, as individual member states would approve a number of products as being "substantially equivalent" to their conventional counterparts.[156]

Again significantly in terms of later developments, the regulation (like the earlier Directive 90/220) contained a safeguard clause allowing member states, "as a result of new information or a reassessment of existing information" to "temporarily restrict or suspend the trade in and use of the food or food ingredient in question in its territory" (Article 12). Once again, any member state invoking such safeguards would be required to inform the Commission, giving the grounds for its decision. Again the Commission would rule on whether the safeguard was legal, working in cooperation with the Standing Committee on Foodstuffs consisting of member-state representatives.

In comparison with the US system, the regulatory structure established by Directive 90/220 and Regulation 258/97 was more complex, more decentralized, and more politicized than the US system. It was more decentralized because of the key role of member states to start, oppose, and reject (through the imposition of safeguards) the approval of a GM seed or food. It was more politicized because of the involvement of politicians in the approval process. And it was more complex in that it created more institutional "veto points," where the approval of new GM varieties or the release and marketing of EU-approved varieties could be blocked. We do not use the term "politicized" in a derogatory way. Indeed, many argue that the European approach is more "democratic," as opposed to a US technocratic approach. What we will show is that, from an institutional perspective, the regulation of GM foods and crops in Europe has been channeled through a political process more than a purely administrative one.

2.3.3. From legislation to implementation

The adoption of these new EU regulations was not, however, the end of the story, for a series of contingent events enabled opponents of GM products to make use of their multiple veto points, triggering a member-state revolt and a complete breakdown of normal EU decision-making. The implementation of the new regulations quickly became inextricably and controversially linked to a series of food-safety scandals that rocked the Union during the second half of the 1990s, most notably the BSE scandal that struck in 1996. In March of that year, the British government of Prime Minister John Major revealed a possible connection between Creutzfeldt-Jacob Disease, a fatal disease for humans, and bovine spongiform encephalopathy (BSE) or "mad-cow disease," a disease spread among cattle through their consumption of contaminated feed (including ground meat and bone meal). The feed had been approved by regulatory officials, even though the public was taken aback to learn that cows had become carnivores.[157] The BSE outbreak infected some 150,000 cattle in the UK, triggering their wide-scale slaughter, a Community ban on the export of British beef, a plummet in beef sales throughout Europe, and a loss of consumer confidence in regulatory officials. The crisis did not abate quickly, moreover, with France and Germany reporting new outbreaks of the disease in 2000 and 2001, and with new food scandals emerging over dioxin-contaminated feed and sewage sludge in feed, among others.[158]

Perhaps most importantly for our purposes, the BSE scandal raised the question of risk regulation "to the level of high politics, and indeed of constitutional significance."[159] It generated extraordinary public awareness of food-safety issues and widespread public distrust of regulators and scientific assessments. Prior to the admission of the BSE risk, "the European Commission had relied on the advice of the [EU's] Scientific Veterinary Committee, which was chaired by a British scientist and primarily reflected the thinking of the British Ministry of Agriculture, Fisheries, and Food—advice which subsequently proved flawed."[160] There was thus considerable skepticism as to the political independence of the committee's "scientific" advice. Within Britain itself, UK scientists continued to reassure the public regarding the lack of BSE risk with information that was soon contradicted. The British Minister of Agriculture even had himself photographed with his four-year-old daughter eating hamburgers to demonstrate that British beef was "perfectly safe" when it turned out that the risk was real and other children died from it.[161] Following the

BSE scandal, comparative polling found that only 12 per cent of Europeans expressed trust in national regulators, compared with around 90 per cent of US citizens who believed the USDA's statements on biotechnology, even though (as we have seen) the USDA can be subject to considerable political influence.[162] As a result, it was much easier for anti-GMO activists to harness "public outrage" and overcome "collective action" challenges to mobilize a large opposition movement.[163] As empirical studies have shown, "trust in risk management is easier to destroy than to create."[164]

It was in this socio-political context that GM crops were first commercially introduced in the US and Europe. In April 1996, within a month of the ban on British beef, the Commission approved the sale of a GM soy product over member-state objections. In November 1996, the GM soy was imported from the US to the EU, spurring widespread protest by Greenpeace and other groups. Soybeans are ingredients in more than half of processed foods, and the US shipped between 25 and 40 per cent of its soybeans to the EU.[165] Many activists feared that GM soy would replace feed that had been banned in response to the BSE crisis. In short, widespread media coverage and public debate about GM foods began just as the BSE food crisis struck, which helped link the two issues before the European public.[166] As behavioral studies show, people's assessments of risks are not equally weighed, but rely on certain heuristics, or in common usage, rules of thumb. A central one is the *availability heuristic* according to which one assesses the extent of risk by recalling examples that come recently to mind.[167] This tendency can result in "probability neglect" in which individual judgments focus on worst-case scenarios, neglecting to address the probability that they would occur.[168] In thinking about the risks of GM foods, European consumers recalled recent experiences with the "mad cow" and other food safety scandals in which scientists' and regulators' advice turned out to be wrong. Precisely at the time that GM products came to market, food risks were salient in the European public's consciousness, unlike in the US.

Two other events occurred in late 1996 that add important context to the contestation that was to engulf EU decision-making over GMOs. In December 1996, a Scottish scientist announced to the world the first successful reproduction of a cloned mammal, a sheep named "Dolly," suggesting that the cloning of humans could follow shortly. The announcement spurred further ethical challenges to biotechnology research. Also in December, the US and Canada lodged complaints before the WTO challenging the EU's ban on hormone-treated beef on the grounds that the EU ban constituted a disguised barrier to trade and was not scientifically justified. The WTO judicial bodies subsequently held against the EU, and, when the EU failed to comply with the ruling, authorized the US and Canada to adopt retaliatory tariffs on EU farm products, leading in turn to widespread protests among European farmers and anti-globalization activists. The WTO case, in particular, rallied a federation of smaller French farm producers that fervently opposed GMOs, the *Confédération Paysanne*, led by Jose Bové.[169] Bové quickly became a symbol for anti-globalization and anti-WTO movements worldwide, and a French national figure, embraced by the leaders of each major French political party.[170]

The close succession of these events illustrates how the popular understanding of GM products in Europe became associated with consumer anxieties related to food safety crises, distrust of regulators and scientific assessments, disquiet over corporate control of agricultural production, ethical unease over genetic-modification techniques, environmental concerns, and anger over the use of international trade rules by the US to attempt to force "unnatural" foods on Europeans. A widespread cross-sectoral movement organized to oppose GMOs in Europe, bringing together environmentalists, consumers, and small farmers. This movement raised the political profile of GM policy across the EU in a common manner, although with different characteristics in discrete national arenas.[171] The movement operated at multiple levels, working the media, and local and national political processes, coordinating transnationally, and lobbying the Commission and EP.[172] The British media dubbed GM products "Frankenstein" foods, playing off fears that scientists and public agencies could not control the release of GM products. European negative attitudes toward GM crops and foods rose rapidly. In early 1996, 46 per cent of the French were against GMOs, a figure that rose to 65 per cent in 1999, and 75 per cent in 2002. Over 80 per cent of Germans expressed negative opinions about GMOs by late 1998.[173]

In the midst of the fray, the Commission approved in January 1997 the sale of another GM food crop (a Bt corn variety owned by Novartis) over the objection or abstention of all but one of the fifteen member states.[174] The Commission was able to do so because of the approval procedure set forth in Directive 90/220 pursuant to which the Council could only reject or amend the Commission's proposal by a unanimous vote. France, which had initially approved the variety, announced that it supported the Commission's authorization.[175] As a result, even though fourteen member states either opposed or abstained from supporting the Commission at this point, the approval went forward. Soon even France opposed commercialization of this GM variety following shifts in French domestic politics.[176]

The member states did not simply accept the Commission's decision. They actively undermined its implementation, invoking the safeguard clause of Directive 90/220 which permitted a member state to prohibit an approved GM variety in its territory if it had "justifiable reasons to consider that [the] product...constitutes a risk to human health or the environment." Austria was the first to act, promptly prohibiting the cultivation and marketing of the GM maize variety on February 14, 1997. Luxembourg followed suit on March 17. Over time, more member states deployed safeguard bans, undermining the central purpose of Directive 90/220 to create a single market for GM crops under a harmonized regulatory system. By January 2004, nine member-state safeguards, applied by Austria, France, Greece, Germany, and Luxembourg, were in effect.[177]

Opponents of GMOs worked not only the political process: they took their battle to the marketplace as well. Non-governmental organizations such as Greenpeace and Friends of the Earth campaigned against GMOs, pressuring retailers not to carry GM products, informing consumers about GM products in stores, and even distributing magnifying glasses to consumers in Germany to help them search for GM products on the labels.[178] Under pressure from potential consumer boycotts of their foods, many large European brand companies and retailers refused to buy or sell GM foods. Danone, Nestle, and Unilever decided not to use GM varieties in 1998, fearing consumer boycotts of their brands. Large UK and French supermarket chains, such as Sainbury's and Carrefour, pledged that their chain-labeled foods would be free of GMOs, wary of consumer boycotts of their stores.[179] The EU trade association EuroCommerce demanded that US producers and distributors segregate and label GM soybeans. Major purchasers of soybean imports into Europe, such as subsidiaries of Unilever, simply refused to buy US soybeans.[180] Monsanto organized a media campaign to raise support for GM products, but the campaign backfired, serving primarily to increase public awareness that GM food products had arrived or were on their way. Surveys in the UK and France indicated that negative perceptions of GMOs rose following the Monsanto advertising campaign.[181]

Thus, although GM soy and maize varieties had been validated by risk assessments conducted by EU scientific committees and been legally authorized for marketing throughout the EU, they were subject to member-state bans and were barely commercialized at all.[182] In fact, the Scientific Committee on Plants had issued sixteen favorable opinions on applications for placing GM plant varieties on the market under Directive 90/220, and only one unfavorable opinion "due to an insufficient risk assessment," resulting in the withdrawal of the application.[183] The scientific committee recommendations were simply ignored.

In response to the popular backlash against GMOs successfully stirred by non-governmental groups and captured in national media, a group of member states (Denmark, France, Greece, Italy, and Luxembourg) pronounced in June 1999 the need to impose a moratorium on all approvals of GM products, pending the adoption of a new and stricter regulatory system. In an annex to the press release of the Environment Council meeting in Luxembourg on June 24/25 1999, the Danish, French, Greek, Italian, and Luxembourg delegations declared:

> The Governments of the following Member States, in exercising the powers vested in them regarding the growing and placing on the market of genetically modified organisms (GMOs), [...] point to the importance of the Commission submitting

> without delay full draft rules ensuring labeling and traceability of GMOs and GMO-derived products and state that, pending the adoption of such rules, in accordance with preventive and precautionary principles, they will take steps to have any new authorizations for growing and placing on the market suspended.[184]

For nearly six years, from October 1998 (when two GM varieties of carnations were approved) to May 2004, no GM varieties were authorized for sale in the EU market, the only exception being for foods derived from GM varieties deemed "substantially equivalent" to traditional foods and subject to the simplified procedure under the 1997 Novel Foods Regulation. Rather, member governments within the Council would obstruct the authorization of any new GM variety, pending the adoption of a revised EU regulatory framework (examined in Chapter 6).

2.4. The sources—and subsequent stability—of regulatory polarization

By the late 1990s, then, the US and the EU had adopted significantly different regulatory *standards* and distinct regulatory *systems*, summarized in Table 2.4. In contrast with the US system, predicated on what the US calls "science-based" decision-making conducted by specialized regulatory agencies with minimal legislative interference, the EU system featured decision-making by political bodies (the Commission, the Council, and its regulatory committees) based on the criteria that included social and economic considerations alongside scientific assessments. Of course, biotech skeptics argue that the US government uses the term "sound science" for political ends and they distrust US government claims. We do not evaluate these claims from a scientific perspective. Rather, we simply note that, in practice, these distinctive regulatory systems gave rise to different regulatory standards for GM products, with more rapid approvals and less onerous restrictions imposed in the US than in the EU.

The *causes* of this regulatory polarization, we argue, are multiple, reflecting a combination of interests, institutions, ideas, and contingent events.[185] More specifically, we offer two arguments in this section about the causes of regulatory polarization and about its persistence over time, respectively. First, we question the ability of any single factor to explain the differences between the US and European approaches to GM foods and crops, and we dispel the notion than any of these factors determined US or EU regulatory polarization from the outset. Rather, after examining the strengths and weaknesses of each explanation, we put forward a multicausal explanation that looks at cultural predispositions and institutional choices in terms of the opportunities and tools that they provide to different interest groups in their political struggles

Table 2.4 Comparison of US and EU approaches to biotechnology regulation

	Aspect	US	EU
Philosophical	View of biotech	Substantially equivalent	Inherently different
	Approach to risk management	"Sound science"	Precautionary principle
Institutional/regulatory	Decision-making style	Administrative	Political
	Pre-release notification	Field tests—mandatory	Mandatory
		Pesticides—mandatory	Mandatory
		Foods—voluntary if GRAS	Mandatory
	Approval required	Field tests—yes	Yes
		Pesticides—yes	Yes
		Foods—no if GRAS	Yes
	Labeling	Only in specific instances	Mandatory

Source: Abbreviated and modified from Young 2003: 466.

over the framing of the issues. We maintain that politicians and interest groups both in the US and in Europe were active in struggles over the framing of the perception of the risks and benefits of agricultural biotechnology, and these perceptions were facilitated by different contingent events on each side of the Atlantic. It was, therefore, not inevitable that US regulators would adopt a product-based approach to GMO regulation, nor was it obvious from the outset that the EU would adopt the strict, politicized, and highly precautionary system that emerged over the course of the 1990s.

Secondly, however, we argue that both political actors and interest groups on each side of the Atlantic have subsequently been constrained by the differential policy frames and institutional choices examined above. In line with historical institutionalist analysis, we argue that early decisions have produced "self-reinforcing" systems on both sides of the Atlantic which, although subject to reversals in response to major shocks, are now difficult to change. In the US, agricultural biotech has been defined as an evolutionary development in plant breeding technology that has resulted in new opportunities, and whose risks can be technocratically managed as any other plant variety by specialized agencies under existing legislation. In the EU, in contrast, agricultural biotechnology has been viewed as presenting a new form of technology and raising broad social concerns so that scientific risk assessment should be just one input into a political decision. These policy frames, in turn, have informed the legally binding institutional frameworks adopted by each side, both of which have proven resistant to change. To the extent that change has proven possible (a subject we address in Chapter 6), it has demonstrated a clearly path-dependent character, introducing incremental changes at the margins of regulatory systems that have remained remarkably stable over time. For this reason, US/EU regulatory polarization, although not inevitable from the outset, has proven robust and enduring, as have the trade disputes that it has generated.

2.4.1. Sources of regulatory polarization

Existing explanations of US or EU regulatory polarization typically emphasize one of the four key factors: interest-group pressures, institutional differences, cultural differences, and contingent events. To start with an interest-based approach, we agree with Thomas Bernauer and others who argue that the structure of *interests* and interest-group pressures have varied between the US, where biotech firms, farmers, and food distributors acquired an early advantage in the adoption of the new technology and pressed for an accommodating regulatory structure, and the EU, where producer interests have been weaker (as a result of the later development of the industry) and less cohesive.[186]

In Europe, the biotechnology industry could have become quite strong and competitive, as a number of leading biotech firms are headquartered in the EU.[187] Over time, however, agricultural biotech investment fled Europe and concentrated in the US on account of the inauspicious European regulatory environment. North American inventors, for example, received 3,035 patents for "agbiotechnologies and crop genetics" compared to 774 patents granted to European inventors during the period from 1982–2002.[188] In 2007, the Commission's Enterprise and Industry Directorate General reported that "public and private investments into agro-biotechnology research have decreased in the EU during the past decade and there has been a delocalization of private R&D to countries outside Europe."[189]

Most importantly, the biotech industry in the US was able to get farmers to adopt GM soybeans, corn, and cotton relatively quickly following their development, creating what one might call *facts in the ground*.[190] Once GM varieties were in the ground in large quantities, US farm association interests became a key source of support for GM crops and foods, so that a broad array of Congressional representatives had little incentive to tighten legislation.

In Europe, in contrast, the biotech industry failed to get early approval and early adoption of GM crops by farmers, so that European agricultural lobbies have never had a large stake in fending off regulatory restrictions. Indeed, given the increasingly strict regulations on the growing and marketing of GM foods and crops as well as adverse public opinion toward GMOs, European farmers have generally approached GM crops with trepidation, fearing that they might not be able to sell them because of responses in the European marketplace. European farmers are wary of distributor and wholesaler reluctance to buy GM foods in response to consumer demands and the threat of anti-GMO activist pressures.[191] Moreover, since US farms are generally larger than European farms and are less likely to be located next to densely settled or protected landscapes, they can adapt to requirements to create buffer zones between their fields and those of adjacent organic (or GM-free) farms at a lower marginal cost.[192] Many European farmers have thus benefited from the challenges confronting US grain producers and traders because of EU legislation, either because their US competitors grow GM varieties or because US non-GM varieties may be inadvertently mixed with GM varieties in the US grain distribution system.[193]

If European farmers were at a significant competitive disadvantage on account of EU restrictions on GM varieties, and if there were supermarket and consumer acceptance of GM foods in Europe, one would likely see greater activism of European farm associations in the political and regulatory process. Political pressure from concentrated producer interests could help to secure a more permissive regulatory system. It appears that this indeed has taken place in the EU's treatment of GM enzymes, which are used in the production of cheese, beer, and biofuels. In contrast with other GM foods and crops, European companies are among the world's leaders in GM enzyme production, and it is striking that GM enzymes have been expressly exempted from the EU's approval and labeling requirements for GM products.[194] Foods derived from the use of GM enzymes (such as cheese and beer) are not labeled, so that European consumers are not in the position to refuse to buy them on this basis. Yet labeling of other GM-derived foods is required in the EU (see Chapter 6), and there is no serious political prospect of this legislation being relaxed.

Moving from producer interests to consumer interests, the fragmented nature of the US consumer movement appears to have contributed to its relatively weak response to GMOs in the 1990s, while the centralization of EU consumer interests allowed consumer groups a greater influence on national and EU decision-making. European skeptics of GM technology were also able to use "direct action" campaigns, such as the uprooting of GM fields, to obtain widespread press coverage that highlighted their views on the risks and uncertainty of the new technology.

Despite these contrasts, we do not believe that the regulatory differences were determined solely by different configurations of interests. European governments and the European Commission wish to promote the competitiveness of EU research and technology, and the EU's largest agribusiness trade associations would more actively support access to biotech products were European consumer and supermarket responses similar to American ones. There is little evidence that Europe is simply being protectionist in constraining imports of GM products from the US, since there are plenty of non-GM varieties grown elsewhere that can fill the gap. Although the biotech industry may have been better organized in the US and although GM skeptics better so in Europe, there remain strong advocates of the technology in Europe and strong skeptics in the US.

Differences in US and EU *institutions* have also played a key role. In the EU, relatively decentralized decision-making, with a particularly strong role for the individual member states, has created a system with multiple veto players, facilitating a ratcheting-up of standards and the imposition of a *de facto* moratorium on new GM approvals imposed by a minority of member governments. By contrast, the more centralized US federal system has produced a more streamlined and politically insulated approval process undertaken by long-standing and more publicly trusted federal agencies like the FDA, the EPA, and the USDA. Such an institutional perspective fits with Sunstein's notion that independent agencies tend to be "more deliberative, calculative, slower and more likely to be error free," resulting in a more "science-based" approach that is less responsive to populist sentiment mobilized by anti-GMO activists. In Sunstein's view, independent agencies are better able to weigh probabilities of both costs and benefits, as opposed to banning new technologies on the grounds of what he calls an "incoherent" precautionary approach.[195]

We agree with institutionalist theorists that institutions can structure politics and channel how issues are framed, and we shall argue presently that the US and EU regulatory frameworks for GMOs, once established in the 1980s and early 1990s respectively, have proven highly resistant to change. However, we do not find it credible that the different approaches to GM regulation in the US and Europe were determined by the characteristics of a more centralized US system of specialized agencies compared to a more fragmented EU system involving a heightened role for politicians. Although specialized regulators may have different approaches to risk regulation than politicians responding to populist concerns, institutional differences did not determine the different approaches of the US and the EU to the emergence of GMOs as a public-policy issue. As we have shown, US regulation could have gone the other way in the 1980s had the EPA taken the lead, or had Congress intervened, or had President Reagan not been in the White House. In short, the key US policy choices of the 1980s were made in the context of political struggles; they were not preordained by the institutional structure of the US federal system.

Similarly, we are skeptical of the claim that the US regulatory system, with its emphasis on regulation by independent expert regulators, makes the US more inherently risk-acceptant than the EU. The US is *more* precautionary than the EU in many other fields, such as the regulation of carcinogens and nuclear energy, among others.[196] The US even appears to be more precautionary than Europe in respect of biotechnology used for medical research, which has raised significant US political debate around ethical issues, even though the FDA, EPA, and USDA are also involved in its regulation to varying extents. Sheingate thus correctly concludes that "the United States is neither uniformly supportive of biotechnology nor does industry dominate all facets of the biotechnology policy process. Rather, there is significant variation across biotechnology applications as well as over time."[197] Even today, were US public concern regarding GMOs to rise, Congress could pass new statutes to increase precautionary regulatory controls, as it has in other areas in the past.

In sum, institutional differences between the US and the EU did not, in themselves, determine the different approaches to biotechnology regulation taken by the two sides. Even in the US, while the debate has been less "politicized" in the sense that politicians have been less involved, politicians (the Reagan White House) were quite engaged in setting up the Coordinated Framework in 1986; and US biotech proponents remain active in promoting the technology in domestic debates. US and EU institutions *mattered*, to be sure, in the sense of offering private interests and public officials distinctive opportunities and constraints, but the general institutional structure of each system did not constrain actors so narrowly as to determine in any straightforward way the regulatory frameworks for agricultural biotechnology adopted by each side.

In addition to the importance of material interests and institutions, we believe that a full explanation requires consideration of *ideational and cultural differences*. The regulatory contrast between the EU and the US can be attributed, in part, to the impact of such differences, summarized by Marsha Echols under the rubric "different cultures, different laws."[198] European governments and their consumers have generally been willing to accept the safety of traditional foods, such as raw milk cheeses and cured meats, while challenging the adoption of new technologies for food production and preservation such as irradiation and genetic modification. Americans, by contrast, have generally been skeptical of traditional European methods, while remaining more open to the use of new technologies in food production and preservation. Similarly, Europeans, who live in a much more densely populated continent, tend to view agriculture as associated with land preservation in an age of urban sprawl, while Americans, benefiting from much larger tracts of nature reserves, are more likely to view agriculture solely in terms of food production.[199] Because Europeans' experience of the environment is more directly linked to agricultural landscapes, they may be more sensitive to GM risks. European publics were thus more likely to be receptive to arguments regarding the risks of these new crops and foods than were Americans.

We also agree that perceptions of risk are influenced by cultural predispositions. Mary Douglas and Aaron Wildavsky in their book *Culture and Risk* tie perceptions of risks to competing cultural views that can be categorized as egalitarian, individualistic, or hierarchical.[200] According to this cultural theory of risk, egalitarians tend to be more concerned with environmental risk, especially to the extent that it affects social inequality; individualists tend to be more dismissive of claims of risk, relying more on market-based approaches; and hierarchists tend to be conservative and focus on the defense of status. To the extent that American culture is relatively more individualistic and European cultures more egalitarian or hierarchical, it is indeed plausible that

Americans and Europeans can approach choices involving risk with different predispositions.

Many cultural theorists go further in their conclusions, maintaining that, *"culture is prior to facts."*[201] In the context of the regulation of GM products, they would argue that regulatory choice is not simply a matter of neutral, technocratic, science-based assessment, but implicitly a reflection of cultural *values*.[202] For instance, in their review of Sunstein's book attacking the coherence of the precautionary principle (and implicitly of European regulation of GM products), four leading cultural theorists of risk offer the example of nuclear power regulation as a reflection of cultural values. They state:

> The difference between French and U.S. attitudes toward nuclear power, and the resulting differences in the regulations of the two nations, are hardly a matter of "coincidence" or chance. In contrast to members of the public in the United States, those in France are much more likely to hold a *hierarchical* worldview. This difference not only disposes the French to be more accepting of nuclear power risks, but also to be more confident in the ability of technical and governmental elites to manage any such risks.[203]

The story in France regarding GM regulation indeed started similarly. At first, GM technology was largely viewed in France as a "matter of progress and national competitiveness to be managed by the elite."[204] As late as 1997, "France was the second most popular country—only after the US—for GMO field experiments."[205] As a result, companies most frequently applied in France for approvals of GM varieties under the 1990 directive, and it was France alone in 1996 that supported the Commission's famous approval of a variety of Bt corn, in a vote where all other member states either opposed or abstained.[206]

However, longitudinal studies demonstrate the problem with purely "culturalist" approaches. For example, "the image of nuclear power" significantly changed over time in the US "from solidly positive to overwhelmingly negative," with interest groups active on both sides in their attempts to frame perceptions of the issue.[207] Similarly, the story of GMO regulation in France is more complicated, showing why a cultural explanation is not sufficient. After France supported the Commission's approval of the Bt corn, it banned the cultivation of that same corn by blocking it from being listed in France's national seed catalogue. Paradoxically, it was a conservative government, under Prime Minister Alain Juppé, which did so in February 1997. This decision was reversed in November 1997 when, again paradoxically, a new socialist government, under Lionel Jospin, took power, with the leader of the Green Party as Minister of the Environment.[208] However, the socialist government soon reversed course yet again in its policy. The French parliament initiated a series of initiatives that triggered public and private hearings and a "Citizens' Conference" that drew significant media attention to GMO concerns. In 1998, the government created a new French Food Safety Agency (*Agence*

Française de Sécurité Sanitaire des Aliments) that, together with its Commission on Biomolecular Engineering (*Commission de Genie Biomoléculaire*, or CGB), which had been quite supportive of the technology, was to review applications of GM varieties. The government then expanded the membership of the CGB to include GM skeptics and created a *Comité de Biovigilance* (under the Ministries of Agriculture and the Environment), whose membership included NGOs such as Greenpeace and José Bové's anti-globalist, anti-agribusiness *Confédération Paysanne*. The French government became one of the backers of an EU moratorium on new approvals of GM products, and it imposed a national safeguard against two varieties of GM oilseed that had received EU approval, in defiance of the Commission. GM field trials in France quickly dropped from 1,100 in 1998 to below 50 in 2002, half of which were destroyed by anti-GMO activists.[209] The company, Novartis, that held the rights to the original Bt corn approved by France finally issued a statement that it would not sell the crop in France until its consumers were ready for it.

Thus, while we accept that the French have different attitudes toward foods than Americans and that they may often be more "hierarchical" in their attitudes toward government and society (including in the weighing of risks), something else happened that *changed* French attitudes toward agricultural biotechnology. In other words, culture is not static, and it was not deterministic of French or European approaches to agricultural biotech regulation. In our view, cultural predispositions rather offered opportunities in political struggles over the framing of the issues, which has varied over time and across issue-areas.[210]

Related to cultural explanations of divergence, the differential use by Americans and Europeans of heuristics (or mental shortcuts) in assessing risks could help to explain the polarization of regulatory approaches. Considerable media attention on food safety scandals in Europe meant that the food safety risks were a more salient issue in Europe at the time that GM varieties were first commercialized. Europeans were thus more likely to rely on what has been called an "availability heuristic" in weighing the risks of GM products because these risks were salient—that is, they were readily available to them cognitively at this critical moment. In addition, we have seen processes of social polarization on both sides of the Atlantic, with GM foods being branded as US attempts to force Europeans to change their mode of life in the context of debates over globalization, and Europeans being dismissed as irrational in their approaches to new food technologies. In this way, "local convergence can generate global polarization."[211]

Finally, *contingent events* have played an especially important role in the EU regulatory process. The BSE and other food-safety scandals of the 1990s led to a dramatic increase in public concern about GM foods and crops in Europe, and to a dramatic decline in public trust in scientists and government regulators compared with the much higher level of such trust in the US. In a similar fashion, the US's successful WTO complaint in the beef-hormones case contributed to the rise of anti-American and anti-globalization sentiment in Europe, and to a conviction in public opinion and among some officials that the US should not be allowed to force GM foods down the throats of European citizens. Here again, however, these contingent events did not "cause" US or EU policies; rather, they acted as *exogenous shocks* within each political system, presenting various actors with new resources, or new constraints, in the ongoing struggle to set domestic policies. In the EU, strategic actors ranging from NGOs to member-state governments mobilized to use the "mad cow" and other food scares, as well as the resulting decline in European public trust of health officials, in their political struggles. They did so in order to shape popular perceptions of GM foods and the resulting EU rules that govern their approval and marketing.

In short, we find that explanations based on any one of these factors (interests, institutions, ideas, or contingent events) are insufficient to explain the stark regulatory differences that have arisen between the US and Europe. We believe that the best explanation for these enduring differences is that US and European interest groups have pursued their interests within existing institutional and cultural contexts that, together with contingent events, provided opportunities as well as constraints, to define issues and frame perceptions of GM foods in different ways. In the US, interest groups, with the support of a Reagan Administration skeptical of strong public regulation, sought a regulatory framework that would treat new GM varieties as substantively equivalent to their conventional counterparts. In doing so, they were able to draw on a supportive institutional and cultural context, including a regulatory system featuring strong and specialized government regulators, a diverse consumer protection movement, and a tradition of accepting the use of new technologies in food production. In this context, GMO supporters and their political and administrative allies were able to "foster a more controlled" and "optimistic image" of agricultural biotechnology, as opposed to the risks of "Frankenstein foods" or "mutant organisms gone wild."[212]

By contrast, pro-GMO interests in Europe were weaker, particularly within the agricultural community. European farmers never got GM crops in the ground to a significant extent before activist groups generated popular resistance to US imports, and they thus never had a major stake in them and never emerged as a champion of the new technology. Pro-GMO interests, rather, encountered an institutional and cultural framework that provided multiple veto points and multiple sources of opposition to GMOs on environmental, food-safety, and ethical grounds, resulting in the more demanding, politicized, and process-oriented EU regulatory system. The subsequent evolution of the EU regulatory process, however, as well as the French U-turn on GMOs and the subsequent declaration of a moratorium in the second half of the 1990s, is in large part a direct result of contingent events, namely the BSE crisis and other food-safety scandals of the 1990s which undermined public support for the technology and trust in regulators at a crucial time for the introduction of GM foods and crops.

Hence, however much we might wish for a simple, monocausal explanation of the observed regulatory polarization between the US and the EU, the most convincing explanation is one that combines an interest-based approach with institutional and cultural constraints, and one that acknowledges that, in the absence of continent events, things might have turned out differently.

2.4.2. Historical institutionalism, increasing returns, and self-reinforcing institutions

Neither the US nor the EU, we have argued, was preordained by its interest-group, institutional or cultural characteristics to adopt the precise regulatory framework that it did when GM foods and crops emerged as a public-policy issue in the 1980s and early 1990s. By the same token, however, the respective US and EU regulatory frameworks, once adopted, have proven remarkably resilient in their essential characteristics—a claim that we substantiate empirically in Chapter 6.

The explanation for this resilience, we argue, can be found in historical institutionalist theory, which examines the effects of institutions on politics *over time*, maintaining that institutional choices taken at one point in time can persist, or become "locked in," thereby shaping and constraining actors later in time.[213] In perhaps the most sophisticated presentation of this strand of historical-institutionalist thinking, Paul Pierson has suggested that political institutions and public policies are frequently characterized by "positive feedbacks," insofar as those institutions and policies generate incentives for actors to stick with and not abandon existing institutions, adapting them only incrementally to changing political environments. This positive feedback may reflect constraints *from above*, in the form of legally binding rules that are difficult or costly for political actors to change, or *from below*, as societal actors adapt to and develop a vested interest in the continuation of specific public policies.[214]

Insofar as political institutions and public policies are characterized by positive feedbacks, Pierson argues, politics will be characterized by certain interrelated phenomena, including *inertia*, or *lock-ins*, whereby existing institutions may remain in equilibrium for extended periods despite considerable political change; *a critical role for timing and sequencing*, in which relatively small and contingent events that occur at *critical junctures* early in a sequence shape the institutional context for the reception of events that occur later; and *path-dependence*, in which early decisions provide incentives for actors to adapt to and thus perpetuate institutional and policy choices inherited from the past, even when the resulting outcomes are manifestly inefficient. With regard to this last concept of path dependence, perhaps the most influential notion in recent historical-institutionalist work, Pierson cites with approval Margaret Levi's definition:

> Path dependence has to mean, if it is to mean anything, that once a country or region has started down a path, the costs of reversal are very high. There will be other choice points, but the entrenchments of certain institutional arrangements obstruct easy reversal of the initial choice. Perhaps the better metaphor is a tree, rather than a path. From the same trunk, there are many different branches and smaller branches. Although it is possible to turn around or to clamber from one to the other—and essential if the chosen branch dies—the branch on which a climber begins is the one she tends to follow.[215]

At the extreme, institutions and policies can become *self-reinforcing*, such that the operation of the institution or policy not only resists change, but also bolsters its societal support base in such a way that the institution becomes more difficult to change, and more stable in the face of external shocks, over time.[216]

In this context, we argue that the US and EU regulatory frameworks for agricultural biotechnology, while not themselves determined by preexisting institutional constraints, have since generated significant positive feedback, lending each system considerable resistance to change, and indeed making each system self-reinforcing. In each case, timing and sequencing have proven vital, as the initial regulatory frameworks were adopted at critical junctures that shaped subsequent developments. In the US case, the critical juncture occurred in the mid-1980s, when the introduction of agricultural biotechnology and of GM foods and crops presented policy-makers with a crucial set of choices. The US biotech industry at this time was strong and well-organized, had influential allies in the Reagan Administration and among government regulators, and faced little contestation regarding GMOs from consumer groups, environmental groups, or the general public. In this context, the Reagan Administration was able to establish a comprehensive regulatory framework within existing statutory authority, hence obviating the need for Congressional approval. The results came close to the preferences of the biotech industry.

The critical juncture in the EU came a few years later when the EU prepared to adapt a harmonized EU approach to the laboratory testing and large-scale cultivation of GM crops. It did so in the context of an EU with relatively weak interest-group support for GMOs, diverse preferences among EU governments and a decision-making system with a large number of veto points. As a result, the EU's initial regulatory framework, codified in Directive 90/220, laid down a more demanding regulatory procedure closer to the preferences of GM skeptics. In addition, a further critical juncture came during the second half of the 1990s when GM foods were first ready for importation and sale in the EU. At this time, the BSE and other food-safety scandals strengthened the position of those actors who sought to make the EU's regulatory framework more restrictive, resulting in the post-1998 moratorium and the subsequent strengthening of the regulatory framework early in the following decade (see Chapter 6).

Just as importantly, in each case the regulatory frameworks adopted have generated positive feedback, both by creating institutional rules that could be changed only with difficulty, and by generating adaptations among interest groups and public opinion that contributed to the stability of the two frameworks. In the US case, the early adoption in 1986 of a relatively welcoming regulatory framework contributed to the rapid growth of the biotech industry and the equally rapid adoption of GM crops by American farmers. These farmers have represented the bulwark of political support for the existing framework in the face of ensuing environmental and food-safety contention about GMOs. Similarly, the approval of numerous GMOs by the relatively trusted FDA, the public framing of GMOs as being "substantially equivalent" to their conventional counterparts, and the lack of any labeling requirement in the US have blunted the mobilization of US public opposition against GMOs, which remains muted by comparison with the EU.

In the EU case, by contrast, the early adoption of a relatively restrictive regulatory framework, together with the turn against GMOs and the subsequent declaration of a *de facto* moratorium, discouraged farmers from planting GM crops, prompted retailers to resist GM foods, and led to the flight of agricultural biotech investment from Europe, further undermining societal support for GM foods and crops. At the same time, the EU's convoluted legislative process, requiring qualified majorities among the member states in the Council of Ministers and an absolute majority of the European Parliament that has turned largely against GMOs, has created a huge institutional hurdle to the relaxation of the EU regulatory framework.

Finally, where external shocks have led to pressures for change on each side, the change has proven to be strongly path-dependent. Thus, as we shall see in Chapter 6, pressures for a tightening of US regulatory standards now occur in a context in which industry and agricultural support for existing standards is strong, and where a fundamental change would require either an act of Congress (supported by both houses, and with the possibility of presidential veto) or a striking about-face by government regulators. Changes to the US regulatory system have thus been incremental and path-dependent, adjusting particular regulatory requirements at the margins without fundamentally altering the approach adopted in the 1986 Coordinated Framework. Similarly, any change in the EU regulatory framework now faces widespread public opposition to GMOs and weak industry and agricultural support, as well as multiple veto points in the Commission, Council, and European Parliament. In that context, and despite considerable pressures for change from the US, fundamental changes in the EU regulatory framework have proven impossible, and those that have taken place have amended Directive 90/220 in one direction—towards greater precaution, greater politicization, and higher standards. We return to the domestic stories in the US and Europe in Chapter 6 after examining the two sides' strategies and interactions in bilateral and multilateral fora.

2.5. Two-level games: Domestic politics and the prospects for bilateral and multilateral cooperation

The entrenchment of the respective US and EU regulatory systems, finally, *matters* in the context of the transatlantic and global dispute over GMO regulation, because the ability of EU and US officials to negotiate, compromise, and deliberate is profoundly affected by the nature and intensity of preferences of their respective constituencies back home.[217]

To understand why, consider the "two-level games" model of international negotiations put forward by Robert Putnam.[218] In Putnam's model, all international negotiations take place simultaneously at two levels: at the international level (Level 1), chief negotiators (also known as statesmen, chiefs-of-government, or COGs) bargain with their foreign counterparts in an effort to reach mutually beneficial agreements; while at the domestic level (Level 2), the same chief negotiator engages in bargaining with her domestic constituencies, who must ultimately ratify the contents of any agreement struck at Level 1. Like traditional, unitary-actor models, Putnam's model assumes that chief negotiators or COGs retain a monopoly on the external representation of the state, and negotiate international agreements on behalf of their domestic constituents. Unlike unitary-actor models, however, Putnam's model posits that the negotiating position of a given COG is constrained by her "win-set," i.e., "the set of potential agreements that would be ratified by domestic constituencies in a straight up-down vote against the status-quo."[219] More specifically, Putnam argues that the nature and size of a given state's win-set is determined by several factors, including the preferences and bargaining power of various producer and consumer groups.

At first blush, the chief negotiator or COG would seem to be in an unenviable position, caught between a rock (her own domestic constituency) and a hard place (her fellow COGs). Yet the COG may exploit her privileged position at both tables, and a specific set of strategies specified by Putnam and others, to manipulate her own and other countries' win-sets, and thereby achieve some autonomy in the pursuit of her domestic and international goals. More specifically, the COG may manipulate her domestic win-set in various ways. She might, for example, manipulate the domestic ratification procedure to make it more or less demanding (as in the case of US "fast track"); or she might increase or decrease opposition to an agreement through the provision of side-payments to important swing groups, or through the manipulation of information about the terms of the agreement.[220] Alternatively, a COG might try to manipulate the domestic win-set of a foreign country, such as by targeting concessions or threats to specific groups within that country. For example, the US has targeted retaliatory sanctions in international trade disputes at particular groups and legislative constituencies within foreign countries.

Through the use of such tactics, COGs may increase their bargaining power in international negotiations, as well as the likelihood of domestic acceptance if an agreement is reached. Thus, for example, when confronting other COGs over a difficult distributional issue, a COG may increase her international bargaining leverage through a strategy of "Tying Hands," deliberately constricting her own win-set in order to provoke concessions from other governments. Alternatively, if a COG is eager to reach agreement, she may engage in a strategy of "Cutting Slack," expanding her own win-set in order to secure domestic ratification of an international agreement. Using her agenda-setting power as the negotiator of an international agreement, a COG may negotiate and propose for ratification an agreement that—among the universe of ratifiable agreements in her own win-set—lies closest to her own policy preferences, whatever those may be. Perhaps most intriguingly, Putnam and his followers argue that a COG may even exploit her presence at both tables to secure *domestic* policy reforms that might otherwise prove impossible to attain because of opposition from particular interest groups.

Thus, while Putnam's model is best known for combining the domestic and international arenas into a single bargaining model, it turns out that the strategies and preferences of individual statesmen, or COGs, are central to determining the outcome of any international bargain. The specific choice of strategies, however, depends on the COG's *preferences*, which Putnam characterizes in terms of the relationship between the domestic win-set and the COG's own "acceptability set," that is, the set of agreements preferred by the statesman to the status quo. As Moravcsik points out, "The possible configurations can be divided into three categories: the statesman-as-agent, the statesman-as-dove, and the statesman-as-hawk.[221]

In the case of the "statesman-as-agent," the COG's preferences are identical to those of her median domestic group, and her aim is simply to negotiate an agreement as close as possible to her median constituent's ideal point. In the case of the "statesman-as-hawk," by contrast, the COG should be more reticent and more recalcitrant than her own domestic constituency, and we would most likely expect a hawkish COG to attempt to manipulate her constituency's win-set toward a more hard-line position. Finally, in the case of the "statesman-as-dove," the COG is more eager than her constituents to reach an agreement, and may be expected to "cut herself slack" by expanding her own domestic win-set to secure ratification of her preferred agreement. In addition, when two "dovish" statesmen enter into negotiations, "incentives are created for COG collusion" whereby COGs conspire to undermine domestic opposition to their preferred outcomes in their respective countries.[222]

Putnam's model is based on a series of simplifying assumptions, including the notion that chief negotiators or COGs enjoy a monopoly of representation for their respective jurisdictions. As we shall note in the next chapter, this assumption is increasingly invalid in the contemporary world, when individual units of government (such as government departments, regulatory agencies, or Commission directorates-general) strike up direct "transgovernmental relations" across national boundaries.

Nevertheless, Putnam's two-level games model does point to several crucial features of international negotiations that inform our analysis of the international law and politics of GM foods and crops. We emphasize three. First, and most obviously, the US federal government, which serves as the chief negotiator for the US, and the European Commission, which serves as chief negotiator for the EU on the trade-related aspects of GM foods, are both limited in their negotiating ability by their respective constituencies, which determine the size of their win-sets. As we have seen, the respective EU and US regulatory frameworks are both supported by strong and entrenched domestic coalitions, dramatically limiting the size of each side's win-sets, tying the hands of negotiators on both sides, and reducing the space for potential agreement. It has also, as we shall see in Chapters 3 and 4, reduced the scope for open, deliberative decision-making on the issue of GM foods and crops.

Second, while the entrenched positions of domestic constituencies are crucial, the preferences of chief negotiators are also important in a two-level games model, with each COG potentially playing the role of "hawk," "agent," or "dove" relative to domestic constituencies. As we shall see, the US government has been relatively hawkish, taking a strong pro-GM stance in international forums consistent with the views of the US biotech industry, despite much weaker support for GM foods among US public opinion. By contrast, the European Commission has been relatively dovish, generally preferring to resume approvals of GM crops and come into compliance with international trade law, notwithstanding tremendous resistance in European public opinion. This position has put the Commission in a consistently uncomfortable position between US and international pressure on the one hand, and domestic European resistance on the other.

Third and finally, we should expect to see both the US government and the European Commission, in their capacities as chief negotiators, engage in the types of negotiating strategies outlined by Putnam, including manipulation of their own win-sets, manipulation of other countries' win-sets, and the exploitation of international pressure to push through desired domestic reforms. Thus, for example, we shall see the Commission consistently attempting to use international pressures to secure reforms to domestic European legislation, restart the European approval process, and remove national safeguard bans on specific GM foods and crops. Once again, however, the Commission's ability to do so has been dramatically limited by the size and intensity of the anti-GMO coalition within the Union.

2.6. Conclusions

The US and the EU both faced the challenge, in the course of the 1980s and 1990s, of developing a regulatory framework for agricultural biotechnology. The two sides adopted starkly different regulatory frameworks. These frameworks reflected divergent interest-group pressures operating within institutional and cultural constraints in light of contingent events. The US emphasized a product-based approach of science-based decision-making by regulators, with only passive oversight by political actors. The EU enacted a process-based, precautionary, and more politicized approach. Once adopted, these two regulatory systems generated positive feedback, becoming self-reinforcing and remarkably resilient in the face of pressures for change.

The existence of these two distinctive and self-reinforcing regulatory frameworks is of more than academic interest because each side's "domestic" regulation of GM foods and crops has external implications for the other and for the world. The friction between the two systems has generated a decade-long series of trade and regulatory disputes. The impact of the EU's moratorium and its stricter standards was felt almost immediately in the US, where around two-thirds of all GM crops were grown.[223] US exports of soy to the EU were valued at $2.3 billion in 1996, about twenty times the value of lost beef sales to the EU in the earlier WTO dispute, but these sales plummeted to just over $850 million in 2004.[224] US corn exports to the EU almost totally collapsed from $400 million in 1996 to less than $10 million in 2004, dropping to less than $3 million in 2005.[225]

The EU's more stringent process-based regulatory approach soon spread to other countries. Yet the use of GM varieties developed and patented in the US spread as well, including the leading agricultural producers such as Argentina, Brazil, India, and China. As a result, the transatlantic dispute over biotechnology regulation threatened to escalate into broader trade conflicts, spurring multiple attempts by the two sides either to resolve their differences or to defend and advance their perspectives in different bilateral and multilateral fora.

3

The Promise and Failure of Transatlantic Regulatory Cooperation through Networks

The dispute over the regulation of agricultural biotechnology presented the US and the EU with a dilemma: on the one hand, both sides sought to regulate GM food and crops in line with their respective regulatory traditions and the demands of their citizens, while on the other hand, both sides had a clear interest in preventing the GMO issue from escalating into a full-scale trade war.[1] Faced with such tensions, US and EU officials attempted bilateral cooperation, enlisting networks of scientists, non-governmental groups, and, especially, government regulators from both sides to deliberate together, and to understand and overcome their regulatory differences. In doing so, they drew on the positive experiences of transatlantic cooperation among regulators in other fields such as competition policy, reflecting calls for transnational governance through regulatory networks, and the creation of transatlantic policy communities. It was hoped that such networks could foster genuine deliberation, a joint search for the best solution to the challenges of regulating agricultural biotechnology, among US and EU regulators. Despite these hopes, the record of US/EU regulatory cooperation in the area of agricultural biotechnology has been disappointing, with little evidence of genuine deliberation or joint problem-solving among US and EU regulators, whose approaches remain distinct and who remain under intense pressure from their respective legislatures, interest groups, and public opinion.

This chapter examines the prospects and expectations of bilateral regulatory cooperation, and explains the failure of bilateral cooperation to produce any significant formal or informal agreements or evidence of genuine deliberation between the US and EU regulators, other than scientists regarding risk-assessment practices. This chapter is organized in five parts. In the first, we review and extend the literature on international governance through "transgovernmental networks" of national officials, while in the second, we examine the prospects for deliberation in such networks. As we shall see, international relations theorists have increasingly posited that international negotiations may follow not a "logic of consequentiality," in which instrumentally rational state actors bargain for advantage, but rather a "logic of arguing," in which governmental officials and policy experts from many different nations engage in an open-minded and deliberative search for truth and for the best policy in any given issue-area. As we shall also see, however, even proponents of such "deliberative supranationalism" concede that deliberative behavior or arguing is most likely under a fairly restrictive set of conditions, which may be difficult to meet in international politics, especially where powerful states disagree. In the third section, we briefly examine evidence from over a decade of transatlantic regulatory cooperation across a broad range of issue-areas such as competition policy, data protection, financial services, and mutual recognition agreements, noting the considerable variation in successful cooperation across these fields and the numerous obstacles to productive and deliberative regulatory cooperation. In the fourth section, we turn specifically to risk regulation, noting the relative strengths, weaknesses, and prospects of deliberation through transgovernmental technocratic networks and in public fora as a means of evaluating and managing novel risks. The fifth section, finally, reviews the abortive attempt, during the latter half of the 1990s, to build systematic transgovernmental networks for biotech regulation between the US and the EU, and examines the reasons for the disappointing outcome of these efforts. This failure, we argue, demonstrates the difficulty and fragility of deliberation in international politics, and sets the stage for the more extensive and marginally more successful efforts to cooperate in multilateral regimes, examined in Chapter 4.

3.1. Governance through transgovernmental networks

The rise of the transatlantic dispute over GMOs during the 1990s coincided with a revival, among practitioners and scholars, of the concept of governance through so-called "transgovernmental" networks of lower-level government officials cooperating directly with their counterparts in other jurisdictions to address overlapping regulatory concerns and to resolve and accommodate regulatory differences. This recent focus on regulatory networks has been complemented by the creation of transnational (non-governmental) networks and mixed public–private networks to address transnational regulatory challenges in an open-ended, collaborative, and transparent manner, again focused on joint "problem-solving" as opposed to the defense of distinct "interests."

The term "transgovernmental relations" was coined in the 1970s, by Robert O. Keohane and Joseph Nye, who defined the term as "sets of direct interactions among sub-units of different governments that are not controlled by the policies of the cabinets or chief executives of those governments."[2] Crucially, this definition suggests, governmental subunits below the level of chief executives, and from "domestic" rather than "foreign" ministries, may interact directly with their domestic counterparts in other countries, and these interactions may be monitored and controlled at best imperfectly by chiefs of government and by foreign-policy bureaucracies that have traditionally had a monopoly on external representation. Such autonomy is more obvious among regulatory agencies, central banks and courts with some *de jure* guarantee of independence, according to Keohane and Nye, but it may also be found in ordinary government ministries and bureaucracies whose control of specialized expertise and information provides them with some degree of *de facto* autonomy in the making of national policies.

Transgovernmental relations have always existed, according to Keohane and Nye, but have become increasingly significant since World War II, when the social and economic role of governments expanded along with their increasing economic interdependence. Under these circumstances, "bureaucracies find that to cope effectively at acceptable cost with many of the problems that arise, they must deal with each other directly rather than indirectly through foreign offices."[3] International organizations can play a key role in fostering these interactions, both as fora within which national bureaucrats can have direct and sometimes regular contact with their foreign counterparts, and as catalysts and members of transgovernmental networks. Thus, a combination of increasing interdependence and increasing institutionalization of international relations meant that national subunits of government would face increasing pressures—and increasing opportunities—to interact regularly and independently of their traditional foreign-policy apparatus.

Furthermore, as David Vogel has pointed out, the demand for direct transgovernmental relations has been magnified as trade interdependence increases and "behind-the-border" regulations of states become subject to scrutiny as potential barriers to trade. Because these national regulations are typically made and enforced by lower-level regulators in domestic governments and specialized regulatory agencies, the removal of such regulatory barriers would seem to demand direct cooperation among regulators in areas such as competition policy, food safety, and consumer protection.[4]

International law and international relations scholars have recently revived the study of transgovernmental networks, which Anne-Marie Slaughter characterizes as a "new world order" governed in large part by transborder networks of regulatory agencies ("the new diplomats") and courts, who interact and cooperate to provide joint governance of issues ranging from anti-trust and financial markets to food safety and the protection of the environment. While recognizing the problems of democratic oversight of these groups, Slaughter maintains that these new transnational mechanisms are normatively desirable because they are more likely to provide a "fast, flexible, and effective" means of transnational governance. In her words:

Stop imagining the international system as a system of states—unitary entities like billiard balls or black boxes—subject to rules created by international institutions that are apart from and "above" these states. Start thinking about a world of governments, with all the different institutions that perform the basic functions of government—legislation, adjudication, implementation—interacting both with each other domestically and also with their foreign and supranational counterparts. States still exist in this world, indeed, they are crucial actors. But they are "disaggregated."[5]

Because they are composed of specialist or sectoral officials rather than the type of generalist officials traditionally found in foreign ministries, transgovernmental networks are more likely to reach a consensus based on their shared knowledge and interests, including their interest in policy learning, yet this consensus is likely to be relatively narrow, in isolation from related issue-areas and from the broader foreign-policy goals of the states in question. Transnational policy networks of professionals that hold particularly homogeneous cognitive orientations have been viewed as forming "epistemic communities" (or knowledge communities) to stress how they share core sets of beliefs, principles, goals, and methods for validating claims that facilitate collaboration. The term, taken from the Greek word *"episteme"* meaning "knowledge," refers to a network of professionals who share at least four attributes which facilitate collaborative problem-solving: "(1) a shared set of normative and principled beliefs...; (2) shared causal beliefs...; (3) shared notions of validity...; and (4) a common policy enterprise...."[6] Scientists are a primary example of a community that shares such common cognitive orientations so that, when discussing a policy challenge, they are more likely to engage in deliberation. Scholars have demonstrated, for example, their central role in shaping understandings and policy responses to the threat to the earth's ozone layer and other environmental issues such as climate change.[7] Policymakers can, under the proper conditions, also engage in such reasoning, in particular when they are granted relative autonomy from their political masters. These relatively autonomous horizontal networks of lower-level governmental representatives can also work with members of international secretariats in specific policy areas to form vertical networks (between the international and national levels) to define policy options and affect policy outcomes.[8]

Perhaps the best examples of these transgovernmental networks and the most extensive scholarship on them are to be found in the EU, where member-state government officials have long interacted through networks such as the Committee of Permanent Representatives (Coreper, which is charged with negotiating the details of the Council's drafts for EU legislation) and "comitology" committees (which are charged with overseeing the Commission's implementation of EU regulations, including individual decisions and further rule-making).[9] Members of Coreper and their numerous working groups play a crucial legislative role in the EU, shaping much of the legislative output of the Union, yet its members come largely from domestic ministries in the various member states, and are not elected officials. Similarly, the members of the comitology committees tend to come from specialized agencies within the executive branches of the member states, with specific policy portfolios.

In recent years, moreover, these traditional EU networks have been joined by others, such as the European Competition Network (bringing together the EU Commission and the national competition regulators in each of the Union's 27 member states) as well as similar networks of utilities and financial regulators.[10] In many cases, these networks of regulators are even formally independent of their own domestic governments. They generally meet out of the public eye, seeking common solutions to common problems. In addition, the EU has experimented with the so-called "Open Method of Coordination" (OMC), an essentially transgovernmental network of national officials who meet regularly to compare and coordinate public policy in a range of issue-areas such as employment and pensions policy.[11] Within the area of agricultural biotechnology, an attempt at governance by networks is reflected in the EU's comitology committees for GM seeds and foods, as well as in the new European Food Safety Authority (EFSA), both discussed in Chapter 6. Whether these networks live up to their promise depends in large part on how they operate in practice—the question to which we now turn.

3.2. Logics of social action, deliberation, and "arguing"

Perhaps most interestingly for our purposes, an increasing number of scholars have identified these new and emerging networks as a promising venue for what Christian Joerges has called "deliberative supranationalism," an efficient and normatively desirable system in which national government officials meet and *deliberate* in search of the best solution to common policy problems.[12] This emphasis on deliberation, reiterated in an exploding body of literature, derives largely from the work of Jürgen Habermas, whose theory of communicative action has been adapted to the study of international relations and EU governance.[13] Under this Habermasian conception, actors are able to agree on a common policy because they are willing to yield to the force of the better argument and find reasoned consensus on basic validity claims and their implications.

The starting point for such deliberative approaches is the claim, made most clearly by Thomas Risse in the field of international politics, that there is not one but three "logics of social action," namely (1) the logic of consequentiality (or utility maximization) emphasized by rational-choice theorists, (2) the logic of appropriateness (or rule-following behavior) associated with sociological institutionalist and constructivist theory, and (3) a logic of arguing (or deliberation) derived largely from Habermas' theory of communicative action.

The first of these approaches, the "logic of consequentiality," derives from the expected-utility assumptions of most rational-choice theories, namely that actors (be they individuals, firms, or states) possess specific preferences over states of the world, and act systematically to maximize their respective utility under physical and social constraints. Such assumptions underlie much of the literature on international conflict and cooperation, including not only "hard" rational choice and game-theoretic models, but also most regime theory, rational-choice institutionalism, a considerable amount of international law scholarship, and much of the literature on historical institutionalism discussed in Chapter 2.

These core assumptions of rational-choice, and the theories based on them, have recently been challenged by the growing number of constructivist and sociological institutionalist theorists in international law and international relations.[14] Constructivism itself, like rational choice, is not a substantive theory of international relations *per se*, but a broader "meta-theoretical" orientation with implications for the study of international politics. As Thomas Risse explains:

> ... [It] is probably most useful to describe constructivism as based on a social ontology which insists that human agents do not exist independently from their social environment and its collectively shared systems of meanings ("culture" in a broad sense). This is in contrast to the methodological individualism of rational choice according to which "[t]he elementary unit of social life is the individual human action." The fundamental insight of the agency-structure debate, which lies at the heart of many social constructivist works, is not only that structures and agents are mutually co-determined. The crucial point is that constructivists insist on the *constitutiveness* of (social) structures and agents. The social environment in which we find ourselves, "constitutes" who we are, our identities as social beings.[15]

For constructivists, then, institutions are understood broadly to include not only formal rules but also informal norms, and these rules and norms are expected to constitute actors, i.e., to shape their identities and their preferences. Actor preferences are not exogenously given and fixed, as in rationalist models, but *endogenous* to institutions, and individuals' identities shaped and re-shaped by their social environment. Taking this argument to its logical conclusion, constructivists have generally rejected the rationalist conception of actors as utility-maximizers operating according to a logic of consequentiality, in favor of March and Olsen's conception of a "logic of appropriateness."[16]

According to March and Olsen, institutions, understood broadly to include "the routines, procedures, conventions, roles, strategies, organizational forms, and technologies around which political activity is constructed,"[17] do not simply provide a set of strategic constraints within which actors seek to maximize their individual utility. Rather, institutional rules, routines, and roles are internalized and followed "even when it is not obviously in the narrow self-interest of the person responsible to do so."[18] March and Olsen argue that actors do not necessarily calculate the expected utility of alternative courses of action given their specific preferences and choose the optimal one when given a choice or a social situation, but rather seek to undertake the action most appropriate to their social role and to the nature of the situation. In a logic of consequentiality, therefore, a given actor will answer and act upon the following script:

1. What are my alternatives?
2. What are my values?
3. What are the consequences of the alternatives for my values?
4. Choose the alternative that has the best consequences.[19]

By contrast, an actor behaving in accordance with the logic of appropriateness would follow this alternative script:

1. What kind of situation is this?
2. Who am I?
3. How appropriate are different actions for me in this situation?
4. Do what is most appropriate.[20]

In essence, March and Olsen argue that human behavior is essentially rule-guided rather than utility-maximizing, with actors seeking to "do the right thing" in light of their socially learned roles and values. This conception of social action has proven influential in constructivist international relations theory as well as normative legal theory, in which international institutions are posited to "teach" norms to states and their representatives, who behave "appropriately," given their socially learned rules and roles.[21]

Drawing on Habermas' theory of communicative action, however, Risse argues for a third logic of social behavior, which he calls the "logic of arguing," derived largely from Habermas's theory of communicative action and emphasizing the interrelated concepts of argumentation, deliberation, and persuasion.[22] In Habermasian communicative action, or what Risse calls the logic of arguing, political actors do not simply bargain based on fixed preferences and relative power; they may also "argue," questioning their own beliefs and preferences, and being open to persuasion and the power of the better argument:

> Arguing implies that actors try to challenge the validity claims inherent in any causal or normative statement and to seek a communicative consensus about their understanding of a situation as well as justifications for the principles and norms guiding their action. Argumentative rationality also implies that the participants in a discourse are open to being persuaded by the better argument and that relationships of power and hierarchy recede into the background. Argumentative and

deliberative behavior is as goal oriented as strategic interaction, but the goal is not to attain one's fixed preferences, but to seek a reasoned consensus. Actors' interests, preferences, and perceptions of the situation are no longer fixed, but subject to discursive challenges. Where argumentative rationality prevails, actors do not seek to maximize or to satisfy their interests and preferences, but to challenge and justify the validity claims inherent in them, and they are prepared to change their views of the world or even their interests in light of the better argument.[23]

At the extreme, Risse argues, we can distinguish two distinct types of social interaction: *bargaining*, in which actors with fixed preferences negotiate and exchange threats and promises in an effort to maximize their respective preferences, and *arguing*, in which open-minded participants seek to discover the truth, and indeed their own preferences, through a collective process of deliberation, argumentation, and persuasion.

This deliberative or argumentative conception of social action resembles the logic of appropriateness insofar as it rejects the rational-choice notion of individual actors with fixed preferences and a simple utility-maximizing function, but it goes beyond the role-playing of the logic of appropriateness to emphasize the collective search for truth (although without implying that there exists an "absolute" truth), the deliberative process through which actors seek to approximate the truth, and the prospects for argumentation and persuasion within the deliberative process.[24]

In the view of many political and legal theorists, moreover, these processes lead to the promise of a normatively desirable "deliberative democracy," in which societal actors engage in a sincere collective search for truth and for the best available public policy, and in which even the losers in debates accept the outcome by virtue of their participation in the deliberative process and their understanding of the principled arguments put forward by their fellow citizens.[25]

Habermas and his followers concede that genuine communicative action, or argumentative rationality, is likely under a fairly restrictive set of three preconditions. First, the participants must share a "common lifeworld...A supply of collective interpretations of the world and of themselves, as provided by language, a common history, or culture." Such a common lifeworld, it is argued, can be "provided by a high degree of international institutionalization in the respective issue-area... [or through] conscious efforts by actors to construct such a common lifeworld through narratives that enable them to communicate in a meaningful way," and is a vital precondition for any collective search for truth.[26]

Second, deliberation is most likely in the presence of "uncertainty of interests and/or a lack of knowledge about the situation," as for example in policy-making that involves scientific and technical considerations.[27] This element of uncertainty is doubly important for theorists of deliberation. On the one hand, it provides all actors with an incentive to deliberate collectively to benefit from each other's knowledge and to join in the collective search for truth, and increases the likelihood that each participant will approach discussions with a genuinely open mind. On the other hand, the presence of uncertainty also serves to obscure one's own preferences as well as distributional questions about who wins and loses from a given policy. If actors are unable to assess potential winners and losers, or indeed their own preferences in a given issue-area, then bargaining from fixed positions makes little sense, and deliberation emerges as the optimal means of ascertaining one's own preferences and the best outcome.

Third and finally, international deliberation is most likely within "international institutions based on nonhierarchical relations enabling dense inter-actions in informal, network-like settings." Such settings, according to Risse, are most likely to approximate the Habermasian construct of an "ideal speech situation" in which participants regard each other as equals with equal access to the discussion.[28] Within this third precondition, however, we find significant disagreement in the literature about the impact of openness, transparency, and politicization on officials' ability to engage in genuine deliberation. While some scholars argue that it is the public nature of deliberative democracy that requires actors to limit their appeals to naked self-interest and argue in terms of commonly accepted normative principles,[29] other scholars posit that individual participants are more likely to leave aside preconceptions and fixed interests, and join in the collective search for truth, in closed, *in camera* settings where compromise will not be second-guessed by governmental leaders, powerful constituencies, or public opinion "back home."[30]

These hypothesized preconditions are by no means satisfied everywhere in international politics and law; but where they are present, Habermasian and constructivist scholars predict that international actors will engage in arguing rather than bargaining, presenting their arguments in a common language, such as those of law or science, and proceeding to decisions on the basis of "the better argument" rather than the bargaining power of the respective actors. Such Habermasian assumptions, moreover, have guided an increasingly sophisticated empirical literature in political science, in which scholars have attempted to articulate and operationalize clear scope conditions for deliberation, including the aforementioned preconditions of a common lifeworld, complexity and uncertainty, and ongoing discussions in an informal institutional setting.[31] The promise of deliberation and arguing also resonates in international legal studies, particularly in the "new governance" and legal pluralism literatures which examine the role of deliberation and discursive engagement in supranational fora, such as the EU, the WTO, and the UN, as well as through transnational regulatory networks.[32]

Empirical studies of deliberation face significant methodological hurdles in distinguishing between arguing and bargaining or between genuine communicative action and "cheap talk."[33] Despite these obstacles, the promise of deliberation has received extraordinary attention in the study of the EU, whose dense institutional environment and networked forms of governance are seen as a particularly promising place to look for evidence of international deliberation.[34] This analytical claim, moreover, has been married to a normative case for Joerges' "deliberative supranationalism," which he claims offers a potentially compelling solution to the challenge of democratic legitimacy within the EU.[35]

In empirical terms, EU scholars have examined deliberation in three EU-related fora: comitology committees, the Constitutional Convention of 2003–2004, and the "new governance" mechanisms of the Open Method of Coordination. With regard to the first of these, Joerges and Jürgen Neyer draw on Habermasian accounts of deliberative democracy as well as constructivist analysis in political science to argue that EU comitology committees provide a forum in which national and supranational experts meet and deliberate in a search for the best or most efficient solutions to common policy problems. In this view, comitology is not an arena for hardball intergovernmental bargaining, as rationalists assume, but rather a technocratic version of deliberative democracy in which informal norms, deliberation, good arguments, and consensus matter more than formal voting rules, which are rarely invoked. In support of their view, Joerges and Neyer present evidence from their study of EU foodstuffs regulation, where they find that the importance of scientific discourse limits the ability of delegates to discuss distributional issues, particularly in scientific advisory committees, which in turn focuses debate and deliberation onto scientific questions. In addition, the authors point out, delegates not only meet regularly in comitology committees but have also often met as part of advisory committees and working groups involved in the adoption of the legislation in question—an ideal setting for long-term socialization into common European norms. In this way, the authors argue, comitology committees pass from being institutions for the strategic control of the Commission to being fora for deliberative interaction among experts for whom issues of control and distribution as well as the carefully contrived institutional rules of their respective committees recede into the background in favour of a collective search for the technically best solution to a given policy problem.[36] Joerges and Neyer's claims remain controversial, however, with rational-choice scholars arguing that EU member states design and utilize comitology committees systematically as instruments of control, and that evidence of deliberation in such committees remains partial and sketchy.[37] Other critics, moreover, question the normative value of "deliberative supranationalism" in comitology committees, noting that such expert deliberation takes place largely outside the public eye and thus presents a tradeoff between deliberative decision-making on the one hand and transparency and accountability on the other.[38]

A second EU arena often identified as a promising venue for deliberation was the Convention on the Future of Europe, which met in 2002–2003 to consider changes to the EU Treaties and concluded with a concrete proposal for a draft Constitution for Europe. Composed of representatives of EU institutions, national governments, national parliaments, and representatives from candidate countries, the Convention was explicitly conceived as a deliberative body, in contrast with the intergovernmental conference that had produced the contentious and inelegant Treaty of Nice in 2000.[39] The actual meeting of the Convention, Paul Magnette suggests in a careful and theoretically informed study, illustrated elements of both arguing and bargaining. On the one hand, Magnette argues, the public nature of the debates and the imperative of achieving consensus among the participants compelled participants to publicly justify their positions in terms of broad constitutional principles and the common good and to refrain from overt threats or horse-trading. On the other hand, Magnette concedes, representatives of national governments did, on occasion, present fixed national positions in debate, and there is at best mixed evidence that participants in the debate were genuinely open to persuasion and to changing their preferences as in a Habermasian "ideal speech situation." Furthermore, "the *conventionnels* knew and acknowledged that their experience would be followed by a classic IGC, and tended to anticipate it. They deliberated, but under the shadow of the veto."[40]

Finally, the promise of deliberation has also been emphasized by students of the Open Method of Coordination (OMC). Based on previous EU experience in areas such as economic policy coordination and employment policy, the OMC was codified and endorsed by the Lisbon European Council in March 2000, and is characterized as an intergovernmental and legally non-binding form of policy coordination based on the collective establishment of policy guidelines, targets, and benchmarks, coupled with a system of periodic "peer review" in which member governments present their respective national programs for consideration and comment by their EU counterparts. By and large, the OMC has been utilized not as a replacement for but as a complement to the traditional "Community method" in areas where member governments have been reluctant to adopt binding regulations, as in the areas of employment policy, social inclusion, and pensions reform.

The OMC remains controversial, both politically and in the academic community. For many commentators, the OMC offered a flexible means to address common policy issues without encroaching on sensitive areas of national sovereignty, representing a "third way" between communitarization and purely national governance. In addition, the basic elements of the OMC—institutionalized cooperation, iteration within non-hierarchical networks, and emphasis on exchange of information and learning—all suggested that OMC networks were potentially promising arenas, and potential test cases, for Habermasian deliberation.[41]

Recent empirical work, however, has at least tempered the more far-reaching claims put forward by the supporters of the OMC. On the one hand, in-depth studies of the European Employment Strategy (EES) and other OMC processes suggest that while the OMC has indeed led to some sharing of experiences and the creation of a common language and common indicators for the analysis of public policy, and some scholars suggest that bargaining power in OMC committees depends at least in part on the strength of one's arguments and not (or not solely) on the size of one's country. "Strategic bargaining" according to one study, "is not the general mode of interaction in the committees."[42] By the same token, however, a number of scholars have argued that, when it is time to negotiate politically sensitive provisions, detailed targets, or public recommendations to the member governments, national representatives revert to a presentation of fixed national positions, engaging clearly in bargaining rather than arguing behavior, and demonstrating few signs of having been persuaded to change their basic approach to employment or other issues.[43] In sum, while EU institutions and policy procedures such as comitology, the Convention, and the OMC might seem to be most-likely arenas for Habermasian deliberation, evidence for such behavior remains at best partial, and the EU's status as a deliberative democracy open to question.

3.3. Transatlantic networks and deliberation:
A mixed record of success

In addition to the EU, the transatlantic (US/EU) relationship has been identified by scholars and practitioners as a potentially promising setting for both transgovernmental networks and deliberative decision-making. While traditionally less interdependent and more thinly institutionalized than the EU, the US/EU relationship moved toward greater economic interdependence and greater institutionalization during the 1990s, reflecting the growth in transatlantic trade and foreign direct investment, the maturation of the EU and the European Commission as a political actor, and the eagerness of policy-makers on both sides to manage their interdependence as well as the trade disputes that have arisen regularly over the past several decades.[44] For Slaughter and others, this deepening transatlantic relationship provides both an illustration and a testing ground for the increasing importance of governance through transgovernmental networks.[45]

More specifically, the expectation that the US/EU relationship would become the site of transgovernmental networks and deliberative decision-making was encouraged by an empirical development, namely the establishment of the New Transatlantic Agenda (NTA) in 1995 and the Transatlantic Economic Partnership (TEP) in 1998, both of which represented efforts by Washington and Brussels to deepen their economic ties and to address long-standing trade-related regulatory tensions.[46] From the beginning, much of the thrust of the NTA has been directed at the removal of regulatory barriers to transatlantic trade and investment. The TEP, in turn, committed the two sides explicitly to an ambitious program of regulatory cooperation designed to reconcile, if not eliminate, regulatory barriers to trade and regulatory conflicts between the two sides. Largely in response to this mandate, the US and the EU have signed over a dozen formal regulatory cooperation agreements in the past decade, in areas as diverse as competition policy, data privacy, customs procedures, veterinary standards, and the mutual recognition of testing and certification procedures (see Table 3.1). These formal regulatory agreements, moreover, represent only a fraction of the many informal contacts that occur between US and EU regulators, both bilaterally and in various multilateral fora.

How successful has this experiment been? A thorough survey of US and EU regulatory cooperation is beyond the scope of this chapter, but cross-sectoral work on transatlantic regulatory cooperation reveals dramatic differences in the success or failure of regulatory cooperation across issue-areas.[47] Consider, very briefly, the observed outcomes in the following five sectors:

- **Competition Policy.** Transatlantic cooperation in the area of competition policy actually predates the NTA and the TEP, having begun with a formal cooperation agreement between the EU Commission and the US Justice Department and Federal Trade Commission in 1991. Particularly in the area of mergers and acquisitions, antitrust regulators on both sides of the Atlantic have actively sought each other out, sharing information and coordinating their investigations of over 600 proposed mergers. Relations between the two sides have not always been harmonious—witness the transatlantic disagreement over the proposed GE/Honeywell merger, approved by US officials but rejected by the EU Commission in 2001—but with very few exceptions, cooperation among competition policy officials has been productive, limiting the potential for regulatory disputes, improving the quality of regulation in both jurisdictions, and fostering a gradual convergence of regulatory approaches across the Atlantic.[48]

- **Mutual Recognition Agreements.** In the absence of genuine regulatory harmonization, the EU and other international bodies have engaged in the "mutual recognition" of standards, with each jurisdiction retaining its own regulatory standards but accepting the equivalence of other countries' standards and hence the safety of products produced in conformity with them. The US and the EU took a similar approach in their 1997 decision to negotiate and sign a Mutual Recognition Agreement (MRA), which provides for the mutual recognition of testing and certification standards for six types of industrial products. This transatlantic MRA is less ambitious than the EU's efforts, in that the two sides agreed to recognize only

Table 3.1 Major transatlantic regulatory cooperation agreements

Agreement	Targeted Regulations	Year
Competition Policy Agreement	Competition regulations	1991
EC/US Agreement on Drug Precursors	Illicit drug regulations	1997
EC/US Customs and Cooperation Agreement	Customs certifications	1997
EU/US General Mutual Recognition Agreements	Conformity assessment testing in six sectors: telecommunications equipment, electromagnetic compatibility, electrical safety, recreational craft, medical devices, and pharmaceuticals	1997
EU/US Positive Comity Agreement	Competition relations	1998
EU/US Veterinary Equivalence Agreement	Animal export certifications	1999
EU/US Safe Harbour Agreement	Data protection regulations	2000
EU/US Agreement on Mutual Recognition of Certificates of Conformity for Marine Equipment	Conformity assessment testing for marine equipment	2001
EU/US Guidelines on Regulatory Cooperation and Transparency	Non-binding guidelines for cooperation among EU and US regulators regarding technical barriers to trade	2002
Financial Markets Regulatory Dialogue	Informal dialogue devoted primarily to avoiding disruption to trade in financial services resulting from distinct EU and US regulations	2002
Mutual Recognition Agreement on Marine Equipment	Mutual recognition of standards for marine equipment	2004
EU/US Agreement extending customs cooperation to container security	Agreement implementing US Container Security Initiative in EU	2004
EU/US Passenger Name Recognition Agreement	Agreement allowing US Customs and Border Protection to collect passenger data relating to flights from EU to US	2004
Policy Dialogue on Trade and Border Security	Semi-annual meetings among senior US/EU customs/border officials	2004
Roadmap for EU/US Regulatory Cooperation and Transparency	Sector-by-sector plans of action for EU/US regulatory cooperation	2004 2005
Transatlantic Economic Council	Overarching structure for regulatory cooperation in targeted sectors.	2007

Source: Updated from Pollack 2003a: 93.

each other's testing and certification procedures, rather than substantive standards. Even so, the implementation of transatlantic MRAs has proven difficult in several sectors, most notably as US specialized agencies like the Food and Drug Administration (FDA) and the Occupational Safety and Health Administration (OSHA) have been slow to recognize the equivalence of European laboratories' assessment of pharmaceuticals, medical devices, and electrical safety equipment.[49]

- **Data Privacy.** The case of data privacy illustrates once again the threat to international trade posed by differing domestic regulations. In this case, the EU's stricter and more comprehensive standards on the protection of data privacy threatened to lead to a moratorium on the transfer of European data to US firms, insofar as those firms failed to offer data protection standards equivalent to those of the 1995 EU Data Privacy Directive. Both harmonization and mutual recognition of the two very different US and EU systems proved impossible, leading to an innovation in the form of the Safe Harbor Agreement signed by the Commission and the US Department of Commerce in 2000. Under the agreement, the US and the EU retain their respective regulatory systems, but US firms may voluntarily comply with a negotiated set of data privacy standards that the EU Commission has recognized as "adequate" under the EU directive, and such firms thus become eligible for transfer of data from the EU. The Safe Harbor agreement has been credited with defusing a potentially explosive transatlantic trade conflict and with ratcheting up US data privacy standards in practice, but questions nevertheless remain about the level of protection and the policing of US firms' compliance with the Safe Harbor principles.[50]

- **Financial Services.** The US and the various countries of the EU have long adopted differing regulatory standards for financial services, including issues such as corporate accounting standards and auditing requirements. Until several years ago, the US, with its large, unified financial market and powerful regulators such as the Securities and Exchange Commission (SEC), closely guarded its regulatory sovereignty and paid relatively little attention to European concerns. This tendency was reinforced in the wake of the Enron and other US financial scandals, when the US Congress quickly and with little consultation adopted the 2002 Sarbanes–Oxley Act, which would require many EU firms to meet two distinct, and in some cases incompatible, sets of regulatory standards. Over the past several years, however, US regulators, led by the SEC, have demonstrated a remarkable willingness to work with their European counterparts, establishing a Financial Markets Regulatory Dialogue in 2002 and interpreting US legislation in ways that would avoid placing undue regulatory burdens on European companies. This shift, attributed by Elliot Posner to the completion of the European regulatory framework for financial services, demonstrates the ability of US and EU regulators to work constructively in an area where the US had previously proceeded with little consideration of its European counterparts.[51]

• **Trade and Border Security.** The terrorist attacks on the US in September 2001 precipitated a wave of new Congressional legislation and regulation regarding the protection of US borders, including a Customs Security Initiative (CSI) to monitor incoming container ships, and new rules on Passenger Name Recognition (PNR) requiring foreign airlines to provide information about passengers on incoming flights. These requirements, adopted unilaterally by Congress, created significant difficulties for the EU, which was forced to adapt on short notice to US policies at odds with its own procedures. Here again, however, US and EU regulators have responded pragmatically and for the most part effectively, establishing a Policy Dialogue on Trade and Border Security in 2004 and reaching bilateral agreements on the implementation of both CSI and PNR. Border security remains a sensitive area, with continuing EU concerns about the Congressional requirement for EU countries to issue biometric passports for visa-free entry into the US, but thus far the new Policy Dialogue shows promising signs of addressing these problems proactively, rather than responding to crises after they occur.[52]

The record of transatlantic regulatory cooperation, then, is a mixed one. In some areas such as competition policy, financial services, and trade and border security, US and EU regulators have cooperated effectively, preventing or at least mitigating economic disputes between them, and in some cases, showing evidence of convergence in regulatory approaches. In other areas, however, agreements like the MRAs and the Safe Harbor Agreement have run into serious difficulties at the implementation stage. Reflecting these difficulties, the two sides agreed in April 2007 to the creation of a new Transatlantic Economic Council, which would meet regularly, coordinate regulatory cooperation activities, and report on progress to the annual US/EU summit meetings. The Council met twice in 2007 and early 2008 and issued an optimistic report to the heads of state and government,[53] but its first year of operation yielded few significant breakthroughs, and notably failed to resolve the dispute over the EU's ban on chlorine-washed poultry from the US, which had been put forward by American officials as an important test of the Council's ability to deliver.[54]

The causes of success and failure in these areas are complex, and a complete discussion is beyond the scope of this chapter, but recent studies have begun to identify a number of such factors, including: (1) sharp and persistent differences in regulatory procedures or philosophies on each side of the Atlantic; (2) domestic provisions limiting the transfer of confidential information to foreign regulators; (3) the multi-leveled nature of both the EU and US polities, which can necessitate the participation of central and state legislators, regulators, and private actors for successful cooperation; (4) the limited nature of the NTA and the TEP as executive agreements, which fail to bind either legislators or state regulators on either side; and (5) the domestic politicization of regulations in areas such as data privacy, which can sharply limit the ability of regulators to engage in substantive compromises with their foreign counterparts.[55]

All of these factors have indeed impeded regulatory cooperation in one or more issue-areas, and deserve further study. In addition to these factors, however, the decade-long record of transatlantic regulatory cooperation points to two less-cited but arguably more fundamental factors that might impede effective regulatory cooperation, and that suggest important caveats to Slaughter's compelling image of international governance by regulator-diplomats. First, much of the existing literature on transgovernmental relations is premised on the notion that domestic regulators voluntarily seek each other out in an effort to improve the quality of their own domestic regulations and eliminate tensions caused by incompatible regulatory standards. While this image does indeed seem accurate in areas like competition policy, in many other areas we find that domestic regulators can be *resistant* to cooperation that they perceive to be a distraction from their core regulatory mandates. Indeed, several of the transatlantic regulatory dialogues surveyed above resemble not so much the voluntary romances undertaken independently of governments as much as "shotgun marriages" undertaken grudgingly under pressure from heads of government and trade officials.

Second, transatlantic regulatory cooperation is not a purely technical exercise, but poses important questions about the distributive consequences of different regulatory standards and the role of the power in determining the outcomes of regulatory conflicts[56]—a point to which we will return at greater length in the next chapter. Across a number of cases, we find that the US and EU regulators do not approach transatlantic negotiations with an open mind and uncertain preferences, but rather with strong and clear preferences for retaining and indeed exporting their domestic standards to the international level. In such circumstances, the "power of the better argument" may not win the day, and the outcome of regulatory cooperation, and the pattern of gains and losses for each side, depends in large part on the bargaining power enjoyed by each side.

The importance of power has been largely ignored in the literature on transatlantic economic relations, which tends to equate bargaining power roughly with market size, and hence, to assume a rough equality between the US and EU. As Elliot Posner has argued in the case of financial services, however, bargaining power can also vary as a function of the domestic *institutional characteristics* of each side, such as the EU's newly established competence to establish rules for the EU's market in financial services. More generally, the US and EU are both complex, multi-level polities, which allocate veto points to a range of actors, including regulators with varying degrees of independence, legislative bodies, and sub-federal states, each of which can resist demands for compromise and effectively tie the hands of their respective negotiators.[57] In the transatlantic MRAs, for example, powerful US regulators like the FDA and the Occupational Safety and Health Administration have used their independence to delay implementation of provisions to which they remain opposed.[58]

Against this background of transatlantic regulatory cooperation, both scholars and practitioners have advocated ambitious proposals for transgovernmental, transnational, and deliberative decision-making in the area of agricultural biotechnology. Sean Murphy, for example, has called for the establishment of a global "epistemic community of knowledgeable state and non-state actors representing a wide range of affected interests, common perspectives and bargaining positions" which could "develop convergent policies and expectations," and which ultimately could lead to a stronger "transnational legal regime" with "more complete and precise rules" that would be "more likely to promote state compliance."[59] Indeed, the context of GM food regulation would seem to provide the very attributes that Risse identified as favoring a resort to transnational deliberation, including a "lack of knowledge about the situation," "uncertainty of interests," "a high degree of international institutionalization in the issue area," and the fact that "apparently irreconcilable differences [could] prevent them from reaching an optimal rather than a merely satisfactory solution for a widely perceived problem."[60] And indeed, the latter half of the 1990s witnessed a concerted effort by the US and the EU to engage in extensive bilateral consultations involving both transgovernmental networks of government regulators as well as "transnational" networks of private actors on both sides the Atlantic.

Nevertheless, the experience of transatlantic regulatory cooperation in other issue-areas suggests caution, given that many of the factors impeding cooperation in other issue-areas, including incompatible regulatory philosophies, politicization of issues, the limited ability of regulatory cooperation to bind other governmental actors, and the presence of distributive conflicts, appear to be present in the area of agricultural biotechnology as well.

3.4. Deliberation and risk regulation

Furthermore, in addition to the other obstacles to successful regulatory cooperation, it is worth noting that the attempt to resolve regulatory disagreement by bringing together government officials, scientists, and civil society representatives reflects fundamental differences among leading theorists regarding the respective roles of technocratic elites, politicians, and mass publics in risk regulation. Cass Sunstein, for example, argues for the delegation of risk regulation to experts because mass publics are more subject to heuristic biases in risk analysis that can lead to serious error.[61] Sunstein, building from the work of behavioral economists and psychologists, finds that engaging mass publics over complex risk issues can exacerbate their fears, leading to errors in judgments and incoherent regulatory policy.[62] Organizing a series of town meetings and citizen dialogues to discuss the risks of GMOs, for example, may lead citizens to conclude that these varieties are more risky than conventional varieties, because why else would a meeting be necessary? This phenomenon may explain why Monsanto's advertising campaign and UK and French regulators' attempts to reassure their citizens of the safety of GM foods through citizen conferences appear to have backfired.[63] Sunstein, rejecting "populism," thus finds that the best response of regulators to mass publics regarding risk regulation is to *"change the subject."*[64] The problem with mandatory labels for GM foods, he notes, is that they "suggest a danger that might not exist."[65] Sunstein's approach to the role of deliberation over risk regulation reflects the US model of delegating risk decisions for agricultural biotech approvals to specialized technocratic agencies. As he concludes, "Democratic governments care about facts as well as fears... [t]hey take careful steps to ensure that laws and policies reduce, and do not replicate, the errors to which fearful people are prone."[66]

In contrast, theorists who address the role of cultural factors in risk perception stress the importance of deliberation among all stakeholders and find that Sunstein's proposals of how to deal with "fear and risk" are elitist, reflective of a "fear of democracy" and "an attempt to exclude the regulation of risk from its ambit."[67] As James Boyle writes, "by withdrawing political questions from the public sphere and giving them over to expert decision making, technocratic rationality actually diminishes the possibility of democratic debate over the ends, in the name of an improved analysis of means."[68]

Regarding risk evaluation itself, cultural theorists maintain that the public has a different and richer conceptualization of risk than do technocrats, taking into account whether risks are voluntarily assumed and controllable, as well as the distributional implications of risk decisions.[69] They state that there are reasons not to completely trust scientific assessments, which are also affected by cognitive biases, and thus "rival" public conceptions should be heard.[70] Moreover, they contend that an important way of gaining citizen trust is by making regulatory decisions more transparent and open to public participation.[71] As Dan Kahan et al. write, "Just as consultation breeds trust in expert risk regulators, the perception that such officials are remote and unaccountable erodes it."[72] This analysis supports an EU approach that provides for direct political input into risk management decisions in order to reflect divergent cultural affiliations and beliefs. The function of deliberation in such a context is that "deliberation can alter individuals' understandings of the relationship between their cultural affiliations and particular beliefs."[73] Because of "the limitations of risk science, the importance and difficulty of maintaining trust, and the subjective and contextual nature of risk," Slovic calls for "a new approach—one that focuses on introducing more public participation into both risk assessment and risk decision-making to make the decision process more democratic."[74] Not surprisingly, European policymakers tend to stress this conception of deliberative democracy in agricultural biotechnology regulation, focusing on European cultural values and the need to engage mass publics.

These different perspectives on the respective roles of technocratic networks and public deliberation in risk regulation have led to different models of decision-making over risk in the US and EU, as we saw in the last chapter. Sunstein highlights the importance of narrowly defined networks of regulatory officials, as reflected in the US approach. Cultural theorists stress the need for broader networks open to all affected parties, as reflected to a greater extent in the EU approach.[75] Hence, among both scholars and practitioners we find fundamental disagreement about who should participate in decisions about risk regulation, and what sorts of factors should be considered—a challenging starting point for any effort at joint, deliberative decision-making.

3.5. Transatlantic networks for agricultural biotechnology

Notwithstanding these caveats, policy-makers were quick to identify agricultural biotechnology as an area in which a structured dialogue among regulators might build mutual understanding and trust, provide early warning of disputes, and perhaps, in time, contribute to a gradual convergence of regulatory approaches to GM foods and crops. Throughout the late 1990s, the US and EU established numerous working groups of scientific experts to exchange information relating to GMOs. The US Congress even amended the Federal Food, Drug and Cosmetics Act to instruct the FDA "to move toward the acceptance of mutual recognition agreements relating to the regulation of drugs, biological products, devices, foods...between the EU and the US."[76]

In 1995, the New Transatlantic Agenda established a High Level Environment Consultation Group which was to work in conjunction with a Permanent Technical Working Group on biotech issues. The latter group was to bring together representatives from the EU's Environment and External Relations Directorates-General and from the Office of the US Trade Representative (USTR), the Environmental Protection Agency and the US Department of Agriculture. Then, in connection with the launch of the Transatlantic Economic Partnership at the US–EU Summit in December 1998, the two trading partners set up a new overarching group—the TEP Biotech Group—to coordinate discussions and information exchange "with a view to reducing unnecessary barriers to trade."[77]

The TEP Biotech Group was to oversee two pre-existing technical groups focused on information exchange over biotech issues: (1) the Agrifood Biotech Group (which consisted of governmental representatives from the Commission's agricultural and industry directorates-general and from the US Department of Agriculture and USTR), and (2) the US–EC Task Force on Biotechnology Research (which coordinated workshops bringing together US and European scientists). The Agrifood Biotech Group, founded in 1996, focused mainly on regulatory and technical issues that could affect "the commercialization of biotechnology-derived plants and food products." It addressed, in particular, the application of the EU's Novel Foods Regulation. Industry representatives were invited to participate in its meetings. The research task force, founded in 1990, held workshops and symposia on biotech research and risk assessment.[78] The TEP Biotech Group even launched a pilot project, in which, following a first phase of comparing molecular data requirements for scientific assessments, the FDA and its EU counterparts in DG Industry were to consider a process of "simultaneous application for product approval."[79] FDA and Commission officials also held bilateral meetings (in person and by video) in which they exchanged information on "existing scientific guidance for biotech food safety assessment" and "the structures available for using scientific advice." The FDA reported that they "found similarities in the underlying principles for safety assessment and thought that regulatory cooperation in this area would be useful."[80]

The creation of these transgovernmental groups, however, in the words of a US government official, can also be viewed as a means "for politics to try to smooth things over."[81] The groups were only created when it appeared that the US and EU were going in different directions, as became apparent in discussions within the OECD (discussed in Chapter 4). Although both sides spoke of creating a "joint application process" for GM products, an idea that was strongly promoted by the biotech industry, those within the regulatory agencies "knew that it would never happen." In the words of an FDA official, "We haven't been able to do it in any other area, why think we could do it in this one, which was so contested?"[82]

Of all the working groups, the research task force was the most successful, meeting relatively actively and reportedly in a deliberative manner.[83] The participating scientists shared methodologies and testing results, resulting in transatlantic convergence in how risk assessments are to be conducted.[84] As a former member of DG Research stated, "Its success is seen as attributable to a clear decision from the start to avoid getting drawn into discussion of policy issues," but rather to focus on scientific and technological ones.[85]

The output of these working groups, however, ultimately disappointed both sides. The overarching TEP Biotech Group only met a few times.[86] The two sides never progressed to a formal cooperation agreement in the biotech sector, nor was there any clear sign of convergence in US and EU regulatory approaches as a result of these exchanges. In 1999, for example, Commission officials consulted widely with their US counterparts in designing the fledgling European Food Safety Authority, closely examining the structure and procedures of the FDA as a potential model. The Commission's eventual proposal for the EFSA, however, drew only selectively from the FDA model, accepting the role of a specialized agency in scientific risk assessment, but insisting that risk management continue to be undertaken by political bodies such as the Commission, Council, and European Parliament.[87] When fully operational, EFSA is expected to employ only around 370 people and have a budget of 70 million euros, a tiny agency compared to the US FDA that employs over 9,000 people and has 2,100 scientists working for it.[88] Moreover, EFSA operates far from decision-making in Brussels as it is based in Parma, Italy, and has even had difficulty getting staff to move there. The FDA and the Commission therefore remained very different types of regulators guided by distinctive mandates, with the FDA in particular refusing to consider any serious challenge to its model of "science-based" decision-making, and with the Commission unable or unwilling to move towards a more specialized regulatory agency model along the lines of the FDA.

As little progress was made among US and EU governmental representatives, the US and EU decided to form two transatlantic civil-society groups in November 1999. One was a transatlantic scientific advisory committee consisting of scientists specifically working on genetically modified organisms. It was to provide counsel to the Commission until its proposed Food Safety Authority was approved and operational, thereby facilitating a common US and EU approach to risk assessment. The second was the EU–US Biotechnology Consultative Forum composed of twenty independent experts, including research scientists from industry, university biotech centers, and representatives of NGOs, including to the GM-skeptic Friends of the Earth, a fervent opponent of GM products. It was charged with examining the issues posed by biotechnology for regulators on both sides of the Atlantic. In the optimistic words of acting EU ambassador to the US, John Richardson, these groups could help forge "a transatlantic consensus on where we go with biotechnology in the future."[89]

The Consultative Forum, which was to make policy recommendations, issued its final report, by consensus, in December 2000, although there was apparently some severe contention within the group.[90] Both sides officially welcomed the report, which offered a partial endorsement of each side's views, although the report was much more supportive of the EU than the US would have liked, and biotech advocates within the Commission (such as in DG Research) found it to be an abomination, causing "great irritation and anguish in the interested scientific communities."[91] For example, the report called for mandatory pre-market examination by regulators of all genetically modified foods and feed prior to their approval (recommendation 1) and endorsed a version of the precautionary principle as a core element of biotech regulation (recommendation 12)—both core elements of the EU regulatory procedure. It also called for "mechanisms that will make it possible to trace all foods, derived from GMOs" (recommendation 8) and for "content-based mandatory labeling requirements for finished products containing novel genetic material" (recommendation 15)—proposals that the EU was to transpose into a regulation in 2003. In addition, it endorsed regulatory procedures for risk assessment and risk management that are transparent and open to "a broad range of specialists and stakeholders (e.g., social scientists, ethicists, representatives of civil society)" (recommendations 13 and 14).[92] It arguably went even further than the existing EU regime, calling, on the one hand, for mandatory approvals and labeling of those GM foods that are "substantially equivalent" to conventional varieties and thus subject to the simplified procedure under the EU Novel Foods Regulation (recommendation 5),[93] and, on the other, for an international liability regime (recommendation 11) when the EU did not itself have one and the Commission argued that one was not necessary.

Despite the report's calls for a process-based approach to agricultural biotechnology, with mandatory labeling, the US State Department (still working under the Clinton administration) diplomatically welcomed the report's emphasis on the potentially beneficial effects of biotechnology, including for sustainable development in developing countries, as well as its discussion of openness and transparency in the regulatory process—"many of [which] are already a part of the U.S. regulatory system."[94]

Following these initial formal expressions of support, however, the report was essentially shelved by both sides, which have scarcely ever made reference to it in the subsequent years of conflict over the issue.[95] Given, in particular, that the Bush administration took office just after the report's release, and that it showed little inclination to deliberate and compromise with the EU on a number of policy matters, from the Kyoto Protocol to the UN Framework Convention on Climate Change to the International Criminal Court, the idea of a consensus approach to the regulation of agricultural biotechnology appeared to become less rather than more promising over time.

Along similar lines, both the US and the EU fostered during the 1990s the development of official "civil-society dialogues" among business, consumers, environmentalists, and labor unions. Although the latter two groups met only a few times and are currently defunct, the Transatlantic Business Dialogue (TABD) and the Transatlantic Consumer Dialogue (TACD) have met periodically since the mid-1990s and have taken clear stances on the regulation of biotechnology. The TABD established an Agri-Food Biotechnology Group to conduct studies and prepare recommendations, and throughout the 1990s, the TABD generally adopted a position aligned largely with that of the US, calling for compatible standards between the US and the EU to promote transatlantic trade in GMOs and arguing that mandatory labeling of GMOs "unfairly discriminates among identical or like products."[96] Faced with increasing public opposition to GM products as well as organizational difficulties among the CEOs who constituted its membership, the TABD has taken a much lower profile on the issue of biotechnology regulation since 2002, with no mention of GMOs or biotechnology in its more recent annual reports.[97]

By contrast, the Transatlantic Consumer Dialogue has been consistently active on the issue of biotechnology regulation since its creation in the 1990s, with US consumer organizations joining their European colleagues in calling for mandatory pre-market approval of all GMOs as well as mandatory labeling and traceability provisions. Clearly closer to the EU position than the US position on biotech regulation, the TACD has also called on the US on several occasions to abandon its WTO complaint against the EU, warning of "a 'pyrrhic victory' by the U.S. given the reaction of European consumers."[98] TACD representatives, however, grew increasingly frustrated at their apparently limited influence upon and access to US leaders in particular; hence, the main importance of the TACD thus far appears to be its impact on US nongovernmental groups, which have been regularly exposed to the more critical views of their European counterparts.

By the end of the 1990s, it was clear that the attempt by the US and EU to work through networks to better understand each other's approaches and learn from each other's experiences, possibly leading to harmonization, had failed. US officials found that politicians in Europe have no incentive to engage in these dialogues given the possible backlash against them. Moreover, because of the politicization of agricultural biotech in Europe, as one USDA official states, European "technocrats have stepped back and left the issues for politicians."[99] In some member states (such as Italy), politicians have blocked even the conduct of any trials of GM varieties so that national technocrats lack information even to engage in deliberation over risks and benefits.[100] As the WTO trade conflict intensified, agency scientists specializing in agricultural risk regulation increasingly lamented the lack of organized transatlantic dialogue and information exchange.[101]

Overall, the US and EU regulators did not come to a reasoned consensus on the assessment or management of risks from NGOs, nor did US and EU policies converge as a result of joint deliberations. Where bilateral negotiations did lead to deliberation and consensus, these negotiations tended to be among like-minded groups of private or civil-society groups, and there is little or no evidence that the outcomes of their discussions had any significant effect on policies on either side of the Atlantic.

3.6. Conclusions

The story of bilateral cooperation over the regulation of GM foods and crops, in short, is largely a story of failure, and provides a lesson in the limits of transgovernmental networks and of "deliberative supranationalism." This result, while perhaps disappointing in policy terms, should not be surprising to many Habermasian scholars of deliberation, who have always conceded that deliberation as a "logic of social action" is likely only under a restrictive set of conditions, including a "common lifeworld"; actors' uncertainty about their own preferences and hence about the distributive implications of their collective decisions; and nonhierarchical relations in informal network settings. Agricultural biotechnology was initially believed by some scholars and practitioners to meet these conditions, given the importance of science to regulators on both sides of the Atlantic, the uncertainty about the emerging technology of genetic engineering, and the possibility of constructing informal bilateral networks of government regulators and civil-society groups.

In practice, however, these postulated preconditions for deliberative decision-making have been largely absent in the area of biotechnology, and the results of bilateral negotiations have been correspondingly disappointing. Despite the fact that the US and the EU are both advanced industrialized democracies, a "common lifeworld" could not be taken for granted, either among US and EU regulators who proceeded from sharply distinctive regulatory frameworks and philosophies,[102] or among civil-society groups who differed both across the Atlantic and across the producer–consumer divide.

Similarly, the presence of scientific complexity and uncertainty does not appear to have promoted deliberation in the ways anticipated by students of communicative action. Habermasian theorists of deliberation, we have seen, often posit complexity and uncertainty as providing favorable conditions for deliberation, on the plausible reasoning that uncertainty places a premium on truth-seeking while obscuring distributional conflicts associated with hardball bargaining. Our findings, however, suggest that the positive role of uncertainty can be limited, and may in some instances cut *against* deliberation. Uncertainty's role can be limited, we argue, because despite the scientific complexity of agricultural biotechnology as an issue, the various actors in the debate including biotech companies, farmers, and the US and EU officials who represent them appear quite able to articulate their own interests vis-à-vis GM foods and crops, and fight for their preferred positions in subsequent bargaining. Just as importantly, we find that uncertainty can cut both ways in areas of risk regulation such as agricultural biotechnology. Faced with environmental and food-safety risks that cannot be measured with absolute certainty, many European consumers, governments, and EU institutions themselves have opted in favor of a rather extreme interpretation of the precautionary principle, in which uncertainty yields not a collective search for truth, but an uncompromising rejection of GMOs, with little or no regard for the causal arguments (e.g., scientific risk assessments) in their favor.

Finally, despite the best efforts of the architects of the NTA, the informal networks established to discuss GMO regulation were consistently hampered by the rampant politicization of the issue, with European public opinion and member governments placing pressure on Commission negotiators to defend EU policies, and US negotiators under corresponding pressure from biotech firms and agricultural associations with a stake in the US regulatory system. Genuine deliberation, in which participants put aside preconceived notions and submit to the "power of the better argument," was at best difficult in such a context. In this sense, our analysis lends further support to the view that transparency and politicization *decrease* the prospect for deliberation in transnational bodies, which appear to function most effectively in closed, *in-camera* settings.

To the extent that we do see evidence of deliberation, most notably among scientific experts and single-issue interest groups like the TABD and the TACD, these outcomes appear to reflect two characteristics of the negotiations, namely (1) the nature of the participants, including their diversity and the relative presence or absence of a pre-existing "common lifeworld" among them; and (2) the terms of the discussion, which could be restricted to scientific issues of risk assessment or take in more explicitly political issues of risk management.

The most successful of the networks (at least for a time) was the research task force. As predicted by students of epistemic communities, the US and European scientists were much more likely to share, in Habermas' terms, a common lifeworld and thus a common way of discussing the risk assessment issues raised by GMOs. The narrowing of the scope of debate to mutually agreed terms—technical risk assessment—further facilitated deliberation. The participants also had mutual interests in learning from each others' work to improve their own professional practices. Indeed, there is reason to believe that bilateral negotiations, if left entirely to scientific experts and restricted to technical questions, could result in significant deliberation and joint discovery. For example, as we shall see in Chapter 6, the EU's scientific bodies (now coordinated under EFSA) have repeatedly reached agreement that the GM varieties notified to them are no more risky than conventional varieties,[103] much as FDA scientists have found in the US. Were US and EU regulatory decisions left to these scientists, there would likely be little regulatory conflict because the scientists in EFSA and the FDA and USDA generally see the technology the same way.

Among policymakers, deliberation is generally more likely to occur among those working in a single regulatory area, as witnessed by the transnational competition and financial networks that have emerged. In the GMO context, deliberation would more likely occur if a network were limited to environmental or food safety specialists, even though institutional and political factors would constrain them more than scientists. Yet, transnational coordination over agricultural biotechnology is even more challenging because it is a *multi-sectoral regulatory issue* involving officials from many different domains including food safety, the environment, agriculture, scientific research and development,[104] and trade. In addition, different US and EU agencies took the lead role on the biotech file so that parallel agencies were not in control on both sides. While environmental regulators took the lead in the European regulatory framework, the USDA did so in the US, with both agencies holding very different policy mandates and priorities.[105] Moreover, even where parallel agencies came to the network, their priorities also differed in reflection of the divergent interests of their underlying stakeholders. For example, the USDA and DG Agriculture held differing priorities since genetically engineered seeds have generally not been used by European farmers but are widespread in the US for crops such as soybeans and corn, and GM foods are generally not available in the EU, unlike in the US. The EU's Directorate-General for Agriculture thus did not face pressure from European farmers to open EU or global markets to GM foods. Rather, European farmers benefit from the increased market risks that US growers face in the EU (and possibly global markets) because of US adoption of GM varieties.

Finally, trade officials, in particular from the US, assumed central positions in the transatlantic networks, affecting the dynamic of discussion. The United States Trade Representative was a member of the 1995 Permanent Technical Working Group on biotech, the 1996 Agrifood Biotech Group, and the 1998 TEP Biotech Group. The participation of USTR lawyers refocused discussions from regulatory protection to the impact of regulatory barriers on trade. Their presence made clear that the threat of a legal claim before the WTO lurked in the shadows. Once the US brought its WTO case, even the transatlantic scientific networks became less active. This experience reflects the challenges of implementing the earlier 1997 transatlantic MRA, but in an exacerbated way. The USTR's leading role in designing the MRA framework agreement and negotiating its annexes for specific regulatory fields impeded the sense of ownership of US regulatory officials (such as FDA) of a process that affected their regulatory mandate, so that they were less committed to the agreement's implementation.[106] In the agricultural biotech case, however, an actual WTO legal filing loomed in the background so that discussions were not open-ended, but constrained by the prospects of trade litigation and the potential instrumental use of any statements in such litigation.

As regards non-governmental networks, anti-GMO activists can be viewed as operating in quite a different lifeworld than do executives from agricultural biotechnology companies such as Monsanto. Not surprisingly, therefore, the like-minded participants within the Transatlantic Business Dialogue and the Transatlantic Consumers' Dialogue, respectively, tended to operate according to a "common lifeworld" internally, and the two groups each came to reasoned—and diametrically opposed—consensus positions. Where a more diverse group of stakeholders was brought together, as in the Biotechnology Consultative Forum, the resulting "consensus" position simply papered over differences among the various participants, and was in any event publicly praised and then quickly put aside by both US and EU authorities.

As we can see, there is a tension between the inclusiveness of a deliberative network and the quality of the deliberation that occurs. The more that one moves to "global deliberative equality" in which "all relevant and affected parties in processes of transgovernmental deliberation" are included (in Slaughter's words),[107] and in which actors "recognize each other as equals and have equal access to the discourse" (in Risse's),[108] the more difficult deliberation becomes. Here again we encounter a trade-off between effective deliberation and inclusiveness: when a policy domain is multi-sectoral, as with GMOs, inclusion of different types of sectoral officials, and in particular trade officials, seriously limits the prospects for deliberative decision-making.

In practice, then, starkly distinctive trading interests resulting from the application of different regulatory systems and regulatory philosophies have repeatedly impeded a genuine deliberative dialogue, calling into question the existence of a common transatlantic "lifeworld" with respect to the regulation of GMOs. Even if regulators (as opposed to trade officials) from the two sides were able to bridge their different approaches and deliberate together, they would and already do find themselves operating in a highly politicized issue-area characterized by strongly mobilized interest groups and by often volatile public opinions limiting their ability to engage in significant substantive compromise. For all of these reasons, the bilateral experiment in regulatory cooperation on biotechnology turned out to be short-lived.[109] With the failure of bilateral cooperation, both actors turned their attention to negotiations and (potential) deliberation in wider multilateral fora, in the hope of either persuading or pressuring the other side to change its policies.

4

Deliberation or Bargaining? Distributive Conflict and the Fragmented International Regime Complex

In the absence of successful bilateral cooperation, both the US and the EU attempted to find cooperative solutions to the problems of biotech regulation in various multilateral fora. Such multilateral regimes, it is often argued, can help states to cooperate by lowering the costs of negotiation, promoting deliberative decision-making, facilitating learning and the formation of common norms, monitoring compliance with the resulting international agreements, and providing a forum for international dispute settlement.[1] Consistent with this view, the US and the EU have attempted, over the past two decades, to engage in multilateral cooperation through a variety of international regimes. However, the multilateral cooperation in the area of agricultural biotechnology has faced two interrelated challenges: conflicts over the distribution of costs and benefits; and the problem of overlapping and conflicting regimes in a fragmented international system.

First, while both sides in the transatlantic GMO dispute had a clear incentive to avoid an all-out trade war, each had a strong preference to retain and to "export" its own regulatory system to the international level, placing the political and economic burdens of adaptation on the other side. International cooperation on the regulation of GMOs is thus characterized not only by joint gains but also by *distributive conflicts* when the parties differ on the *terms* or the content of possible international regulations. As we will see, the US has not wanted to overhaul and segregate its grain distribution system at a significant cost for what it sees as an incoherent EU labeling and traceability regime for products containing or derived from GMOs. The EU, for its part, has not wanted to accept such US products without more demanding EU internal scrutiny, subject to the "precautionary principle," as well as EU labeling and "purity" threshold requirements for approved varieties. Such a distributive conflict, as we shall see, has led both parties to engage in hardball bargaining and interfered with the development of deliberative problem-solving.

Second, the regulation of GM foods and crops is the subject of what has been termed a *"regime complex"*—a set of partly overlapping regimes regulating various aspects of GM foods and crops that can reflect distinct, and sometimes conflicting, normative principles and rules.[2] International cooperation in such regime complexes is likely to be characterized by several inter-related phenomena: inter-state negotiations in one regime will be limited by precedents in others; states will "forum-shop," seeking the most favorable regimes within which to advance their interests; and legal inconsistencies are likely to arise among regimes, creating new problems of inter-regime coordination. The regulation of agricultural biotechnology, we show, falls within such a regime complex, being subject to a number of regimes relating to trade, environmental protection, and food safety, among others. This chapter therefore focuses on the promise as well as the difficulties of multilateral cooperation in the area of agricultural biotechnology, with a focus on four major international regimes: the Organization for Economic Cooperation and Development (OECD), which addresses cross-cutting issues relating to agricultural biotechnology among advanced industrialized countries; the WTO and its Agreement on Sanitary and Phytosanitary Measures (SPS Agreement), which addresses trade-related aspects of GM foods and crops; the Biosafety Protocol signed at Cartagena in 2000, which deals with the environmental implications of GMOs; and the Codex Alimentarius Commission, which handles food-safety standards.

By contrast with the rather bleak story of bilateral negotiation in the previous chapter, we do find some evidence here of deliberative decision-making, most notably in the technical committees of the OECD and the Codex Alimentarius Commission. Consistent with Risse's predictions, we find that US, EU, and other experts were able to deliberate about the risks of GM crops and find common ground on a number of issues, particularly where negotiations were limited to scientific and technical questions of risk *assessment* and where these technical discussions were insulated from questions of risk *management* and technical barriers to trade. By the same token, however, in each regime setting, where the discussion turned from scientific issues of risk assessment to political issues of risk management, and where outcomes in one regime were linked to trade disputes in other regimes, the prospects for deliberative decision-making receded dramatically.

Through a close examination of these multilateral regimes, we are able to show how "hard" and "soft" law mechanisms in these regimes interact with each other in ways that have often been overlooked in international law scholarship. Each side of the agricultural biotech dispute has attempted to export its own regulatory approach to the global level, seeking allies among third parties, engaging in "forum-shopping" to find particularly hospitable regimes, and producing awkward compromises within, as well as inconsistencies among, the various international regimes. In the process, "soft" and "hard" law mechanisms have not interacted in a complementary and progressive manner, as often theorized, but rather served to constrain and undercut each other. As a result, more flexible "soft law" regimes, like the Codex Alimentarius Commission, have been "hardened" by concerns over the implications of their decisions in the "hard law" WTO regime, while the "hard law" WTO regime has been "softened," being made more flexible and less predictable as its judicial process has sought means to avoid deciding the substantive issues in dispute. Yet we also suggest that tensions and even conflicts among the various "hard" and "soft" international law regimes should not necessarily be lamented. The tensions we observe simply reflect underlying differences among states and state constituencies, and in particular among powerful ones, in a diverse, pluralist world. In such a context, overlapping, fragmented regimes can also provide service to each other, signaling states and international decision-makers to tread softly in applying their particular rules, taking account of developments in other spheres of international law and politics.

The chapter is organized in three parts. In the first, we provide the theoretical background to multilateral cooperation in biotech regulation, first noting the purported benefits provided by multilateral regimes and then addressing two interrelated impediments to multilateral cooperation—those of distributive conflict and regime complexes—which are often overlooked, or not sufficiently stressed, in international law and international relations scholarship. We examine, in particular, the impacts of forum-shopping among multilateral regimes and the resulting challenges of fragmentation and conflicts of law among them. The second and longest section of the chapter provides detailed process-tracing accounts of US/EU cooperation and conflict over GMO regulation in the OECD, the WTO, the Convention on Biodiversity (CBD), and the Codex Alimentarius Commission, analyzing the different "hard" and "soft" law mechanisms available in these regimes, and how they have developed and interacted in this field. The third and final section builds from these case studies to summarize our findings regarding the prospects and the limits to international cooperation over agricultural biotechnology policy through multilateral regimes.

4.1. Multilateral regimes and GMOs: Theoretical background

4.1.1. GMO regulation as a global, not just a transatlantic, issue

The pressures for the US and the EU to turn to multilateral regimes were due not only to the failure of bilateral negotiations, but also to the very nature of agricultural biotechnology, which by the late 1990s had emerged as a *global* political issue, with other OECD members and less-developed countries increasingly enmeshed in transatlantic debates over the adoption of GM foods and crops. Early signs of the globalization of GM issues appeared when the anti-GMO movement spread, to different degrees, from Europe to much of the rest of the world. Japan and Korea, two WTO members traditionally raising barriers to US agricultural exports, announced that they would tighten approval procedures for GM varieties and require mandatory labeling of GM seeds and foods in response to constituent demands.[3] In the Japanese market, prices for GMO-free varieties were surging; companies and department store chains were advertising GMO-free foods; and a new GMO-inspection industry was developing.[4] Even Australia and New Zealand, both large agricultural exporters, announced in 1999 that they would require labeling of all GMO-derived foods, and all but one Australian state placed a ban or moratorium on the planting of GMOs.[5]

The majority of developing countries, generally concerned over the expansion of patent and other rights over seeds and plant varieties, also supported a move toward a restrictive new treaty on trade in GMOs, while also criticizing the monopoly rights that large US and European firms hold over new seed technologies. Some countries, such as Thailand, welcomed new price premiums available for their GMO-free varieties. The Brazilian state of Rio Grande de Sul declared itself a GMO-free zone in the hope of assuring and attracting foreign buyers of its crops.[6] In 2002 the African countries of Zambia, Zimbabwe, and Mozambique rejected donated food from the US that was produced using biotechnology.[7] The US responded vociferously to these developments. In a June 2003 speech to the Biotechnology Industry Organization (BIO), the US biotechnology trade association, President George Bush denounced EU limits on the imports of GMOs based on "unfounded, unscientific fears...For the sake of a continent threatened by famine," Bush continued, "I urge the European governments to end their opposition to biotechnology."[8]

The US was not alone in these debates, as its call for open markets for GM food products was soon supported by other major agricultural producers, including Canada and Argentina in particular. Farmers in developing countries began to adopt some GM varieties despite government bans, which resulted in eventual government authorizations. In Brazil, farmers adopted GM soybeans notwithstanding regional bans on GMO cultivation, triggering a grant of amnesty from the Brazilian President to those farmers who illegally planted them, followed by government authorizations.[9] In China and India, farmers adopted GM cotton varieties, including those that had not been authorized, and both governments invested heavily in agricultural biotech research and development. China and India are viewed as pivotal for the global adoption of GM crops because of the size of their populations and economies, and also because of the impact that their policy choices may have in other developing countries. Overall the total acreage of GM soy, corn, and cotton continued to increase, particularly in six key countries—the US, Canada, Argentina, India,

Brazil, and China—reaching around 102 million hectares in 2006 according to one study.[10]

In this context, neither the US nor the EU limited its attention to bilateral discussions, but sought to advance their interests through multilateral regimes. Because agricultural biotech regulation implicates so many policy issues, multiple multilateral regimes were available. These included the WTO and its SPS Agreement, which addresses trade-related aspects of GM foods and crops; the CBD and its Biosafety Protocol signed at Cartagena in 2000, which deals with the environmental implications of GMOs; the Codex Alimentarius Commission, which handles food-safety standards; and the OECD, which addresses multiple trade and regulatory issues facing its developed-country members.

4.1.2. Regime theory: How regimes can foster multilateral cooperation over GMO regulation

International regimes were famously defined in a collective research project as "sets of implicit or explicit principles, norms, rules, and decision-making procedures around which actors' expectations converge in a given area of international relations."[11] Regime theorists contend that these regimes can help states cooperate by performing a number of useful functions, both at the negotiating stage and during the subsequent implementation of international agreements. At the negotiating stage, international regimes can reduce the transaction costs of negotiating international agreements, by providing fora for regular meetings of officials, and by providing secretarial, logistical, informational, and deal-brokering support for international negotiations.[12] The various trade rounds of the WTO, the regular meetings of the parties to the UN CBD, the task forces of the OECD, and the working groups of the Codex Alimentarius Commission have all performed this basic function in the area of agricultural biotechnology, although as we shall see, these various fora are characterized by different memberships, different voting rules, and differing modes of representation. Once a given agreement is adopted, international regimes can also play a key role in their implementation, most notably by monitoring member-state compliance with their provisions, reporting instances of non-compliance, and—in select cases such as the WTO Dispute Settlement Body—providing third-party dispute resolution.[13]

The rationalist variant of regime theory examines why and how states create international regimes in the rational pursuit of their interests.[14] Its central arguments build from game theory, a branch of applied mathematics that social scientists use to study different scenarios (or games) in which players interact. In regime theory, the players are typically states (with the EU in the trade context being treated as a single actor). Each player is self-interested and rationally attempts to maximize its own utility, but in doing so, affects the others' choices. Different scenarios involve different sorts of interactive games. Some games are zero-sum (one player's gain is equal to another's loss), and others are not (as when both players can gain through cooperation). While zero-sum games entail pure conflict, nonzero-sum games mix conflict with at least some common interest.

The Prisoner's Dilemma (PD) scenario is a classic nonzero-sum game which serves as the backdrop for much regime theory. In this game, each of the two actors will benefit from cooperation by pursuing their mutual interests together, but they both face short-term (or "myopic") temptation to cheat (or "defect") from their agreements. If each of them follows a myopic self-interest by betraying the other, then they both are worse off than if they had cooperated.[15]

The strategic structure of PD can be represented as a simple payoff matrix, as in Table 4.1. Within each cell of the table, we see the ordinal payoffs for players 1 and 2, arrayed from most preferred (4) to least preferred (1) outcome. As in the story that illustrates the PD game, both the players would prefer a cooperative outcome (3,3) to the non-cooperative equilibrium (2,2) in Table 4.1. In game-theoretic terms, the equilibrium outcome of the game is "Pareto-suboptimal"—that is, both players could benefit from mutual cooperation.[16] Yet, in the absence of mutual trust and/or a clear enforcement mechanism, each player has a rational incentive to defect, no matter what the other player does.[17]

As applied to international politics, in this scenario, each of two states can benefit from cooperation if its negotiating partner complies with its agreement; but each state also faces a strong post-decisional temptation to renege on the agreement because the reneging state's payoff is greatest when the other state complies and it does not. Yet if both states renege, then both are worse off than they would have been if both complied. The trick is how to get self-interested parties to collaborate for their mutual benefit when they face incentives to betray the other party.[18]

International trade theory and international environmental regimes—both of which, as we will see, have come into play in the GMO context—provide

Table 4.1 The Prisoners' Dilemma

		Player B	
		B1 (cooperate)	B2* (defect)
Player A	A1 (cooperate)	3,3	1,4
	A2* (defect)	4,1	2,2 **

Source: Adapted from Stein 1982: 306.
4 = most preferred, 1 = least preferred.
* Player's dominant strategy.
** Equilibrium outcome.

examples of a multi-state extension of the PD game. First, under the international trade theory of comparative advantage, each state benefits from trade because it can specialize in the production of goods in which it has a relative advantage (compared to what else it can produce) and trade those goods for ones in which other states have the relative advantage. However, a state may be tempted to renege on its free-trade commitments for political and economic reasons. If the other states retaliate, however, by reneging on their commitments, then the exporting interests will be harmed and the first state's welfare reduced, politically and economically. The goal of the trade regime is, thus, to provide a mechanism to reassure each state that the other will comply with its trade commitments.[19]

In the environmental context, the classic "tragedy of the commons" provides another example of the PD. In the tragedy of the commons situation, each player benefits from using a common environmental resource (e.g., fisheries, the global climate) at no fee. If all players use the commons without any form of rationing, however, then the commons will be depleted or polluted, so that all the players are worse off. Once again, each player faces both an incentive to cooperate to protect a commonly valued resource and an incentive to "free-ride" on the conservation efforts of others. But this free-riding can induce other states to renege, resulting in the commons' depletion or pollution.[20]

In these contexts, regime theorists have argued that international regimes can reduce the prospects of reneging and thus increase the effectiveness of international agreements in facilitating the achievement of national goals. They do so by providing mechanisms for monitoring state compliance and reporting instances of non-compliance to the members of the regime, increasing both the likelihood of being caught and the costs of reneging, and affecting states' concerns over their reputational interests before a broad audience. When such a game is repeated over time (so that it is an iterative and not a one-shot game) and when monitoring by the regime reassures states of each other's compliance, then regimes can facilitate states' willingness to engage in mutually beneficial cooperative arrangements. In Keohane's terms, these regimes can "empower governments rather than shackling them... because they render it possible for governments to enter into mutually beneficial agreements with one another."[21]

International regimes vary in the services that they provide. Regimes designed to protect global environmental commons exemplify the use of monitoring and reporting functions. For example, the Framework Convention on the Protection of the Ozone Layer (1985) and the subsequent Montreal Protocol (1987) have created a secretariat whose primary function is to report on member-state compliance with the provisions of the agreement.[22] Some international regimes, such as the trade regime of the WTO, go beyond this reportorial function, establishing courts and other dispute-settlement bodies designed to adjudicate disputes over the meaning of the relevant agreement and the responsibilities of the various parties, and sometimes to authorize sanctions for non-compliance. Most basically, international regimes can provide the crucial element of iteration, changing the game from a one-shot to a repeated-play game in which cooperation can be rewarded and defection punished in subsequent rounds of play. Put differently, international regimes lengthen the "shadow of the future," facilitating cooperative outcomes.

As we saw in the previous chapter, constructivist scholars go beyond these transaction-cost approaches, arguing that the institutionalization of a given issue-area in international politics can also encourage the sorts of ongoing, face-to-face interactions that can be conducive to deliberative decision-making, and in the process, facilitate learning and the adoption of normative understandings, which can affect the very perception of rational self-interest. Constructivist scholars like Risse do not assume that deliberative behavior is automatic in international politics, but they do claim that international regimes, which foster close and iterated cooperation among national officials, can create the conditions under which deliberation about GMO regulation might take place. Put differently, international regimes can at the very least increase the efficiency of decision-making, and at best transform international decision-making from a logic of bargaining to one of arguing that can transform actors' perceptions of the national interest.[23]

Our focus in this chapter, as in much of the regime-theory literature, is on common and competing interests and on the potential for distributive conflict among those interests. However, we do not mean to suggest either that the dispute is fueled only by interests, at the expense of normative concerns, or that such interests are narrowly economic or protectionist. First, as we have shown in Chapter 2, norms are embedded in the two different systems, which exhibit strong path-dependencies. When policy negotiations move to the bilateral and international levels, each party can be viewed as advocating not only its own interests, but also the normative beliefs underlying its regulatory approach. Second, we wish to make clear that when we refer to distributive costs and benefits, we are referring not only to economic costs and benefits, but also and perhaps especially to political ones. Government negotiators are likely to incur severe political costs if they negotiate outcomes at the multilateral or bilateral levels that compromise domestic norms as embedded in domestic legal systems. In short, both norms and interests are bundled into the positions that both US and EU negotiators defend in transatlantic and global negotiations.

In the area of agricultural biotechnology, all of the regimes that we examine in this chapter provide for a great deal of face-to-face interaction, significant elements of iteration as well as some degree of reporting on member-state compliance, while the WTO goes further with the establishment of a third-party dispute-settlement system for the adjudication of claims among members. Therefore, we might expect multinational regimes such as the OECD, the WTO, the CBD, and the Codex Alimentarius to facilitate the adoption and implementation of cooperative outcomes in areas like agricultural biotechnology. Indeed all four regimes *have* witnessed extensive negotiations and the adoption of formal agreements about (or directly implicating) the regulation of GMOs. As we will see in this chapter, however, the existence and creation of new international regimes do not necessarily facilitate (or even signal) efforts at cooperative and deliberative decision-making but can simply provide stages for strategic maneuvering in which underlying conflict continues to play out in new fora, for two primary reasons that we now examine.

4.1.3. A first obstacle to regime-based cooperation: The distribution of (absolute) gains

Despite the general promise of regimes in fostering cooperation under anarchy, international relations scholars have identified a number of potential obstacles to successful regime-based cooperation. The realist literature, for example, has emphasized the dual challenges of (*a*) cheating and (*b*) relative gains as the primary obstacles to successful cooperation, focusing on concerns over national security and the balance of power.[24] When faced with a collective-action or public-goods problem under anarchy, realists have long argued, states acting under anarchy face a strong temptation to renege on their commitments or cheat, and the spectre of such cheating is expected to undermine cooperative efforts or forestall them entirely.[25] However, as neoliberal institutionalists have demonstrated, the temptation to renege can be effectively managed by international institutions that monitor state compliance and authorize decentralized enforcement, as in cases like the WTO, which features well-developed dispute-settlement mechanisms.[26]

Other realists point to the so-called "relative gains" problem, in which states will avoid engaging in cooperation that would make them *absolutely* better off because it may nevertheless make them worse off *relative* to other states, affecting the balance of power among them. In an anarchical international system, this argument goes, states face a constant threat to their very existence, and so concerns about relative position are likely to trump opportunities for absolute gain.[27] However, as scholars like Duncan Snidal and Robert Powell have demonstrated, relative gains concerns are not a constant in international politics but vary as a function of factors such as the number of states in a cooperative agreement, the costs of using force, and the nature (threatening or benign) of the international environment. Put differently, relative-gains situations emerge as a special case of interactions among states, such as where the numbers of relevant states is small, the costs of using force is low, and the security environment is unforgiving.[28] In the case of agricultural biotechnology, we did find some actors (in Chapter 2) concerned about the relative positions of the US and EU resulting from regulatory policy choices, although in this case the concerns arise out of a desire to establish a leading position in an emerging technology rather than out of fears for military security.[29] But there is little evidence that concerns over such relative gains have been crucial in driving the conflict over biotech regulation, and indeed EU biotechnology policy has in practice voluntarily ceded global technological leadership to the US, a clear violation of realist expectations.

Rather than focusing on these traditional realist fears, therefore, we identify two additional, distinct impediments to international cooperation and deliberation in areas such as agricultural biotechnology: (*a*) the impact of distributive conflict, and (*b*) the existence of multiple, overlapping political and legal regimes or "regime complexes." Neither of these impediments, we hasten to add, arises from pessimistic realist assumptions about the ubiquity of cheating or the importance of relative gains for the balance of power. Even if we adopt the relatively optimistic neoliberal assumptions that institutions can monitor compliance (and hence reduce incentives to cheat) and that states are concerned exclusively (or primarily) with absolute gains, the problems of distributive conflict and regime complexes still emerge as substantial obstacles to successful international cooperation in areas such as biotechnology regulation. We consider each of these impediments, briefly, in turn.

The first, distributive, challenge to regime theory focuses on the appropriateness of the PD game as the proper model for many instances of international cooperation, because it fails to capture the potential for distributive conflicts among the participants. In the PD model, states are assumed to have a common interest in reaching a cooperative outcome, and the primary impediment to successful cooperation is the fear that other states will cheat on their agreements. In PD models of international relations, these problems are typically addressed by creating mechanisms for the monitoring of state behavior and the sanctioning of states that violate the terms of the agreement. If PD is an accurate description of the situation facing states, then international regimes should indeed facilitate cooperation by monitoring compliance and (in the case of dispute-settlement bodies) providing for decentralized enforcement as well.

However, the PD game deemphasizes another important obstacle to successful cooperation, namely conflicts among states with *different interests* over the *distribution* of the costs and benefits of cooperation—even if these states are primarily concerned about "absolute gains" for themselves, and not about "relative gains" from a balance-of-power perspective.[30] That is to say, when states cooperate in international politics, they do not simply choose between "cooperation" and "defection," the binary choices available in PD games, but rather they choose specific *terms* of cooperation such as the specific level of various tariffs in a trade regime or the precise levels of different armaments in an arms control regime, and so on. As James Morrow notes, "There is only one way to cooperate in PD; there are many ways to cooperate in the real world."[31] In game-theoretic terms, there may be multiple equilibria—multiple possible agreements that both sides prefer to the status quo—and states face the challenge of choosing among these many possible agreements.

Different terms of cooperation can have different distributive implications, affecting states' calculation of costs and benefits, both economically and politically. In an international trade agreement, for example, one side may prefer to drastically reduce tariffs on industrial goods, while another may place a stronger emphasis on reducing agricultural tariffs. In this situation, states face not only the challenge of monitoring and enforcing compliance with a trade agreement, as in the PD model, but also of deciding on the terms of cooperation—in this case, the combination of agricultural and industrial tariff reductions in the agreement. Yet PD models, with their binary choice of cooperation or defection and their emphasis on Pareto-improving outcomes, fail to capture these elements of international cooperation.

In light of these problems with PD models, an increasing number of scholars turned toward a "situation-structural" approach to cooperation, in which different problems or issue-areas are characterized in terms of distinct 2 × 2 games, of which PD is the only one possibility. Arthur Stein, for example, offered a distinction between what he called "collaboration" and "coordination" games. According to Stein, cooperative dilemmas in international politics can be classified into two categories. The first of these categories, which Stein calls "dilemmas of common interests," arises when two or more states would benefit from cooperation, but face incentives to renege on their agreements, as in the PD and collective action models referred to above. The solution to such dilemmas of common interests, according to Stein, is "collaboration," in which states can agree to mutual cooperation and rely on formal regimes to provide iteration, monitoring, and enforcement of agreements, and hence reduce the temptation to renege.[32]

The second of Stein's categories, "dilemmas of common aversion," arises when states seek to avoid a particular outcome and need to coordinate their behavior in order to do so. In game-theoretic terms, dilemmas of common aversion are categorized by multiple equilibria, any of which would allow the players to avoid their mutually undesirable outcome. Two neighboring states, for example, might wish to facilitate rail travel between them, in which case they would wish to avoid a situation in which the gauge of their railroad tracks are incompatible; yet there are many different potential standards that the two states might adopt, and so the states will need to coordinate to select one among the multiple possible equilibria, and hence to avoid their mutually undesired outcome of incompatible rail lines. Turning to the GMO case, the US and EU clearly desired to avoid a trade war—creating a dilemma of common aversion—but the question remained, *on what, or whose, terms*?

Dilemmas of common aversion, in turn, come in two variants. In the first and simpler variant, referred to variously as "common indifference" or "pure coordination" situations, states prefer to coordinate their behavior but have no preference among the multiple possible equilibria. In this type of coordination game, illustrated in Table 4.2, cooperation is relatively easy, because both actors are indifferent about the specific equilibrium outcome chosen and need simply to decide between themselves on one of them. In such situations, states may rely on conventions, or they may delegate the choice to third parties such as international organizations that provide "constructed focal points" around which states can coordinate their activities.[33]

More difficult are what Stein calls "dilemmas of common aversions with divergent interests," also known as the Battle of the Sexes game. In this situation, illustrated in Table 4.3, both players agree on the least preferable outcome or outcomes to be avoided, and both agree to coordinate their behavior to avoid such an outcome, but each one prefers a specific (equilibrium) outcome. In Table 4.3, for example, Player A prefers outcome A1B1, whereas Player B prefers A2B2. A more concrete example is provided by the so-called Battle of the Sexes game, in which two players (say, a husband and wife) agree that they want to take a vacation together, but disagree on the destination (he prefers

Table 4.2 Dilemmas of common aversion and common indifference

		Player B	
		B1	B2
Player A	A1	1,1 **	0,0
	A2	0,0	1,1 **

Source: Adapted from Stein 1982: 310.
1 = most preferred, 0 = least preferred.
** Equilibrium outcome.

Table 4.3 Dilemmas of common aversion and divergent interests

		Player B	
		B1	B2
Player A	A1	4,3 **	2,2
	A2	1,1	3,4 **

Source: Adapted from Stein 1982: 310.
4 = most preferred, 1 = least preferred.
** Equilibrium outcome.

the mountains, she the beach). In such a game, the primary challenge is not the threat of cheating (since both players prefer *some* joint vacation to being alone), but rather of deciding which of two possible equilibrium outcomes (the mountains or the beach) will be selected.[34]

As in pure coordination games, any agreement in Battle is likely to be self-enforcing once adopted, with little need for monitoring or enforcement mechanisms, since both players prefer either cooperative outcome to uncoordinated behavior.[35] By contrast with both PD and pure coordination games, however, the Battle game is characterized by a strong distributive conflict over the terms of cooperation. Turning once again to the GMO case, both the US and EU sought to avoid a potential trade war over GM foods and crops, but the choice of international regulatory standards was fraught with distributive implications, since either side would face significant technical, economic, and political costs in moving from its preferred regulatory framework toward that of the other side.

Stein's distinction between collaboration and coordination games would later be taken up by scholars like Duncan Snidal, Kenneth Oye, and Lisa Martin, who would develop a more general "situation-structural" approach to international cooperation, in which different empirical situations could be modeled using a series of 2 × 2 games such as Harmony, PD, Pure Coordination, Battle, Assurance, Chicken, and Deadlock.[36] For our purposes here, the most important implications of this literature have to do with the impact of distributive conflict on international cooperation, both generally and in our specific issue-area of agricultural biotechnology. We therefore offer four specific, overlapping arguments that are relevant to understanding the international law and politics of agricultural biotechnology and related areas: namely, the importance of distributive conflict and power in affecting regime outcomes; the pervasiveness of distributive conflict in nearly all cooperation scenarios, where the terms of cooperation must be agreed prior to the design of monitoring and enforcement rules; the particular impact of distributive concerns in standard-setting, in which the choice of a particular international standard is likely to have significant distributive implications for the states concerned; and finally the impact of distributive conflict on the use of information and the prospects for deliberation.

First, regime theory, with its emphasis on PD and collective-action models, has under-emphasized both distributive conflict and the role of state power in determining the outcome of regulatory conflicts. This, in turn, has affected much international law scholarship, some of which has welcomed regime theory for its validation of international law's role without critically examining the role of distributive conflict and state power in affecting international law's terms and implementation. In international politics, as Stephen Krasner argues, efforts at cooperation often take the form of a Battle of the Sexes game, in which different states have clear preferences for different international standards. Even if all states benefit from a common standard, raising the prospect of joint gains, the distribution of those gains depends on the specific standard chosen, and the primary question is whether and how states can secure cooperation under their preferred terms.

The most important question, Krasner argues, is not whether to move toward the "Pareto frontier" of mutually beneficial cooperation, but *which point* on the Pareto frontier will be chosen. Under such circumstances, he suggests, outcomes are determined primarily by the use of state power, which may be employed in one of three ways: (*a*) to determine who may play the game; (*b*) to dictate the rules of the game, including the possibility of a single state moving first and imposing a *de facto* standard on others; and (*c*) to employ issue-linkages, i.e., the application of threats and promises in related issue-areas, to change the payoff matrix for other states and induce those states to accept one's preferred standards.[37] Like Stein, Krasner views such coordination regimes as stable and self-enforcing, since weaker states would only hurt themselves by failing to accept common international standards; yet this self-enforcing nature of the regime should not obscure the fact that the regime produces winners (who secure cooperation on terms closer to their preferences) and losers (forced to cooperate on terms favorable to others), and that state power plays a key role in determining the shape of the regime and the standards adopted under the regime.[38] We should, therefore, expect the outcome of a Battle of the Sexes game to be determined in large part by powerful states, with weaker players being excluded from negotiations, or forced to accept a *fait accompli*, or induced to accept powerful states' terms through threats and promises in related issue-areas. In the GMO regulatory context, developing countries will be placed in a difficult situation where powerful players—the US and EU—clash.

Second, distributive conflict is not unique to Battle games, but emerges as a generic and nearly ubiquitous feature of all international cooperation. By contrast with the approach of situation-structuralists like Stein and Krasner, James Fearon has argued in a landmark article that it is misleading to attempt to characterize international cooperation over any given issue as being exclusively *either* a PD *or* a Battle game. Rather, Fearon maintains,

> ... understanding strategic problems of international cooperation as having a common strategic structure is more accurate and perhaps more theoretically fruitful. Empirically, there are always many possible ways to arrange an arms, trade, financial or environmental treaty, and before states can cooperate to enforce an agreement they must bargain to decide which one to implement. Thus, regardless of the substantive domain, problems of international cooperation involve first a bargaining problem (akin to the various coordination games that have been studied) and next an enforcement problem (akin to a Prisoners' Dilemma game).[39]

More specifically, Fearon models international cooperation as a two-stage game in which states first agree on the terms of cooperation, and then establish any monitoring and sanctioning provisions necessary for enforcement. Linking these two stages into a single game provides useful insights into the significant challenges of successful international cooperation. For example, he demonstrates, a long "shadow of the future" can lessen problems of enforcement, by reassuring the players that the game is an iterated one and that compliance will be rewarded and noncompliance punished over the long haul. By the same token, however, a long shadow of the future can *exacerbate* distributive conflicts: If states know that the rules and standards they adopt will bind them and their successors for many years to come, they will have a greater incentive to bargain hard and to hold out for their preferred standard, knowing that it will shape the patterns of gains and losses well into the future.[40] Put differently—and in stark contrast to the socialization and deliberation hypotheses of the constructivist literature presented in Chapter 3—we should expect states that are highly interdependent and embedded in stable international regimes to bargain harder and more obstinately, if and insofar as those states face distributive conflicts and expect the results of their negotiations to influence outcomes well into the future. Such has been the case in the GMO context in which bargaining over the terms of cooperation in Codex, for example, have been adversely affected by the enforcement mechanisms under the WTO. The impact of distributive conflict on regime outcomes will nonetheless vary in relation to a number of factors including the issue area and type of game at stake.

Third, the setting of international standards, as for GM foods and crops, is particularly prone to distributive conflicts, and international standard setting can and should be theorized as a coordination game that often creates incentives for parties to engage in strategic bargaining. Some standard-setting negotiations may take the form of a "pure coordination" game, in which the various participants are entirely indifferent among the possible standards to be adopted: indeed the constructivist or "world society" literature depicts international standard-setting as an essentially technocratic and deliberative process in which calculations of interest and power recede into the background.[41] However, as Walter Mattli and Tim Büthe have argued convincingly, almost any potential international standard is likely to have varying distributive implications for states and firms, and so we can expect actors to attempt to "export" their domestic standards to the international level, minimizing their adaptation costs, while their trading partners and competitors are forced to adapt and adjust to a new and different standard.[42]

Whether such standardization negotiations are, strictly speaking, Battle of the Sexes games is open to question. For example, if a particular state or private firm faces sufficiently high costs of adjusting to a new international standard, it might prefer to block a common standard rather than accept coordination on a standard that leaves it at a substantial disadvantage vis-à-vis its competitors, resulting not in a Battle of the Sexes game but in Deadlock.[43]

Indeed, as we shall see, the clash of starkly different US and EU positions has frequently resulted in blockages and failures to reach agreement within the various regimes where that conflict has played out. In an economically interdependent world, however, even the US and EU cannot ignore a standard being adopted by a critical mass of states within a regime, complicating a negotiator's calculations as to her position. Our case studies thus show intensive and ongoing US and EU engagement over standards in multiple regimes. More generally, however, as Fearon rightly suggests, the significant issue is not the specification of a particular 2 × 2 game for each negotiation (which we will see are sometimes issue-specific within a particular regime), but the fact that international standards can have significant distributive implications, creating potential winners and losers, providing all players with incentives to engage in strategic bargaining, and sharply discouraging any attempt at open, deliberative decision-making. Negotiating environmental, health, and safety standards, where they have significant trade implications, can be particularly difficult because of the distributive stakes, as we will see.[44]

Fourth and finally, distributive conflict affects the use of information, and reduces the prospects for deliberative decision-making in regimes. In an insightful article, James Morrow notes that international cooperation is plagued not only by problems of monitoring and sanctioning, as emphasized in PD models, but also by problems of distribution and information. As Morrow argues, any Battle of the Sexes scenario creates two interlinked problems over which the players must cooperate to solve:

> The first is the problem of distribution: which solution will actors adopt in the face of divergent preferences over the possible solutions? The second is the *problem of information*: is there a solution that best suits all of them and if so, which solution? Because each actor has some, but not perfect information on the value of different solutions, all actors have a joint incentive to pool their information to determine whether they agree that one solution is preferable to another. These two problems work against one another. Distributional problems create incentives to misrepresent one's private information in the hope of gaining what is likely to be a more favorable solution; yet, the actors require accurate communication messages to solve an informational problem.[45]

This inherent tension between distributional and informational problems, Morrow argues, creates a trade-off: States can solve one or the other of these problems, he suggests, but they cannot solve both simultaneously. Just as importantly, the greater the distributional conflict between the players, the greater the incentive for both players to misrepresent information rather than sending truthful signals to their negotiating partners.[46]

Morrow's findings, although derived from a formal model of international cooperation, dovetail with some constructivist accounts of deliberation (reviewed in Chapter 3). Coming from different starting points, both agree that the presence of distributive conflict, and the salience of power relationships among participants, increases the incentives for actors to bargain hard and treat information strategically, and hence decreases the likelihood of deliberative decision-making in international politics. Hence, to the extent that international cooperation is characterized by distributive conflicts, the prospects for deliberative decision-making are dramatically worsened. Actors in such contexts will tend to deploy information strategically to advance their preferred policy outcomes, as opposed to being transparent in common pursuit of better policy.

In the case of agricultural biotechnology, we show in this chapter that the negotiation of international standards for the regulation of GM food and feed is indeed characterized by stark distributive conflict between the world's two major economic powers, and by an effort by both sides to export their domestic standards to the international level and shift the burden of economic and political adjustment to the other side. On the US side, as we shall see, the government, farmers, and biotech firms have been concerned that overly rigid or protectionist EU standards have already damaged US economic interests, and they fear the spread of those standards to additional countries, which would further shrink the potential market for GM foods and crops. For this reason, US officials have consistently attempted to export their own, more permissive, and "science-based" approach to the various international regimes promulgating biotech standards, while at the same time seeking to undermine the adoption and application of EU standards, both diplomatically and through the threat of WTO litigation.

On the European side, EU representatives begin with a domestic regulatory system that is highly restrictive of GM foods and crops and fear the imposition of a US-style global standard that could leave EU farmers and biotech firms at a relative disadvantage, subject the EU to potential environmental and food-safety risks, and subject EU and national politicians to substantial political costs associated with changing EU regulation in the face of strong anti-GMO public sentiment. As we shall see throughout this chapter, therefore, EU representatives have consistently attempted to export the Union's more precautionary approach to the global level, while at the same time safeguarding the EU regulatory system from legal challenge under the WTO. Indeed, this EU effort to export its own standards is part of a broader and more explicit strategy promulgated by the European Commission to "promot[e] European standards internationally through international organization and bilateral agreements," which, the Commission argues, "works to the advantage of those already geared up to meet those standards."[47]

In sum, although the US and the EU share a common interest in avoiding a global trade war over the regulation and marketing of GMOs (and over non-tariff barriers in the form of environmental and health regulations generally), they have differed sharply in their preferred terms for a solution in this case. The US prefers what it terms a "science-based" approach that leads to liberalized trade in GM seeds and foods, and the EU prefers a "precautionary" one that is more restrictive of trade. Each believes that adapting to the other's approach will impose significant economic and political costs on it. Under these circumstances, we expect each side—the US and the EU—to attempt to export its own model to the international level, using positive and negative issue-linkages to influence the positions of other states, and we expect them to rely less on deliberative decision-making and more on bargaining from fixed positions.

4.1.4. A second obstacle: The challenge of regime complexes and the fragmentation of international law

This brings us to a second challenge to traditional regime theory, one that suggests that a proliferation of regimes may *not* lead to joint problem-solving, convergence, or accommodation, but rather to tensions, inconsistencies, and conflicts among the various regimes. From the beginning, regime theorists have acknowledged a multiplicity of regimes designed to deal with different issues: indeed, the classic definition of regimes, offered by Krasner, identifies regimes as "principals, norms, rules and decision-making procedures around which actor expectations converge *in a given issue-area.*"[48] Yet an increasing number of real-world problems—including agricultural biotech regulation—do not fall neatly within the jurisdiction of a single regime, but rather lie at the intersection of multiple regimes. These overlapping regimes result in a *regime complex*, which Raustiala and Victor have defined as: "an array of partially overlapping and nonhierarchical institutions governing a particular issue-area." As they state,

> Regime complexes are marked by the existence of several legal agreements that are created and maintained in distinct fora with participation of different sets of actors. The rules in these elemental regimes functionally overlap, yet there is no agreed upon hierarchy for resolving conflicts between rules. Disaggregated decision making in the international legal system means that agreements reached in one forum do not automatically extend to, or clearly trump, agreements developed in other forums.[49]

Decision-making in these regime complexes is characterized by several distinctive features, of which we emphasize three. First, negotiations in a given regime will not begin with a blank slate, but will typically demonstrate "path dependence," taking into account the developments in related international regimes.

Second, individual states, responding to domestic political contexts and seeking to advance their interests, will engage in "forum-shopping," selecting particular regimes that are most likely to support their preferred outcomes. More specifically, states will select regimes based on characteristics such as their membership (e.g., restricted or universal), voting rules (e.g., one-state-one-vote vs. weighted voting, and consensus vs. majority voting), institutional characteristics (e.g., presence or absence of dispute-settlement procedures), substantive focus (e.g., trade or environment or food safety), and predominant functional representation (e.g., by trade or environment or agricultural ministries), each of which might be expected to influence substantive outcomes in more or less predictable ways.[50]

Third, the dense array of institutions in a given regime complex will create legal inconsistencies among them. States will respond to these inconsistencies with efforts either to demarcate clear boundaries among various regimes or to assert the primacy or hierarchy of one regime over the others, in reflection of a state's substantive preferences. States will engage in "strategic inconsistency," attempting through one regime to create conflict or inconsistency in another in the hopes of shifting the understanding or actual adaptation of the rules in that other regime in a particular direction.[51] Powerful states are likely to be particularly adept at such forum-shopping.[52]

Raustiala and Victor's empirical analysis focuses on the issue of plant genetic resources, which they argue exists at the intersection of intellectual property, environmental protection, agriculture, and trade. In a similar fashion, the regulation of agricultural biotechnology is subject to a regime complex, including regimes addressing trade, environmental protection, food safety, and agriculture, among other issue-areas. Within this regime complex we find that the expected characteristics of path-dependence, forum-shopping, and legal inconsistency arise.

The political science analysis of overlapping regimes is complemented by a growing legal literature about the "pluralism" and "fragmentation" of international law.[53] In 2000, for example, the International Law Commission (ILC) included the topic "Risks ensuing from the fragmentation of international law" into its work program, and in 2002, it created a Study Group to make recommendations concerning the topic, renamed "Fragmentation of international law: difficulties arising from the diversification and expansion of international law."[54] For many international lawyers, the result of such fragmentation is legal uncertainty and potential conflict between international legal regimes.[55] As the 2006 ILC report states:

> [W]hat once appeared to be governed by "general international law" has become the field of operation for such specialist systems as "trade law," "human rights law," "environmental law," "law of the sea," "European law" ... each possessing their own principles and institutions. The problem, as lawyers have seen it, is that such specialized law-making and institution-building tends to take place with relative ignorance of legislative and institutional activities in the adjoining fields and of the general principles and practices of international law. The result is conflicts between rules or rule-systems, deviating institutional practices and, possibly, the loss of an overall perspective on the law.[56]

Scholars disagree regarding the causes of such fragmentation and whether fragmentation is positive or negative for international law, but they largely concur on its development.[57] Many legal scholars view this development as a manifestation of the rise of a global legal pluralism, which refers to "the presence in a social field of more than one legal order."[58] As a theory or analytic framework, legal pluralism differs from much of regime theory in that it challenges monist conceptions of the state and of state interests and rather emphasizes the interaction between distinct normative orders—state and non-state—while de-emphasizing the role of formal texts. In this sense, legal pluralism has a "radically heterogeneous" conception of law and social order, taking a post-modernist constructivist orientation that focuses on social diversity and informality more than on formal rules and hierarchic authority.[59] As Francis Snyder writes, "the various sites [for governing globalization, whether public or private, domestic, regional or international] are not all necessarily hierarchically ordered in relation to each other. Instead, they demonstrate many other types of interrelationships, sometimes hierarchical, sometimes not, sometimes competing, sometimes collaborative."[60]

Although legal pluralism and theories of regime complexes have quite different starting points, in particular regarding their conceptions of the role of states, they both raise the question of how legal regimes interact and potentially constrain one another where there is no central authority. As Roderick Macdonald writes from a legal pluralist perspective, "[d]ifferent legal regimes are in constant interaction, mutually influencing the emergence of each others' rules, processes and institutions."[61] These regimes are not "self-contained,"[62] in a way that some legal commentators fear, but rather exercise normative pressure on each other, as we will demonstrate below. Lines of communication between regimes exist, but, crucially, there is no hierarchy imposing a particular discipline. We are particularly interested in the interaction of "harder" and "softer" forms of international law within a regime complex, which calls into question much of the previous theorizing about them.

4.1.5. Regime complexes and the interaction of international hard and soft law

Legalization in international politics has been defined, in formal terms, as consisting of a spectrum of three factors: (i) precision of rules; (ii) authority or obligation of rules; and (iii) delegation to a third-party decision-maker.[63] International regimes vary in the extent of their legalization along these dimensions, which can give them a "harder" or "softer" legal character. In this respect, hard law "refers to legally binding obligations that are precise (or can be made precise through adjudication or the issuance of detailed regulations) and that delegate authority for interpreting and implementing the law."[64]

As an institutional form, hard law features many advantages, allowing states to commit themselves credibly to international agreements, monitor and enforce such agreements, and solve problems of incomplete contracting.[65] Yet hard law also entails significant costs, restricting the behavior of states, infringing on national sovereignty in potentially sensitive areas, and encouraging states to bargain fiercely and at length in light of the ongoing and legally binding commitments into which they are entering.

In contrast with this ideal-type of hard law, soft law is often defined as a residual category: "The realm of 'soft law' begins once legal arrangements are weakened along one or more of the dimensions of obligation, precision and delegation."[66] Hard and soft law, therefore, are not binary categories but can be arrayed on a continuum, and "soft law" agreements can feature varying degrees of obligation, precision, and delegation—they need not, that is, be uniformly low on all three counts.

Soft law agreements have obvious disadvantages, in that they create little or no legal obligation and provide no third-party dispute settlement to resolve disputes among the parties and fill in the details of incomplete contracts. For this reason, soft law is viewed by many scholars as a second-best alternative to hard law, either as a way-station on the way to hard law or as a fall-back when hard-law approaches fail.[67] Defenders of soft law, however, argue that soft-law agreements offer significant offsetting advantages compared to hard law, in that they are often easier to negotiate; impose lower "sovereignty costs" on states in sensitive areas; provide greater flexibility for states to cope with uncertainty and learn over time; and allow states to be more ambitious and engage in "deeper" cooperation than they would if they had to worry about enforcement.[68] For this reason, a growing number of scholars in law and political science have advocated a pragmatic approach, selecting hard-and soft-law approaches depending on the characteristics of the issue in question.

Although those with constructivist leanings may question this positivist characterization of law, because it distracts from how law operates normatively,[69] we find this conceptualization to be useful to make a central point about how legal regimes with different characteristics affect each other within a regime complex. Our primary interest here is not in the respective advantages and disadvantages of these characteristics encapsulated by the terms hard and soft law, which have been well examined elsewhere, nor do we wish here to enter into the academic debate as to whether soft law (or for that matter international law) actually constitutes "law."[70] We focus, instead, on the coexistence and interactions among regimes with hard-law and soft-law characteristics in the regime complex for agricultural biotechnology.

While the respective costs and benefits of hard and soft law remain a subject of contention, both legal and political science scholars have moved increasingly toward a view that "soft" and "hard" international law can build upon each other as complementary tools for transnational problem-solving, with soft law serving either as a way-station to hard law or as an alternative where hard-law options are unavailable.[71] Dinah Shelton, for example, examines how "combinations [of hard and soft law] may be essential to achieve specific goals."[72] Similarly, a leading US international law casebook addresses in its introduction to soft law how "treaties and state practice [i.e., binding hard law] give rise to soft law that supplements and advances treaty and customary norms;... [while] soft-law instruments are consciously used to generate support for the promulgation of treaties or to help generate customary international law norms [i.e., binding hard law]."[73]

By contrast, we demonstrate how hard and soft laws are not necessarily mutually supportive, but rather can counteract and undermine each other when multiple regimes with different functional orientations overlap. As we shall see, the regime complex for agricultural biotechnology comprises both soft-law regimes, including the non-binding guidelines and recommendations issued by the OECD and Codex, alongside hard-law regimes such as the Cartagena Biosafety Protocol and the WTO, the latter of which combines relatively detailed, legally binding rules with a particularly strong system of third-party dispute settlement to interpret and enforce them. Simplifying somewhat, we argue that the coexistence of these hard and soft-law regimes has led to some "hardening" of the soft-law regimes like Codex, and to some "softening" (and more flexibility and less predictability) of the hard-law WTO dispute-settlement system. More precisely, we contend that the OECD and especially Codex have lost some of their traditional advantages as soft-law regimes, growing more contentious, more difficult, and less deliberative over time because states are concerned about how these regimes' decisions can be used in the hard-law WTO dispute-settlement system. By contrast, the quintessential hard-law regime of the WTO dispute-settlement system has been softened somewhat, as panelists and Appellate Body members need to take into account not only political pressures from the member states, but also the growing overlaps, tensions, and conflicts between the WTO legal order and the provisions—both hard and soft—of neighboring international regimes.

4.1.6. Two final theoretical points

Before we move to the specific regimes at issue, we make two final points in an effort to use our empirical work to advance theorizing on the prospects and limits of international cooperation. First, *the two challenges to international regulatory cooperation that we have identified—distributive conflict and fragmented regime complexes—do not operate in isolation but interact and build on each other, making cooperation still more difficult.* Where existing regimes have distinctive implications for substantive outcomes on a given issue, distributive conflicts encourage states to engage in forum-shopping among the various regimes available, seeking to negotiate or litigate in the regime most likely to secure their preferred outcome. These distributive conflicts create incentives to strengthen existing international regimes that are more receptive to a state's position and to undermine those that are not. Where an existing regime complex is unfavorable, a state may attempt to create a new countervailing regime to support its position and change the status quo. We see powerful states, in particular, well-positioned to use these strategies, as Benvenisti and Downs have pointed out.[74] As we will show, the distributive conflict between the US and EU over the terms of international regulation pertaining to GMOs has provided each side with an incentive to forum-shop among regimes within the regime complex, with the US finding the WTO and OECD to be more favorable venues. We will further see the EU deploy the strategy of creating a new countervailing regime in its entrepreneurial role behind the 2002 Biosafety Protocol.

Second, and finally, building from our two-level game approach introduced in Chapter 2, *domestic politics within states also affect international regime developments*. In Chapter 6, we will address the impact that international regime developments have had within the US and EU respectively, including through their provision of tools for constituencies in domestic political and legal struggles. Here, in contrast, we note how politics within states reciprocally affects developments within a regime complex. In both the US and the EU, the regulation of GM foods and crops is a responsibility shared and disputed among multiple agencies, ministries, Directorates-General, and levels of government, which respectively have primary responsibility for different issues, such as the environment, food safety, agriculture, and trade. In practice, as our interviewees have repeatedly pointed out, particular ministries, agencies, and DGs favor developments in one regime over another, which would provide them with more authority in domestic arenas or produce outcomes closer to their own preferences.[75] Indeed, our research into the workings of various international regimes reveals domestic turf wars among ministries, because international developments affect their relative power domestically. Environmental ministries took the lead in the Biosafety Protocol, which, in turn, has resulted in new national legislation around the world which has granted environmental ministries new domestic regulatory authority. These developments have sometimes upset their colleagues in agricultural ministries, because the Protocol affects matters within that ministry's traditional domain, and, in particular, the control of the entry of pests and the support of agricultural exports. As one US agricultural ministry official states, "the problem is that countries are not monolithic. When the Biosafety Protocol was negotiated, the agricultural ministries were largely absent either by design or by default."[76]

Conversely, ministries of agriculture can bolster their authority domestically through their participation in organizations in which they play dominant roles such as the Food and Agricultural Organization (FAO), International Plant Protection Convention (IPPC), and World Organization for Animal Health (OIE) that we examine below. IPPC officials, for example, have spoken against the adoption of a new international standard for pest risk analysis under the Biosafety Protocol, because it may conflict with the existing one under the IPPC, which responds to agricultural ministries.[77] Trade authorities likewise can use WTO processes to accomplish domestic regulatory goals, as can private actors, including multinational corporate ones, who are often behind WTO legal cases.[78] Non-governmental organizations also play this game, with environmental NGOs favoring the Biosafety Protocol since it can give rise to new environmental regulation, both internationally and nationally.[79] In each case, these various players see the respective secretariats of these international organizations as potential allies. In the words of one international bureaucrat, secretariats often "fight for their organization to be more influential and gain resources."[80] The choice among international regimes, therefore, as well as the interactions of regimes within a regime complex, is shaped in important ways by domestic bureaucratic politics, as we shall see presently.

4.2. Multilateral regimes and agricultural biotechnology regulation: Four case studies

In the rest of this chapter, we examine the record of international cooperation on agricultural biotechnology in four overlapping multilateral regimes that have played primary roles in the creation of standards and rules applying to agricultural biotechnology and that vary in their "hard" and "soft" law characteristics: the OECD (cross-sectoral, but limited to advanced industrialized democracies), the WTO (trade), the CBD and its Biosafety Protocol (environment), and the Codex Alimentarius Commission (food safety standards). On the whole, we argue, efforts at multilateral cooperation in the regulation of GM foods and crops have been characterized by the dual problems of distributive conflict (which has provided each side with an incentive to export its domestic system through hardball bargaining, rather than engage in a deliberative search for the "best" system) and regime complexes (which have provided each side with an incentive to forum-shop and to create strategic inconsistencies among regimes).

Interestingly, we do find some evidence of deliberative decision-making in some multilateral regimes, particularly within the OECD and to a lesser extent within Codex. As many proponents of deliberation readily concede, however, such decision-making practices thrive only under certain, restrictive conditions. In the case of agricultural biotechnology, we demonstrate, negotiators in the OECD and the Codex *have* been able to engage in some amount of deliberative, technically oriented decision-making, but only *under certain conditions*. More precisely, US, EU and other officials have been most likely to engage in real deliberation where the negotiations focus on scientific issues of risk assessment (as opposed to the more political question of risk management), where the negotiators involved are themselves representative of scientific and technical departments or ministries, and where these deliberations are insulated both from domestic public opinion and from entanglement in trade disputes under WTO law. By and large, these conditions have been satisfied within the OECD, and within the Codex except where the latter system was linked to potential disputes under WTO law, such as the US/EU ongoing dispute over GMOs, and we find some evidence of deliberative decision-making in these cases.

By contrast, however, where the subject has turned from risk assessment to risk management, where trade negotiators have been included in national delegations, and where negotiators have been constrained by an attentive public opinion or by linkages to WTO law and dispute resolution, negotiators have reverted to a clear "logic of consequentiality," reflecting the clear distributive stakes of the conflict for policy-makers and for key constituencies in Europe and the US. Within each of these regimes, we show that the US has sought to promote its more "science-based" approach to biotechnology regulation, while the EU has sought to secure international recognition for its regulatory approach and, in particular, for its interpretation of the precautionary principle. The result in each case has been a series of untidy compromises, representing not a deliberative consensus on the best policy but a series of carefully negotiated bargains allowing each side to claim partial victory without resolving the fundamental issues in dispute.

Moreover, within the complex of regimes relevant to the regulation of GM foods and crops, both the EU and the US have engaged in systematic forum-shopping, pressing their cases in those regimes most receptive to their respective views, and seeking to ensure the hierarchical dominance of their preferred regimes. Thus, as we shall see, the US has demonstrated a particular preference for reference to, and dispute resolution in, the WTO, whose legal provisions most closely support its views, while the EU has pressed its case most energetically within the CBD, whose environmental mandate, promoted by representatives of environmental ministries, has been most receptive to the Union's precautionary approach. These efforts, have, in many cases, resulted in potential legal inconsistencies among the various regimes (reflecting, for example, different issue-areas or different memberships), with no clear hierarchy among the various regimes or the standards they establish.

Of course, many other international organizations have assumed some role in overseeing and pronouncing on agricultural biotechnology in light of the multi-sectoral issues at stake and the publicity and political confrontations that have arisen over ongoing technological developments.[81] These organizations include, to name a few, the FAO,[82] the World Health Organization (WHO), the International Plant Protection Convention (IPPC), the World Organization for Animal Health (OIE), the United Nations Environmental Programme (UNEP),[83] the International Standards Organization,[84] the UN Educational Scientific and Cultural Organization (UNESCO), the UN Industrial Development Organization (UNIDO),[85] the UN Economic Commission for Europe (ECE) and other regional UN commissions, the International Programme on Chemical

Safety (IPCS), the International Labour Organization (ILO), the World Bank, the Consultative Group for International Agricultural Research (CGIAR),[86] the International Centre for Genetic Engineering and Biotechnology (ICGEB), and the International Council of Scientific Unions (ICSU). The FAO, for example, boldly issued its 2003–04 report on *The State of Food and Agriculture* to address, in the words of its subtitle, *Agricultural Biotechnology: Meeting the Needs of the Poor?* The report examined and assessed the state of knowledge of the benefits and safety record of adopted agricultural biotech varieties in developing countries, providing potential impetus and political cover for developing country governments to explore their adoption. As we will examine further in Chapter 7, much of the conflict between US and EU approaches to agricultural biotechnology will play out in the ministries, fields and markets in developing countries, with the larger countries such as Brazil, China and India taking the lead.[87]

The conflicts among US, European, and increasingly other states have been repeated in these overlapping fora, only further demonstrating the challenge of international cooperation in a world of multiple international institutions with their proliferating committees, working groups, task forces, and expert consultation publications. There are many fora to "shop," creating challenges for even the most well-resourced states and regional organizations, such as the US and EU. Where negotiations have remained narrowly focused on technical issues such as risk assessment, these organizations have been able to produce consensus documents based on a relatively deliberative approach, as we will show regarding the OECD and Codex. However, in all fora, including in the OECD and Codex, once issues have shifted toward policy such as the "legitimate factors" to take into account in "risk management" or the labeling of products derived from agricultural biotechnology, parties turned to hard bargaining from fixed positions, resulting in either no consensus between the US and European positions or compromise language that provides little practical guidance for decision-makers or the public.

4.2.1. The OECD and biotech: A node for networks

The OECD is an international organization based in Paris which facilitates policy harmonization primarily through soft law mechanisms and consensus decision-making. It was one of the first international organizations to play a major role in attempting to harmonize approaches to agricultural biotech regulation. In this way, it was a locus for transatlantic regulatory networks in this area.[88]

The OECD is the successor to the Organization for European Economic Cooperation (OEEC), which was created in 1948 to promote economic cooperation, distribute funds, and help to coordinate the reconstruction of Europe as part of the Marshall Plan. In 1961, the organization changed its name to the

OECD, and it has gradually expanded its membership to thirty countries today, which represent the most developed countries, complemented by a growing number of observer nations. The OECD Council acts as the central body made up of one representative from each member country and one from the EU. The OECD has a secretariat of around 700 professional staff divided into different functional directorates, and the organization itself works through functional committees that bring together relevant officials from the OECD's members.[89] "In all there are approximately two hundred committees, working groups and expert groups, involving the combined participation of thousands of senior officials from member-country governments."[90]

Although international treaties are sometimes negotiated under the OECD's auspices,[91] the OECD has been most effective in operating as a high-powered think tank for developed countries, providing information and analysis to help them implement and coordinate more effective national polices in a wide variety of areas including economic policy, labor policy, environmental policy, and the development of science and technology. In this capacity, it "has acted as a policy coordinator, information broker, and forum for policy learning in the emerging field of biotechnology."[92] Indeed, the OECD has been held up by many scholars as an ideal-typical case of deliberative decision-making and as the model for the EU's own Open Method of Coordination. As James Salzman writes, "In the little that has been written on the OECD, it is often held up as a prime example of an organization that operates on the basis of cooperation and informal networks, relying on 'soft law'—recommendations and guidelines—rather than hard rules."[93] Indeed, the OECD has issued a number of important recommendations, guidelines, and other consensus documents regarding agricultural biotechnology.

The OECD secretariat has followed biotechnology since the technology's earliest days but has been particularly active since the early 1980s, playing an important coordinating, brokering, and information-providing role. At the beginning, when application and commercialization of the science lay in the distant future, the OECD's Directorate for Science, Technology and Industry took the exclusive lead. However, as the technology moved toward application and commercialization, an increasing number of OECD committees became involved in light of the cross-cutting nature of biotechnology issues, including the Environment Committee, Industry Committee, the Committee for Agriculture and the Trade Committee.[94] In order "to facilitate co-operation" regarding biotechnology issues between OECD Directorates that service these committees, the OECD created an Internal Coordination Group for Biotechnology (ICGB) in 1993, which meets three to five times a year.[95] The ICGB publishes the *Biotechnology Update* newsletter, which provides up-to-date information on the OECD's activities in this field.

The OECD has developed important guidelines for biotech policy over the last several decades, as regards both environmental protection and food safety.

It created a Committee for Science and Technology Policy in 1972 prior to the famous Asilomar meeting (of 1975) and the committee's secretariat identified genetic research as an area that it should pursue. It was not, however, until 1981 that the committee added biotechnology to its work program. Once it did, the OECD began to play an important global role. In 1982, for example, the OECD issued a broad report on scientific trends in biotechnology and their policy implications, which spurred the committee to follow-up on the issues of biotech safety, patent protection, R&D, and the technology's long-term economic effects.[96] In 1986, the committee issued its "Blue Book" on *Recombinant DNA Safety Considerations*, adopted by the OECD Council as a recommendation, constituting, in Cantley's words, "a first step in the harmonization process of safety principles and practices among the Member countries of the OECD."[97] In the early 1990s, the OECD's Group of National Experts developed a set of "Good Developmental Principles" (GDPs) "to guide researchers in the design of small-scale field experiments with genetically modified plants and microorganisms."[98] The result was a new 1992 OECD publication, *Safety Considerations for Biotechnology*, which established separate good development principles for plants and microorganisms and specifically addressed environmental safety. These GDPs were later adopted by OECD member nations.

The OECD Directorate for Science, Technology and Industry, working with the Environment Directorate, then turned to developing guiding principles for large-scale releases of GMOs, covering plants, microorganisms, and animals, which led to a series of reports. In 1993, the OECD published *Safety Considerations for Biotechnology: Scale-up of Crop Plants* which publicized the concept of "familiarity" which would become important for the work of the Codex Alimentarius Commission. The concept "is based on the fact that most genetically engineered organisms are developed from organisms such as crop plants whose biology is well understood." The idea is to facilitate the ability of risk assessors to draw on such knowledge and experience when evaluating the introduction of GM plants and microorganisms into the environment.[99]

The Working Group on Harmonization of Regulatory Oversight in Biotechnology, established in 1995 when the approval of the first commercialization of GM crops was being considered, continues work in this area. It has since published twenty-nine consensus documents, with others under development, regarding the biology and introduced traits of specific GM varieties, including wheat, rice, canola, soybean, and maize, among others.[100] Its primary goal is "to ensure that the information used in risk/safety assessments, as well as the methods used to collect such information, are as similar as possible."[101] One or two OECD members typically take the lead in preparing a first draft of the document which is then brought to the full working group. Through creating a common understanding of the host plants and introduced traits at issue, it can ease the process for companies preparing files for a risk assessment, reduce costs by avoiding duplicative work, and help to facilitate dialogue if there are any divergent findings in the actual assessment. The Working Group is now expanding its work to study the molecular characteristics of the genes inserted and the impact of the receiving environment on the risks posed.

The Working Group, which focuses on environmental impacts, is complemented by an OECD Task Force for the Safety of Novel Foods and Feeds, which focuses on food and feed safety. The Task Force was created in 1999 and builds from the previous work of an *Ad Hoc* Group of National Experts.[102] It consists of "individuals nominated by the governments of OECD Member countries [who,] for the most part, . . . work in ministries or agencies with responsibility for ensuring the safety of . . . genetically modified foods and feeds."[103] Other international organizations contribute to the Task Force's work, including the FAO and the WHO.[104] The main OECD achievement in the area of food safety was the 1993 "Green Book," *Safety Evaluation of Foods Derived by Modern Biotechnology—Concepts and Principles*. This publication developed the concept of "substantial equivalence,"[105] which, as we saw in Chapter 2, was incorporated in different ways into both US and EU regulatory policy and in particular the EU Novel Foods Regulation. The concept creates a common baseline—that of the conventional counterpart—against which the food safety risks of a GM variety can be assessed. The concept was also subsequently adopted at the international level, including in the Codex Alimentarius Commission's 2003 *Principles for the Risk Analysis of Foods Derived From Modern Biotechnology*.[106] The OECD continued to develop the concept, confirming in 1997 that "the determination of substantial equivalence provides equal or increased assurance of the safety of foods derived from genetically modified plants, as compared with foods derived through conventional methods."[107]

Building on this work, the Task Force has likewise published eleven "consensus guidance documents," with others under development, regarding the characteristics of specific GM varieties. These documents are again used to help national and international practitioners assessing the safety of GMO food and feed in a common manner, applying the concept of "substantial equivalence."[108] More recently, some leading developing countries, such as Brazil, China, India, Thailand, and South Africa have participated in the Task Force. Thailand and South Africa, for example, have taken the lead in preparing the first draft of OECD consensus documents, with the assistance of the US, regarding risk assessments of GM varieties of interest to them, such as papaya and cassava.[109]

The OECD's work can be viewed as a successful application of deliberation to reach common conclusions that can, in turn, operate as a form of "soft law" guidelines for national implementation. In this way, the organization exemplifies the role of global transgovernmental networks and the hope of policy-making through the logics of appropriateness and argument, as put forward by Habermas, Risse and others. In fact, in commenting on our theoretical discussion of deliberation through networks in Chapter 3, a former official at OECD wrote in the margins, "[w]hat is described here sounds very similar to the working practice of the OECD since many decades."[110] As Cantley writes regarding the OECD's work in the early and mid-1980s on biotechnology, "although consensus was difficult to achieve, given wide divergences both in the perceptions of risk in biotechnology, and in the state of scientific development in different Member counties, there remained a persistent and common interest in mutual learning and a willingness to compromise."[111] As he further states regarding the OECD's earlier *Ad Hoc* Group of National Experts on Safety and Biotechnology, "[t]he government representatives in the group came from a number of different government ministries and agencies, all having some direct interest in biotechnology safety; science and technology, environment, public health, government and others. Thus, the main task, but also the chief difficulty, of the group was to reconcile the different perspectives of these agencies and to promote an interdepartmental approach—as well, obviously, as an international one."[112] In this regard, he finds the networks to have been successful in deriving and promulgating policy guidelines and principles, based on shared understandings and assessments of the state of the science, which national governments could implement, potentially leading to overall harmonization of policies ensuring worker, consumer, and environmental safety.

The main constraint, however, has been the politicization of the issues in Europe, which resulted in the EU forsaking the OECD's earlier guidelines. The OECD's 1986 Blue Book, for example, concluded:

- any risks raised by rDNA organisms are expected to be of the same nature as those associated with conventional organisms. Such risks may, therefore, be assessed in generally the same way as non-recombinant DNA organisms...

- there is no scientific basis for specific legislation to regulate the use of recombinant DNA organisms.

Scientists and other experts within the European Commission's Directorate-General for Science, Research and Development pointed to these principles in attempting to forestall what they viewed as unnecessary, technology-stigmatizing, GM-specific EU legislation. As we know, however, the EU's DG Environment, working with the Council and Parliament, went the other way. As the issues became increasingly salient in European politics, the OECD's influence, based on deliberation among experts from different policy domains, waned. As one former member of the Commission who also worked in and with the OECD laments,

> once "GM" had been politicized as a category, and had become the basis for passionately advocated demands for regulation (in particular, by Green political parties and NGOs in Europe), more scientific, technical and sector-specific arguments

and circles found they were effectively ignored. It was a disheartening, demoralizing and embittering experience in which to participate as an impotent witness.[113]

The persistent conflicts between the US and EU over agricultural biotechnology have, in short, been replicated in the OECD, affecting its work program. The studies that the OECD prepares are a function of member demands developed by consensus, supplemented by *ad hoc* member funding of specific studies. To give an example, upon a proposal by the EU, the OECD created and reported on a survey in 2003 answered by twenty OECD members regarding their approaches for monitoring, detecting and identifying GMOs (with the US notably not participating).[114] These national approaches are pursuant to GM-specific legislation enacted first within the EU and now, as we will see, pursuant to an international environmental treaty. As the OECD reports,

> [t]he project was intended to identify programs/systems that are currently used or developed to detect or monitor products or organisms. The aim was to appraise the frameworks of such systems as a foundation to develop potential systems for the detection/identification of transgenic organisms, and to better understand the different approaches between member countries.[115]

The survey showed that "many respondents mention that they are working toward the implementation of the Cartagena Protocol on Biosafety" (covered below), with "the most striking common ground" being the "new EU Directive 18/2001" which "will likely prove the most influential initiative toward the goal of harmonization of regulatory standards and procedures among European countries."[116]

In parallel, US officials have attempted to advance their perspectives through OECD initiatives, repeating the "mantra" that regulation of agricultural biotech should be "science-based," and that "there is no scientific basis for specific legislation to regulate the use of recombinant DNA organisms," such as by reference to this language from the 1986 OECD Blue Book.[117] In 2007, the US pushed for a new initiative in the OECD Working Group regarding the treatment of "low-level presence" of transgenic varieties in bulk shipments, in parallel with a new initiative it provided in the Codex Alimentarius Commission, as we will see below.[118] Through these multilateral initiatives, the US is attempting to create pressure for the EU to change (or otherwise soften the impact of) the EU's zero-tolerance thresholds for varieties that have been approved in foreign markets but not yet in the EU, even where the EU's scientific authority has found the varieties to be safe for human and animal consumption. This issue is particularly important for the US because of the impact of the EU's zero-tolerance thresholds for not only international trading markets, but also for US approvals for domestic consumption as well, because of the difficulty and costs of segregating grains, as we will see in Chapter 6.

Although both sides have attempted to use the OECD strategically to advance their interests, the OECD's work is nonetheless best viewed as an attempt to facilitate harmonized approaches where possible through soft law mechanisms. The organization continues to develop, compile, and publish biotechnology-related guidelines, statistics and policy briefs. The Working Group and Task Force publish "consensus documents" concerning the characteristics of specific GM varieties to facilitate a common approach to their risk assessment by national and EU regulators, where possible. Regulators may take different risk management decisions, but the consensus documents at least permit them to do so based on common risk assessment understandings and principles. In 2006, the OECD published a two-volume *Safety Assessment of Transgenic Organisms* that included all consensus documents to date, with the express aim of identifying "elements of scientific information used in the environmental safety/ risk assessment of transgenic organisms which are common to OECD member countries... [in order] to encourage information sharing and prevent duplication of effort among countries."[119]

The OECD's information-gathering function is replicated in the organization's BioTrack online database, created in 1995, which holds information on "biotechnology products tested and approved in member countries." Researchers can view this information by searching the database, and the information can be organized by company and organism.[120] The OECD's website also provides links to an inventory of biotechnology statistics for all its member countries.[121] These statistics include the types of biotechnology that each nation uses, the tests each performs, and the publications resulting from these tests. In 2007, the OECD Biotrack Online programme was developing a unique identification system for biotech crops which can help to facilitate application of the Cartagena Protocol and its Biosafety Clearing House (discussed below). While proponents of agricultural biotechnology may find the "tracing" of GMOs based on unique identifiers to be a harmful stigmatization of the technology, the OECD's technical work nonetheless can facilitate a harmonized approach that in turn facilitates trade in these varieties, at least compared to the alternative of no common identification system.

Finally, the OECD provides informational services aimed to facilitate the development of agricultural technology. For example, the OECD is working to facilitate cooperation for the development of biological resource centers (BRCs) whose primary function is to gather and store information about organisms, collecting them and maintaining databases regarding their genetic information. The OECD maintains that these BRCs are a "vitally important element" of a "sustainable international scientific infrastructure," which is meant to support the delivery of the benefits of biotechnology.[122] The OECD Working Party on Biotechnology, finding that biotechnology "is a key driver for sustainable economic growth,"[123] has advocated the establishment of a Global Biological Resource Center Network and a framework of common operating standards to help regulate the BRCs.[124]

The OECD, in short, demonstrates both the promise and the limits of technocratic, deliberative networks of government officials. On the one hand, the organization, composed of mostly like-minded advanced industrialized countries and engaging in technical, soft-law analyses, has been a model of deliberative decision-making for both scholars and practitioners, including in the area of biotechnology where its work has proven influential in risk assessment and in the detection of GM products. On the other hand, however, the limits of the OECD have also been apparent in two ways. First, the OECD has proven more harmonious in scientific discussions of risk assessment, while remaining divided or silent on the more difficult and politicized issues of risk management. Second, as we have seen, the OECD has increasingly been riven by the US–European conflict over how GM seed and crop varieties should be treated. The US has viewed the OECD as an institutional culture relatively more conducive to its perspectives on agricultural biotechnology, especially compared to the Biosafety Protocol (addressed below), and it allegedly pushed to further OECD's work in response to the Protocol.[125] The US wished to ensure continuation of OECD work programs that have, in particular, resulted in consensus documents for individual GM varieties, including those of interest to non-OECD developing countries such as Thailand (for papaya) and South Africa (for cassava). The US has continued to advocate its "science-based" approach and has referred to OECD documents such as the 1986 Blue Book and 1993 Green Book as evidence of an international call for such an approach. Yet in light of the considerable divergence in US and European views on such issues as the definition and application of a "precautionary principle" (discussed below), this member-driven organization is limited in what it can accomplish. What it has done more recently in the area of agricultural biotech is to create consensus documents for risk *assessments* of specific varieties. Risk *management* decisions, however, remain at the national and EU level where considerable differences remain. The other three international organizations that we cover include rules and standards that have greater implications for these risk management decisions.

4.2.2. The WTO and the SPS Agreement

The WTO, founded on January 1 1995, is known primarily as a "hard law" regime based on binding rules that are enforced through a rather remarkable international trade dispute-settlement system, although it also includes soft law components as we will see. It is the successor organization to the General Agreement on Tariffs and Trade, GATT, and, as of January, 2008, consisted of 151 members, including the EC.[126] The organization's central mission is to facilitate the negotiation, monitoring, and enforcement of international trade liberalization.[127] These three functions (negotiation, monitoring, and enforcement) are fulfilled through various organs within the WTO, assisted by a secretariat.[128] The WTO is particularly noteworthy in international law for its binding dispute-settlement mechanism, governed by its Understanding on Rules and Procedures Governing the Settlement of Disputes ("Dispute Settlement Understanding") and administered by its Dispute Settlement Body. This body, which meets about once every two weeks, oversees the formation of dispute-settlement panels, the adoption of dispute settlement reports, and the authorization of sanctions against countries that fail to comply with WTO dispute-settlement rulings.

Since the creation of the GATT in 1948, tariff rates have generally plummeted around the world. Yet while tariffs have fallen, differences among national regulations, and in particular risk regulations, can create significant non-tariff barriers to trade (NTBs), especially in agricultural and food products. For example, the percentage of US food imports governed by non-tariff barriers rose from around 50 to 90 per cent from 1966 to 1986, the year when the Uruguay Round was launched.[129] The problem of non-tariff barriers was first addressed in the Standards Code, a "side agreement to the GATT" that thirty-two (out of 102) GATT contracting parties signed at the conclusion of the "Tokyo Round" of trade negotiations in 1979.[130] The code, however, was largely ineffective on account of the weakness of its dispute-settlement provisions (the creation of panels could be blocked), the lack of detailed provisions regarding risk regulation, and the fact that most GATT members did not sign it.[131] In fact, "[n]ot one SPS measure was successfully challenged before a GATT dispute-settlement panel after the Tokyo Round, and several other prominent disagreements over SPS measures remained unresolved."[132]

Accordingly, under US initiative, a new agreement, the WTO Agreement on the Application of Sanitary and Phytosanitary Measures, was negotiated as part of the Uruguay Round of Trade Agreements that created the WTO to discipline members' sanitary and phytosanitary (SPS) measures.[133] It covers all governmental measures which are applied to protect human, animal, or plant life or health from a list of risks, such as from pests, diseases, and toxins.[134] The SPS Agreement was complemented by the Agreement on Technical Barriers to Trade (TBT Agreement) which covers regulations that lay down mandatory technical product and process requirements that lie outside of the SPS Agreement's scope—that is, requirements other than SPS measures. For example, requirements for the labeling of GM foods to provide information to consumers should be considered non-SPS measures, and thus subjected to the TBT Agreement, as discussed in Chapter 5.

The SPS and TBT Agreements reflect a shift of attention toward NTBs. From a trade perspective, the former system did not sufficiently counter domestic protectionist measures blocking agricultural trade. From a regulatory perspective, however, the trading system now implicates itself much more deeply into national regulatory processes. The SPS Agreement does not establish international standards for biotechnology or other food-safety questions (a role left to the Codex Alimentarius Commission, examined below). However, the Agreement does incorporate and promote the adoption of international standards in a manner that could be interpreted in a highly constraining way. Article 3.1 provides that, "To harmonize sanitary and phytosanitary measures on as wide a basis as possible, Members shall base their sanitary and phytosanitary measures on international standards, guidelines or recommendations, where they exist," subject to exceptions where "there is a scientific justification" or a different level of protection desired. The Agreement further provides that measures "which conform to international standards, guidelines or recommendations shall be deemed necessary to protect human, animal and plant life or health," and thus consistent with the SPS Agreement's requirements.[135] In this way, the WTO has significantly increased the stakes of negotiations in "voluntary" standard-setting bodies such as the Codex Alimentarius Commission to which it refers.[136]

In addition, the Agreement not only requires the adoption of non-discriminatory national food safety regulations, but also establishes rules that limit the ability of states to adopt trade-restrictive regulations without "scientific justification."[137] The Agreement requires members to "ensure that any [SPS] measure...is based on scientific principles and is not maintained without sufficient scientific evidence," regardless of whether it is applied equally to domestic and foreign products.[138] Article 5.1 requires, in particular, that measures be based on risk assessments, prescribing: "Members shall ensure that their sanitary or phytosanitary measures are based on an assessment, as appropriate to the circumstances, of the risks to human, animal, plant life or health, taking into account risk assessment techniques developed by the relevant international organizations."[139] The only exception is "where relevant scientific evidence is insufficient," in which case "a Member may provisionally adopt...measures on the basis of available pertinent information," provided that it "shall seek to obtain the additional information necessary for a more objective assessment of risk and review the...measure accordingly within a reasonable period of time." The SPS Agreement thus places the onus on a state that would restrict trade through national regulations to demonstrate that its regulations are based on a scientific risk assessment. These terms are binding and enforceable before WTO dispute-settlement panels and the WTO Appellate Body.[140]

For many commentators, the regulatory requirements in the SPS Agreement are highly problematic. The SPS Agreement can be read to require that "science" always trumps politics in national (and, in the EU case, regional) regulatory policy.[141] Such a reading raises concerns about a "democratic deficit" in the design and application of WTO rules.[142] As the late Robert Hudec pointed out:

Traditionally, trade agreements have focused on eliminating discrimination against foreign trade by disciplining governmental measures that impose competitive disadvantages on foreign goods vis-à-vis domestic goods with which they compete. In the recent Uruguay Round trade agreements, however, it appears that the draftsmen...added another goal, one that can be described as the prevention of unjustified regulation per se, whether or not such a regulation creates a competitive disadvantage for foreign goods vis-à-vis domestic goods. Thus, for example, a food safety measure that is not based on scientific principles would be a violation of Article 2 of the [SPS Agreement], whether or not it discriminates against foreign goods.[143]

In a similar vein, David Wirth writes that the SPS Agreement's requirements create "an opportunity or a temptation to accomplish substantive goals similar to those in the domestic regulatory reform debates through international processes in the face of domestic obstacles to achieving those same aims at the national level."[144] These opportunities could be of interest to domestic constituencies (such as a segment of business), government agencies that have lost in internal regulatory debates, or an executive whose wishes are being constrained by parliament (or vice versa). In each case, these domestic players will have interests in common with foreign traders negatively affected by the domestic regulation. Moreover, as Hudec continues, a WTO rule that requires regulatory "rationality" can provide these "foreign traders...a greater set of legal rights than is given to the domestic producers with whom they compete."[145] SPS rules are thus subject to the challenge that democratic governments, in response to constituent demands, may decide for multiple reasons to regulate in a manner that is not rationally justified according to most scientists and is not determined by some international standard-setting body in Geneva, but is nonetheless not discriminatory. For this reason, an amicus curiae brief submitted by a group of academics in the GMO case points out that "the GMO dispute implicates not only technical concerns about barriers to trade but also political concerns about a democratic deficit in the design and operation of the WTO itself."[146]

On the other hand, risk regulation adopted with no scientific risk analysis suggests that protectionist motives could lurk behind it, or that most of the costs imposed by the regulation are possibly being shifted to non-represented foreign parties. Even if the motive for the measure is not protectionist, the measure can have the greatest adverse impact on foreign producers (and not domestic ones), because they were not taken into account in the domestic decision-making process. The requirement of a risk assessment can serve, in Howard Chang's words, "a prophylactic purpose."[147] It creates a procedural mechanism that requires that domestic regulators must at least weigh scientific evidence before adopting non-discriminatory regulations that have disparate adverse effects on foreign traders. Howse makes the related point that, from the perspective of "deliberative democracy,"

democracy...requires respect for popular choices, even if different from those that would be made in an ideal deliberative environment by scientists and technocrats, *if* the choices have been made in awareness of the facts, and the manner that they will impact on those legitimately concerned has been explicitly considered.[148]

The requirement of a risk assessment procedurally helps to ensure that regulatory decisions more likely respect *"real* choices," after taking into account scientific evidence.[149]

4.2.2.1. WTO "SOFT LAW" DISPUTE-SETTLEMENT MECHANISMS

The WTO is best known for its justifiably renowned dispute-settlement system, pursuant to which a panel, subject to appeal before the WTO Appellate Body, issues legal decisions. The WTO, however, also includes many other mechanisms that can facilitate dispute settlement, which can be viewed as forms of "soft law" governance. As we mentioned above, the WTO helps states to monitor each others' fulfillment of their existing commitments. It does so through a Trade Policy Review Mechanism (TPRM), complemented by a web of councils and committees. The TPRM reviews national trade-related policies on a member-by-member basis, with the US and EU being subject to a TPRM review and report every two years. In addition, WTO councils and committees oversee implementation of each of the WTO's substantive agreements, with the SPS and TBT committees being respectively responsible for overseeing implementation of these agreements. A member may raise concerns over another's SPS-related measures through the TPRM or the SPS or TBT committees.

More SPS disputes are in fact settled within the "soft law" committee process than under the WTO dispute settlement understanding, a point missed by many legal analysts. Article 7 and Annex B of the SPS Agreement, entitled "Transparency," require members "to notify changes in their SPS measures" to the SPS committee, which holds two to four general meetings and periodic special meetings each year. Between 1995 and 2005, the committee heard 330 complaints, over half relating to human health measures, many of which have been resolved without litigation before the dispute-settlement system.[150] In some cases, a member changes or withdraws its measure; other countries are provided technical and even financial assistance to adapt to it. Focusing discussion around scientific assessments, including jointly conducted ones, has served to resolve many of these disputes.[151] As Roberts and Unnevehr conclude, "The establishment of the SPS Committee has provided a forum for airing grievances and made it easier to identify and track contentious regulations. These mechanisms have facilitated the resolution of disputes between countries at every level of development."[152] The US has repeatedly challenged the EU's agricultural biotech regulations before the SPS committee to keep pressure on the Commission. For example, at the time when EU member states pressed to block EU approvals, the US stressed before the committee, "The principle that national health measures and international SPS measures must be based on science is fundamental to the effective implementation of the SPS Agreement."[153]

In sum, the SPS committee process establishes a potentially useful "soft law" forum in which both sides to the dispute can discuss and air grievances, albeit in the "shadow" of the hard-law dispute-settlement process, and the committee process has demonstrated its effectiveness in other disputes. In the case of agricultural biotechnology, however, the committee process offered little hope for resolving the dispute, leading ultimately to the WTO's hard-law alternative, dispute settlement.

4.2.2.2. WTO "HARD LAW" DISPUTE-SETTLEMENT: THE MEAT-HORMONES PRECEDENT

If matters are not resolved in the SPS committee (or bilaterally following notice to it), then they may be brought to the dispute-settlement mechanism and its "hard law" enforcement. In this sense, the "soft law" mechanisms can be viewed as operating in the shadow of potential complaints under the WTO Dispute Settlement Understanding.

The US brought six of the first seven SPS complaints respectively against Korea, Australia, the EU, and Japan.[154] As of February 2007, the US had been involved in eleven SPS disputes, nine times as a complainant, whereas the EU had likewise been involved in eleven SPS disputes, but eight times as a defendant.

The US challenge to the EU's ban on the sale of hormone-treated beef has received the most media and scholarly attention, providing an important precedent for the GMO case. The hormone dispute actually began in 1989, when the EU (acting under the terms of a 1988 directive) instituted a ban on the use of synthetic growth hormones in beef cattle and prohibited the import of animals or meat from animals that had been treated with these hormones. Although the EU directive had been adopted primarily on the grounds of European consumer concerns about the safety of hormone-treated beef, the ban had an immediate and dramatic impact on beef producers in the US, where some 90 per cent of all beef cattle are treated with synthetic growth hormones, and where FDA studies have consistently shown that the growth hormones in question are safe for human consumption.

In 1995, after the entry into force of the SPS Agreement and the Dispute Settlement Understanding, the US initiated legal action against the EU, alleging that the EU ban was inconsistent with the terms of the SPS Agreement, because it was not based on scientific evidence, a risk assessment, or agreed international standards, and it arbitrarily differentiated between products. The EU, by contrast, argued that the SPS Agreement acknowledges the right of states to determine the appropriate level of health protection for their consumers, and that the ban was justified under the precautionary principle.[155] A WTO dispute-settlement panel was established in May 1996 and issued its report in favor of the US in August 1997. The EU appealed the panel's decision, and the WTO Appellate Body issued a second report in January 1998, once again in favor of the US.

Both the WTO panel and appellate decisions were complex, involving hundreds of pages of legal reasoning and scientific testimony. The Appellate Body overrode the panel's assessment on several issues, including the panel's finding that the EU's regulation was required to, and in fact failed to, conform to a Codex Alimentarius Commission standard, and that the EU ban represented "a disguised restriction on international trade," because it was "inconsistent" with other EU regulations regarding the use of natural and synthetic hormones.[156] After significantly narrowing the reach of the panel's decision, however, the Appellate Body agreed that the EU had failed to base its beef-hormone ban on a scientific risk assessment, undermining the EU's claims that the ban was adopted to protect human health. In response to the EU's invocation of the precautionary principle, both the panel and the Appellate Body found that the precautionary principle—whatever its status under international law on which they declined to rule—could not override the express provisions of the SPS Agreement, in particular the requirement of a risk assessment under article 5.1 of the Agreement. In accordance with the Appellate Body's findings, the Dispute Settlement Body ruled in February 1998 that the EU ban was inconsistent with the terms of the SPS Agreement and instructed the EU to bring its regulations into compliance by no later than 13 May 1999.

Facing continuing pressure from its own consumers, however, and hopeful of producing additional scientific findings that might justify the ban, the EU failed to act, and the US retaliated on 17 May 1999, applying tariffs in the amount of $116.8 million targeted against specific EU products such as *foie gras*, Roquefort cheese, and Dijon mustard. These US tariffs in turn sparked a wave of protests among French and other European farmers, including an attack in August 1999 by a group of French farmers on a McDonald's restaurant, selected as the symbol of the threat of globalization and, more specifically, American food culture to French traditions. "McDonald's encapsulates it all," said one commentator. "It's economic horror and gastronomic horror in the same bun."[157] Although the leader of the farmers' group, Jose Bové, was jailed for his part in the attack, he was also hailed as a hero in the French press for his opposition to American efforts to force upon Europeans foods that were widely seen as unwanted and unsafe. As of October 2007, both the EU's beef-hormone ban and the US punitive tariffs remained in place, in spite of the expressed desire of both sides to reach a negotiated settlement. The EU also challenged the US retaliatory tariffs before the WTO on the grounds that new risk assessments now justify the EU's ban, leading to a follow-up WTO case.[158]

The transatlantic dispute over the EU's policies regarding GM varieties is analytically similar to the dispute over beef-hormones. EU trade-restrictive regulations on GMOs were adopted once again without being based on a scientific risk assessment, leaving the Union open to WTO legal challenge. US governmental authorities again sided with US producers and repeatedly protested to the EU bilaterally and before relevant WTO committees, to no avail. Until 2003, the Clinton and then the Bush administration refrained from taking legal action before the WTO, fearing that such a case could prompt a European consumer backlash against GMOs and harm the prospects for new WTO trade negotiations. The US eventually lost patience and brought a WTO complaint in May 2003 which resulted in a WTO panel report adopted in November 2006. We address relevant WTO jurisprudence under the SPS Agreement and the GMO panel decision itself in Chapter 5. In the interim, the EU and US continued their regulatory dispute in other international fora, each seeking to alter the terms of the international debate in its own favor.

4.2.3. The Convention on Biodiversity and the Cartagena Biosafety Protocol

From the perspective of the EU, the most promising of the alternative fora for addressing the regulation of agricultural biotech was the 1992 CBD, one of a series of framework agreements adopted at the 1992 UN Conference on Environment and Development at Rio de Janeiro, Brazil. By contrast with the case of hormone-treated beef, where no other international treaty existed to support EU claims regarding its right to restrict imports of hormone-treated foods, the CBD offered a forum within which the EU could press for an international environmental agreement supporting its precautionary approach for biotech regulation.

The CBD is characterized by nearly universal membership, including 190 parties as of January 2008, but the US is only an observer before it.[159] Although the US signed the CBD, the Senate has not ratified it. The ethos of the CBD is quite different than that of the WTO, and the lead representatives of states before it tend to come primarily from environmental ministries and departments, rather than trade and agricultural ministries, who (as we have seen in Chapter 2 and will see below) have their own stakes in inter departmental policy debates.

In the 1992 Convention, the parties agreed to "consider the need for and modalities of a protocol setting out appropriate procedures" in relation to the transfer and use of "any living modified organism [LMO] resulting from biotechnology that may have an adverse effect on...biological diversity."[160] The vast majority of parties supported the adoption of a protocol. By contrast, the US was more reluctant about negotiating in that forum, and indeed a reticent US and a small number of grain-exporting countries, known as the "Miami Group,"[161] were initially able to block the signature of a protocol in February 1999 in Cartagena, Columbia. However, all countries eventually compromised and the Biosafety Protocol was finally adopted on January 29, 2000. At least to one US negotiator, the US was at a disadvantage in the negotiations, because it was only an observer and not a party to the CBD, and thus to the Protocol.[162]

Many US negotiators at the time felt that this was "not a real environmental treaty aimed at alleviating environmental problems, but one to provide protection to the EU in any WTO litigation over GMOs."[163] In this sense, negotiations involving the US and EU were less about joint problem-solving under uncertainty and more about negotiating an environmental treaty in the shadow of potential trade litigation. From the perspective of one US negotiator, "the EU was willing to give in to developing-country environmental concerns to get what it wants for a GMO case."[164] Whether this perception is correct is less important than the perception itself, showing deep wariness of EU intentions and skepticism about the prospect of deliberative problem-solving through the Biosafety Protocol as opposed to instrumental strategizing.

As we will see, the negotiations have been characterized by a logic of bargaining, as opposed to mutual problem-solving involving deliberation, on account of their distributive implications. The three central issues that divided the US from the EU and most of the world were: (*a*) the application of the precautionary principle to decisions to ban imports and require labeling; (*b*) whether the Protocol should cover bulk commodities intended for consumption (e.g., crops) or be limited to organisms intended for direct introduction into the environment (e.g., seeds); and (*c*) the relation of the Protocol to WTO rules, with the US seeking to establish the primacy of WTO law and the EU using the CBD forum to strengthen the precautionary principle in international law. The parties compromised on all three issues, although the Protocol can be viewed as countering some US positions under the WTO's Sanitary and Phytosanitary Agreement.

First, the parties negotiated over the issue of the integration of the precautionary principle into the Protocol, with the final language clearly favoring the EU position. Although article 15 of the Protocol provides that countries will undertake "risk assessments...in a scientifically sound manner," the agreement favors precaution in risk management decisions. Right at the start, the Protocol's preamble "reaffirms" the "precautionary approach contained in Principle 15 of the Rio Declaration on Environment and Development" (fourth recital), as does Article 1 of the Agreement entitled "Objectives." In addition, Article 10 expressly incorporates a version of the precautionary principle, providing that a country may reject the importation of "a living modified organism for intentional introduction into the environment" where there is "lack of scientific certainty regarding the extent of the potential adverse effects... on biological diversity in the Party of import, taking also into account risks to human health." A similar provision applies to a country's rejection of bulk GM commodities (such as soybeans, corn and cotton) for food, feed, or processing (Article 11). While the US government authorities may maintain that this reference to "human health" is only made in the context of a treaty on biodiversity that does not address food safety *per se*, clearly the US would have preferred such broad statements to be excluded. In addressing human health, the Protocol arguably extended the scope of the underlying CBD.

At first glance, the US largely prevailed as regards the second issue in having the Protocol's mandatory pre-shipment notification and consent provisions limited to GMOs intended for release into the natural environment (e.g., planting), so that these provisions do not apply to bulk crops intended for food processing and mass consumption (Article 5). Rather, as for bulk shipments of crops, a party is only obligated to notify a Biosafety Clearing-House once it approves a GMO for the national market, and the US is participating in this system. However, the Protocol leaves it to each country to decide whether to permit the importation of such products and provides that they may apply the precautionary principle in making this decision (article 11). In addition, such shipments must be clearly labeled that they "may contain" LMOs (Article 18), a requirement that was subsequently made more stringent against US wishes, as further described below.

Third and finally, as for the relation of the Protocol to WTO rules, the US demanded a "savings clause" to preserve its WTO rights, because otherwise there would be an argument under international law that conflicting provisions in a treaty signed last in time prevail over those in a prior treaty (known as *lex posteriori*).[165] The US did obtain such a clause, but it failed to obtain a clear reservation of its WTO rights.[166] Rather, references to other "international agreements" are only made in the Protocol's preamble, and these references are in tension with each other. The preamble provides that "this Protocol shall not be interpreted as implying a change in the rights and obligations of a Party under any existing international agreements." The next phrase, however, states that "the above recital is not intended to subordinate this Protocol to other international agreements." As an EU representative stated, the two clauses effectively "cancel each other out," leaving the legal relationship between the two regimes unclear and allowing both sides to claim a partial victory.[167] The EU, therefore, could point to the Biosafety Protocol as evidence of an international consensus (involving over 140 parties) regarding the application of the precautionary principle to the regulation of biotechnology. It could (and, as we will see, did) modify its existing legislation to comply with its international commitments under the Biosafety Protocol, pointing to these obligations in its defense against the US' WTO challenge to the EU's biotech regime in 2003. The US, by contrast, could claim that nothing in the Biosafety Protocol compromised US rights under the SPS Agreement, which it would invoke in its WTO complaint, including because the US is not a signatory to the Protocol. Moreover, the Protocol contains only weak, voluntary dispute-settlement provisions, reducing the prospect of a judicial body making pronouncements under it, decisions which could place even greater normative pressure on a WTO dispute-settlement panel.[168]

154

The Protocol took effect on September 11, 2003, and, as of January 2008, had been ratified by 143 parties, including the European Community.[169] The US, by contrast, is not a party to the CBD and hence has not signed or ratified the Biosafety Protocol. Nevertheless, the US participated as an observer in the first three conferences of the parties in February 2004, June 2005, and in March 2006, respectively, and has indicated that "as a practical matter, firms in non-Party countries wishing to export to Parties will need to abide by domestic regulations put in place in the importing Parties for compliance with the Protocol."[170]

Key issues, however, have been left open for negotiation in subsequent meetings of the parties. These meetings have demonstrated the fundamental tensions in national approaches to GM regulation, especially as biotech varieties are produced and exported by more countries. At the first conference of the parties in February 2004, the parties to the Protocol agreed to stricter labeling requirements for trade in GM seeds and bulk crops and paved the way toward new liability rules, but important details remained at issue.[171] Some additional progress was made on implementation at the second conference of the parties in June 2005, where the parties established the work program for the new biosafety clearing house, agreed on measures to promote capacity-building among less-developed countries and approved rules prepared by the Protocol's committee on compliance. The 2005 meeting, however, also demonstrated the stark divide that remains between, on the one hand, the EU and other countries (including many less-developed countries in Africa), which continue to advocate a highly precautionary approach, and, on the other hand, GM producers and other agricultural exporters (including most notably Brazil and New Zealand, as well as the US and Canada as observers) that advocate a less restrictive approach.

According to the terms of the Protocol, the Parties were to agree, by September 11, 2005, on the detailed documentation requirements for bulk shipments of LMOs, supplementing the original requirement that such shipments bear a label indicating that they "may contain" LMOs. These details are central to the Protocol's impact on international commodities trade. At the June 2005 meeting in Montreal, the Parties split into three competing factions. Many African countries, led by Ethiopia, argued for a substantial tightening of the labeling requirements, replacing the "may contains" language of the Protocol to a less equivocal "contains," together with a list of all LMOs contained in a given shipment. These countries, lacking the technical capability to test for LMOs in imported shipments, sought to place as much of the burden as possible on agricultural exporters. The EU, by contrast, took a more flexible position on the "may contains" language, but sought binding provisions on the thresholds for adventitious presence of GMOs, in keeping with the EU's own legislation on labeling and traceability (see Chapter 6). These proposals were rejected, however, by Brazil and New Zealand, who argued that strict rules on adventitious presence would impose an unreasonable burden on exporters. The June meeting ended amid mutual recriminations and without any agreement on labeling requirements.[172]

A compromise was reached in March 2006, after further contentious debate, at the third meeting of the parties.[173] The compromise decision "requests Parties" and "urges other Governments" to take "measures ensuring that documentation accompanying LMOs intended for direct use as food or feed, or for processing" complies with the importing countries' requirements and clearly states that the shipment "contains" LMOs where the LMOs' identity is "known through means such as identity preservation systems." In contrast, where the identity is "not known," then the term "may contain" will continue to be used. However, since the text does not provide further specification, the labeling decision, in practice, may remain in large part with the exporter. In addition, the text expressly confirms that "the specific requirements set out in this paragraph do not apply to…movements" of LMOs between parties and non-parties to the Protocol. Furthermore, no agreement was reached regarding the adventitious presence of GM varieties except to state that "the expression 'may contain' does not require a listing of living modified organisms of species other than those that constitute the shipment." The parties also agreed to discuss phasing out the "may contain" labeling option by 2012.[174] These controversial issues will thus remain on the agenda for subsequent session.

The ongoing debate among the parties over the labeling of LMOs reflects the distributive implications of any agreed standard. Compliance costs will vary significantly in light of the definition of the label (e.g., "may contain," "contains," or required listing of specific LMOs), and the number, type, and place of sampling and testing. One study shows that just the annual testing costs of corn export cargoes could vary between $1 million and $87 million depending on whether one or twenty samples are required per cargo, while noting that these calculations do not include other significant compliance costs such as handling and overhead charges and shipping delays.[175] These negotiations over labeling requirements highlight the stark differences between agricultural exporters and importers—even in the absence of the US as a party—because of their asymmetric distributive implications.

Overall, the EU has found a more favorable forum in the Biosafety Protocol to fashion international rules and norms that contain its "fingerprints," coinciding with and supporting its regulatory approach to agricultural biotechnology. In particular, the Protocol has created new rules providing for the application of the precautionary principle in national decision-making and the requirement of labeling of LMOs in conformity with an importing country's requirements. In addition, discussion continues regarding risk assessment and risk management principles, including the taking into account of "socio-economic considerations"[176] as well as liability rules.[177] In this way, the

Protocol can serve as a counterweight to the WTO SPS Agreement's narrower focus on science-based regulatory measures.

From a law-in-action perspective, the Biosafety Protocol is affecting national regulation world-wide, as countries adopt new regulations to implement its requirements. According to a USDA official, as the EU was the driver behind the Protocol, the EU can also be seen as the driver of national regulatory change. In his words, the Protocol is now "the default international legal instrument governing agricultural biotech."[178] Moreover, an FAO economist notes, the Protocol's impact is not only on bulk commodities but also on authorizations for the importation of seed to be used in field trials to develop locally adapted GM varieties, adversely affecting the development of local biotech capacity, and thus the use of agricultural biotech in countries around the world.[179]

The negotiation and implementation of the Biosafety Protocol thus exemplifies the reciprocal multilevel nature of international law and international relations. Environmental ministries took the lead in its negotiation and are empowered by it vis-à-vis agricultural and trade ministries. However, trade and agricultural ministries also have their favored international organizations which, in turn, provide them with tools in policy debates at the national level. Trade ministries have the WTO, and agricultural ministries have the FAO in which the Codex Alimentarius Commission and IPPC are housed.

Where these ministries (particularly in large countries) feel affected by the operations of another international regime, they frequently monitor developments in that regime. At the 1999 WTO ministerial meeting in Seattle, the environment ministers of the EU member states "publicly contradicted their own trade spokespersons and prevented the issue of LMO regulation from forming any part of the WTO framework," as had been proposed by the EU's Trade Commissioner at the time, Pascal Lamy.[180] In turn, as the potential impact of the Protocol on commodities trading became clearer to agricultural exporting nations, they sent "representatives of trade and finance ministries, sometimes replacing more familiar faces from environment and agriculture ministries," upsetting any notion of a "club-like" or "epistemic" community.[181] As a US official states, the US tried to convince countries at the time of negotiation of the Protocol of the implications for trade, but the negotiators "did not have a clue about how agricultural trade worked," as "agricultural ministries were largely absent either by design or by default."[182] Now, as more countries produce and export biotech crops and as the Protocol's potential impact on international commodity shipments becomes clearer to agricultural producers and agricultural ministries, further elaboration and implementation of its requirements may become more contentious. As the USDA official states, "now that the practical issues are becoming more apparent, agricultural ministries are playing more of a role, though it remains hard to tell how this will play out over time."[183]

Our study of the Biosafety Protocol thus also exemplifies the politics to which a fragmented international legal system gives rise. On the one hand, these different regimes are not closely coordinated in a clear and consistent manner. For example, the SPS and the other WTO agreements contain no reference to the Biosafety Protocol or the CBD. The SPS Agreement rather refers to international standards set by three other international bodies, to which we now turn. In fact, EU efforts to have the CBD recognized as an observer to SPS and TBT committee meetings have been blocked, despite the CBD's formal requests.[184] Similarly, as we will see, the EU unsuccessfully pointed to the Biosafety Protocol's provisions relating to the precautionary principle before the WTO panel in the *EC-Biotech* case. On the other hand, however, WTO decision-makers are aware of the Protocol's existence, which can indirectly affect their decision-making, as arguably occurred in the *EC-Biotech* case that we analyze in Chapter 5.

4.2.4. The Codex Alimentarius Commission and related standard-setting bodies

The WTO SPS Agreement officially recognizes three entities as international standard-setting bodies, each of which can set standards relevant to GMO regulation. They are the International Office of Epizootics (or OIE, which now also calls itself the World Organization for Animal Health),[185] the IPPC, and the Codex Alimentarius Commission (Codex). The OIE addresses animal health, the IPPC plant health, and the Codex food safety, all three of which are implicated by biotechnology. As a result, all three organizations engage in work programs respectively addressing genetically engineered plants, animals, and foods, meaning that any country wishing to follow and participate in international norm-making in this area must dedicate resources to multiple international agencies with their various agendas. All three have adopted norms and standards to which parties and dispute-settlement panels refer in support of their legal and normative positions and determinations in WTO trade disputes, as would occur in the GMO case (see Chapter 5). We only briefly address the OIE and IPPC before turning to the political negotiations in Codex in more detail, since Codex is responsible for food safety standards, it has addressed the broadest range of biotech issues and has done so for a longer period of time. The debates in the OIE and IPPC somewhat replicate those that we will examine in greater detail within Codex.

4.2.4.1. THE INTERNATIONAL OFFICE OF EPIZOOTICS (WORLD ORGANIZATION FOR ANIMAL HEALTH)

The OIE collects, analyzes, and disseminates information on animal diseases and provides technical assistance to countries for animal disease control. Based in Paris and having a membership of over 170 states, it is the only one among the three standard-setting bodies that is not a UN organization. The OIE (like the IPPC) is more intergovernmental in nature than the Codex and does not include a formal process for recognizing non-governmental organizations as observers. The central actors are government veterinary specialists, typically from within the ministry of agriculture.

The OIE's work in the area of agricultural biotechnology is relatively recent, as the development of transgenic animals has developed more slowly than for plants. In 2005, however, it created an *Ad Hoc* Group on Biotechnology that is addressing the cloning of animals and the use of biotech vaccines and diagnostic tests on them.[186] The Codex Task Force on Foods Derived from Modern Biotechnology (described below) has asked the OIE to liaise with the Codex secretariat and report to the Task Force regarding OIE's work in this area.[187]

4.2.4.2. THE INTERNATIONAL PLANT PROTECTION CONVENTION

The IPPC addresses the protection of plants and natural flora, aiming, in particular, "to prevent the spread and introduction of pests of plants and plant products, and to promote appropriate measures for their control."[188] The initial convention was adopted in 1951 at an FAO conference, but the entity remained sleepy and did not operate functionally as a standards-setter until the creation of the WTO. It then suddenly became one of the world's three recognized standard-setting bodies under the SPS Agreement—the one responsible for the development of *international phytosanitary standards*.[189] The IPPC was thus substantially amended in 1997 to adapt to its newly prescribed role. The revised 1997 text enlarged the IPPC's focus from national defensive responses against pests to the coordination of international plans of action. It also extended the IPPC's scope of concern beyond agricultural plants to include natural flora and the environment. The IPPC, in this way, again exemplifies the catalytic effect the WTO can have on the development of other international regimes, whether in reaction to it (as in the case of the Biosafety Protocol) or in furtherance of its trade liberalizing aims (as in this case).

The 1997 convention went into effect in 2005 (after ratification of two-thirds of its contracting parties) and (as of January 2008) had 166 members, including the EU. It is governed by a Commission on Phytosanitary Measures (consisting of all the parties) that adopts International Standards for Phytosanitary Measures aimed at controlling the introduction and spread of plant pests. It is housed at the FAO in Rome and has a tiny secretariat of five professional members who must respond to and service over 160 member countries. Like the OIE, it is intergovernmental in nature, the primary actors this time being plant protection specialists who tend to come from national agricultural ministries.

IPPC standards are adopted through a complex and evolving process involving technical panels, working groups, and a standards committee, before being presented to the full Commission. Decisions are generally made by consensus, although standards can be adopted by a two-thirds vote at the following meeting of the Commission. This IPPC work is linked to the size of the organization's budget. When, for example, the US suddenly paid its arrears, the work program went from "five to six standards to around ninety."[190]

The parties were careful in the 1997 IPPC to attempt to safeguard their rights under the SPS Agreement, providing in the second phrase of the preamble "that phytosanitary measures should be technically justified, transparent and should not be applied in such a way as to constitute either a means of arbitrary or unjustified discrimination or a disguised restriction, particularly on international trade," and expressly noting the SPS Agreement at the preamble's conclusion. Article III of the IPPC further specifies that "[n]othing in this Convention shall affect the rights and obligations of the contracting parties under relevant international agreements," with the WTO agreements clearly in mind. Moreover, the IPPC's substantive provisions regarding certification, consignment, and quarantine measures make reference to trade concerns including the need "to minimize interference with international trade."

Given concerns that agricultural biotech varieties could pose risks to conventional plant varieties, natural flora, and ecosystems, the IPPC has begun to address biotech issues, although they have yet to become an important part of its agenda for standard-setting. In 2004, the interim Commission (consisting of the parties to the treaty) adopted a standard for "*Pest risk analysis for quarantine pests including analysis of environmental risks and living modified organisms*," which supplements an earlier IPPC standard on risk analysis for quarantine pests. The key issue debated was how to determine if a LMO is a "pest" which is subject to national regulatory controls of pests. The parties eventually compromised in agreeing that national authorities should assess the risks of those LMOs that may present a phytosanitary risk, while noting that "other LMOs will not present phytosanitary risks beyond those posed by related non-LMOs and therefore will not warrant a complete pest risk analysis."[191] In light of the standard's focus on the details of technical risk analysis, as opposed to broader risk management policies, it was easier for the parties to reach consensus in a deliberative manner, as we will see regarding Codex.[192] Although the "precautionary principle" has "reared its head" in IPPC discussions, it has yet to be included in the IPPC's general principles.[193] Overall, the IPPC has been much less active in engaging with agricultural biotech-related issues than Codex, in part because of its narrower mandate (focusing on pests), and in part because states have provided it with fewer resources than Codex.

Because the 1997 revisions amended IPPC's scope of concern beyond agricultural plants to include natural flora and the environment, its work overlaps with that of the CBD and Biosafety Protocol. These two organizations have significantly different institutional cultures, with lawyers much more present in Biosafety Protocol discussions, which tend to be more adversarial, whereas IPPC has a more "cooperative culture."[194] Moreover, environmental ministries tend to play the lead role in the Biosafety Protocol while agricultural ministries dominate the IPPC. As a result, there are turf wars. For example, various parties to the Biosafety Protocol (represented by officials from environmental ministries) have raised the need to establish a risk assessment procedure for LMOs as pests under its mandate, and the IPPC officer has kept responding that the IPPC already has one.[195]

The IPPC also includes a soft law non-binding dispute-settlement system that is available to address biotech issues involving pests, which we will see were litigated extensively in the *EC-Biotech* case. The IPPC indeed states that its system "is designed to provide an alternative to, or complement that of the World Trade Organization."[196] The IPPC manual provides for the establishment of "a committee of experts" consisting of five individuals, two designated by each of the parties and three independent ones. This committee will decide, from a technical standpoint, such issues as whether a country conducted a pest risk analysis, whether it did so properly, and whether it rendered a justifiable conclusion. This formal process, however, has not been used since 1997, and IPPC officials conjecture that, at least in one case, a country decided not to invoke the formal procedure because of "political concerns" raised by the foreign affairs ministry.[197] The IPPC has nonetheless handled around six dispute-settlement matters since 1997 on an informal basis, most of them involving pest risks associated with coconut exports.

To give an example of how the process can work in a "soft law" manner, in one case a South American country blocked imports from a Caribbean country on phytosanitary grounds. The IPPC arranged for FAO officials to help provide capacity for the Caribbean country to address the pest risk. It then provided the information to the South American country to assure it that its import ban was not needed. Acting as a trusted intermediary, the IPPC was able to diffuse and ultimately resolve the conflict, although resolution took around two years before imports were permitted.[198] In this way the IPPC can facilitate dispute settlement, in part (according to one IPPC official) because "there is no involvement of lawyers."[199]

One option in the GMO dispute would be for the parties to resort to this non-binding technical process. However, although this "soft" form of dispute settlement may seem quite attractive, it would likely be placed under considerable strain if it were used in conjunction with a contentious WTO "hard law" complaint. One can imagine that a country could attempt to build a technical dossier for a WTO complaint, such as the US regarding the EU's measures on GMOs, through use of the technical findings from an IPPC "non-binding" procedure. The IPPC manual indeed contemplates that the "report may also be made available on request to competent bodies of international organizations responsible for resolving trade disputes."[200] A WTO panel could simply refer to the IPPC's technical findings as to whether a country's risk assessment was adequate and its phytosanitary measures appropriate. Were such to occur, there would likely be considerably greater scrutiny in international law circles of this relatively unknown dispute-settlement process, subsequently affecting its "soft law" operation.

The FAO's response to the WTO GMO case provides an example of other international organizations' concerns about being implicated by the WTO dispute-settlement process. Following NGO protests against the FAO's favorable assessment of the prospects of agricultural biotechnology for developing countries in the 2003–04 *The State of Food and Agriculture* report, the FAO's leadership became extremely wary of appearing to "take sides" in the WTO case.[201] When the WTO panel asked the IPPC to recommend phytosanitary experts for the panel to consult, an IPPC official prepared a list, but senior FAO management blocked the IPPC from providing it (at least formally) to the WTO panel. If FAO management did not want to provide even an official list of experts out of concern for being implicated, one can imagine its reaction to the prospect of IPPC technical findings being used by claimants in an agricultural biotechnology-related WTO dispute. The IPPC soft law system, as a result, has lain in abeyance, once more suggesting how hard law systems can adversely affect soft law ones, as opposed to complementing them in a progressive manner.

4.2.4.3. THE CODEX ALIMENTARIUS COMMISSION

The final multilateral organization that we cover is the Codex Alimentarius Commission (Codex), an intergovernmental body established in 1962 by the UN FAO and the WHO to facilitate international trade in food through the adoption of international food-safety standards. The Codex has 178 members (at least formally, since many countries do not actively participate) and permits non-governmental organizations to be observers.[202] It has a surprisingly small secretariat in light of its notoriety as a global standard-setter, consisting of only ten professionals in 2007, three of whom were seconded to it by governments, two from Japan, and one from Korea.[203] Leading states appear to have purposefully kept the organization small. The Codex secretariat is lodged, however, within the FAO, the largest of the UN's specialized agencies, with around 3,000 employees, of whom around 1,200 are professional staff.[204] As we will see, the nature of Codex has been significantly affected by the WTO, which has both heightened concerns over the distributive implications of Codex standard-setting, and helped catalyze moves to make Codex more transparent in the process.

The Codex works through different committees and task forces in which individual Codex members play the lead organizing and financial role as "chair." Constraining the secretariat's role in this way arguably helps wealthier states to retain greater control of Codex standard-setting agendas. Where matters raise more difficult issues, these bodies in turn refer them to "working groups," composed of Codex members, as well as to outside groups of experts, designated jointly by the FAO and WHO, for consultation. In December 2006, for example, the Codex Task Force on Foods Derived from Biotechnology referred questions for an expert consultation concerning, among other issues, whether non-antibiotic resistance marker or reporter genes "have been demonstrated to be safe to humans in food products."[205]

The operating costs of Codex committees and task forces are borne by the member who accepts to be the "chair" of such body, which helps to reduce the strain on Codex's limited budget.[206] As a result, most Codex meetings are not held at its secretariat's headquarters in Rome. For example, France is the chair of the Codex Committee on General Principles whose meetings are held in Paris; Canada is chair of the Committee on Labelling, and most of that committee's meetings are held in Canada, organized by Health Canada, the Canadian federal health department; Japan is chair of the Codex Task Force on Foods Derived from Biotechnology, whose meetings it has hosted mainly in Japan, organized by the Japanese health ministry. However, working group meetings under these committees are now being co-chaired by developing countries, and their meetings sometimes held in them, representing an attempt to make Codex more accessible to developing countries.[207]

Traditionally the US and EU have driven Codex agendas and still do so to a large extent. Each works to find allies for its own positions. Conventional wisdom is that the EU has a significant advantage not only because of its large membership (now comprising 27 votes in Codex) but also, in particular, because of its close relations with ACP (Africa-Caribbean-Pacific) countries with which it has had a series of partnership agreements and even longer historical ties. The US, however, has worked hard to develop relations with other Codex members, including through technical assistance programs.

Yet Codex is far from being a mere tool of the US and EU. Its membership has expanded over time, and its aims have broadened. From the point of view of leading agricultural exporting nations (such as the US, Canada, and Australia), the Codex system initially represented, in particular, a means of counteracting the possibility of exclusion from wealthy (and expanding) European markets following the creation and development of the European Community and the adoption of an increasing number of EU standards.[208] Since the 1970s, an increasing number of developing countries have joined Codex, seeking to promote "agricultural development generally and the use and management of agricultural chemicals in particular."[209] The FAO and WHO, together with the Codex framework, helped to provide them with a means to acquire necessary resources, including technical assistance to enhance their food safety systems.[210] As other countries' stakes in international agricultural trade grow, they have become more proactive on certain issues. Overall, Codex provides a global forum for the discussion and adoption of food standards in which foreign exporter interests have a greater voice than within purely national political bodies, facilitating market access for them.

Codex's work can be classified into three main types: (i) standards that define particular commodities (e.g., mineral water); (ii) standards that define acceptable levels of food additives and pesticide residues; and (iii) codes of conduct, principles, and guidelines such as for risk assessments and labeling.[211] The Codex has developed over 3,000 standards including for around 1,000 food additives and 200 pesticide residue standards, as well as over 200 standards for food commodities and over 40 codes for food safety practices.[212] The Codex system has been particularly noted for its success in harmonizing and elaborating pesticide standards, eventually becoming "the main international forum on pesticides regulation."[213] As we will see, most of Codex's work on agricultural biotechnology consists of principles and guidelines prepared initially by a Codex committee or a task force.

Codex traditionally represented a form of "soft law," since the standards were not binding and, by definition, there was no need for a dispute settlement system to enforce them.[214] The process for producing Codex standards involves committees of experts which are established to deliberate over the appropriate standards. A designated Codex committee elaborates a draft standard or guideline subject to comments by member governments and interested international organizations, as part of either a five- or eight-step process.[215] Once the process is completed, standards are approved by the full Codex Alimentarius Commission which meets once every two years.

Voting in Codex is by simple majority of the votes cast, except for changes in the rules of procedure, which requires a two-thirds majority. Any Codex member can call for a vote. In practice, however, standards are generally adopted by consensus, with no vote actually being taken. In fact, no vote over a Codex standard has been taken since 1997.[216] Moreover, because the standards are technically non-binding, a country retains the choice not to adopt it nationally. Prior to the WTO, countries thus did not feel as constrained by the Codex system.

Overall, Codex became a sort of gentleman's club—or epistemic community—of food specialists, with strong industry representation, based on the following characteristics:

> (1) the position of Codex as relatively isolated from international hard law and politics, (2) the voluntary nature of Codex activities and output, (3) agreed-upon norms, which restrained members from both obstructing the process of elaborating new Codex standards and from letting trade considerations override all other considerations, and (4) lack of sanctions in situations where [standards are] not followed.[217]

Those attending Codex meetings were (and the majority remain) food safety experts, often with technical scientific backgrounds from national administrations and industry.[218] Although only governments can vote, the process has often been driven by industry, which seeks to reduce compliance costs resulting from multiple national regulations.[219] The stated goal has been progressive harmonization of food safety standards, resulting in enhanced consumer protection and facilitation of international trade.

The situation of the Codex, however, changed considerably with the creation of the WTO and the adoption of the SPS Agreement in 1995. Under the SPS Agreement, implementation of a Codex standard creates a presumption of compliance with "harder" WTO law provisions, subject to binding dispute settlement. More precisely, article 3.1 of the SPS Agreement provides that WTO members shall base their food safety standards on international standards, guidelines, and recommendations (specifying those of Codex), subject to certain exceptions. Article 3.2 further states that a member's conformity with these international standards shall be presumed to comply with WTO law. These SPS provisions have significantly increased Codex's notoriety and affected how Codex operates in some areas. More concretely, WTO dispute resolution dramatically increased the significance of Codex standards, not only providing a significant impetus to harmonization activities, but also "hardening" Codex decision-making by providing US and EU negotiators with an incentive both to export and to protect their respective regulatory standards within Codex committees.

The effect of Codex standards became clear in the US–EU trade dispute over the EU's ban on beef produced with growth hormones. In 1995, at the first Codex session following the creation of the WTO, the US strategically "forced a vote and the adoption of Codex standards" covering five bovine growth hormones, winning the vote by a 33–29 margin, with seven abstentions.[220] It was hardly a consensus decision, but it was enough to establish a "voluntary" international Codex standard under the organization's voting rules. Shortly afterwards, the US initiated its WTO complaint against the EU, contending that the EU's ban was not "based" on an international (Codex) standard as required by the SPS Agreement.

Since then, the US and European countries have placed increasing importance on the negotiation of new regulatory principles and standards within Codex, since these principles and standards may be invoked (and already have been invoked) in the decisions of WTO panels and of the Appellate Body. As a European Commission representative before Codex concludes, "In the past, if we disagreed with Codex standards or Code of Practice, we could ignore it and take our own legislation. Now we can't."[221] In response, states began sending more than food experts and food agency officials to Codex meetings, complementing them with "delegates from the diplomatic services and ministries of trade, industry, finance, and foreign affairs." In an empirical study, Veggeland and Borden note an increase of such representatives to the Codex Committee on General Principles from ten in 1992, to thirty-two in 2000, to forty-one in 2001.[222]

The enhanced importance of Codex led to a major campaign by the EU to gain full membership in Codex, culminating in the amendment of Codex's rules to permit regional economic integration organizations to become members (which the US opposed) and the EU's formal accession to the organization in November 2003.[223] The Commission now speaks and votes on behalf of the EU in Codex "where an agenda item deals with matters of exclusive Community competence."[224] In practice, this appears to mean that the Commission will often represent the member states in Codex on biotech matters, since annex II of the Council Decision provides: "As a general rule, the European Community has exclusive competence for agenda items dealing with harmonization of standards on certain agricultural products, foodstuffs..., including labeling, methods of analysis and sampling, as well as codes and guidelines."[225] Pursuant to Codex rules, the EU must notify all Codex members before a Codex meeting whether it or the EU member states alone have competence to speak, and if applicable to vote, on the agenda items.[226]

4.2.4.4. BIOTECH REGULATION IN CODEX

The subject of biotechnology regulation first came before Codex during the early 1990s, following the creation of a joint WHO/FAO expert consultative group that provided a report in 1990 on biotechnology and food safety. Since then, a number of Codex committees and a special Task Force on Foods Derived from Biotechnology have addressed biotechnology related food-safety concerns. The major issues raised in these various Codex sub-groups include the conduct of risk assessments of foods derived from biotechnology, the use of "other legitimate factors" besides science in risk management, the use of the precautionary principle in food safety regulation, and the use of labeling and traceability requirements for bioengineered foods. Of all the biotechnology related issues discussed in Codex, the issue of how to conduct risk assessments, handled by the Task Force, was most often characterized by deliberation in light of the technical, science-based issues in question. In contrast, discussions of risk management policy issues involved hard bargaining from relatively fixed positions.

As with the work of the OECD, much of the work of the Codex *does* reveal evidence of genuine deliberation over technical, science-oriented matters. For example, the Codex Commission established a Task Force on Foods Derived from Biotechnology in 1999 for a four-year period, which was quite successful and thus reestablished for a further four-year period in 2004. Discussions in the Task Force during its first mandate led to the adoption of agreed *Principles for Risk Analysis and Guidelines for Safety Assessment of Foods derived from Modern Biotechnology*, including two technical guidelines and a special annex for the conduct of food safety assessments. The principles reflect and recapitulate what was negotiated for general Codex principles for risk assessment and risk management, applying them to the case of modern biotechnology. The guidelines and annex, in contrast, go into considerably more detail regarding technical issues involving the conduct of risk assessments of foods derived from biotechnology, building on the concept of substantial equivalence.[227]One guideline addressed the assessment of foods derived from genetically engineered plants and the other foods derived from genetically engineered microorganisms. The annex covered food safety assessments for allergenicity. Now that many countries are devising regulations on agricultural biotech varieties for the first time, these Codex food safety assessment guidelines will become a first point of reference, helping to shape national regulations.

The Task Force's work demonstrates how deliberation toward consensus can work within Codex. The Task Force delegated drafting to three sub-working groups to develop texts respectively on principles, analytical methods, and allergenicity, and it "relied on three well-funded joint FAO/WHO expert consultations" on plants, allergenicity, and microorganisms, working toward consensus in a "relatively straightforward" manner.[228] The 2002 Codex evaluation report highlights that, in contrast to other Codex biotech work,

> [t]he development of these texts on principles, plants and micro organisms is an example of consensus building in a very short time period.... The scientific basis of risk assessment means that the established methods could be easily transferred to both the guidelines for plants and micro organisms.... [T]he overall success of [the Task Force] is undeniable and may be attributed to the restricted terms of reference, science-base of the discussion, established international definitions and methods of risk assessment, technical input resources and focus on food safety for human health.... [P]art of the explanation is undoubtedly that the issue was one of science not culture.[229]

In light of this success, the Task Force recommended that its mandate be extended for another four years to continue its work on guidelines for evaluating the safety of foods from biotechnology. At its first meeting, the Task Force agreed to prepare a guideline for food safety assessments of foods derived from r-DNA animals, complementary to the existing guidelines prepared by the Task Force on foods from r-DNA plants and foods from r-DNA microbes, and an annex to the Plant Guideline on foods from r-DNA plants modified "for nutritional or health benefits.[230] At the second meeting of the Task Force, the US, pressed by the US biotech industry, was able to add to the Task Force agenda a project to address the handling of low-level content of GM varieties that adventitiously appear in traded goods. The Codex guideline, in fact, is scientifically similar to the early food safety guidance developed by FDA, which also focuses on evaluating the safety of new proteins in GM plants. The development of the FDA guidance was part of a broader US government effort to address GM crops grown in research trials, both to minimize the likelihood that material from such crops would enter the food supply and to avoid any adverse food safety consequences if such material were to enter the food supply at low levels. This effort was prompted in part by the Starlink scandal discussed in Chapter 6, even though that scandal involved commercial GM corn approved only for animal feed adventitiously appearing in corn chips and tortillas.

Given that the US has typically been the first to authorize GM varieties for commercialization (before they have been approved elsewhere), which varieties are often mixed inadvertently in low quantities by grain distributors with other varieties, the US was eager to promote international standards to help avoid or facilitate resolution of the ensuing international trade conflicts. The EU and other countries initially resisted, noting that such guidelines could be viewed as "legitimating accidents."[231] To minimize these implications, the EU insisted that the guidelines would be limited only to GM varieties that had been authorized for consumption in a Codex member (as opposed to those that had been authorized nowhere for human consumption, as in the Starlink case) and would focus on whether the risk assessment could be expedited in these cases. The EU made clear that any adventitious presence of unauthorized GM material "at any level" was illegal under EU law and that the working group should also focus on "strengthening data and information systems."[232]

After much discussion, the working group, which was co-chaired by the US, Germany, and Thailand, prepared draft guidelines regarding "food safety assessment in situations of low levels of recombinant-DNA plant material in food," where the GM variety is commercially authorized in one country, but not in the recipient country. The result was a compromise that did not meet US aspirations, as the guidelines only slightly simplify the recommended approval process for GM varieties that have been approved elsewhere, recommend no minimum threshold to facilitate trade, and are only voluntary. Moreover, on the EU's insistence, these simplified safety guidelines are to be accompanied by a new Codex database that will help countries identify low-level content of GMOs that they have not approved so that they can more easily block their importation.[233] Importing countries will now more easily be able to obtain the material and tests to detect the presence of GM varieties in bulk shipments that they have not approved for consumption.

The Codex process has encountered particularly severe difficulties in addressing issues that implicate risk management policy over transgenic varieties. Three Codex sub-groups have addressed them: the Committee on General Principles (regarding the use of the precautionary principle and "other legitimate factors" besides science in risk management); the Committee on Labeling (regarding the labeling of GM foods); and the Committee on Food Import and Export Inspection and Certification Systems (regarding the issue of traceability). Here, in particular, we find little evidence of deliberative decision-making. Rather, we find arduous negotiations between the US and EU, each of which again put forward distinctive and sharply opposed proposals for international standards and guidelines on issues that could have a direct bearing on the application of the SPS Agreement to national regulatory measures and in particular in the WTO biotech case initiated in 2003.

As regards the application of the precautionary principle, the Codex Commission charged the Committee on General Principles to devise principles on risk analysis within the Codex standard-setting framework, with the EU and the US each advocating their traditional approaches to the question. The EU sought an expansive definition of the precautionary principle as applying to both risk assessment and risk management. The US delegation, in contrast, noted pointedly that, "a precautionary approach was already built into risk assessment; this concept should not be used by risk managers to overrule risk assessment."[234] After nearly four years of debate, the Codex Commission finally agreed to a compromise text that acknowledged that "precaution is an inherent element of risk analysis" within the Codex framework, while offering little clarification about its use at either the national or international levels.[235] The committee is still engaged in difficult discussions regarding the use of precaution by governments in risk analysis, but this discussion remains at the "second stage" of Codex standard-making, and there is little sign of consensus.[236]

The US and the EU also clashed in predictable fashion on a second cross-cutting issue affecting GM regulation, namely the guidelines for the invocation of "other legitimate factors" (dubbed OLFs) besides science that could be invoked by Codex in establishing international standards. Following domestic EU law and practice, EU members argued that the Codex should consider a range of OLFs such as consumer concerns and animal welfare. Consumer concerns lie at the heart of the EU's regulatory response to GMOs, and European concerns over the welfare of transgenic animals could further restrict US food exports in the future. The US thus vociferously argued that "giving consideration to...these factors could open a Pandora's box," and therefore sought to restrict Codex decision-making to scientific considerations.[237]

Once again, the members eventually reached a vague compromise on an addition to the Codex Manual of Procedure of "Criteria for the Consideration of Other Factors" in Codex decision-making. However, the criteria did not indicate what constituted an OLF, noting instead that "only those other factors that can be accepted on a world-wide basis, or on a regional basis in the case of regional standards and related texts, should be taken into account in the framework of Codex." The statement further constrained the use of OLFs by stating that other relevant factors "should not affect the scientific basis of risk analysis" or create "unjustified barriers to trade," with a footnote noting that it should be done "according to the WTO principles, and taking into account the particular provisions of the SPS and TBT Agreements."[238] In effect, the compromise reached within the committee allowed for the invocation of unspecified OLFs in Codex decision-making, but proceeded to limit that invocation with a series of vaguely stated restrictions. When the Task Force on Foods Derived from Biotechnology likewise addressed the use of "other legitimate factors" in risk management, the Principles simply reiterated that risk management should be proportionate, based on a risk assessment, and, "where relevant, [take] into account other legitimate factors in accordance with the general decisions of the Codex Alimentarius Commission, as well as the Codex Working Principles for Risk Analysis."[239]

Finally, the US and EU engaged with other Codex members in contentious negotiations over the issues of labeling and traceability of GM foods. The US, EU, and other members have battled in the Committee on Food Labelling and elsewhere within Codex for over 15 years regarding the labeling of foods derived from biotechnology, but to no avail.[240] The US argues that Codex should limit its focus to voluntary labeling covering only foods that are not "substantially equivalent" to conventional counterparts. It contends that mandatory labeling of foods derived from biotechnology may lead to consumer deception regarding the foods' safety. European delegations, in contrast, argue in favor of mandatory labeling guidelines for all biotech-derived foods, whether or not they are "substantially equivalent" to conventional varieties. They maintain that such labeling respects a consumer's right to know.[241] When members agreed to the 2003 *Principles and Guidelines on Foods Derived from Biotechnology*, the document conspicuously did not address the labeling of GM foods because US and EU positions proved impossible to reconcile.[242]

The US and EU still battle over these issues within the Committee on Food Labelling, and the Committee remains deadlocked. The US, supported by Argentina, Australia, Mexico, Paraguay, the Philippines, and Thailand, again maintained in 2006 that labeling of GM-derived foods should be voluntary and should be limited to foods that are substantially different from their conventional counterparts in terms of composition, nutritional value, or allergenic content. The EU, supported by Japan and many other countries, again called for process-based labeling in which all GM-derived foods would be labeled. The Canadian chair of the talks proposed a compromise in which "mandatory" labels would apply to GM foods substantially different from their counterparts, while "optional" labels could apply to all other GM foods. This proposal was put to a Codex working group, which reported back to the committee in 2006.[243] The parties, however, could not reach consensus, and the US and Argentina recommended that Codex discontinue its work on this issue.[244]

Not surprisingly, the Codex evaluation report points to GM labeling as "one of the most difficult issues Codex has addressed, where there appear to be intractable differences between country positions...and progress (if any) has been painfully slow...." It points out that "[t]his particular issue reflects a broader difficulty in international harmonization when cultural differences among countries mean that consumers have different interests and priorities." As a result of this experience and in light of Codex resources, the report recommended that Codex prioritize work that has "an impact on consumer health and safety," and put the lowest priority on "informational labeling relating to non-health and non-safety issues."[245]

Similarly, as regards traceability as a risk management tool, the US and EU again clashed. The EU and European members maintained that traceability had a two-fold purpose: to facilitate product recalls where foods have been found to be hazardous, on the one hand, and to facilitate verification of the authenticity of labeling statements, on the other. The US "strenuously opposed the European approach to traceability," maintaining that traceability for consumer protection purposes in respect of labeling was not appropriate for biotech-derived foods, arguing that traceability systems are disproportionately costly for such a purpose.[246] On these grounds, it blocked within the Task Force on Foods Derived from Biotechnology language that would support the use of traceability as a tool to inform consumers of the ingredients of food. In the ensuing 2003 *Principles and Guidelines on Foods Derived from Biotechnology*, the Codex members were only able to agree on vague language that acknowledges the use of "tracing of products" as a risk-management tool. The parties, however, again added a footnote that any tracing requirements for risk management "should be consistent with the provisions of the SPS and TBT Agreements," indicating once more the reach of the WTO in decision-making in other fora.[247]

European countries were successful, however, in including the concept of traceability for consumer protection in the adoption of general Codex *Principles for Traceability/Product Tracing as a Tool within a Food Inspection and Certification System*, adopted in 2006. These principles provide that such a tool "may apply to all...stages of the food chain," including to "reinforc[e] confidence in the authenticity of the product and the accuracy of information provided on the products" such as for "organic farming."[248] Although the US was able to block specific reference to biotech products, the general principle is now included, representing at least a partial victory for the EU. However, as part of a compromise, the US and other agricultural exporting states were able to specify in the text that a traceability system "should not be mandatory for an exporting country to replicate," "should not be more trade restrictive than necessary," and "should be practical, technically feasible and economically viable."

4.2.4.5. CONCLUSIONS ON CODEX

The results of these negotiations have thus far proven to be disappointing in terms of producing any consensus. As Sara Poli states, "After lengthy discussions, these conflicts led to poor compromises, which do not have practical impact on the activity of the Codex Commission."[249] Like the paragraphs of the Cartagena Protocol dealing with the relation between Cartagena and WTO law, much of these Codex texts simply paper over rather than settle the differences among the parties, potentially delegating clarification of these issues, if at all, to the WTO dispute-settlement system. Rather than hard law and soft law working in coordination toward genuine "problem-solving," the hard law of the WTO has constrained and to some extent "hardened" what was supposed to be a flexible, "voluntary" process for harmonized rule-making and guidance to facilitate trade in agricultural products. Strategic bargaining in defense of trade interests has often replaced technical discussions. As Victor writes, we are now more likely to see "dueling experts," reflecting US adversarial legalism, than "independent expert panels" working collaboratively to "synthesize complex technical information" as in Joerges' image of deliberative supranationalism.[250] As Veggeland and Borgen add, we now see a "replication of WTO coalition[s] and positioning pattern[s] in the Codex."[251] An organization in which decision-making was formerly based primarily on a "logic of arguing" or deliberation has been transformed to one more frequently based on a "logic of consequentiality" or bargaining.

Yet by raising the profile of Codex, the WTO has not only made countries more circumspect of negotiations in it, but it has also pressed Codex to become more accountable to an array of participants outside the traditional epistemic food standards community, from trade representatives to consumer groups. As Victor writes, "the impression that international standards are now more relevant has also entrained new actors—such as consumer protection organizations—into the process."[252] Michael Hansen of Consumer's Union confirms that Codex has indeed "opened up from being a backwater where industry dominated."[253] The addition of these new voices may result in consensus breaking down, especially in controversial cases, where votes used to matter less, but when reached, the consensus is more likely to take account of broader constituencies, both in terms of countries and interest groups.

On less politicized matters, the organization has been able to continue its technical work, as shown by the Codex Principles for the Risk Assessment of Foods Derived from Modern Biotechnology. As the Codex evaluation report states, "comparison with the pre-Uruguay Round level of standard setting activity does not show that there is actually any slowing down. Rather, the expectation of accelerated decision-making has not been realized and there is frustration with the normal pace of activity."[254] Less controversial technical work appears to be proceeding as before, albeit with greater awareness of the stakes.

Moreover, officials within the WTO and Codex regimes have increasingly worked together to ensure greater coherence so that one should not view the systems as completely "fragmented." As Joanne Scott notes, the SPS committee's periodic review of national measures has helped forge "a horizontal access between the regimes," serving "to enhance the transparency of their activities, and critical oversight thereof."[255] Codex and other standard bodies attend SPS committee meetings as observers. The secretariat that services the SPS committee also provides feedback to Codex on behalf of WTO members, bringing to its attention issues that they believe should be clarified, as well as standards that may need to be updated.[256] A good example of recent coordination between the SPS committee and Codex is the dispute over sulphur dioxide (SO_2) residues in cinnamon between the EU and Sri Lanka. The EU had banned Sri Lankan imports of cinnamon with residues of SO_2, although it permitted low levels of residues in other spice imports. Sri Lanka brought its concerns to the SPS committee whose secretariat notified Codex that it had no applicable standard. Codex members responded by agreeing within three months to a new standard for SO_2 residues in cinnamon. The EU in turn lifted its measure and provided some technical assistance to Sri Lanka.[257]

Codex principles also continue to be important for panels in WTO cases. The WTO panel in the GMO case, which we examine in Chapter 5, repeatedly referred to Codex (as well as some IPPC and OIE) standards, principles, and guidelines. In total, the WTO panel cited them in thirty-six separate paragraphs of the panel report (not including numerous citations in footnotes). It did so when interpreting the meaning of such terms used in the SPS Agreement as "additive," "contaminant," and "risk assessment." It also did so to support one of its central findings regarding the relation of "uncertainty" to a "risk assessment." The panel cited the Codex *Working Principles for Risk Analysis* which maintains that "[t]he responsibility for resolving the impact of uncertainty on the risk management decision lies with the risk manager, not the risk assessors."[258] This principle supported the panel's finding that the documents cited by the EU in support of the EU member state safeguards did not constitute a "risk assessment," and thus the safeguards violated article 5.1 of the SPS Agreement that measures be based on one, as we will see in Chapter 5.

Deliberation has by no means disappeared within Codex on account of the rise of the WTO. Where Codex members retain a focus on matters involving scientific risk assessment, working with the assistance of expert consultative reports and through committees and working groups, the system continues to work relatively well. At times, decisions can be made quite quickly, including through Codex–SPS committee coordination, as in the above example regarding residue limits on Sri Lankan cinnamon. Governance through networks of government officials and members of different international secretariats *can* work.

Moreover, even where the WTO's legal system has significantly affected Codex, one cannot simply conclude that it has been for better or for worse. Rather, from a normative perspective, increased participation and contention in Codex can be viewed as positive when these same issues are controversial within and between states. Indeed participation of nation-states and NGOs has increased within Codex.[259] Some decisions in Codex may be less flexible than in the past, but they are also likely to be made in a more transparent manner because of state and broader interest-group demand. Codex's soft law system has simply had to respond to a new international context that includes a more legalized international trading system. This has informed the updating and focus of Codex's technical work, as well as its handling of matters that are particularly politically controversial. Any "consensus" that might otherwise have been reached in Codex's pre-WTO days would have been both less informed in its decision-making process and misleading in what such consensus might suggest.

Nonetheless, as regards the norm of deliberation, where decisions within Codex regard politically salient policy-oriented issues, such as the "risk management" of food products derived from biotechnology through labeling, traceability, the use of precaution, and the taking into account of non-science factors (such as socio-economic concerns), we see little or no evidence that negotiations are governed by a "logic of arguing" among like-minded technocrats. As one FAO official summarizes, the SPS Agreement has resulted in a "slowing [of] the Codex process for some controversial standards," and "the Codex tradition of arriving at consensus has...broken down in controversial cases."[260] Codex's engagement with the risk management of foods derived from agricultural biotechnology is *the* prime example of this breakdown.

4.3. Conclusions: Distributive conflict, regime complexes, and the limits of cooperation

Just as with the US–EU bilateral networks that we examined in Chapter 3, US–EU cooperation in multilateral fora has delivered largely disappointing results from the perspective of conflict resolution. Here again, it was hoped that multilateral regimes might lower the transaction costs of negotiations, enhance monitoring and enforcement, and thus improve compliance with international agreements. Just as importantly, both scholars and officials hoped that institutionalized cooperation in networks of national and supranational experts would result in deliberative decision-making, policy learning, and a reasoned consensus about the central issues in the regulation of agricultural biotechnology. Yet while the issue of agricultural biotechnology was taken up within several multilateral regimes, including the OECD, the WTO, the CBD, and the Codex Alimentarius Commission, effective cooperation and deliberation that might lead to consensus were impeded by distributive conflicts among the world's two economic powers (the US and the EU) and by tensions among the overlapping regimes themselves.

As we have seen, the intense politicization of the issue, the entrenchment of two sharply divergent regulatory systems governing the world's two largest economies, and the prior failure of deliberation at the bilateral level meant that the various multilateral negotiations on agricultural biotechnology resembled a Battle of the Sexes game, in which each side sought common international standards *on its own terms*. While we did find some evidence of deliberative decision-making, most notably within the OECD and the Codex Alimentarius

Commission, we also found that such deliberation was most likely where negotiations focused on issues of scientific risk assessment, where the negotiators were themselves scientific and technical experts, and where committee or network deliberations were insulated from domestic public opinion and from the WTO dispute-settlement system.

Once again, however, where the subject has turned from risk assessment to risk management, where international trade negotiators played a key role in negotiations, and where negotiations took place in the shadow of domestic public opinion or WTO dispute settlement, negotiators reverted to a logic of consequentiality, bargaining from essentially fixed positions and seeking international codifications of established domestic positions. Furthermore, consistent with Krasner's analysis of Battle of the Sexes, the negotiating power of the various actors was clearly important, with the US and the EU playing dominant, as well as diametrically opposed roles in each of the four regimes surveyed in this chapter. Given the roughly equal economic power brought to bear by the two sides, however, the result in each case was not a clear victory for either side, but a series of muddled compromises that papered over rather than resolved the fundamental transatlantic differences.[261]

In addition to the constraints arising from distributive conflicts, effective cooperation was further hampered by the difficulty of negotiating simultaneously in the various regimes governing agricultural biotechnology. As we have seen, both sides have actively "forum-shopped," with the US clearly preferring the WTO with its emphasis on trade and its strong dispute-settlement system, and the EU preferring cooperation within the CBD, whose broad membership and strong environmental emphasis appeared to favor the EU's more precautionary approach. These differences led in turn to inconsistencies among the various regimes and to efforts by each side to assert the primacy of its own preferred regime. In no case, however, was a clear hierarchy of regimes established, leaving enormous uncertainty among governments, producers, and consumers about the definitive global rules for the regulation of GM foods and crops.

The ability of states to bring their distributive conflicts to multiple regimes within a fragmented international law system, in turn, calls into question the characterization of hard and soft law as progressively interacting. They can do so, as witnessed by our discussion of the resolution of disputes within the SPS committee, of the committee's successful reference of matters to Codex for standard-setting, and of the WTO dispute-settlement panel's use of Codex definitions of risk assessment terms for the interpretation of WTO obligations. But when matters become highly politicized and their resolution has distributive consequences, hard and soft law regimes can work at cross-purposes. Hence, as we have seen in the agricultural biotech case, soft-law regimes like the Codex can become effectively "hardened" through linkage to a hard law regime like the WTO, losing many of the traditional advantages of soft law such as flexibility, speed, and the potential for deliberation. The hard law regime of the WTO, by contrast, has arguably been "softened" by its linkage to overlapping regimes (e.g., the Biosafety Protocol), constraining the operation of what looked to be hard, binding WTO rules, as we will address further in Chapter 5.

Yet the fact that different international law regimes may be at odds with each other should not necessarily be lamented. The tensions simply reflect underlying differences among states and social constituencies in a diverse, pluralist world. In such a world, overlapping regimes can provide an important service to each other. The operation of softer law regimes, for example, can signal states and WTO judicial decision-makers to tread softly when applying WTO law, and in particular, the SPS Agreement to disputes over national risk regulations. They prompt internal responses within the WTO regime to preserve its own social and political legitimacy. Although Benvenisti and Downs are correct that the existence of multiple regimes favors powerful states which are best positioned to forum shop effectively,[262] regulatory disputes, as in the GMO case, also frequently pit powerful states against each other, putting international regimes in a particularly delicate situation.

In the context of agricultural biotechnology regulation, the CBD's Biosafety Protocol has provided a counter-voice which can protect the WTO system from the trade law regime's relative insularity from global politics.[263] Even if WTO panelists do not directly invoke the Protocol's provisions or even if they state that the provisions do not apply because the US is not a party to the Protocol, panelists can implicitly take the Protocol into account. They can do so through their appreciation of the political stakes of alternative interpretations of WTO rules (from a rationalist perspective) and through the Protocol's impact on the framing and broader social understanding of the issues (from a constructivist one). The existence of the Protocol can affect the interpretation of WTO legal provisions and the appreciation of the underlying facts of the case to which WTO law is applied.[264] Moreover, it is simply not in the interest of the WTO as an organization to ignore the content of an international environmental agreement, especially one having over 140 parties.[265] As we will examine in Chapter 5 when we address the WTO panel report's response to the US challenge to the EU's biotech regime, the WTO's legitimacy will be under greater threat if it makes decisions in a closed environment oblivious to non-trade norms and perspectives. In such a case, the WTO will receive feedback, but only *after* the panel issues the decision. That feedback may be much more deleterious to the WTO judicial system's legitimacy and authority. Overlapping and interacting international regimes reduce this risk by flagging concerns to which WTO judicial bodies might otherwise be blind.

5

WTO Dispute Settlement Meets GMOs: Who Decides?

Throughout the trade-related conflicts described in the previous chapters, the US consistently brandished, and the EU girded itself for a possible US challenge to the EU's agricultural biotechnology policy before the WTO.

For many years, both the Clinton and George W. Bush administrations had resisted bringing a WTO challenge over GMOs, hoping to address the dispute diplomatically and without the risks of global litigation. In 2003, however, the US, joined by co-complainants Canada and Argentina, finally filed suit against the EU before the WTO, challenging the EU's *de facto* moratorium and its various safeguard bans, and in November 2006, the WTO Dispute Settlement Body adopted the WTO panel's ruling in the case, which was largely in favor of the complainants.

In this chapter, we examine the WTO case and the panel ruling in detail, asking why the US brought its case after waiting so long; on what grounds the US chose to challenge EU policy; how the WTO panel adjudicated this politically charged and legally complex case; and what difference the ruling may make on the law and politics of ongoing GMO disputes. We focus in large part on the decision of the panel, asking not only who won the case, but how the panel chose to interpret WTO law within broader political, institutional, and normative contexts. We show that the WTO panel was placed in a difficult position in light of the political saliency of GMOs and challenges to the legitimacy of its authority. We nonetheless address the positive role of WTO dispute settlement even in this politically charged area, and examine the impact of the WTO legal process in the handling of the US–EU (and now global) dispute over the regulation of agricultural biotechnology.

This chapter proceeds in six parts. In part 1, we assess why the US delayed bringing a complaint before the WTO for so long despite the financial implications of the EU regulatory restrictions which have been much more significant than those in the earlier US challenge regarding the EU's ban on the sale of meat produced with growth hormones. Choices to use WTO law, as we show, need to be seen in political and social context. In part 2, we present the complainants' claims and the EU's defenses in the GMO case under applicable WTO rules. In part 3, we summarize how the SPS Agreement had been previously interpreted in relevant WTO jurisprudence, arguably softening the impact of some of its provisions. In part 4, we turn to the complex and (for many) controversial interpretive moves made by the WTO panel in a decision of over 1,000 pages that was adopted without appeal by the WTO Dispute Settlement Body on November 21, 2006. As we shall see, the panel largely took a procedural approach in its decision, finding that the EU had engaged in "undue delay" in its approval process and that member states had failed to base their national bans on a risk assessment as required not only under WTO law, but as confirmed by the EU's own risk assessment body EFSA. Despite these elements, which are often (and correctly) cited as a victory for the US, the panel made interpretive moves that enabled it to avoid examining the EU's moratorium under the substantive obligations imposed by the SPS Agreement, in effect "deciding not to decide" these questions.

In part 5, we examine the panel's decision within the analytic framework of comparative institutional analysis, showing how the interpretive choices at the panel's disposal have significant institutional implications. In this section, we take a normative turn, and in doing so, highlight the difficult institutional choices faced by the panel from a *governance* perspective. We show how judicial bodies (and legal scholars) in interpreting WTO texts implicitly make institutional choices, although they are typically not explicit about them. This section examines how these judicial interpretive choices effectively allocate institutional authority for balancing competing policy concerns to different institutional processes, such as the market, political and administrative bodies and courts, and how they involve allocations at different levels of social organization, from the local to the global. Each of these institutional decision-making processes has its attributes and deficiencies in terms of the dynamics of participation within it, ultimately affecting *who decides*. In other words, ultimately the WTO judicial process does not alone decide these issues (what we term to be an illusory judiocentric perspective), but does so through its impact on other decision-making processes. In the case of the GMO decision, we argue, the panel largely took a *procedural turn*, acknowledging the role of WTO's members in regulating risks, but subjecting these members to various obligations pursuant to which they must justify their decisions where they have impacts on foreign parties.

Finally, in part 6, we conclude the chapter by assessing the potential role, if any, of the WTO dispute-settlement system in helping to resolve the US–EU regulatory conflict. We contend that, despite the evident risks of further politicization of the conflict and of potential EU noncompliance with the adopted panel report (even though the report decided only in part in favor of the complainants), the WTO case offers the possibility not so much of convergence or harmonization, than of clarification of the two sides' legal obligations under WTO law and potential mutual accommodation.[1] We show how WTO law and its dispute-settlement system have served to channel political disputes into defined legal parameters, empower domestic constituencies with an interest in compliance with the WTO law and legal decision in question, and encourage domestic regulators to operate more transparently, taking into greater account the effects of their actions on foreign parties.

5.1. Explanation of the delay and initiation of the US complaint

The US government, responding to pressure from US farm associations and agricultural biotechnology companies, was long frustrated with the EU's restrictions on GM crops and foods. Although the US often threatened to bring a complaint before the WTO—its preferred international forum for dealing with GM issues—it delayed doing so for years. US forbearance finally gave way, however, in May of 2003 when following a slight further delay in light of US–European tensions over the 2003 invasion of Iraq, the Bush administration initiated requests for WTO consultations over the EU's *de facto* moratorium on new approvals and over member-state bans on EU-approved varieties.[2] The US filed its request for WTO consultations twelve days after President Bush declared the end of "major combat operations" in the second Iraq war, on May 1, 2003, from the aircraft carrier *U.S.S. Abraham Lincoln*. In August 2003, the US requested the establishment of a WTO dispute-settlement panel, which was formed in March 2004.

What first interests us is why the US declined for so long to bring a legal case against EU restrictions on GM products. Unlike in its earlier challenge against the EU's ban on hormone-treated beef under the General Agreement on Tariffs and Trade (GATT) and the WTO, the US refrained from bringing a WTO claim against the EU for years, preferring to undertake the bilateral and multilateral negotiations examined in Chapters 3 and 4.[3] It appeared to question the usefulness of pressing its case before the WTO judicial process, wary of wielding law, arguably drafted in its favor, to advance its aims. From a review of this dispute over time, we find that the US initially chose these less aggressive diplomatic routes for four primary reasons: (i) US authorities, in consultation with US industry, understood that EU authorities were severely constrained by EU consumers' and member-state politicians' demands, and believed for a long time that bringing a WTO legal case might be counter-productive in Europe; (ii) there was some risk that media coverage of a WTO dispute over GMOs could raise the notoriety of GMOs among US consumers and, in doing so, adversely affect their commercial prospects within the US itself; (iii) following the mass demonstrations at the WTO ministerial meeting in Seattle, US authorities were reticent to initiate a controversial new trade case involving consumer health and environmental protection that could undermine the WTO's authority; and (iv) European concerns over GM foods appeared to have growing political support around the world, as represented by the January 2000 signature of the Biosafety Protocol and new labeling requirements imposed by other countries. The US waited to see the fallout of these developments. A WTO case, especially one over such a politically charged issue, does not arise autonomously of social and political context. We examine each of these issues below, together with the developments that eventually led the US to bring the WTO case.

5.1.1. US recognition of European consumer opinion and the potential of a populist backlash

US authorities and affected US industries recognized that the EU stance was a populist one and that EU authorities' hands had been tied. US industries did not want to be seen as forcing genetically modified foods on European consumers since the market backlash could be severe, with an increasing number of brand food companies and retail chains forsaking products with genetically modified ingredients. US authorities and companies rather hoped to work with EU authorities and EU scientists over time to convince the European public that genetically modified foods are safe and can be beneficial to human health and the environment. Yet US commercial constituencies became increasingly concerned over the slow pace of change in the EU, and increasingly pressed the US government, in particular through Congressional representatives, to bring the case.[4] Patience had not paid off, as the EU moratorium continued. There seemed to be little to lose from bringing a case, at least in terms of the prospects of opening the EU market to GM products.

5.1.2. Spillover effects of EU policy in the US

It at first appeared that the media attention given to the US–EU dispute over European restrictions on GM foods could affect the political and commercial playing fields within the US. US consumers seemed to be unaware that they were eating foods derived from GM varieties, and a major international trade case could increase their awareness. Such a change could, in turn, potentially trigger demands for greater US regulation of GMOs. As we have seen, however, despite the increased media attention, consumer attitudes did not fundamentally change in the US, and concerns about an anti-GM backlash in the US eased. The policy space within the US was thus relatively clear to bring a WTO case.

5.1.3. The fallout of the anti-WTO Seattle demonstrations

The November 1999 demonstrations in opposition to the launching of a "millennium round" of trade negotiations in Seattle, undercut US strategies to place legal pressure on EU authorities through the WTO. Among the protestors, anti-GMO demonstrators dressed in monarch butterfly costumes and sported other monarch butterfly images in reference to the dangers, they felt, pesticide-expressing GM varieties pose to this and other species.[5] The riots in Seattle forced a cancellation of the ministerial's first day of meetings and ultimately the collapse of the ministerial without a mandate for a new round of negotiations. By joining an anti-WTO coalition that rallied a number of constituencies, anti-GMO activists could help foment opposition to the Clinton administration's hopes to obtain "fast-track" trade negotiating authority from Congress, and, ultimately, adversely affect Vice-President Gore's candidacy in the November 2000 presidential elections. Anti-GMO activists' votes could go to Ralph Nader, the Green party's candidate. Moreover, US commercial interests—including telecommunications, industrial property industries, and other "new economy" sectors—did not wish the US to trigger further opposition to trade liberalization endeavors on account of another WTO lawsuit over food safety and environmental protection. This was not an opportune time politically for a US Democratic administration to commence a legal challenge against the EU's trade restrictions on genetically modified products.

By August 2002, however, the new Bush administration had received Trade Promotion Authority, the Doha negotiating round had been launched, and anti-WTO protests had somewhat mollified in the wake of the September 11, 2001 terrorist attacks.[6] Politicians, activist groups, and the media switched their primary attention from economic globalization to security issues, and in particular to the US-led invasion and occupation of Iraq and the "war on terror." Moreover, were a GMO case to have any effect in the US presidential election in 2004, it would likely again draw votes toward a Nader candidacy and away from the Democratic party, to the benefit of President Bush's reelection. Although the political orientation of the US government may ordinarily play little role in the bringing of a WTO dispute, the GMO case was a high-level one with clear political ramifications in which the Bush administration was less fearful than the Clinton White House that a WTO case could upset its other trade policy and political goals.

5.1.4. Foreign and international developments

Finally, as we have seen, the anti-GMO movement was moving beyond the EU to both advanced industrialized and less developed countries. Among the former, Japan, Korea, New Zealand, and Australia joined the EU in adopting more stringent agricultural biotech regulations, while the majority of developing countries, generally concerned over the expansion of patent and other rights over seeds and plant varieties, supported a move toward trade-restricting regulation of GM products, resulting in the Biosafety Protocol. Yet by 2002, it was clear that the US also had a number of allies in its challenge against European trade restrictions on GMOs. Notwithstanding the controversies over GM foods and crops in Europe and elsewhere, global plantings of GM crops increased dramatically, rising by around 15 per cent in 2003 and 20 per cent in 2004, as well as an additional 11 per cent in 2005 and 13 per cent in 2006.[7] In 2005, it is reported that "the billionth acre was planted, and growth rates remain in double digits."[8] Just as importantly, the US, while still the largest grower of GM crops, was joined by an ever-growing number of other major producers.[9] These new producers included a number of less developed countries (including China, India, Argentina, Brazil, and South Africa), as well as advanced industrialized countries (in particular Canada, and with around 60,000 hectares of Bt maize being planted in Spain in 2006, within the EU itself). Many of these countries came to support the US view that exports of GM foods and crops should not be impeded by overly strict national regulations. When the US ultimately brought a case before the WTO, it was joined by Canada and Argentina as co-complainants, with Egypt dropping out at the last moment.[10]

Overall, it appears that the US finally initiated the WTO complaint because of a combination of increased frustration of US commercial constituents over lost sales to Europe; US commercial concerns over the impact of EU regulatory restrictions on regulatory developments in third countries; US commercial concerns over future GM varieties that companies hoped to introduce in the US and to market abroad; and hopes that a WTO case could favorably influence regulatory debates within the EU and globally over GMOs toward more "science-based" determinations. Since anti-WTO demonstrations had somewhat subsided, since the US administration had obtained trade promotion authority, since the Doha negotiating round was in process, and since a Republican administration had little to lose (and potentially could gain) politically within the US, there seemed to be much less of a downside from a WTO case compared to the risks of continued European intransigence and a proliferation of foreign trade restrictions on biotech products. Perhaps most importantly, by 2003 the administration had despaired—understandably, in light of the evidence presented in Chapters 3 and 4—of finding a solution suitable to it through bilateral or multilateral cooperation, or through developments in domestic EU politics. As USTR Robert Zoellick stated in January 2003, "We've tried to be patient, we've tried to work the system, we tried to pay attention to European political sensitivities. It's not moving. It's not being solved."[11]

5.2. The WTO complaint

In their May 2003 requests for consultations, the complainants Argentina, Canada, and the US limited their WTO claims to a challenge of the EU's *de facto* moratorium on approvals, and the EU member-state "national marketing and import bans" on those biotech products that had been approved. Agricultural trade associations within the US, led by the American Soybean Association, pressed the USTR to initiate a WTO challenge against the EU's labeling and traceability rules as well.[12] USDA officials maintained that EU labeling and traceability rules could cost US producers $4 billion in lost exports, in addition to billions in new infrastructure costs.[13] Law firms in Washington had a legal case ready to go.[14] The filing of such a complaint, however, appeared to depend on the outcome of the initial case.

The three complainants made their initial request for consultations under the SPS Agreement, the Agreement on Agriculture (Agriculture Agreement), the Agreement on Technical Barriers to Trade (TBT Agreement), and the General Agreement on Tariffs and Trade 1994 (GATT 1994). The US' written submissions focused on the provisions of the SPS Agreement, although the US "reserved the right" to bring claims under the TBT Agreement. Canada and Argentina also focused on the SPS Agreement, but they set forth cumulative and alternative claims under the TBT Agreement and under article III.4 of GATT 1994.[15] They did so because the TBT Agreement covers all "technical regulations" and "standards," including "labeling requirements," except for "sanitary and phytosanitary measures."[16] As we will see, it was unclear how the WTO panel would apply these agreements, especially where different aspects of the EU measures might be covered by both of them. For example, the EU measures might aim to protect both consumer health risks (covered by the SPS Agreement), and a consumer's right to information (covered by the TBT Agreement).[17]

The parties' claims were set forth in three parts, in which they respectively challenged the EU's "general moratorium," its "product-specific moratoria," and EU member-state marketing and import bans applied to biotech seeds and food. That is, they challenged the application of both EC Directive 2001/18 and its predecessor Directive 90/220 governing "the deliberate release into the environment of [GMOs]," and EC Regulation 258/97 regulating "novel foods and novel food ingredients."

The US made four primary claims against the general moratorium and the product-specific moratoria. First, the US maintained that the EU imposed "undue delay" in its product and marketing approvals, in violation of article 8 and annex C of the SPS Agreement. Second, it contended that the EU failed to "publish promptly" its "moratorium" in violation of article 7 and annex B of the agreement. Third, it argued that the general moratorium and product-specific moratoria are not based on *risk assessments* as required under article 5.1, thus also resulting in a violation of article 2.2 of the SPS Agreement.[18] Fourth, the US claimed that the EU applies arbitrary or unjustifiable distinctions in the levels of protection required for GM products in violation of article 5.5 of the agreement. In particular, the US noted the EU's less stringent regulatory treatment of products produced with "biotech processing aids," such as enzymes used in the production of European cheeses, which do not require regulatory approval under GM-specific legislation.[19] Interestingly, there apparently was a group within the US government that wished to limit the US challenge to procedural issues under articles 7 and 8 of the SPS agreement because of concern regarding US vulnerability to legal challenge, especially under the US Bioterrorism Act of 2002.[20] Some commentators, including the EU itself, indicate that the Bioterrorism Act lacks a risk assessment as required under the SPS Agreement and results in discrimination against foreign food imports.[21]

The US then challenged the nine "safeguard" measures adopted by six EU member states which ban the importation or marketing of biotech products that have been respectively approved under Directive 90/220 or Regulation 258/97.[22] The US maintained that these member-state measures were also not based on a risk assessment, as required under article 5.1. Moreover, in each case, the "EU scientific committees considered and rejected the information provided by the member States."[23] Finally, the US specifically challenged Greece's import ban under article XI:1 of GATT 1994. Article XI prohibits the use of quantitative restrictions, subject to the exceptions set forth in GATT article XX. Greece's measure expressly "prohibits the importing into the territory of Greece of seeds of the genetically modified rape-plant line bearing reference number C/UK/95/M5/1."[24]

5.3. Relevant WTO jurisprudence

As we addressed in Chapter 4, many commentators find that the SPS Agreement is problematic because of its constraints on national regulation which is non-discriminatory, but is not sufficiently based on international standards determined within the Codex, IPPC, and OIE (which are otherwise voluntary), or is not based on a sufficient risk assessment. The prevailing norm in the GATT had been that of non-discrimination but the WTO SPS Agreement's provisions extend beyond this norm, requiring, in Hudec's terms, national regulators to act "rationally" and have their legislative decisions assessed on these grounds by a supranational judicial body. Prior to the GMO case, however, the WTO Appellate Body had responded in its jurisprudence to concerns over the SPS Agreement by interpreting it to provide greater discretion for domestic regulatory policymakers than many had expected. In particular, the Appellate Body's interpretations have reduced the potential constraints of provisions of the SPS Agreement that require WTO members to "base" national measures on international standards, and to respond to risks in a consistent manner.

Regarding the first requirement, article 3.1 provides: "To harmonize sanitary and phytosanitary measures on as wide a basis as possible, *Members shall base their sanitary or phytosanitary measures on international standards*, guidelines and recommendations, where they exist, except as otherwise provided for in this

Agreement, and in particular in paragraph 3."[25] The Appellate Body's early juris-prudence appears to have circumscribed the potential reach of article 3.1 as an independent legal obligation. In the *EC-Meat hormones* case, the Appellate Body overruled the panel in respect of its interpretation of the phrase "shall base their [SPS] measures on international standards." The Appellate Body found that a party does not need to "conform to" these standards, and that such an interpretation (of the panel) would inappropriately vest international standards with "obligatory force and effect," transforming them into "binding norms."[26] As Joanne Scott writes, because the article 3 provisions on harmonization no longer appear to constitute independent, autonomous obligations, "the bite of international standards in the WTO is shown once again to be less fierce than many had anticipated...So sensitive is the AB to the sovereignty concerns of Member States in the face of international standards that, arguably, it has strained the meaning of the relevant texts in downplaying the authority of such standards."[27] Victor likewise concludes that "far from imposing a strict harmon-ization between national and international standards—which was the main fear of the Agreement's detractors—the Agreement," because of the Appellate Body's interpretations, has "actually allow[ed] diversity to flourish."[28]

Similarly, the Appellate Body, through its interpretation in the *EC-Meat hormones* case, appears to have constrained the potential reach of article 5.5 of the SPS Agreement which addresses the consistency of a state's regulation of risks. Article 5.5 provides:

> With the objective of achieving consistency in the application of the concept of appropriate level of sanitary or phytosanitary protection against risks to human life or health, or to animal and plant life or health, each Member shall avoid arbitrary or unjustifiable distinctions in the levels it considers to be appropriate in different situations, if such distinctions result in discrimination or a disguised restriction on international trade.

In the *EC-Meat hormones* case, the panel found that the EU violated this pro-vision on account of its differential treatment of natural and synthetic hor-mones when used for growth purposes as compared to "natural hormones occurring endogenously in meat and other foods."[29] Likewise, the panel held that the EU violated this provision when the EU banned the use of natural and synthetic hormones for growth purposes in beef, but did not restrict the use of carbadox and olaquindox (anti-microbial agents) mixed in feed given to piglets.[30] The Appellate Body, however, overruled the panel. First, in strong terms, the Appellate Body found the panel's ruling regarding naturally occur-ring hormones to be "an absurdity," since "there is a fundamental distinction between added hormones (natural or synthetic) and naturally-occurring hor-mones in meat and other foods." Second, although the Appellate Body agreed that the EU's differential treatment of carbadox and olaquindox was arbitrary, it overruled the panel and found in favor of the EU because the EU had not engaged in "discrimination or a disguised restriction of international trade" which is a separate requirement for an article 5.5 claim. In this respect, the Appellate Body noted "the depth and extent of the anxieties" in the EU concerning the use and possible abuse of growth hormones in meat, and in the process showed considerable deference to EU decision-making.[31]

Although these other provisions certainly remain relevant, a central issue in an SPS case has become how demanding the Appellate Body and panels will be on a member's basing its SPS measure on a "risk assessment," pursuant to article 5.1 of the agreement—a third requirement. Article 5.1 provides:

> Members shall ensure that their sanitary or phytosanitary measures are based on an assessment, as appropriate to the circumstances, of the risks to human, animal or plant life or health, taking into account risk assessment techniques developed by the relevant international organizations.

In the *EC-Meat hormones* case, the Appellate Body limited the panel's holding under article 5.1 as well, while this time finding against the EU. The Appellate Body stated that members may rely on minority scientific opinions, including in risk assessments conducted in other countries, and can take account of "factors which are not susceptible of quantitative analysis by the empirical or experimental laboratory methods commonly associated with the physical sciences."[32] In the *EC-Asbestos* case, the Appellate Body found that a country may decide to reduce the risk to zero through a ban where a substance is carcinogenic and thus life-threatening (as France had done). The Appellate Body also found that risk factors alone can differentiate products that are otherwise similar.[33] In the *EC-Meat hormones* and *Japan-Varietals* cases, the Appellate Body settled on a rational relationship test, stating that "whether there is a rational relationship between an SPS measure and the scientific evidence is to be determined on a case-by-case basis and will depend upon the particular circumstances of the case, including the characteristics of the measure at issue and the quality and quantity of the scientific evidence."[34]

A key issue for national regulators has thus become how stringent the Appellate Body will be in determining what constitutes a sufficient scientific risk assessment. Where the definition of a risk assessment is more stringent, such judicial review can become substantively intrusive in its effects.[35] The Appellate Body test for a risk assessment set forth in the *Australia-Salmon* case can be viewed as a relatively strong, substantive-oriented requirement because of the degree of specificity that it required. There, the Appellate Body found that a risk assessment in respect of pests and diseases must:

> (1) identify the diseases whose entry, establishment, or spread a Member wants to prevent within its territory, as well as potential biological and economic consequences associated with the entry, establishment or spread of those diseases;
> (2) evaluate the likelihood of entry, establishment or spread of these diseases, as well as the associated potential biological and economic consequences; and

(3) evaluate the likelihood of entry, establishment or spread of these diseases according to the SPS measures which might be applied.[36]

It then went onto state that "likelihood" means "the probability" of entry, establishment or spread. Although these requirements can be viewed as "procedural" ones in that panels are to focus on the modalities of the risk assessment, and not on the assessment's substantive findings, the line between procedural and substantive can be a fine one. As Alan Sykes has explained the Appellate Body's standard may be extremely hard to meet in many cases, having clear substantive implications for risk regulation.[37] A key question in the GMO case thus was—How demanding would the panel be regarding EU risk assessments of genetically engineered varieties?

5.4. The September 2006 panel decision: Deciding not to decide

In September 2006, the panel finally circulated its decision to WTO members, which was adapted without appeal in November 2006.[38] The decision was 1,087 pages in text, and over 2,400 pages when including annexes. The panel expressly avoided deciding (or in its words, examining) many issues, including

—whether biotech products in general are safe or not
—whether the biotech products at issue in this dispute are 'like' their conventional counterparts, [a]lthough this claim was made by the Complaining Parties.[39]

The panel found in favor of the US, Canada, and Argentina, but largely on procedural and not substantive grounds. In particular, in respect of the EU moratorium and product-specific moratoria, the panel only found that the EU engaged in "undue delay" in its approval process in violation of Article 8 and Annex C of the SPS Agreement. The panel avoided determining whether the EU had based a decision on a risk assessment or whether the assessments showed actual risks or greater risks than for conventional plant varieties. It did so by holding that the moratoria did not constitute "an SPS measure within the meaning of the SPS Agreement, but rather affected the operation and application of the EC approval procedures, which are set out in the relevant EC approval legislation and which [the panel] had found to be SPS measures."[40] In other words, the panel found that the EU legislation, which the complainants did not challenge, constituted an SPS measure, but that the EU practices, which they did challenge, did not. On this legalistic distinction, the panel decided against all of the complainants' claims against the EU moratoria other than the Article 8 claim for "undue delay." In this way, the panel could effectively *decide not to decide* as regards the substance of any regulatory measure at the EU level.[41]

In contrast, the panel found that all of the EU member-state safeguards against EU-approved varieties constituted SPS measures, and that these member-state measures failed to be based on a risk assessment and thus were inconsistent with the EU's substantive obligations under articles 5.1 and 2.2 of the agreement. The panel similarly found that the EU member states failed to comply with the SPS Agreement's version of a precautionary principle in article 5.7, which the panel characterized as providing a "qualified right" to implement temporary measures in situations of uncertainty, subject to certain requirements. In doing so, the panel implicitly supported the Commission's earlier position regarding the legality of the member-state bans under internal EU law, providing leverage to EU central authorities within the multi-level EU governance context.

We now examine each step in the panel's interpretation of the SPS Agreement's text to reach these results, highlighting their institutional implications in light of the interpretive options available.

5.4.1. Applicability of the SPS Agreement

The first key interpretive choice confronting the panel having institutional implications was whether the SPS Agreement applied. This first threshold issue was critical to the case because of the different legal requirements contained in WTO agreements. The SPS Agreement arguably contains more stringent provisions than the other potentially applicable WTO agreements in that it alone explicitly requires that measures be based on a scientific "risk assessment." In consequence, if the panel found that the SPS Agreement did not apply, then the panel likely would show greater deference to EU decision-making processes and thus have less input into them.

In order to demonstrate this point, we need to review briefly why GATT and TBT claims are likely to be less intrusive. GATT requirements focus primarily on whether a measure is discriminatory. For example, the EU would not have engaged in any discrimination in violation of GATT article III so long as GM and conventional varieties are found *not* to be "like products"—that is, so long as GM varieties are considered to be different than conventional varieties under a number of criteria, including consumer perceptions.[42] This is the case because the EU treats European-developed and foreign-developed GM varieties the same. Although the panel denied making any decision as to whether biotech and non-biotech varieties are "like products," the panel suggested that they were *not* in its analysis of Argentina's GATT article III.4 claim. The panel wrote, "it is not self-evident that the alleged less favourable treatment of imported biotech products is explained by the foreign origin of these products rather than, for instance, a perceived difference between biotech products and non-biotech products in terms of their safety, etc."[43] Moreover, even if a panel found that the EU's measures were inconsistent with GATT article III.4, the EU would have an article XX defense. Article XX provides, in general language, that measures must be "necessary to protect human, animal or plant life or health," and not "constitute a means of arbitrary or unjustifiable discrimination between countries where the same conditions prevail, or a disguised restriction on international trade." Its more open-ended language would make it easier for the EU to justify its measures.

The TBT Agreement also arguably provides greater grounds for state regulatory intervention than does the SPS Agreement. For example, the TBT Agreement contains general language that regulations "shall not be more trade-restrictive than necessary to fulfill a *legitimate objective*, taking account of the risks non-fulfillment would create."[44] The list of what constitutes a "legitimate objective" is an open one, and includes protection of the "environment" and "the prevention of deceptive practices." Moreover, Article 2.2 of the TBT Agreement provides that, "[i]n assessing such risks, relevant elements of consideration are, *inter alia*, available scientific and technical information, related processing technology or intended end-uses of products." In other words, "available scientific and technical information" appears to be just one element of consideration among others (*"inter alia"*) to be taken into account in applying the TBT Agreement. Because of the more open-ended language of the TBT Agreement, a party should more easily be able to raise non-science-based rationales to justify a measure under it. Overall, since neither the TBT Agreement nor the GATT contain a provision requiring that technical regulations be "based" on a risk assessment, EU measures based on non-SPS objectives would have stronger grounds for being upheld as consistent with the EU's WTO obligations. The EU moratoria and member-state safeguard bans would, as a result, more likely withstand WTO scrutiny.

The panel faced three choices in determining whether the SPS Agreement was applicable. It could interpret that it was applicable, in which case the TBT Agreement would not apply. It could find that it was not applicable, in which case the TBT Agreement and/or perhaps the GATT would apply. Or it could determine that the EU legislation contained both SPS and non-SPS objectives so that claims and defenses could be raised under SPS and TBT Agreements (as well as the GATT).

The panel first addressed, in abstract terms, the EU's defense that a measure could have multiple aims, some of which fall within the SPS Agreement's scope and others which do not, in which case the SPS Agreement would not apply to them.[45] The EU argued that if the rationale for a regulatory measure includes both an SPS objective resulting in an infringement of the requirements under the SPS Agreement and "also a non-SPS objective," then the infringing member would have to correct the SPS aspect "by removing the SPS objective and the elements of the measure therefrom." It would not, however, otherwise have to terminate the regulatory action if it remained consistent with other WTO requirements, such as those under the TBT Agreement or the GATT.[46] The US and Argentina, in contrast, maintained that only the SPS Agreement applied.

From a formal legal perspective, the panel agreed with the EU that a single legal requirement could have two different purposes, one covered by the SPS Agreement and another falling outside of the SPS Agreement's scope.[47] However, it found that all of the risks of concern under the EU legislation fell within the scope of the SPS Agreement. To determine whether the SPS Agreement applied, the panel turned to article 1.1 of the agreement and the definition of SPS measures in Annex A. Article 1.1 provides that the agreement "applies to all [SPS] measures which may, directly or indirectly, affect international trade." Since the EU's measures clearly "may" affect international trade, the key issue was whether they constituted "SPS measures." In a long section involving 73 pages of analysis, the panel parsed through the meaning of almost every word used in Annex A, frequently referring to the *Shorter Oxford English Dictionary* and other dictionaries, looking at the words' ordinary meanings in their broader "context."[48] As regards an SPS measure's "purpose," the annex defines SPS measures as "any measure applied to" protect against a list of enumerated risks, and in particular risks to human, animal, or plant life, or health arising from pests, diseases, disease-carrying organisms, additives, contaminants, and toxins, as specified in four sub-paragraphs (a) through (d).

The issue of whether the EU legislation contained one or more "purposes" falling outside of the SPS Agreement's scope was heavily litigated, resulting in endless linguistic analysis. The complainants contended that the SPS Agreement applied since the EU maintained that its measures are needed, on the one hand, to protect humans from such risks as toxicity, allergenicity, contamination, horizontal gene transfer, and antibiotic resistance, and on the other hand, to protect the environment from such risks as the invasiveness of new species, the development of resistance in pests, impacts on non-target species, and other unintended effects arising through the use of GMOs. In support, they cited text from the applicable EU directives and regulation, as well as the information required from applicants in the EU approval process.[49] The EU, in contrast, maintained that its directives and regulation also aimed to protect broader ecosystem concerns, including as regards "non-living components in the environment, such as biogeochemistry," and thus also involved non-SPS objectives.[50]

The panel sided with the complainants, and disagreed with the EU's contention that because the legislation aimed to protect biodiversity, the legislation also expressed a purpose that was not covered by the SPS Agreement. The panel arrived at this result by broadly interpreting the coverage of particular terms used in Annex A such as "animal," "plant," "pest," "additive," "contaminant," "arising from" and "other damage."[51] The panel concluded that all of the potential adverse effects indicated by the EU arising from the release of GMOs into the environment fell within the scope of the SPS Agreement.[52]

The panel came to similar conclusions regarding Regulation 258/97, the Novel Foods Regulation.[53] As Christine Conrad states, by relying on the hypothetical and indirect (as opposed to identified and direct) risks of GM varieties, they found that the SPS Agreement had a very expansive scope of coverage.[54] Because the panel found that all of the risks addressed by the EU legislation were covered, directly or indirectly, by Annex A of the SPS Agreement, they found that there was "no basis" for applying the TBT Agreement and found it "not necessary to make findings...under [GATT] Article III.4."[55] Joanne Scott thus raises a concern over SPS "imperialism" in which the SPS Agreement trumps otherwise applicable WTO law.[56] What interests us in particular, are the institutional implications of these panel interpretations. Because the panel determined that only the SPS Agreement applied, it arguably would show less deference to EU decision-making, and as a result have more input into EU decision-making processes for agricultural biotech approvals.

5.4.2. Whether the moratoria constitute SPS measures

The next interpretive issue facing the panel having institutional effects was whether EU general and product-specific moratoria existed, and, if so, whether they constituted "SPS measures" for purposes of the agreement. If the panel found that the moratoria existed but did not constitute "SPS measures," then some of the SPS Agreement's procedural provisions would apply, but its substantive requirements would not. The panel indeed took this route, having important institutional effects.

The panel first found that the EU had engaged in *de facto* general and product-specific moratoria on approvals of GM products.[57] It based its decision on an extensive review of statements and documents issued respectively by the European Commission, the Council, the European Parliament and the member states, and in particular five member states whose formal 1999 declaration stated that they would take steps to suspend all EU authorizations of GM varieties.[58] In addition, the panel painstakingly examined the approval process for each of the twenty-seven varieties (involving "product-specific moratoria") where the EU or lead member-state authority took no action for years.[59]

Having determined that the moratoria existed, the panel determined whether they constituted "SPS measures." Here the panel agreed with the EU that the EU's general and product-specific moratoria did *not* constitute "SPS measures" because the moratoria constituted neither "requirements" nor "procedures" within the framework of the SPS Agreement. It noted that "the mere fact that the decision in question related to the application, or operation, of procedures does not turn that decision into a procedure for the purposes of Annex A."[60] The panel distinguished the procedures under the EU legislation which were SPS measures, and "the procedural decision to delay final substantive approval decisions," which was not an SPS measure.[61] Under this casuistic reasoning, the panel concluded that "the moratorium was not itself an SPS measure, ... but rather affected the operation and application of the EC approval procedures."[62]

Thus, while the panel found that the SPS Agreement had a broad scope of coverage in terms of the "purpose" of a measure, it found a narrower one in terms of the measure's "nature." In this way, the panel both avoided addressing claims under the TBT Agreement and avoided examining substantive claims against the moratoria under the SPS Agreement, while still taking over a thousand pages to reach this conclusion! It arguably did so in light of challenges to its authority to decide these substantive issues, as examined in Section 5.5.3 below.

5.4.3. Legality of the moratoria

The panel finally turned to the complainants' substantive and procedural claims on page 624 of the report. Because it had determined that none of the moratoria constituted an "SPS measure," the panel would find that none of the SPS Agreement's substantive requirements applied to them. Yet by determining that the EU violated certain procedural requirements, the panel would return the substantive issues to EU administrative and judicial processes that must render their decisions without "undue delay". They must do so in the shadow of a potential future claim under these same SPS substantive requirements.

The panel held that the EU had not acted inconsistently with any of the SPS Agreement's substantive requirements since each requirement arises only when a particular measure constitutes an "SPS measure." On this definitional ground, the panel found that the EU moratoria were not inconsistent with the SPS requirement that a member base its measure on a risk assessment (the claims under articles 5.1 and 2.2).[63] It likewise found that the EU did not apply "arbitrary or unjustifiable distinctions in the levels it considers to be appropriate in different situations, if such distinctions result in discrimination or a disguised restriction on international trade" (under articles 5.5 and 2.3). It found the same with respect to the claim that the EU took measures that were "more trade-restrictive than required to achieve their appropriate level of sanitary or phytosanitary protection" (under articles 5.6 and 2.2). In short, by finding that the moratoria did not constitute SPS measures, the panel avoided having to engage in any substantive analysis of the claims.

In contrast, the panel found that the EU violated procedural requirements in engaging in "undue delay" in approving the GM varieties, in violation of article 8 of the SPS Agreement, which, in turn, refers to Annex C of the agreement.[64] The first clause of Annex C provides that "Members shall ensure, with respect to any procedure to check and ensure the fulfillment of phytosanitary measures, that: (a) such procedures are undertaken and completed without undue delay." Article 8 and Annex C, unlike other SPS provisions, does not refer to "SPS measures," but rather to procedures to fulfill SPS measures. In this case, the procedures were taken to fulfill the relevant EU legislation which the panel had determined was an SPS measure.

In response to the procedural claims, the EU argued that the delays were not "undue" because of the time needed to revise the EU legislative framework to include labeling and traceability requirements, and the changing state of the science. The panel was not persuaded, finding that such arguments could not be used endlessly to delay taking a decision.[65] Otherwise, the panel stated, "Members could evade the obligations to be observed in respect of substantive SPS measures, such as article 5.1, which requires that SPS measures be based on a risk assessment."[66] The EU, in other words, was evading taking a decision as required under article 5.1. The panel returned the substantive issues to EU administrative and judicial processes, but with some guidance pursuant to which public authorities and private actors can refer to WTO requirements as leverage.

In addition, the panel refrained from determining whether the general moratorium had ended, changing its initial finding in a leaked "interim report." The EU approved a biotech product for the first time in six years during the middle of the proceedings, which was one of the specific varieties listed in the US complaint.[67] In its interim report, the panel found that the *de facto* general moratorium had thus ended.[68] In its final report, however, the panel left unresolved whether a moratorium continued to exist. It rather instructed the EU "to bring the general *de facto* moratorium on approvals into conformity with its obligations under the SPS Agreement, if, and to the extent that, that measure has not already ceased to exist."[69]

According to Washington insiders, this switch from the interim to final panel report was important for the US government, US farm associations and companies like Monsanto, on account of the greater leverage that they now have in lobbying within the EU, exhibiting once more how international law has its effects.[70] They now have greater leverage to press the EU to approve GM varieties, including by threatening a potential WTO compliance proceeding under article 21.5 of the Dispute Settlement Understanding (DSU). For the US, a general moratorium still exists. Although the EU has approved a number of GM varieties since 2004 for consumption following EFSA risk assessments, the EU had approved no varieties for cultivation.[71] A general moratorium thus arguably still applied to the EU's application of the deliberate release directive 2001/18.

5.4.4. The member-state safeguard bans

The panel turned finally to the member-state safeguard bans, and again determined step-by-step whether the SPS Agreement applied, whether the safeguard measures were "SPS measures" for purposes of the SPS Agreement, and whether the complainants' substantive claims against the measures were valid. This time the panel produced a different outcome on the substantive claims, an outcome which needs to be viewed in the context of the multi-level structure of EU decision-making.

The panel first determined that each safeguard fell within the scope of the SPS Agreement pursuant to the panel's earlier criteria,[72] and that each constituted an "SPS measure" (because the member states actually took a decision to ban imports). The panel then examined whether each of the safeguards violated the EU's obligations under the SPS Agreement. The panel focused on the claim that the member-state safeguards were not "based on a risk assessment" in violation of article 5.1 of the SPS Agreement, and were not otherwise "consistent with the requirements of article 5.7" for provisional measures. In assessing the applicability of these provisions, the panel's report became rather tortuous.[73] The complainants argued that article 5.7 should be viewed as an "exception" to the requirements of article 5.1 so that both articles would need to be reviewed. The EU contended, in contrast, that articles 5.1 and 5.7 should be viewed as addressing two "parallel universes," one for definitive measures and the other for provisional ones. Since the member-state provisions were provisional, the EU contended that only article 5.7 applied. The panel took a somewhat confusing middle view in which it found that article 5.7 constitutes a "qualified right" of a party to take provisional measures, suggesting that it constitutes a separate track under which the burden of proof lies on the complainants.[74] The panel nonetheless started its analysis of the claims under article 5.1 because it believed (without further explanation) that, "in the specific circumstances of this case, the critical issue in our view is whether the relevant safeguard measures meet the requirements set out in the text of Article 5.1, not whether they are consistent with Article 5.7."[75] The panel thus began its analysis as if article 5.1 were the primary obligation, and only then turned to article 5.7 to see if that article's requirements were met, after which the panel made "final" conclusions.

Applying article 5.1, the panel found that none of the member-state safeguards were based on a risk assessment. Key to the panel's analysis was the definition of a "risk assessment" as set forth in Annex A and as elaborated by the Appellate Body in the *Australia-Salmon* case. The panel repeatedly turned to this Appellate Body report which, as we have seen, defined the required risk assessment relatively stringently, finding that an SPS risk assessment must evaluate "the probability" of entry, establishment or spread of a disease or pest.[76] Many of the EU member states cited outside scientific studies in support of their safeguard measures, but the panel found that none of these studies constituted a risk assessment for purposes of the SPS Agreement because none of them addressed this key issue of "probability."[77]

The EU argued, in the alternative, that the member-state safeguards were based on the risk assessments conducted at the EU level, and that different conclusions could be drawn from these risk assessments. The panel considered these EU evaluations (whether conducted by the relevant EU body or the initial member-state competent authority) to constitute "risk assessments" within the meaning of the SPS Agreement, since no party argued otherwise. However, it found that none of these evaluations supported the member-state safeguards and that no member state explained how or why it assessed the risks differently based on such risk assessment. As a result, none of the safeguards could be viewed as "based" on them. Because, in the panel's view, none of the safeguards bore a "rational relationship" to a risk assessment, the panel found, as a "preliminary" conclusion, that all of them were inconsistent with the requirements of SPS article 5.1, subject to review of the applicability of article 5.7.

In determining whether article 5.7 applied, the panel's findings would again have institutional implications. As the panel stated, "if we were to find that a safeguard measure is consistent with the requirements of Article 5.7, Article 5.1 would not be applicable and we would consequently need to conclude that the European Communities has not acted inconsistently with its obligations under Article 5.1."[78] The panel found, however, that none of the safeguards met the requirements laid out by the Appellate Body from its earlier parsing of the text.[79]

The determinative issue was whether the "relevant scientific evidence was insufficient" for conducting a risk assessment under article 5.1. The parties litigated over whether this determination should be assessed by an objective standard or in relation to the subjective views of the legislator, once again affecting the amount of deference the panel would show to state institutions. The EU contended that the concept must "refer to the matters of concern to the legislator," implicitly raising the issue of the democratic context in which precautionary SPS measures are adopted.[80] The EU argued that members' "level of acceptable risk" varies and must be taken into account. The panel disagreed, stating that "there is no apparent link between a legislator's protection goals and the task of assessing the existence and magnitude of potential risks."[81] The panel thus focused on the technical aspects of risk assessments conducted by "scientists," who "do not...need to know a Member's 'acceptable level of risk' in order to assess objectively the existence and magnitude of risk."[82]

In assessing the merits, the panel found that the relevant scientific evidence was sufficient for a risk assessment in each case. It did so, however, by focusing on risk assessments conducted at the EU level, thereby recognizing the authority of EU scientific risk assessors vis-à-vis EU member-state risk managers. The panel pointed out that the EU's "relevant scientific committees had evaluated the potential risks...and had provided a positive opinion."[83] The panel stressed that "[t]he relevant EC scientific committee subsequently also reviewed the arguments and the evidence submitted by the member state to justify the prohibition, and did not consider that such information called into question its earlier conclusions."[84] The panel thus agreed with the complainants that "the body of scientific evidence permitted the performance of a risk assessment as required under Article 5.1," so that article 5.7 did not apply. Consequently the panel found that each of the member-state safeguards was inconsistent with the obligations under article 5.1 and, "by implication," was also inconsistent with the requirements of article 2.2 that an SPS measure be "based on scientific principles" and "not maintained without sufficient scientific evidence, except as provided for in [article 5.7]."[85] The fact that official EU scientific authorities had engaged in positive risk assessments at the EU level facilitated the panel's interpretive findings. Had they not done so, the panel would have been in a much more delicate position in weighing the sufficiency of the scientific evidence.

Although the panel came out squarely against the member-state safeguards, it nonetheless implicitly pointed to a significant loophole which could facilitate a future panel finding that member-state safeguards are consistent with SPS Agreement obligations, thereby facilitating the EU's ability to comply with WTO requirements. The panel stated that "if there are factors which affect scientists' level of confidence in a risk assessment they have carried out, a Member may in principle take this into account."[86] It declared that "there may conceivably be cases where a Member which follows a precautionary approach, and which confronts a risk assessment that identifies uncertainties or constraints, would be justified" in adopting a stricter SPS measure than another member responding to the same risk assessment.[87] In other words, were there EU-level risk assessment to identify certain "uncertainties or constraints" in its evaluation, there could be grounds for upholding an EU member-state's safeguard measure as being "based" on an EU risk assessment (as required under article 5.1), even though the EU had approved the variety. The panel repeated this same analysis verbatim in assessing whether a member-state safeguard could be found to meet the requirements under 5.7 for provisional measures.[88] In addition, in a letter to the parties annexed to its decision, the panel wrote:

> The Panel's findings relating to Article 5.1 of the *SPS Agreement* preserve the freedom of Members to take prompt protective action in the event that new or additional scientific evidence becomes available which affects their risk assessments. Particularly if the new or additional scientific evidence provides grounds for considering that the use or consumption of a product might constitute a risk to human health and/or the environment, a Member might need expeditiously to re-assess the risks to human health and/or the environment.[89]

These panel dicta could affect EU evaluations (and reevaluations) of GM varieties in the future, exemplifying the reciprocal interactions of international and national laws (and in our case, EU law). The European Commission has already responded by calling explicitly for EFSA to take member-state views into account, as well as to address "more explicitly potential long-term effects and bio-diversity issues" in its risk assessments.[90] If EFSA responds to member-state concerns by indicating greater "uncertainty" in its risk assessments regarding "long-term effects," then EU and member-state measures could withstand WTO scrutiny. In this way, the EU could claim "implementation" of the report without changing the substance of EU or member-state scrutiny. The litigation would have simply constituted a complex, mind game for clever sophist-lawyers. The panel's pointing to ways in which the EU could comply with its judgment reflects its wariness of being viewed as making substantive risk decisions on account of concerns over its own legitimacy, as we examine further at the end of this chapter.

5.4.5. Panel decisions on non-WTO law and Amicus Curiae briefs

Finally, the panel made two other interpretive decisions with broader implications, one regarding the impact of other international law on WTO law, the other regarding the acceptance of *amicus curiae* briefs. The panel's rulings on the impact of other international law, in particular, has significant institutional implications, for here the panel faced a choice of recognizing the authority of other political institutions operating at the international level. In this regard, the panel addressed the EU's contentions that WTO agreements should be interpreted both in light of the 2000 Cartagena Biosafety Protocol to the Convention on Biodiversity, which became effective in 2003, and of the precautionary principle as a general or customary principle of international law.[91]

The panel examined the EU's arguments in light of article 31.3(c) of the Vienna Convention on the Law of Treaties which provides: "There shall be taken into account [in the interpretation of a treaty], together with the context:... (c) any relevant rules of international law applicable in the relations between the parties." The panel interpreted article 31.3(c) of the Vienna Convention narrowly regarding the applicability of non-WTO treaties in WTO disputes. It found that *all* WTO members must be parties to a non-WTO treaty in order for it to be "applicable in the relations between the parties."[92] Because WTO members collectively are parties to very few, if any, other international treaties besides the UN Charter, the panel effectively ruled that WTO panels are not required to take other treaties into account. In doing so, it effectively limited the authority of other international political processes. In this case, since the complainants (as well as other WTO members) had not ratified the Biosafety Protocol, the panel found that the language of article 31.3(c)(3) did not require it to take the Biosafety Protocol into account in the interpretation of the WTO treaty.[93]

The panel had at least two other alternative interpretations available to it with institutional implications. First, it could have found that article 31(c)(3)

197

applies to treaties involving "the parties to a dispute." Such a reading would still have meant that the Biosafety Protocol was not relevant since the complainants had not ratified it, but it would have meant that an international treaty would be applicable in future WTO disputes where the parties to the dispute have ratified that treaty. The panel did note, however, that it "need not, and do[es] not, take a position on whether in such a situation we would be *entitled* to take the relevant rules of international law into account" (emphasis added).[94] In other words, it left open the issue as to whether a panel might have the discretion to take into account another international treaty which all parties to a particular WTO dispute have ratified.

Second, the panel could take other international law into account in interpreting a WTO agreement in order to avoid conflicts among international rules. Here the panel only noted that "other rules of international law may in some cases aid a treaty interpreter in establishing, or confirming, the ordinary meaning of treaty terms in the specific context in which they are used."[95] The panel's finding, however, was quite narrow maintaining that treaties can "provide evidence of the ordinary meaning of terms in the same way that dictionaries do." The panel thus found that it "need not necessarily rely on other rules of international law, particularly if it considers that the ordinary meaning of the terms of WTO agreements may be ascertained by reference to other elements." Although the EU "identified a number of provisions" of the CBD and Biosafety Protocol to be taken into account, the panel found that "we did not find it necessary or appropriate to rely on these particular provisions in interpreting the WTO agreements at issue in this dispute."[96] The panel thus did not examine the provisions of the Biosafety Protocol regarding the exercise of precaution in its interpretation of the SPS Agreement, and in particular SPS articles 5.1 and 5.7, once more limiting the authority of these other international fora.

The panel then turned to the applicability of customary international law in the form of the "precautionary principle." Here the panel followed the Appellate Body's lead in the *EC-Meat hormones* case, declining to "take a position on whether or not the precautionary principle is a recognized principle of general or customary international law."[97] The panel rather noted that there has "been no authoritative decision by an international court or tribunal" which so recognizes the precautionary principle, and that legal commentators remain divided as to whether the precautionary principle has attained such status. It thus "refrain[ed] from expressing a view on this issue," other than declining to apply any such international law principle, if it exists, to the panel's interpretation of the relevant WTO agreements, and in particular, to the SPS Agreement.

Finally, the panel accepted three "unsolicited" *amicus curiae* briefs submitted to it, under its "discretionary authority," thereby potentially opening the WTO judicial process to other participants. The briefs were respectively submitted by a group of university professors who addressed, in particular, the relation of scientific knowledge to government regulation; an NGO group represented by the Foundation for International Environmental Law and Development; and an NGO group represented by the Center for International Environmental Law.[98] Each of the briefs maintained that the panel should find that the EU's regulations and practices complied with WTO law. Each further contended that the panel should grant parties considerable deference in the regulation of agricultural biotechnology in light of the uncertainty of the risks posed, as well as larger democratic concerns. The panel, however, did not "find it necessary to take the *amicus curiae* briefs into account" and thus did not cite them in its reasoning.[99] In this way, the panel again followed previous Appellate Body practice, limiting the direct input of private actors in WTO dispute settlement.

5.5. Institutional choices and legitimacy constraints in WTO judicial decision-making: Who decides?

The WTO panel made a number of interpretive moves in the GMO case which effectively reflect choices over the allocation of institutional authority. Adopting a comparative institutional analytic approach, we examine these institutional choices in this section within the larger context of the legitimacy constraints that the panel faced. We examine and evaluate five radically different institutional alternatives available to the panel through interpretation of the relevant WTO texts, including the one that the panel chose. We also show how the panel, itself operating under institutional constraints, was wary of being forced to make judgments over the EU's regulatory choices in light of existing scientific evidence. We conclude this section by discussing the panel's choices in light of broader legitimacy constraints on WTO judicial decision-making which helps to explain, in our view, why the panel made the interpretive choices that it did. Because of these legitimacy constraints, we find that the panel's decision is not only understandable, but also helpful from the broader perspectives of international dispute settlement and transnational governance.

5.5.1. Institutional choice through judicial interpretation: Comparative institutional analysis and its relation to other analytic frames

Most legal academics examining a case such as the GMO dispute take an interpretive textualist-oriented approach, focused on the WTO judicial process, whether from a formal or a functionalist perspective. They may interpret the relevant legal texts "formally" in terms of their ordinary meaning, or "functionally" in terms of their meaning in light of a normative goal (taking a teleological approach). By doing so, they tend to assume a "judiocentric" perspective as to how these disputes are to be decided, and thus are largely silent as to how these judicial bodies' decisions structurally implicate, on the one hand, *who* ultimately decides these questions, and on the other, how the judicial bodies themselves are affected by the audience that receives and responds to their decisions, which involves states and constituents that wield varying amounts of influence.[100] The approach of many WTO legal scholars is, not surprisingly, one of textual interpretation since this is their domain of relative expertise and comparative advantage.

Such a textualist approach tends to focus on whether disputed facts fall within different *categories* that are often derived from WTO texts and jurisprudence. For example, categories can be extrapolated from terms used in WTO texts, such as "SPS measure," "technical regulation," "like product," and "necessary." The categories can also be constructed in case law and scholarly analysis without the terms being used in the relevant WTO texts, such as "least trade restrictive alternative," "product or process requirement," and "extrajurisdictional" measure. The role of the interpreter and advocate is to match the facts to existing categories or to create new categories for the purpose of analysis or advocacy.

Yet from the perspective of the impact of judicial interpretations, such an interpretive, textualist approach misses what, in fact, the WTO dispute-settlement bodies actually do. Although our approach can also be viewed as "functionalist" in terms of the importance of examining consequences as opposed to applying categorical labels, the analysis is *structural* in that it examines the potential impact of WTO dispute-settlement decisions on other decision-making processes. From this structural perspective, the focus shifts from the question of *what* is being done to the question of *who* is determining it. No longer is the question solely about textual interpretation and the matching of a set of facts to a particular category, but rather about structural relations involving alternative decision-making processes. From a structural perspective, we are interested in the effective allocation of power "from" one institution "to" another. By attempting to shift authority among institutional alternatives, the WTO judicial process can alter relations between *who* decides and affected publics.

The key to a structural perspective is to assess how relations between polities and among constituencies are mediated in different ways through alternative institutional processes. The optic here is to see the WTO judicial process through the broader lens of *governance* and not through a judiocentric perspective focused solely on judicial interpretation and review. We pay considerable attention to judicial interpretation, yet we ground our assessment of what the WTO judicial process in fact does, and what it should do, in an understanding of the effects of a shift in decision-making "from" and "to" alternative institutional processes.

Comparative institutional analysis, as we use it, *is a method of analysis that provides a conceptual framework for comparing the tradeoffs (both the attributes and deficiencies) of real life institutional alternatives for addressing policy concerns in a pluralist world involving constituencies with different interests, priorities, perceptions and abilities to be heard.*[101] From a policy perspective, we cannot meaningfully assess the attributes and deficiencies of one institutional process—beset by resource, informational, and other asymmetries—without reference to other institutions that will be subject to similar (but not identical) dynamics. This analytic approach makes explicit the imperfections and limits of *each* institutional alternative. It recognizes that there may be parallel biases affecting them, but shows why they are never uniform because of their implications for who decides.

This analytic framework is particularly useful in assessing the institutional implications of interpretive choices confronted by international tribunals, and in our case, WTO dispute-settlement bodies. Through this framework, we show that an international dispute-settlement body, such as the WTO panel in the GMO case, does not simply interpret legal texts but, *de facto*, allocates decision-making responsibilities among various governmental and market actors. In doing so, it faces inevitable dilemma in light of the imperfections of each alternative. The purpose of comparative institutional analysis is to make these tradeoffs explicit.

The comparative institutional analytic framework that we use here can be seen in contrast with, and as complementary to, a number of normative analytic frames currently used in international law scholarship, including global constitutionalism, conflicts of laws, and global administrative law approaches. We first briefly summarize each of these analytic approaches and then compare and contrast our comparative institutional analytic framework with them, before applying our framework to the GMO case.

Constitutional law perspectives. As Jeffrey Dunoff has shown, international law scholars of different proclivities deploy different global constitutional law perspectives to address the role of WTO law.[102] These frameworks include those taking a substantive rights-based perspective; an institutional perspective; and a process-based pluralist perspective of constitutionalism. The rights-based constitutional approach, highlighted in the work of Ernst-Ulrich Petersmann, looks at particular constitutional rights, including a right to trade and other "market freedoms" that the WTO is alleged to incorporate.[103] The pluralist process-based constitutional approach, highlighted by the work of Neil Walker, looks at the constitutional principles and discourse that the WTO generates in relation to other constitutional orders.[104] The institutionalist constitutional perspective, as seen in the work of Joel Trachtman, addresses structures of authority within and between different institutions.[105]

The predominant view, when we speak of a WTO constitution, arguably is an institutional one. In the WTO context, some of this work, such as that of John Jackson, focuses on the internal institutions of the WTO and their role in relation to foreign trade restrictions.[106] Much of the trade scholarship also looks at the relation of WTO legal provisions and national regulation in a manner analogous to the dormant commerce clause of the US constitution and the trade provisions of articles 28 and 30, the Treaty Establishing the European Community. These provisions respectively address when US state (or EU member state) restrictions on commerce from other US states (or EU member states) are permissible under the US constitution (or EU constitutive treaty), as the case may be.[107] WTO law is viewed as playing similar constitutional law functions.

The comparative institutional analytic framework used here has much in common with the institutional aspects of constitutional analytic approaches. It fits particularly well with approaches that address how different legal orders interact. Like the constitutional law pluralist and institutionalist approaches, it addresses the reciprocal impact of different institutions on each other.[108] It highlights, in particular, the role that WTO dispute settlement plays in shaping other institutional processes. However, we do not see the need to use the normatively charged term "constitution" as opposed to the more modest term "institution" in the WTO context. The term "constitution," which is used primarily by lawyers and not scholars from other disciplines in addressing the role and functions of the WTO, can be problematic in that it can be *perceived* as one which places the WTO at the top of a global hierarchy, even if this runs directly counter to pluralists' contentions. After all, the term constitutionalism is derived from domestic contexts in which constitutional decisions by courts can trump political ones by legislatures. The comparative institutional analytic perspective thus looks pragmatically at the tradeoffs among different institutional choices that confront the WTO judicial process in a dispute similar to that over the regulation of agricultural biotechnology.

Conflicts-of-law legal pluralist perspective. A second analytic framework that has been proposed for understanding WTO dispute settlement is that of a conflict-of-laws perspective, presented by Christian Joerges. In a compelling stream of papers, Joerges has drawn from legal pluralist insights[109] to address how legal systems can exist simultaneously while paying due respect to one another when they overlap and conflict.[110] Joerges views the WTO dispute-settlement process in terms of how it creates meta-norms to address conflicting national laws, such as the laws of the exporting state and the importing state in a particular trade dispute, as part of a pluralist legal order. These meta-norms are to be applied within states' own jurisdictions. As Joerges writes, "conflicts law seeks to overcome legal differences by dint of meta-norms, which the jurisdictions involved can accept as supra-national yardsticks in the evaluation and correction of their own jurisprudence."[111] For Joerges, these norms serve as a mediating device between conflicting laws in a pluralist world, playing a role between law and politics. Joerges thus characterizes WTO dispute settlement as a form of "comitas" or "comity," constituting "a middle ground between law and politics by advising the latter to take the expertise of the former seriously, and by advising the former to be aware of the limited legitimacy of law that did not originate in a democratic process."[112] A key question for Joerges regarding disputes over risk regulation is what role "science" and "risk assessment" should play as a meta-norm. While Joerges contends that they have a central role to play because of their universalizing character, he strongly argues that they should not be applied by WTO dispute-settlement panels to trump national democratic decision-making in the GMO case, because of scientific uncertainty and because of ethical and other non-scientific concerns.[113]

As with the conflict-of-laws perspective, the comparative institutional analytic approach sees the WTO as a mediating institution. To the extent that the conflict-of-laws approach is an analogy used to address a range of choices in solving trade conflicts, a comparative institutional approach has much in common with it. Unlike the traditional conflict-of-laws approach, however, it focuses on choices involving different institutional processes, as opposed to different "laws." Moreover, it addresses a much broader range of choices than that of the law of the importing and exporting states, finding that the key impact of WTO dispute settlement lies in the role it can play in shaping institutional choices. Finally, while we agree with Joerges that the WTO interjects new norms into transnational governance such as the role of science and risk assessment (what he sees as conflict-of-laws norms), the comparative institutional analytic approach focuses not on the particular norms (though as we will see, they are indeed important), but rather on who applies them and the institutional choices that drive them. That is, a comparative institutional analytic frame focuses not just on what is being applied (the norm), but crucially, on who is applying it. A focus on norms, in this sense, is little different than focusing on textualist–or jurisprudentially–constructed categories. For example, strict scrutiny of whether a national regulatory measure is based on a meta-norm of risk assessment is shifting authority from a national decision-making body to another institution, be it that which defines what constitutes a valid risk assessment (such as the Codex Alimentarius Commission), or that which evaluates the specific risk assessment in question (such as a WTO judicial panel). A focus on the criteria of the norm can obscure the institutional choices that are consciously or unconsciously made. Although Joerges points to the deficiencies of an international dispute-settlement panel relying on science as a "meta-norm," one must pay equal attention to the deficiencies of deferring to a regulatory state, regardless of the impact of its decisions on outsiders, however appealing the regulatory state's articulation of a particular norm may be. In analyzing the GMO case, one must look into the deficiencies of not just one institutional choice, but one must simultaneously (and with equal scrutiny) weigh the tradeoffs of that institutional choice against other (also imperfect) institutional alternatives.

Global administrative law perspective. A third approach that has been well-articulated and that has stimulated a great deal of work is the global administrative law (GAL) project advanced by Richard Stewart and Benedict Kingsbury.[114] This ambitious project has been broad in its focus, and includes within the scope of its analysis the role of transgovernmental and transnational regulatory networks as well as global institutions such as the UN and WTO. The GAL project aims to put forward common principles for administrative decision-making within these different international and transnational processes. In the authors' words, the task is to "identify..., amongst these assorted practices, some patterns of commonality and connection that are sufficiently deep and far-reaching as to constitute, we believe, an embryonic field of global administrative law."[115] In a case such as the GMO dispute, the global administrative approach would look at the role that WTO judicial review can play in reviews of national decisions in terms of their compliance with principles such as transparency, due process, participation of affected stakeholders, proportionality and reasoned justification for regulatory measures.[116] As we saw, these principles were indeed of central concern to the WTO panel in the GMO case, which decided against the EU for the "undue delay" in its application of EU procedures and the lack of a scientific basis for member-state safeguards in light of the EU's own official risk assessments.

The comparative institutional analytic framework used here fits particularly well with a global administrative law perspective, in its focus on the relation of international and national decision-making processes. Nico Krisch aptly describes global administrative law as involving "a constant potential for mutual challenge: of decisions with limited authority that may be contested through diverse channels until some (perhaps provisional) closure might be achieved."[117] In this light, transnational disputes over agricultural biotechnology regulation before the WTO indeed arise in multiple contested sites for its governance. The WTO panel decision in the GMO dispute is thus best seen as part of an ongoing process, both informed by and informing national regulatory practice, transnational regulatory dialogue and developments in multiple international fora. What the comparative institutional analytic approach provides for the GAL project is a tool for evaluating institutional choices for the application of administrative law principles. A focus on GAL principles alone, just as a focus on conflict-of-laws meta-norms or on textual or judicially-constructed categories, will fail to address the inherent institutional choices at stake. Norms, principles, and categories in the abstract do not determine outcomes. Institutional processes do. The comparative institutional analytic approach makes explicit the tradeoffs among these institutional alternatives for decision-making, such as those alternatives which confront a WTO panel in its interpretation of WTO texts.

To summarize then, the comparative institutional analysis used here provides a conceptual framework for assessing alternative interpretive choices in terms of their institutional effects. From a policy perspective, we cannot meaningfully assess the attributes and deficiencies of one institutional process to which authority can be allocated—beset by different biases—without reference to other institutions which would be subject to similar (but not identical) dynamics. A comparative institutional analytic framework helps to situate the implications of judicial interpretation in social and institutional context, recognizing that interpretive choices have effects on different institutional processes in which constituencies of different countries, with varying priorities, perceptions, and abilities to be heard, are able to participate to varying and always imperfect degrees. Our task now is to make more explicit the relative attributes and deficiencies of a WTO panel's interpretive choices in this comparative institutional context. By demonstrating in this chapter how to apply this analysis, we hope to provide a deeper way of understanding WTO judicial decisions, and a better framework for normatively assessing them.

5.5.2. Alternative institutional choices in the GMO case

In deciding the GMO case, the WTO judicial panel implicitly confronted a choice among at least five institutional alternatives. Comparing the institutional choices faced from a normative perspective is not easy as we will see. Any WTO judgment rendered will implicitly be choosing among the relative benefits and detriments of imperfect institutional alternatives. Here are five strikingly different institutional processes to which the WTO panel could attempt to allocate decision-making through its interpretation of the relevant WTO texts in the GMO case. Each should be seen as an ideal type which we examine in order to clarify the institutional implications of legal analysis in this dispute, and generally:

(i) the panel could interpret the agreements to show great deference to EU political decision-makers, finding (for example) that the EU measures included non-SPS objectives, such as the protection of biodiversity, so that they should be interpreted under GATT article III or the TBT Agreement. Under these agreements, the panel could find that the EU's measures are non-discriminatory and reflect a legitimate public policy objective. Alternatively, the panel could reach this result by finding that the EU restrictions were consistent with the SPS Agreement's version of the precautionary principle under article 5.7. Through characterizing the EU's regulatory measures in any of these ways, the panel would *allocate the decision-making to an EU and/or national political process*;

(ii) the panel could stringently review EU decision-making under a relatively clear rule, such as that product bans are presumptively illegal, or that SPS measures must be based on a demanding quantitative scientific risk assessment. Finding that the EU violated its WTO commitments and should thus permit the sale of GM seeds and foods, the panel could effectively *allocate decision-making to the market* through the aggregated decisions of EU consumers. EU consumers could make their decisions on the basis of a labeling system and in response to market advertising. Moreover, a clear rule can spur more efficient bargaining between the parties to resolve their dispute;

(iii) the panel could interpret the agreements to *allocate decision-making to an international political process*. The SPS Agreement refers to international standards set by the Codex Alimentarius Commission and the International Plant Protection Commission which respectively provide (formally) for simple majority or two-thirds majority voting. In addition, other international law, reflective of international political processes, could be deemed relevant, such as customary international law or a treaty governing GMOs such as the Biosafety Protocol, each of which, as we saw, were affirmatively cited by the EU in the case;

(iv) the panel could allocate the substantive decision to itself by balancing the interests and concerns of the parties to the dispute. WTO panels have weighed competing concerns in reviewing other measures, balancing a measure's effectiveness in addressing national public policy objectives against the impact on foreign traders in light of reasonably available policy alternatives. For example, in reviewing whether the EU measure was based on a risk assessment, the panel could weigh the severity and likelihood of the risks posed against the trade impacts of the measure. In this way, the panel would effectively *allocate substantive decision-making to itself, an international judicial process;*

(v) the panel could attempt to focus on the procedures of the approval process as opposed to the substance of the risks posed. For example, the panel could focus on whether the EU approval process was transparent, involved a risk assessment, or resulted in undue delay. In this way, the panel would again allocate decision-making *to EU and/or national political processes, but this time subject to internationally imposed procedural constraints,* whether created through WTO jurisprudence or another international body. In our view, this is the path that the panel largely took through its series of interpretive moves.

Much of the legal scholarship addressing WTO judicial decisions addresses interpretive choices in normative terms. We see this pattern, in particular, regarding the SPS Agreement because of the important public policy issues of health and safety that are at stake. Yet the normative choices, in our view, cannot be only determined based on the substance of the values—such as what health and safety regulation is appropriate. In our view, since people around the world live in vastly different social contexts, resulting in vastly different social priorities, a WTO judicial decision's validity from a *normative* perspective should be assessed primarily in terms of the relative attributes and deficiencies of the alternative decision-making processes available. Since all decision-making processes suffer from imperfections in terms of accountability, the determination of what is a better process needs to be a comparative institutional one—that is, in terms of the relative attributes and deficiencies of real life institutional alternatives. In this section, we ourselves take a normative turn, but one within a larger governance perspective. We show how the SPS agreement can be interpreted, and how in fact the panel's interpretive choices can be read, in light of these broader institutional concerns.

From the perspective of accountability, the dilemma confronting the panel when making its interpretive–institutional choices is that there is no single spectrum of accountability against which institutional decision-making can be assessed since different mechanisms for accountability are themselves in tension. As shown in debates over risk regulation between rationalists (such as Sunstein) and culturalists (such as Kahan), expertise-based accountability mechanisms (focused on effectiveness) are in tension with those of democratic politics (focused on responsiveness). Moreover, in the context of multi-level governance, as Robert Keohane's work shows, internal accountability mechanisms within national democracies are in tension with the external accountability mechanisms of global regimes.[118] In the GMO case, the WTO panel faced the difficult dilemma of addressing the demand to make European internal political and regulatory processes appropriately accountable to affected outsiders (including through reference to scientific risk assessments), while itself remaining appropriately respectful of internal European democratic political processes.

We now assess five different institutional choices that confronted the panel in its interpretation of the WTO legal texts. As we will show, none of these institutional choices are perfect from the perspective of the participation of affected stakeholders. Under each alternative, stakeholder positions will be heard in different and imperfect ways.

5.5.2.1. A POLICY OF DEFERENCE: ALLOCATION OF AUTHORITY TO NATIONAL POLITICAL AND JUDICIAL PROCESSES

One institutional choice favored by many commentators is for the WTO judicial body to show deference to the country implementing the trade restriction, thereby effectively *allocating decision-making authority to a national (or in this case, EU) political process*, subject to judicial review before national courts under national law. For example, a WTO judicial panel could find that the EU national legislation and implementing regulations are in compliance with WTO rules so long as they are non-discriminatory and the regulatory purpose behind them is legitimate, whether the purpose is to protect against potential risks under uncertainty or reflects ethical concerns regarding the technology.

The panel could obtain this institutional result through a number of interpretive moves. First, it could find that the EU restrictions were indeed SPS measures, and were legitimate under the SPS Agreement's version of a precautionary principle, article 5.7.[119] Second, the panel could find that the EU measures included non-SPS objectives, such as ethical concerns regarding the technology or an objective of protecting biological diversity, so that the SPS Agreement does not apply. The panel could then apply the TBT Agreement, finding that the EU regulations are non-discriminatory, reflect "legitimate" domestic policy objectives and are thus WTO-compliant. The panel could have reached similar conclusions by applying the GATT, finding that the EU measures comply with GATT Article III.4 because they do not discriminate among "like products," but rather constitute internal regulations that are enforced against foreign products through an import ban.[120] Some scholars propose more radical means to obtain such a deferential result, maintaining that WTO judicial panels should also be able to decline jurisdiction or apply a political exception doctrine in politically charged cases that implicate trade and other social policies, in which case the national import restriction will not be judicially scrutinized.[121] Through each of these interpretive moves, the panel could respond to legal scholars' contention that WTO rules should be interpreted in deference to the "local values" of the country imposing the trade restriction.[122] In each case, the textual interpretation would result, at least from a first-order analysis, in an allocation of decision-making to EU political processes.

There are strong policy grounds for deferring to domestic political choices for regulating market transactions, given the remoteness of international processes. Participation in democratic decision-making at the national level is of a higher quality than at the international level, because of the closer relation between the citizen and the state, the consequent reduced costs of organization and participation, and the existence of a sense of a common identity and of communal cohesiveness—that is, of a *demos*. National and sub-national processes are better able to tailor regulatory measures to the demands and needs of local social and environmental contexts. They are more likely to respond rapidly and flexibly to new developments. This approach is reflected in the principle of subsidiarity in the EU, as well as by the framers of the US constitution.[123] It is a principle espoused in a variety of scholarly disciplines, from law to political science to institutional economics.[124]

National and sub-national political decision-making processes, nonetheless, can also be highly problematic from the perspectives of participation and accountability. Producer interests are generally better represented than consumers on account of their higher per capita stakes in regulatory outcomes. Producer interests' predominance explains a great deal of protectionist legislation.[125] However, even where national and local procedures are relatively pluralistic—involving broad participation before administrative and political processes that are subjected to judicial review—they generally do not take account of adverse impacts on unrepresented foreigners.

From the standpoint of accountability, if the WTO judicial process showed complete deference to national political processes, permitting them to ignore severe impacts on foreign interests that could be easily avoided through an alternative measure, then it would be effectively delegating decision-making to a process that was not sufficiently accountable to all affected parties. The SPS agreement thus requires members to justify their SPS measures to those affected by them, including on the basis of a scientific risk assessment.

Yet even where a member's regulation appears to lack a "rational basis" in terms of a scientific risk assessment, and yet severely affects foreign traders, WTO judicial intervention raises normative concerns. Although the Appellate Body and panels write in terms of "whether there is a rational relationship between an SPS measure and the scientific evidence,"[126] the underlying concept is that a member's regulation must be rationally supported. Commentators rightly ask, who are the WTO panelists to decide what is "rational." Such a basis for judicial review lies in tension with principles of representative democracy.

However, if one believes in the value of deliberation, whether under the concept of "deliberative democracy" or simply as an important governance principle, then adoption of a trump card that "states have the right to be irrational" is highly problematic when their decisions impose costs on unrepresented outsiders.[127] Because of the EU's market power and its ideational influence in world politics, the EU has a significant impact on what farmers grow around the globe, in particular in ACP (African-Caribbean-Pacific) countries which include the world's poorest. If indeed the technology can offer benefits in increasing plant stability through reducing pests, in raising crop yields, and in reducing the use of pesticides and their risks to farmers and the environment, including to poor developing country farmers (as development analysts contend),[128] then showing broad deference to the EU on the ground that "states have the right to be irrational" is ethically dubious. EU political processes have negatively affected investment in new agricultural biotechnology varieties, whether conducted by private or public bodies, which could (at least potentially) benefit developing country populations.[129]

Of course, such first-order institutional allocation of decision-making to EU political processes does not mean that global markets will play no role. To assess the relation of global market forces to EU policy-making, we must distinguish between two types of EU regulatory intervention—that of restricting the planting of GM varieties and that of restricting their consumption as food or animal feed. There arguably is less need for WTO scrutiny of EU restrictions on the cultivation of GM varieties on environmental protection grounds because of the impact of product market competition. Seed companies would still be able to develop and sell GM seeds to farmers planting them in foreign countries. If the EU permitted the sale of the resulting GM food and feed in the EU market, then these foreign farmers would compete in the EU market with EU growers. Were GM varieties to provide a significant cost advantage to these foreign farmers, then EU farmers would have a strong incentive to lobby for change within the EU political process. Indeed, EU farmers have lobbied EU politicians and have found some support in the European Commission to ease their access to GM animal feed in order to reduce their input costs arising from a feed shortage.[130] In this way, global markets can have an impact on national political processes by activating national interest-group participation. If the EU nonetheless continues to ban the cultivation of GM crops, it would not be to favor protectionist producer interests, but because producer interests were unsuccessful in EU and member-state political processes.

The impact of global markets, however, is arguably different in respect of EU restrictions on the consumption of GM food varieties. Here the restrictions benefit EU farmers from foreign competition in EU food markets. The EU import restrictions thus primarily harm EU consumers, to the extent that they pay higher food prices as a result of the import restrictions,[131] and foreign producers. The EU restrictions, in addition, could harm foreign consumers where the EU's exercise of market power affects regulatory and production choices affecting consumption choices outside the EU. As a result, a second institutional alternative could be considered—that of WTO judicial intervention to press the EU to remove its import restrictions on GM food and feed, where there is no evidence that the GM food or feed varieties in question impose greater health risks than their conventional food and feed counterparts.

5.5.2.2. WTO IMPOSITION OF A CLEAR RULE IN FAVOR OF TRADE: ALLOCATION OF AUTHORITY TO THE MARKET

The WTO panel also had choices for its interpretation of the WTO texts which could result in a constraining review of the EU's risk regulatory measures under a relatively clear rule that would favor international trade and the resulting market competition. The panel, for example, could find that the EU's moratoria on approvals of GM varieties constitute a ban in violation of GATT article XI and are not "necessary" under GATT article XX, because of other reasonably available alternatives such as product labeling. The panel could make an analogous finding under provisions of the SPS Agreement, such as in its interpretation of the article 5.6 requirement that "measures [be] not more trade-restrictive than required." In this way, the EU's *de facto* import bans would be strictly scrutinized. Alternatively, the panel could require a rigorous risk assessment under article 5.1 of the SPS Agreement, one that the panel could closely review with the assistance of outside expert testimony. Some commentators find that the WTO panel took this less deferential approach in respect of the member-state safeguards when it refused to recognize any of the studies indicated by the EU member states as constituting a risk assessment.

Under this second institutional choice, EU constituencies would still be able to buy "GM-free" products on the market. Product labeling could inform consumption decisions (and indirectly, foreign production decisions). Such an approach would effectively shift decision-making over the appropriate balance among trade, environmental, and consumer protection goals from a national (or in this case, EU) political process *to the market* through the aggregated decisions of EU consumers.

This market-based model has many benefits from the perspective of participation. A market-based decision-making mechanism permits for more individualized participation in determining the proper balance between trade, consumer protection, and environmental goals. In this manner, markets enhance democratic voice. Sellers of non-GM products could label their products "produced without GMOs." Consumers, informed through advertising campaigns, could choose which products to buy on the basis of their process of production. In choosing between food products, EU consumers would implicitly choose among alternative regulatory regimes for their production.

Were WTO panels to apply such an approach, they could stimulate not only product competition, but also regulatory competition between jurisdictions.[132] Different jurisdictions could ban or authorize the planting of GM varieties, which (as noted in our review of the first alternative) could be upheld under WTO rules. However, if the EU were to authorize the sale of GM foods for consumption, and ban the sale of GM seeds for cultivation, then EU and foreign regulatory requirements for the production of food would be in competition when EU consumers select which food to eat on the basis of product labeling. In purchasing food, EU consumers would effectively be voting for one regulatory system over another. This market process could, in turn, affect EU political processes. Were GM products cheaper than their non-GM counterparts so that EU consumers could buy them, EU farmers would have a greater incentive to lobby for authorization to cultivate GM varieties themselves.

These market decision-making mechanisms, however, are also imperfect and are subject to skewed participation in the determination of the appropriate balance of policy concerns. Markets are subject to information asymmetries, externalities, collective action problems, and oligopolistic practices. Perhaps most importantly, information costs are high for consumer purchasers, given the complexities of risk assessments. The type of label would affect product pricing, and shape consumer choice, especially were the labels misleading.[133] For example, consumers may react differently to a mandatory labeling regime (imposed on all products that contain or are produced with GMOs) than to a voluntary one (in which producers could label a product as not containing

GMOs). Even if the labels are accurate, many consumers will not take the time to review them adequately. Under a mandatory labeling system, products that must be labeled as "contains GMOs" could be stigmatized as risky without supportive evidence. In addition, where information costs are high for consumers, anti-GM activists can more easily target supermarkets exercising oligopolistic power with threats of boycotts or other negative publicity, so that private standards (such as supermarket requirements on food distributors for GM-free products) may face highly imperfect market competition.

Moreover, the views of concerned EU citizens regarding the alleged environmental impacts of GM crops would be poorly represented in the market process. Some consumers who do not eat the food product in question, whether or not it contains GMOs, would have no impact on the competition between the GM and non-GM products in question, even though they may be quite concerned about the environmental impact of the GM products in question. Other consumers might refrain from buying GM-free products, because they doubt that their purchasing decisions would be effective.

If the cultivation of GM varieties results in environmental costs, these costs might not be internalized in the price charged to consumers, so that the market would not take these costs into account. If EU environmental regulation is more stringent than foreign regulation, resulting in higher prices for EU-grown varieties, then EU farmers may demand that EU environmental requirements be reduced in order for them to compete against foreign producers. EU constituencies opposed to such a reduction in environmental regulation could face collective action problems to counter these producer demands, triggering a "race to the bottom" in GMO environmental regulation.

Finally, we note that the clearer the rule applied by a WTO panel, the more efficiently the EU, US, and other WTO members can negotiate around it. Yet the transaction costs of such negotiations would still be considerable, and the application of the rule would have distributional implications for the negotiation. From a distributional perspective, a policy of clear deference toward EU decision-making would increase what the US and others would have to pay the EU to alter its regulatory policies, while a clear rule imposed against EU trade restrictions would require the EU to pay for the right to retain its regulatory restrictions. Nonetheless, WTO members are constantly engaged in negotiations at the international level, be it within the WTO or in other international regimes which provide alternative decision-making processes to which we turn next.

In short, were the WTO panel to interpret WTO texts in a way that would significantly curtail EU policy discretion, it could help to allocate greater decision-making to the market. This institutional process would provide different opportunities for participation in decision-making that would also entail tradeoffs. These tradeoffs need to be compared with those under the first alternative of deference to an incompletely representative EU political process, as well as with those that follow.

5.5.2.3. THE INTERNATIONAL REGULATORY ALTERNATIVE: ALLOCATION OF AUTHORITY TO AN INTERNATIONAL POLITICAL BODY

Because of their concern over WTO judicial intervention, many legal scholars contend that the weighing of scientific evidence should be left to "the political domain."[134] But if they are right, then this raises the question of which political domain. National political processes are largely unresponsive to those outside of national borders, even though foreigners may be highly affected by national decisions. One institutional alternative which (in theory) is more representative of a broader array of constituents is to *allocate decision-making to a more inclusive political process, an international one.* This third alternative institutional choice, referred to as "positive integration" because it involves the enactment of new supranational regulation, contrasts with "negative integration" promoted through the regulatory competition model (the second one just covered).[135]

A number of legal commentators have advocated the incorporation of consumer protection, environmental, labor, and other regulatory issues into the WTO so that the WTO would become a global regulatory organization, and not just a trade organization with regulatory implications. As international relations scholars have long noted, the clustering of diverse issues within a single regime can facilitate tradeoffs (or side payments) among issues.[136] Andrew Guzman has built on this concept by advocating that:

> the WTO [should] be structured along departmental lines to permit its expansion into new areas while taming its trade bias…Each department would hold periodic negotiating rounds to which member states would send representatives. These 'Departmental Rounds,' however, would be limited to issues relevant to the organizing department.… In addition to the Departmental Rounds, there would be periodic 'Mega-Rounds' of negotiation that would cover issues from more than one department.[137]

In this way, Guzman proposes turning the WTO into a "World Economic Organization."[138]

Regarding the application of this institutional alternative, one can start by looking at the SPS Agreement itself as a political choice. WTO members agreed to the constraints imposed by the SPS Agreement, because they distrusted granting complete discretion to national political processes. In particular, they agreed that national SPS measures must be based on risk assessments in order for them to be justified in light of their trade implications.

Next, one can turn to the three international organizations expressly recognized by the SPS Agreement, for the adoption of harmonized international food, plant and animal health protection standards—the Codex Alimentarius Commission, the International Plant Protection Commission (IPPC), and International Office of Epizootics (OIE). These three bodies each have adopted guidelines and principles providing that regulation be based on scientific risk assessments, to which the panel repeatedly referred.[139]

As we have seen in Chapter 4, all three of these organizations also have programs that specifically address agricultural biotechnology regulation, and they thus can adopt international standards that could settle disputes arising from divergent national regulatory approaches. Their organizational rules respectively provide for adoption of standards by either a simple majority vote (for food and animal health standards under the Codex Alimentarius Commission and OIE) or a two-thirds majority vote (for plant protection standards under the IPPC). The EU (or US) could thus try to force a vote to create a clear international standard—be it on the use of the precautionary principle, the ability to rely on "other legitimate factors" in risk management or otherwise. National or EU regulations that implemented these international standards would then be presumed to be legitimate under the SPS Agreement. In the words of SPS article 3, WTO members' "[s]anitary or phytosanitary measures which conform to international standards, guidelines or recommendations shall be deemed to be necessary to protect human, animal or plant life or health, and presumed to be consistent with the relevant provisions of this Agreement and of GATT 1994." In other words, the EU could work with other countries, such as the ACP countries with which it has a "partnership," to protect itself from WTO judicial challenge through an international political process for standard-setting.

As we have seen, the EU already was successful in having the precautionary principle incorporated into the Biosafety Protocol, which extended the Convention on Biodiversity's scope of coverage to include the protection of "human health." Yet the Biosafety Protocol is not recognized as an international standard in Annex A of the SPS Agreement, and none of the complainants have ratified it. The complainants, in contrast, are all members of the Codex Alimentarius Commission and the IPPC, and these agreements are expressly covered in Annex A.

Finally, the Agreement Establishing the WTO itself provides for majority or supramajority voting, including for interpretations and amendments of the texts of WTO agreements. Thus, in theory, it is possible to interpret and amend the WTO agreements through a political process. These decisions would be made by the WTO General Council or at a WTO ministerial meeting, depending on the issue in question. Article IX:1 provides for a general rule on WTO decision-making that "except as otherwise provided, where a decision cannot be arrived at by consensus, the matter at issue shall be decided by voting," and, in such case, by a simple majority of the votes cast. Articles IX:2 and IX:3 provide respectively for a three-fourths majority vote for authoritative interpretations of the texts and for the waiver of any obligations of a member. Article X contains a specific rule on amendments, providing for a two-thirds majority vote, subject to qualifications depending on whether an amendment would alter substantive rights and obligations.[140]

In practice, however, secondary decision-making by international political bodies is almost always made by consensus.[141] In the WTO, in fact, decisions are always made by consensus and even then they are infrequent.[142] Because of the prevailing norm of decision-making by consensus, the WTO political/legislative system, in contrast to its judicial system, is relatively weak. Similarly, decisions in the Codex and IPPC are typically taken by consensus because the organizations' efficacy would otherwise be undermined. Codex and the IPCC do not constitute chambers within an international parliament and their standards are meant to be "voluntary." If votes were taken against the will of a Codex member, especially a powerful one, it might withdraw or otherwise attempt to disrupt Codex operations. Decision-making over politically contentious matters such as the regulation of agricultural biotechnology by a majority vote within a centralized international political process is generally avoided because member states do not wish to set a precedent which could threaten their autonomy on other matters in the future.

Although international decision-making processes can be more inclusive of affected stakeholders than national political processes, they are quite remote from citizens and thus are subject to severe imperfections in at least five ways. First is the question of which interest groups have access to national representatives that negotiate at the international level. To the extent that parliamentary bodies are relatively disempowered in international fora, interest groups having preferential access to administrative officials will be favored. Second is the question of which interest groups have better direct access at the international level. Those with high per capita stakes in outcomes will invest in following negotiations directly at the international level and sometimes participate in them as observers. Third is the question of the asymmetric power of countries that negotiate at the international level. Countries with large markets tend to wield much greater power in international economic negotiations. Moreover, the bureaucracies of northern countries have greater resources, and larger, more experienced staff. Within Codex, for example, many developing countries have traditionally not attended meetings.[143] Fourth is the challenge of devising appropriate voting rules at the international level. Even if all countries did participate in international economic negotiations in an informed manner, the weighing of votes by country is problematic, where the countries vary in size from small island nations to China. Fifth, even were these centralized international governance mechanisms to facilitate relatively greater voice of a broader array of stakeholders, these mechanisms may be unsuited to respond to local norms, needs, and conditions in rapidly changing environments.

Finally, as we noted in Chapter 4, the current structure of international trade, environmental, and development organizations is fragmented. Rather than moving toward a consolidation of international law, we are seeing the emergence of pluralist regime complexes in which institutions have overlapping jurisdiction, reflecting the *ad hoc* nature of their creation. States sometimes purposefully calculate for the provisions in one agreement to be in tension with, and potentially undermine, those in another which they are unable to change. This brings us back to the question with which we began our discussion of this institutional alternative: which political process should decide?

5.5.2.4. THE JUDICIAL ALTERNATIVE: AN INTERNATIONAL COURT'S BALANCING OF SUBSTANTIVE NORMS AND INTERESTS

Under a fourth institutional alternative, the WTO judicial bodies themselves could "balance" competing preferences for trade, consumer, and environmental protection in their review of the facts of each specific case, assessing the regulatory measure's "proportionality."[144] In contrast to the second approach in which the panel would apply rather bright line rules, under this fourth approach it would apply more open-ended standards to the facts of a case. In this way, the panel could *allocate the substantive decision to itself—an international judicial process*. Under each alternative, a WTO panel is intervening, but under the other alternatives, the panel is effectively allocating authority to some other decision-making process. Under this one, in contrast, it is deciding that it will decide the appropriate balance between competing concerns.[145]

The Appellate Body has explicitly taken a balancing approach in some WTO cases. In the *Korea-Beef* case, involving a Korean requirement that retailers make a choice of selling only Korean or foreign beef (which was allegedly required to ease the government's monitoring of the labeling of the beef's origin so that Korean consumers are accurately informed), the Appellate Body concluded:

> In sum, determination of whether a measure, which is not 'indispensable', may nevertheless be 'necessary' within the contemplation of Article XX(d), involves in every case a process of weighing and balancing a series of factors which prominently include the contribution made by the compliance measure to the enforcement of the law or regulation at issue, the importance of the common interests or values protected by that law or regulation, and the accompanying impact of the law or regulation on imports or exports.[146]

Applying these three listed factors to the specific factual context, the Appellate Body held against the Korean measure.

Similarly, the panel, in applying the SPS Agreement in the GMO case, could have explicitly weighed the severity and likelihood of the risks posed against the trade impacts of the EU's measures, constituting a form of "proportionality" review. It could have done so under any number of SPS provisions, including articles 2.2, 2.3, 5.1, 5.5, and 5.6, as supported by a number of commentators. In the *Japan-Apples* case, a WTO panel engaged in balancing of concerns in applying article 2.2, holding that Japan's "phytosanitary measure at issue is clearly *disproportionate* to the risk identified on the basis of the scientific evidence available" (emphasis added).[147] There is thus a proportionality dimension to SPS article 2.2, and arguably to article 5.1 to which this provision relates.[148] Caroline Foster maintains that panels should use proportionality-type analysis under article 5.6, in determining whether a member's "measures are not more trade-restrictive than required."[149] Alessia Herwig has suggested the same under article 5.5 of the agreement in determining whether member measures constitute "unjustifiable discrimination."[150] Clearly, there is plenty of opportunity for a panel to engage in explicit balancing under the SPS Agreement.

Judicial bodies are sometimes viewed as being better situated than political institutions to weigh expert evidence and facts on a case-by-case basis, because of concerns over potential executive or legislative bias in individual cases. Of course, more informal administrative processes may often be superior to formal judicial procedures for the gathering and weighing of facts. Many commentators thus favor a greater role for "soft law" governance mechanisms such as the WTO committee system, or Codex working groups.[151] Yet where a country can be sanctioned because its regulatory measures fail to comply with an international obligation, then it will likely prefer the use of a more formal dispute-settlement process.

In the GMO case, the panel heard evidence that would permit it to engage in proportionality review. It called on six scientific experts to testify. The panel asked the experts detailed questions in writing and at hearings regarding the risks posed by individual GM varieties and whether the EU member-state bans were supported by risk assessments. In this way, the panel could better assess the concerns at stake.

If WTO panels issue rulings under a balancing test against national and EU regulatory decisions which reflect strongly held values, however, the sociological legitimacy of the WTO judicial process may be strongly challenged, undermining its authority. The WTO judicial body is not only unelected, but also (as any international organization) lies at an extremely remote level of social organization, far from the ordinary citizen. Constituencies may thus find it to be poorly situated to decide substantively whether specific GM products must be authorized because they are safe for human health and the environment according to scientific risk assessments. Moreover, given the history of mistaken scientific judgments, coupled with the possibility of bias in the scientific evidence because the testing of GM varieties is financed primarily by the private sector, WTO panels may wish to avoid being in the position of second-guessing member determinations on "scientific" grounds. Finally, it is much more difficult for WTO members to correct or respond collectively to a WTO judicial decision by amending a WTO rule, compared to in domestic legislative systems, so that members may be wary of WTO panels asserting too much authority in these cases.

The WTO panel was thus reluctant to allocate substantive decision-making authority to itself in the GMO case, under a balancing test. Although the

Appellate Body did so in the *Korea-Beef* case, that case did not involve politically charged environmental issues that would attract the attention of transnational NGOs and the media. In the GMO case, in contrast, the panel likely realized that it lacked the legitimacy to engage explicitly in a delicate balancing on this particular matter. Although, as with any court, WTO panel members are not elected, they are even more subject to legitimacy challenges than domestic courts because of the more fragile social acceptance of their decisions, as we examine further below. Paradoxically, as the need for international judicial review increases because of biases in national political processes, intrusive judicial review can also become more difficult, and judicial panels must weigh the potential adverse reactions to their decisions as a cost to the overall trading system. The WTO panel thus took a proceduralist turn in the agricultural biotech case, in which it could look for allies within the EU political system, an institutional alternative that we now address.

5.5.2.5. THE PROCEDURALIST TURN: INTERNATIONAL JUDICIAL REVIEW OF THE PROCESS OF NATIONAL DECISION-MAKING

Under a fifth institutional alternative, instead of engaging in a balancing of substantive concerns, the WTO panel can review the *procedures* of the national decision-making process to attempt to ensure that national decision-makers take into account the views of, and impacts on, affected foreign parties. As under the first option, the panel would attempt to return substantive decision-making to a national political forum, but unlike under the first option, it would not completely defer to the regulating state. The WTO Appellate Body has adopted this approach in a number of important decisions involving the intersection of trade and social policy.[152]

The panel clearly chose this fifth option in its response to the complainants' challenges to decision-making at the EU level. It avoided addressing the SPS Agreement's substantive provisions by finding that the EU had not adopted a reviewable "SPS measure." By categorizing the EU *de facto* moratoria in this way, the panel avoided examining whether the moratoria complied with article 5.1's requirement that measures be based on a risk assessment, article 5.5's requirement that measures be consistently applied, and article 5.6's requirement that measures be no more trade-restrictive than required to achieve their aims. The panel nonetheless found that the EU had violated its procedural obligations under the SPS Agreement by engaging in "undue delay" in the review process. The EU is now on notice that if it continues to take undue delay in its review of applications for the approval of individual GM varieties, it will be subject to trade retaliation from affected WTO members. The EU's review process is again operating, even though the process remains quite slow and politically charged.

Categorizing the panel's more stringent review of the EU member-state safeguard bans is more complicated. Because the panel held that the safeguards were not based on a risk assessment in violation of article 5.1 of the SPS Agreement, some commentators contend that the panel applied a strict rule concerning risk assessments under the second institutional option, or (alternatively) itself assumed substantive decision-making by balancing competing concerns under the fourth one. However, we view the panel as again returning the issue to the EU political process in light of the two-level nature of EU policy-making. The member-state bans are procedurally subject to EU political challenge and judicial review under EU legislation. Just as the panel held that the EU had engaged in undue delay in taking approval decisions for GM varieties, it implicitly found that the EU was taking undue delay in challenging the member-state bans under the EU's own internal legislation. The EU's scientific bodies had found that the bans were not justified by a scientific risk assessment so that, under the EU's own internal law, the Commission should challenge them.[153]

Had the panel decided in favor of the member-state safeguards, it would likely have been viewed as calling into question the judgments of the EU's official scientific bodies! Thinking counterfactually, such a WTO panel decision would have had a very different impact in internal EU politics. Commentators on the WTO decision have ignored this key institutional aspect of the case. Had EU official scientific bodies not explicitly issued positive opinions on the GM varieties in question, the panel would have been in a much more compromised position and its institutional choice in respect of the safeguard bans could indeed more properly be viewed in terms of the second or fourth alternatives examined above.[154]

Process-based review may seem ideal, since it is relatively less intrusive than substantive review and it directly focuses on the issue of participation of domestic and foreign parties. However, process-based review also raises significant concerns, in particular because strategic actors can manipulate processes to give the appearance of consideration of affected foreigners without in any way modifying a predetermined outcome. Moreover, even if international case-by-case review were possible (which it is not), it will be difficult, if not impossible, for an international body to determine the extent to which a national agency actually takes account of foreign interests. National and EU decision-makers can thus go through the formal steps of due process without meaningfully considering the views of the affected parties.

Process-based review is more likely to be meaningful if the WTO panel can empower actors within existing national political processes that will reduce bias. Judicial actors using this approach recognize that political and administrative processes are not monolithic, but have cracks that can be worked. They understand that for their decisions to be effective, they will need to provide tools that can be used by actors in these processes. If a WTO panel can enlist allies in the EU system to reduce the system's inevitable biases, then process-based review may work. Otherwise, a WTO panel will inevitably have to engage in some form of substantive review if it wishes to have an impact where national measures, responding to majoritarian or minoritarian political demands, prejudice unrepresented foreign traders. In the agricultural biotech context, both EFSA and members of the European Commission are potential allies within EU decision-making processes. The Commission has long been looking for tools to remove the member-state safeguards or at least not have them renewed after they expire by their terms. EFSA will continue to make the risk assessments on which EU decisions are to be based in the future.

In complement, the transparency demands of process-based review can help to activate broader and more informed participation in national and EU political and administrative processes to counter any minoritarian biases in favor of concentrated interests. For example, the conduct of risk assessments has become a focal point in EU decision-making which has become subject to more transparent notice and comment procedures from interested stakeholders.[155] This process also gives rise to an administrative record that can facilitate subsequent judicial review at the WTO level. The prospect of such judicial review, in turn, can create leverage in EU administrative processes so that they are more likely to avoid violations in the first place than would otherwise be the case. Overall, the WTO SPS Agreement, as interpreted in SPS cases, has spurred the EU to adopt authorization procedures that create administrative records that can either justify its measures or subject them to legal challenge.

In sum, to assess whether this institutional outcome through judicial interpretation is normatively desirable, we need to compare it with the implications of other available (and also imperfect) alternatives. We have provided a framework and analysis to do so. Although we may have technically interpreted the WTO agreements differently, we find that the overall thrust of the panel's report was appropriate in its procedural orientation in light of the institutional alternatives. We find so particularly on account of the legitimacy constraints that the WTO judicial process itself faces, to which we now turn.

5.5.3. Institutional choice in context: Sociological legitimacy constraints on WTO decision-making

Assessing the WTO judicial decision in the GMO case should not be done solely in terms of the impact of that decision on other institutions. The relation of international and national law and politics is a reciprocal process. Not only can an international decision affect domestic political processes, but domestic and international politics can also affect the decisions of international judicial actors. Indeed, as creations of states which rely in large part on their members to implement judicial decisions, international courts and tribunals may anticipate the likely domestic and international political reactions to their decisions, and take those reactions into account in their rulings. What appears to be an independent and autonomous judicial decision, therefore, can be subtly influenced by judges' anticipation of the decision's reception among the parties to the dispute as well as the membership of the organization and the broader legal community.[156]

When there is a risk of defiant responses to WTO judicial decisions, especially by powerful members in "hard" cases, the WTO judicial process has an incentive to issue reports that avoid deciding the substantive issues, resulting in what has been termed as a politics of legitimacy.[157] By legitimacy, here, we stress the concept's sociological dimensions in terms of the social acceptance of a judicial decision.[158] In the case of the WTO dispute-settlement system, we refer to whether WTO members and society at large ultimately accept or reject a WTO panel or Appellate Body ruling.[159] For us, many normative aspects of legitimate judicial decision-making are linked to sociological ones in that they point to specific aspirations of the judicial process that can enhance the prospects of social acceptance of its decisions, such as judicial impartiality, transparency, fair access for affected parties, consistency, due respect for political branches at different levels of social organization, and the provision of reasoned and principled decisions.[160]

Panels and the Appellate Body should be concerned with the acceptance of their decisions by the WTO members, as well as by social forces that will place pressure on WTO member governments to defy the panel and Appellate Body decisions. Powerful WTO members such as the US and EU, which are the world's largest traders, are arguably of particular concern. Were the WTO judicial process to come down hard on the EU in the GMO case, the EU would likely not comply with its decision, in response to the demands of EU member states and the larger European public. Moreover, such a ruling could provide fodder to anti-globalist challenges to trade liberalization, and fuel further mass protests against the WTO. The EU's defiance of the WTO decision, coupled with mass protests, could provide a rationale for other WTO members to refuse to comply with WTO legal rulings. One member's noncompliance could trigger other member's tit-for-tat strategies of noncompliance. As McDougal and Lasswell stated about international law around fifty years ago, "[s]ince the legal process is among the basic patterns of a community, the public order includes the protection of the legal order itself, with authority being used as a base of power to protect authority."[161] As Koskenniemi writes in a report from the International Law Commission, "[t]reaty interpretation is diplomacy, and it is the business of diplomacy to avoid or mitigate conflict."[162] The Appellate Body and judicial panels thus have an incentive to write opinions that are slightly ambiguous, leading to different interpretations as to how they can be implemented. In this way, they can shape their decisions to facilitate EU compliance and amicable settlement, and thereby uphold the WTO legal system. Part of diplomacy, of course, involves power variables, and there are justifiable concerns over a pattern of WTO dispute resolution in SPS and GATT article XX cases in which WTO dispute-settlement bodies show greater deference when the US or EU is a respondent.[163]

Our analysis of the WTO panel decision indeed suggests that the WTO judicial process is not independent of politics or strategic action by WTO judicial decision-makers. WTO judges, both panelists and the members of the Appellate Body, have some independent agency. They are not only interpreters and appliers of WTO legal provisions, but the pattern of their jurisprudence suggests that they also assume a mediating role. They can press members to take account of each others' views and interests, and they can spur the settlement of disputes by facilitating compliance with judicial recommendations, including through empowering actors at the domestic level, thereby upholding the WTO legal system. After all, this is a *dispute-settlement system* (not simply a court) whose ultimate "aim," under the Understanding on Rules and Procedures Governing the Settlement of Disputes "is to secure a positive solution to a dispute."[164] As a result, Joerges and Neyer characterize WTO dispute settlement as constituting "a middle ground between law and politics."[165]

The WTO panel demonstrated caution in the GMO case. It decided to tread softly, wishing not to be seen as second-guessing the EU's risk assessment and risk management decisions.[166] Even though the US prevailed, the WTO decision, in our view, left significant discretion to the EU in determining the level of acceptable risk. Moreover, by finding that neither the EU general nor product-specific moratoria were "SPS measures," the panel left a WTO decision over the crucial substantive issue of whether EU-level decision-making was based on a scientific risk assessment for another day, if ever. As regards the member-state safeguard measures, the panel found that they were inconsistent with the EU's substantive WTO commitments to base SPS measures on a risk assessment, but did so by relying on risk assessments conducted by the EU itself in a context where the EU has so far refrained from challenging the safeguards under EU law. The panel can be best viewed as returning the issues to EU political, administrative, and judicial processes in a way that can facilitate compliance with the WTO legal order. The panel even indicated a means for them to do so, which has already elicited a Commission reaction.[167]

The extraordinary length of the WTO panel decision and the significant delay in issuing it provide further evidence of the panel's concerns over challenges to its authority. Ironically, the panel attempted to avoid making substantive decisions, just as the EU had done, in part by using methods which were the basis for its legal holding against the EU. While the panel held against the EU for engaging in "undue delay," the panel itself took over three years from the initial filing of the claim to render a decision, instead of from around seven to ten months as contemplated by the DSU. As a consequence, the panel vastly exceeded WTO guidelines which, under article 12.8 of the DSU, provide, "the period in which the panel shall conduct its examination....shall, as a general rule, not exceed six months." Article 12.9 of the DSU further states that "[i]n no case should the period from the establishment of the panel to the circulation of the report to the Members exceed nine months." Even once composed on March 4, 2004, the panel took over thirty-two months to circulate its decision.[168]

Some might find it a bit presumptuous for the panel to hold that the EU had engaged in "undue delay" in making decisions in this controversial area, and then do so itself. Of course, the panel listed good reasons for the length of its proceedings. It noted in its opinion the inordinate amount of written submissions (which it estimated at 2,580 pages), supplemented by "an estimated total of 3,136 documents," the need to consult with a panel of scientific experts (which provided the panel with an "estimated 292 pages" of responses), the procedural and substantive complexity of the case, and the fact that the three complainants did not consolidate their complaints.[169] The WTO panel faces resource constraints in handling the mass of evidence presented. Yet the panel clearly was in no hurry to make a quick decision, which is one reason that it consulted so many documents and experts. There is a sense as well that the panel purposefully delayed issuing its report until after the WTO Ministerial Meeting in Hong Kong in December 2006, in which intensive bargaining (and demonstrations) took place under (and against) the Doha Round of trade negotiations. None of the parties to the proceeding appeared to object to the delays.

The panel's delay in deciding the substantive claims at the EU level will, in fact, be even longer on account of its decision, if in fact a WTO judicial decision on the substance is ever made. Under the panel's reasoning, only once the EU actually makes a decision which results in an "SPS measure" regarding a GM variety, may a complainant bring a substantive claim. In such a case, the complainant would have to restart the process from scratch. A panel would have to be formed and experts consulted. The actual delay in the panel making a decision on the substance of EU decision-making will thus be much longer than the three and half years that the case formally took (not to count subsequent procedures regarding the EU's implementation of the ruling), if indeed a new claim is ever filed. The panel thereby effectively parried deciding on the substance of EU decision-making.

The length of the decision is also significant. By issuing an opinion that is 1,087 pages, containing 2,187 footnotes, citing the jurisprudence of 60 previous WTO panel and Appellate Body reports, and with more than another thousand pages of annexes, the panel made the decision look both extremely thorough and considerably technical. At the same time, it becomes much more difficult for outsiders to read, understand, and criticize the panel report. Few have the patience to do so. Whether consciously done or not, the thousand-plus page panel decision obfuscates the judicial role, submerging legal conclusions and analysis in a sea of text. The mere translation of the decision into the WTO's other official languages, French and Spanish, resulted in further delay before it could be formally adopted and officially released.

The panel's delay and arguable obfuscation can be viewed in both socio-legal and normative terms. From a socio-legal perspective, the panel was not anxious to make a substantive decision on EU procedures regarding the politically controversial issue of GMOs on account of the likely challenges to its authority. It thus took a tortuous path involving a series of interpretive moves to avoid deciding the substantive issues, as summarized in our earlier step-by-step review of the decision. From a normative perspective, the delay, length, and overall complexity of the panel decision may nonetheless have positive attributes when viewed in light of the interpretive and institutional alternatives that the panel confronted in broader institutional context. The panel was attempting to grant time for the parties to sort out their disputes in the shadow of WTO law, to provide input into EU administrative and judicial decision-making processes (in particular, through giving tools that domestic actors can use under EU law in order to facilitate compliance), to indicate flexible means for the EU to comply with the decision by (strategically) retaining some ambiguity, and to protect its own authority through painstaking textualist justifications for its interpretive moves, given their unstated (but nonetheless significant) institutional implications.

5.6. The potential impact of the WTO decision

The GMO dispute highlights the role—within political and sociological constraints—that can be played by the WTO dispute-settlement system in the governance of transnational regulatory conflicts. To assess its role, one must begin by thinking counterfactually as to what would have occurred had there been no WTO judicial process with compulsory jurisdiction which channeled the US–EU dispute into a legal process.[170] Were there no WTO with compulsory jurisdiction, the EU would have faced considerable pressure from EU member states and other constituents to block formation of a judicial panel or the adoption of a panel decision, as it did in agricultural disputes under the GATT, triggering further US ire.[171] The US likewise was constrained from retaliating unilaterally as it often did under the GATT, especially pursuant to the infamous section 301 of the 1974 US Trade Act. Because there now could be a WTO legal ruling whose adoption the EU could not block, and because the US is not legally authorized to retaliate without authorization of the WTO's Dispute Settlement Body, the US executive is better able to fend off political demands to up the ante through threatening and applying unilateral trade retaliation.[172] If there were no WTO, the US executive could have been under much more pressure from agricultural lobbies and members of Congress to retaliate unilaterally against the EU. In the former GATT days, the US did so numerous times on agricultural matters, including in the "chicken war," the "oilseeds war" and other battles over food and seed products.[173] The WTO legal system and decision thus significantly constrained the US by empowering those in the US who wished to stave off domestic political pressure that could lead to a trade war. Such political pressures can now be more easily channeled through reference to a legal mechanism. Were there no WTO with a DSU with compulsory jurisdiction, it would likely be much more difficult for the US and EU to manage their conflict. The WTO and the DSU, as law and dispute-settlement processes generally, help the parties to "depoliticize" their conflict so that it does not "explode" into "other forms of confrontation."[174]

The WTO panel decision also needs to be understood in terms of how it affects transatlantic intergovernmental bargaining, providing both sides with incentives to return to the bargaining table. EU officials, for example, were concerned about the imposition of WTO-approved trade sanctions, and thus had an incentive to engage in talks to reassure the US, to pursue measures that might bring the EU into compliance, and to seek additional time to comply with US demands in the face of fierce domestic opposition.

On the US side, the office of the US Trade Representative, pressed by the farming and biotech sectors, also had incentives to negotiate with the Commission. On the one hand, the threat of US retaliation provided the US with newly acquired leverage to demand concessions from the EU. On the other hand, however, US stakeholders recognized that "strict panel implementation would not necessarily provide any commercial benefits for U.S. biotechnology firms," since the safeguard bans that were the most obvious indicator of noncompliance were targeted largely at GMOs that were no longer in commercial use.[175] US negotiators and stakeholders therefore articulated a wider set of demands and concerns for the EU, including: speeding up the approval process for pending GMO authorizations; addressing the EU's "zero-tolerance" policy toward the adventitious presence of unauthorized GMOs in US product shipments; reopening the EU market to US long-grain rice, after the Commission had insisted on mandatory testing of US rice shipments in EU ports; and challenging the member-state safeguard bans. US negotiators, in effect, sought a package deal in which the US could declare the EU in compliance with its WTO obligations in return for concessions on these related matters.[176]

Indeed, we do see considerable evidence of an upsurge in frequent, and arguably, constructive bilateral diplomacy since the panel decision. The character of this new diplomacy, however, differs from that of the mid-1990s, which sought to build deliberative linkages directly between US and EU regulators. By contrast, the high-level bilateral diplomacy between the US and the EU since November 2006 has been dominated by trade officials on both sides, and focused primarily on ways to facilitate EU compliance with the WTO decision while minimizing the trade impact of EU policy on US farming and biotech interests. Following the November 2006 panel decision, the European Commission requested a bilateral meeting to discuss the "normalization" of transatlantic biotech trade, and US and EU negotiators held at least five rounds of bilateral talks between February 2007 and the WTO-imposed compliance deadline of November 21, 2007.

In these negotiations, as in earlier meetings, both sides remained under intense pressure from their respective stakeholders to concede as little as possible, yet negotiators repeatedly described the negotiations as substantive and productive. One USDA official, for example, was cited as calling the tenor of the meetings "drastically different" from previous years, which we characterized as "pretty hostile." By contrast, he said, the October 4th bilateral meeting held on the eve of the WTO deadline, "had a lot more give and take and open communication."[177] The meetings, moreover, achieved some substantive results, including a Commission agreement to reopen the EU rice market to US producers, lifting the requirement for testing at the first EU port of entry, in return for a USDA-supervised scheme for testing all EU-bound rice shipments within the US.[178]

In other areas, the negotiations continued to draw out. The Commission, for example, repeatedly promised US negotiators that it would challenge the national safeguard bans, only to have its efforts rebuffed in the Council of Ministers. The Commission finally notified Austria, in May 2008, of its intent to challenge Austria's bans on the sale of two GM varieties were Austria not to remove them, and Austria did remove them at the end of that month.[179] The Commission, however, did not challenge the Austrian safeguard bans on cultivation of these varieties, which remain in effect. Late in 2007, moreover, French President Nicolas Sarkozy announced France's government-ordered suspension of cultivation of the EU-approved MON810 maize variety, creating new challenges for the Commission.[180]

Similarly, US officials at a July 2007 meeting suggested that US industry might agree to follow Argentina's example and cultivate only those GM crops that had been approved in the EU, so as not to fall afoul of the EU's zero-tolerance policy toward the adventitious presence of unapproved GMOs in US shipments. Such an agreement, however, was possible only if the EU's regulatory system became more predictable.[181] As we shall see in Chapter 6, the Commission has not been able to secure the steady flow of new GM approvals in the face of strong member-state opposition, and the member states had (as of this writing) yet to agree to the Commission's proposals (and US demands) to raise the thresholds for the adventitious presence of GMOs that had been deemed to be safe by the EFSA.[182]

Despite its disappointments, US negotiators and stakeholders agreed that the imposition of retaliatory tariffs was likely to be counterproductive, and so the US agreed first to extend the deadline for EU compliance to January 11, 2008, and later announced that it would, in the words of a USTR spokeswoman, "...suspend for a limited period the proceedings on our WTO request for authority to suspend concessions in order to provide the EU an opportunity to demonstrate meaningful progress on the approval of biotech products."[183] In the interim, US negotiators announced that the US would begin to calculate the estimated damage to US exports, both in preparation for any eventual imposition of sanctions and to maintain the pressure on the EU.[184] The EU Commission, for its part, indicated its relief, announcing: "We welcome the measured response by the US, and reiterate out commitment to advance the difficult dossier of biotechnology through dialogue."[185] Whether WTO and US pressure would be sufficient to prompt a substantive change in EU policy, however, remained to be seen.[186]

5.6.1. Domestic effects in EU politics and law

In principle, the WTO panel decision can provide leverage to those in the EU who wish to facilitate the approval of GMOs, thereby mitigating the conflict with the US. This is a central way in which international law can have effects—by being used as a tool in domestic political playing fields where there are divisions within a state.[187] In examining the international law and politics of risk regulation, we should not forget that regulatory debates, such as those over GMOs, arise not just between, but also within, states (and in the EU case, within regions). In federal and other multi-level governance contexts, the WTO tends to empower centralized executives where executive bodies are more pro-trade oriented. WTO rules and judicial decisions provide leverage for executive bodies to fend off protectionist pressures and (more generally) pressures to regulate in ways that adversely affect trading partners. By finding that a moratorium constitutes "undue delay" in violation of WTO commitments, the decision empowers those within the Commission and other advocates of agricultural biotech within the member states to push forward new approvals of GM varieties. Indeed, as we will see in Chapter 6, the Commission approved fifteen varieties for sale as food and feed, from the end of the moratorium in May 2004 through January 2008, although without member-state support, and EFSA has issued a positive assessment regarding other GM varieties in the pipeline. At the EU member-state level, Spain grew around 60,000 hectares of GM corn in 2006, and other member states could follow suit, as six of the member states now grow some GM varieties, although in relatively small quantities.[188]

Within the EU itself, the Commission has been concerned about the technological and competitiveness impacts on European industry from the EU's approach to agricultural biotech regulation, obstructing the EU's aim to spur economic growth through its "Strategy for Life Sciences and Biotechnology," launched in 2002. The Commission's 6th Framework Programme for research, technological development, and demonstration activities proposed that the area of life sciences and biotechnology was its "first priority."[189] Yet the Commission's President at the time, Romano Prodi, lamented that Europe's biotech industry lagged "four or five years" behind and was worth "roughly one-third as much as its U.S. counterpart."[190] A 2004 Communication from Commission President Prodi again noted the "exodus of researchers and rapid decline in GMO field research in the EU and the consequent negative repercussions in innovation and competitiveness."[191] Following the WTO panel decision, EU trade commissioner Peter Mandelson declared in a similar vein, "we must be under no illusion that Europe's interests are served by being outside a global market that is steadily working its way through the issues raised by GM food. They are not."[192]

The Commission had been pressing to end the "delays" in the EU approval process and the WTO decision could help it in this respect. We do not believe that it is simply coincidental that the EU approved a biotech variety in the middle of the WTO case, in May 2004—the first time since the start of the moratorium in 1998.[193] The panel itself noted that "these votes [regarding the authorization of NK603 maize] may have been influenced by the establishment of this Panel."[194] The WTO case provided the European Commission with a sense of justification to approve the GM corn variety over member-state reticence.

The WTO case and decision now also empower the Commission and private parties if they challenge member-state bans on varieties approved at the EU level, most of which have expired or are expiring by their terms. For example, three biotechnology companies (Monsanto, Syngenta, and Pioneer) challenged Italy's ban on food products containing authorized GM maize before the Italian courts. An Italian lower court referred the matter to the European Court of Justice (ECJ) for an interpretation of EU law pursuant to Article 234 of the Treaty. In September 2003, the European Court of Justice ruled that Italy must conduct "a risk assessment which is complete as possible…from which it is apparent that, in light of the precautionary principle, the implementation of such measure is necessary in order to ensure that novel foods do not present a danger," which Italy had so far failed to show. The matter then returned to the Italian courts to apply the ruling based on Italy's presentation of any studies supporting its restrictions. The Italian court ordered the ban removed and the Italian government complied.[195]

More generally, European courts already have adopted similar language to that used by the WTO Appellate Body in examining EU and member-state measures in light of an assessment of the risks posed. As Scott writes, "WTO law may not have direct effect in European law, but its effect in this sphere is palpable nonetheless."[196] The decision thus enhances the leverage of pro-trade and pro-biotech constituencies in member states to press to have the safeguard bans legally removed, at least compared to the previous status quo. In short, although the panel sought to avoid deciding substantive matters itself, its decision (whether intentionally or not) can empower actors in intra-European political processes that can help to mitigate the dispute. We believe that there is evidence of such an intention structurally within the panel report. In terms of assessing the role of international actors such as WTO judicial panels, we thus need to disaggregate the state in terms of competing constituencies, and, through this lens, examine how international institutions can affect politics inside them.

5.6.2. The limits of WTO jurisprudence

Legal victories, of course, do not mean commercial ones, and here we begin to encounter the limits of WTO influence in the EU's domestic law and politics, as we will show in Chapter 6. Italians do not grow or consume GM foods, even though Italy's safeguard ban was removed following Monsanto's court case. Even were the Commission to approve new varieties, approval does not mean commercial acceptance. Moreover, EU member states can again reject the Commission's approval decisions by implementing national safeguard bans. Hungary did so in respect of the GM corn variety approved by the Commission during the WTO proceeding.[197] The Commission can, under the terms of EU law, challenge those member-state safeguards, and has done so in other contexts. As we will see in Chapter 6, however, the EU's member states generally have opposed by lopsided margins any Commission effort to overturn those national bans, making any legal challenge before the ECJ exceptionally sensitive in political terms.

The Commission may also feel constrained from continuing to approve new GM varieties without the support of the Council of Ministers. As Charlotte Hebebrand of the delegation of the European Commission in Washington D.C. pointed out, "although legally the Commission could keep approving these applications under these scenarios, it is a 'bit of a question mark' whether they can keep doing so politically."[198] The Commission may thus attempt to modify application of the EU regulations to ease tensions between EU approvals and member-state resistance. We see signs of this approach in the Commission's October 2006 assessment of implementation of 2003 Directive 1829/2003 on GM food and feed. As we will see in Chapter 6, the Commission, under pressure from EU member states, has asked the EFSA to take account of member-state scientists' views and to broaden its analysis, in line with member-state demands, "to address more explicitly potential long-term effects and bio-diversity issues" in its risk assessments. Since assessing potential long-term effects of GM varieties is extremely difficult, and implicitly raises much greater uncertainty, any EFSA finding in this respect could make EU nonapprovals and member-state safeguards more "challenge-proof" under the SPS Agreement. In addition, assessments of long-term health and biodiversity effects would require applicants to conduct further tests resulting in considerable cost and further delays. As a result, some applicants might (conveniently for the Commission) forgo the EU approval process, especially applicants with fewer resources.

The penalties for noncompliance with WTO decisions, furthermore, are limited, as the Dispute Settlement Body can only authorize the complainant to withdraw trade concessions (such as by raising tariffs) on products from the respondent equal to the amount of trade lost by the complainant as a result of the measures in question.[199] Moreover, such damages are only calculated on a prospective basis, and are not triggered until after the date set for implementation of the panel decision, which can be six months to over a year after adoption of the ruling. For example, in the *EC-Biotech* case, the parties reached an agreement in June 2007 that established November 21, 2007 as the deadline for implementation of the panel decision, which they then extended until January 11, 2008 (almost five years after the initial filing of the WTO complaint), with the US later extending the deadline for a "limited" but unspecified period of time.[200] The concrete steps necessary for implementation remained in dispute.

Given the extraordinary political sensitivity of GMOs in European public opinion, EU member governments may prefer to accept US retaliatory tariffs against European products (were the US to impose them), however painful these may be, rather than back down on the issue of GMO approvals or on the national safeguard bans. Indeed, following the Appellate Body ruling in the *EC-Meat hormones* case, the EU accepted the imposition of US retaliatory tariffs, rather than permit the importation of hormone-treated beef from the US and confront the likely outcry in European public opinion.[201] Put simply, WTO remedies in the *EC-Meat hormones* case, and possibly in the GMO case, even when coupled with the adverse impact on the EU's reputation for its failure to comply with WTO law, may provide an insufficient incentive for EU member governments to change their behavior.[202]

For these reasons, many US officials realize that the transatlantic conflict over agricultural biotechnology will not be resolved through WTO litigation, but only through a change in attitudes of member-state constituents toward GM foods.[203] The fact that neither side appealed the decision suggests that neither was so upset with the outcome—both could declare a sort of victory.[204] In addition, the non-appeal of the decision could suggest that each side saw little prospect of "winning" through further deployment of WTO law.

The WTO panel decision may nonetheless be useful for the complainants as a tool to help ease acceptance of GM varieties in other countries, and in particular in developing countries, as we will explore in Chapter 7. The WTO complaint, in this sense, was never just about the EU. In fact, farmers in India and China are rapidly adopting GM cotton, so that Europeans will increasingly sport GM-cotton T-shirts while eating their GM-free foods. Ironically, Europeans' choice for apparel may largely be between GM cotton and synthetics.

The US, it appears, has little choice but to take a longer-term view that technological advance will eventually win out if it offers real benefits and risk assessments of specific varieties continue to show that there is no evidence that they pose greater risks than conventional counterparts. GM varieties might even be accepted in Europe following their growing use around the world. The EU regulatory process already provides a framework for the approval of GM varieties based on a risk assessment. US companies and USDA officials are in ongoing contact with European farmers about the potential of GM varieties, and some farmers are interested.[205] Over time, if European constituents feel at greater ease with agricultural biotech products, European farmers producing on a larger scale would likely grow them.

5.6.3. Addressing the challenge of the adventitious presence of GM content in trade

In light of our discussion on the role that WTO rules and dispute settlement can play in helping countries to reconcile regulatory differences, we make a final observation as to how the EU could modify its existing legislation in a manner that would enable it to continue with its constraining regulation of GM products (if that is the EU's preference), while being more accommodating of other countries' trade and domestic agricultural and consumer welfare interests as determined through these countries' internal decision-making processes (and in particular, those of developing countries). The EU exercises considerable market power in affecting even US regulatory choices and US agricultural practices on account of the size of the EU's market and the difficulty (if not impossibility) of segregating grain varieties for domestic consumption and trade to the EU. As we will see further in Chapter 6, the EU currently has a zero-tolerance policy for the adventitious presence of GM content in EU imports where a GM variety has not been approved by EU political authorities, even where the EU's own scientific authority (EFSA) has found that the variety in question poses no greater health risks than conventional agricultural varieties. If the EU creates a more expedited process following positive EFSA assessments for sale of varieties as food or feed, then this zero-tolerance threshold for non-approved varieties will matter less. But if the EU combines a zero-tolerance threshold with a refusal to assess without delay and approve varieties for consumption that EFSA has found to be safe, then the EU will not only affect international trade, but also seriously constrain other countries' abilities to make choices over this technology, as we will see in Chapter 7 regarding developing countries.

These other countries may have valid reasons to decide to adopt GM varieties, whether because they find that specific GM varieties reduce the need to use pesticides and herbicides that are harmful to the environment and farm laborers, increase yields and thus reduce prices for consumers and increase income for farmers, or offer other benefits, such as nutritional ones.[206] Different constituencies, confronting different ecological, economic, and social challenges and having different policy priorities may disagree on the relative benefits and costs of adopting transgenic varieties. Yet they are concerned about the EU's ability to constrain (or even determine) their choices through the EU's exercise of market power when the EU's own scientific authorities find that the varieties in question, much less the minute content of them that might be adventitiously imported, pose no additional risk to the health of EU citizens. At a minimum, it can be argued that this particular EU regulatory provision, when combined with a blocked approval process, is disproportionate in its impact on other countries' regulatory choices. Through such an EU legislative change, other countries (and, in particular, developing countries) would be better able to make their own choices about transgenic varieties without being effectively constrained from doing so on account of the externalities of the EU's zero-tolerance threshold in a world of integrated product markets for which it is virtually impossible to segregate grains completely.

We see some prospect for such a change, although it confronts the severe limits that we have identified. As we saw in Chapter 4, Codex and the OECD began to address this issue in 2007, raising its international profile. Later that year, as we will see in Chapter 6, key members of the European Commission called for a modification of this EU rule to permit some tolerance threshold for the adventitious presence of GM content in cases where EFSA has determined that the GM variety in question poses no additional health risks, appearing to do so primarily because of EU farmer demands for animal feed. However, the Commission would have to propose such a legislative change, and the EU Parliament and the Council would each have to approve it, rendering change difficult. In the absence of EU legislative change, we could eventually see a new WTO case initiated against the EU concerning this provision, which would provide some additional leverage for, but by no means any guarantee of, such a policy accommodation.

5.7. Conclusions

To understand the role played by international law and international organizations, one must study the reciprocal relationship between domestic and international processes, which mutually and dynamically affect each other. The WTO judicial process, whose operation we have analyzed in detail in this chapter, will not in itself resolve the differences between the US and the EU, nor will it necessarily even secure compliance of the parties with the decisions of the WTO panel. There are stark limits to the accommodation of deep conflicts, such as over the regulation of GMOs, in particular, given the deeply embedded, path-dependent nature of US and EU regulatory approaches to GMOs (discussed in Chapters 2 and 6), and the largely failed attempts to reach consensus through bilateral networks and multilateral regimes (reviewed in

Chapters 3 and 4). Nonetheless, despite its inherent limitations, the WTO judicial process *can* play a positive role in the ongoing transatlantic and global conflicts over the regulation of agricultural biotechnology, regularizing political conflicts into legal channels, encouraging states to regulate risks in more transparent ways and take into account the interests of affected foreigners, and empowering domestic constituencies that have an interest in complying with the international law and legal decision in question.

The WTO judicial process does not simply judge the legality of national regulatory measures, but also (in the process) can affect the processes through which national measures are determined. In affecting these processes and the measures to which they give rise, it can also, over time, affect national positions in international politics. In the GMO context, by providing a framework of legal rules, the WTO can facilitate dialogue between governments and constituencies concerning the objectives of GMO regulation, the means used to achieve these objectives, and the impact of these choices on different constituencies. As Robert Howse writes, "SPS provisions and their interpretation by the WTO dispute settlement organs...can be, and should be, understood not as usurping legitimate democratic choices for stricter regulations, but as enhancing the quality of rational democratic deliberation about risk and its control."[207]

The WTO SPS Agreement's requirements, and, in particular, the requirement that regulation be based on a risk assessment, cannot guarantee that regulatory policy decisions will be rationally made in a deliberative manner, taking into account the impact on affected constituencies. WTO judicial decisions do not determine procedural or substantive outcomes, especially where issues are politically charged. Far from it. But in light of the alternatives, WTO requirements, such as that regulatory measures be based on a risk assessment, can provide information to national regulatory processes so that regulatory decisions are more likely to be informed and subject to legitimate challenge within the regulating state than in the alternative.[208] In this way, WTO rules can help push WTO members to take into account the impact of their decisions on others, and to justify their decisions in legal and policy terms, and thereby facilitate exchange between governments at the international level and between governments and their constituencies nationally.

WTO rulings can only have limited effects on path-dependent regulatory frameworks and entrenched interests, and we will examine (in Chapter 6) how member-state and public opposition constrains the Commission's ability to comply with the WTO ruling. Nevertheless, we will also see how WTO rules arguably have already made the EU decision-making process somewhat more transparent and more flexible, as they have pressurized EU and member-state authorities directly, and European constituencies indirectly, to justify their decisions over biotech regulatory measures. WTO rules, in particular, have already pressed the EU not simply to ban all GMOs, but rather to engage in case-by-case risk assessments and the adoption of a new labeling regime. In our view, nondiscriminatory labeling rules, in particular, are likely to survive a WTO legal challenge, even though we recognize that labeling can sometimes serve to spread and deepen consumer prejudices, as the ECJ has itself noted.[209] We now return to the EU and US regulatory systems to assess the extent to which they have changed, and possibly converged, in light of the transnational and international initiatives and pressures that we have examined in Chapters 3, 4 and 5.

6

US and EU Policies Since 2000:
Change, Continuity, and (Lack of)
Convergence

The story of GMO regulation thus far has taken us from the domestic to the international level and from regulatory polarization to a so-far futile attempt to agree upon a global regulatory standard for GM foods and crops. Despite this disappointing record of bilateral and multilateral deliberation and negotiation, many observers had hoped throughout this period that the US and EU regulatory systems for agricultural biotechnology might converge, reducing the substantial disparity among the two systems and thereby mitigating the substantial trade tensions and resulting political conflicts between them. The mechanisms hypothesized to drive such a convergence varied. While some analysts placed their hopes in the power of deliberation or persuasion, others emphasized the importance of pressure exerted on one or another side by international organizations like the WTO, by national and transnational interest groups and public opinion, or by market forces. Many American observers, for example, hoped that the EU, under pressure from WTO rules, might move toward what they called a more "science-based" and less "politicized" system of regulation, which would in turn facilitate the resumption of approvals for new GM varieties. Many European observers, by contrast, hoped that either public opinion or market pressures would prompt a process of "trading up" in the US, which might become more precautionary in its own regulations and more demanding in the conduct of its own risk assessments.[1]

In this chapter, we review the development of both EU and US regulatory policies and practices regarding agricultural biotechnology since 2000, in light of bilateral and international pressures. In theoretical terms, this chapter thus falls within the "second image reversed" tradition in international relations theory, and "transnational legal process" theories of law, which respectively examine the ways in which international processes influence domestic politics, and thus laws and practices, in individual states.[2] We identify the direction and the sources of change, and assess the evidence for convergence between the two systems. The first and more detailed section examines the remarkable root-and-branch reform of the EU's regulatory framework. This reform, we argue, was driven partly by the European Commission's concern with safeguarding the EU regulatory system from possible WTO challenge, and included an upgrading of EU risk assessment that superficially resembles US practice and demands. Yet, as we shall see, the reform of the EU system also took place in the face of strongly mobilized public opinion, as well as opposition to GMOs from a sizable group of EU member governments that collectively exercised a veto over any reforms. The result, far from convergence on the US regulatory system, has been a steady and systematic ratcheting-up of EU standards, which are now amongst the strictest in the world. Even after the resumption of approvals in May 2004, the EU regulatory process remains deeply politicized, operating in the face of strong anti-GMO public opinion.

In the second section, we examine the domestic debate over the adequacy and possible reform of the US regulatory framework for GM foods and crops, and the resulting incremental changes to the behavior of market actors as well as US national regulators. In keeping with the "trading up" thesis, we find marginal evidence of movement in US public opinion against GM foods and crops, as well as increased caution by US farmers in adopting new GM varieties not yet approved for sale in the EU and other export markets, and we take note of emerging regulatory challenges such as the development and sale of GM animals. By and large, however, the US regulatory system has proven remarkably stable in the face of these changes, supported by a strong industrial lobby, weak public opinion, and government regulators operating within an institutional framework that has changed very little in its essentials over the past decade.

In sum, we argue, the story of US and EU regulation of GMOs since 2000 has primarily been one of continuity, with the US retaining its preference for a relatively lax, technocratic, and product-oriented regulation, while the EU has retained the essential features of its more precautionary and process-oriented regulatory system involving individual approval decisions by politicians. There have, to be sure, been changes on both sides of the Atlantic, but in both cases these changes have demonstrated a clearly path-dependent nature, with the EU altering its regulatory process in the direction of greater precaution, while US regulators and market actors have made incremental changes to their previous practices in response to market pressures and public opinion. In terms of the larger story of regulatory polarization and international conflict, finally, we find at best modest signs of convergence between the US and the EU which retain distinctive systems in terms of regulatory approaches, procedures, and policy outcomes for agricultural biotech regulation.

6.1. European developments since 2000:
Reform without change

By the late 1990s, the European Commission found itself in a difficult situation, facing competing pressures from above and below. From above, the Commission faced significant pressures from the US and from a potential WTO case that might challenge the EU's *de facto* moratorium or even its regulatory system as a whole. Forestalling such a challenge, and safeguarding the EU system against WTO legal challenge, was therefore a consistent priority of the Commission throughout the 1990s and into the 2000s. From below, meanwhile, the Commission faced equally great but opposing pressure from European public opinion and from GM-skeptical member governments seeking even more precautionary GMO regulations as a prerequisite to any resumption of approvals for new GM foods and crops.

Under the circumstances, the Commission adopted a strategy aimed at satisfying both sets of demands: In the first stage, the Commission would put forward proposals for new legislation that would "complete" the EU's regulatory framework, putting in place additional features (such as a labeling and traceability system) that had long been demanded by GM-skeptical member governments, as well as other features (such as a more systematic use of risk assessment and clearer provisions on national bans) insisted upon by the US. In the second stage, with a "completed" regulatory framework in place, the Commission would seek to resume approvals of GM varieties, thereby resolving the central element of the US's grievance against the Union.

6.1.1. The White Paper on Food Safety and the precautionary principle

Toward this end, in January 2000, the Commission issued a White Paper on Food Safety in which it proposed that the EU would overhaul its food safety system and establish a new centralized EU agency, which was eventually named the European Food Safety Authority (EFSA), to assist with risk regulation.[3] The White Paper set forth the EU's general approach to risk regulation in the food sector, dividing "risk assessment" from "risk management." Specialized scientific committees within the new food authority would conduct risk assessments, and the authority would provide food safety information to consumers and operate a rapid alert system in conjunction with member-state authorities to respond to food safety emergencies. Risk management, by contrast, would remain under the control of the EU's political bodies. In an annexed "action plan" the Commission set forth over eighty new food safety-related measures for adoption, including amendments to Directive 90/220 and the 1997 Novel Foods Regulation.

Table 6.1 Key events in EU biotech regulation, 2000–2008 (November)

2000 (Jan.)	White Paper on Food Safety
2000 (Jan.)	Communication on the Precautionary Principle
2000 (Jan.)	Cartagena Protocol on Biosafety Adopted
2001 (March)	Council and EP adopt Directive 2001/18, replacing 90/220, on deliberate release of GMOs
2002 (Jan.)	Establishment of European Food Safety Authority
2003 (May)	US launches WTO complaint over EU regulation of GMOs
2003 (Sept.)	Council and EP adopt Regulation 1830/2003 on Traceability and Labelling of GMOs
2003 (Sept.)	Council and EP adopt Regulation 1829/2003 on Genetically Modified Food and Feed
2004 (Apr.)	Entry into force of Regulations 1829/2003 and 1830/2003
2004 (May)	Commission ends moratorium with approval of Bt-11 maize
2005 (May)	Commission "orientation debate," decision to press ahead with new approvals, challenge member-state bans
2005 (June)	Environment Council rejects Commission's effort to overturn national safeguard bans
2006 (Sept.)	WTO ruling in *EC–Biotech* dispute
2007 (July)	Member states fail to agree on approval of GM potato for cultivation
2007 (Nov.)	EU Environment Commissioner Dimas proposes to deny approval of two GM maize varieties for cultivation
2008 (Jan.)	France announces new national ban on MON810
2008 (May)	Removal of Austrian safeguard ban on sale of two varieties
	Commission delays approval of seven new varieties pending new studies
2008 (July)	French EU Presidency announces creation of group to consider strengthening EU regulations on GMOs

In February 2000 the Commission issued a Communication on the precautionary principle, indicative of EU authorities' more risk-averse approach in a politicized domain that was raising challenges to the legitimacy of EU law. The Commission declared that the "precautionary principle" would be applied whenever decision-makers identify "potentially negative effects resulting from a phenomenon, product or process" and "a scientific evaluation of the risk...makes it impossible to determine with sufficient certainty the risk in question [on account] of the insufficiency of the data, their inconclusiveness or imprecise nature." It stressed that "judging what is an 'acceptable' level of risk for society is an eminently *political* responsibility."[4] The Commission nonetheless maintained that it hoped to provide some guidance regarding the application of the principle, which it acknowledged was "giving rise to much debate and to mixed and sometimes contradictory views."[5] Building on the case law of the European Court of Justice, the Commission stated that where regulatory decisions are adopted in accordance with the principle, the resulting measures should meet a series of criteria, and in particular, they should be proportionate, nondiscriminatory, consistent, based on cost-benefit analyses where feasible, and subject to review and ongoing risk assessment.[6] When the Council adopted a resolution on the precautionary principle in

December 2000 its evocation granted policymakers greater flexibility. The resolution maintained that risk assessments may not always be possible on account of insufficient data, and that cost-benefit analyses should consider the "public acceptability" of risk management decisions.[7] The EU's evocation of the precautionary principle, already too permissive in the views of US policymakers, had just become more so.

Some supporters of the precautionary principle argue that it represents a more democratic, process-oriented approach to regulation than cost-benefit analysis advocated by Majone, Sunstein, and many others who highlight the importance of regulation based on risk assessments. As Douglas Kysar writes, "A key benefit of the PP, in contrast [to cost-benefit analysis], is that it contains a built-in sensitivity to the need for collective deliberation."[8] However, in the larger political/sociological context, the Commission's use of the precautionary principle can also be seen as a shield from the pressures on it from above and below. The principle provides a first line of defense against a WTO challenge to the EU's regulatory practices, as we saw in our analysis of the WTO case in Chapter 5. The EU's defenders can argue, on the basis of the precautionary principle, that the WTO panelists have no legitimacy to determine the appropriate level of state regulatory precaution based on risk assessments. Simultaneously, the principle provides a means for the Commission to signal to European constituencies and European member-state authorities that it can be trusted in its regulatory decisions, including over new technologies for the production of food. When we hear the Commission saying "precautionary principle," we also hear it saying "trust us." Since a trusted administrator can be left with greater administrative discretion, the precautionary principle can also serve to enhance Commission authority and autonomy vis-à-vis the member states as well as the WTO.[9]

6.1.2. Directive 2001/18: Deliberate release revisited

In response to these competing pressures from above and below, the Commission proposed new legislation in 1998 to govern the deliberate release of GMOs into the environment and the placing of GM food products on the market.[10] Responding to the Commission proposals, both the European Parliament and Council immediately pressed the Commission for further regulatory controls. The majority of the European Parliament insisted on tighter restrictions regarding labeling requirements and thresholds pursuant to which products could contain traces of GMOs and still be sold in the EU. The member states were mixed in their views, with some appearing to do whatever possible to ensure that no GM crops would be grown in their territories (such as Austria and Luxembourg), and others being torn between the demands of GM opponents and those of the biotech sector (such as Germany and the United Kingdom). A "conciliation committee," consisting of the members of the Council and fifteen representatives of the European Parliament, drafted the final text.[11]

The resulting legislation, Directive 2001/18, was finally adopted in March 2001 by co-decision between the Council and the European Parliament.[12] The directive's twin objectives were to protect the environment and human health when GMOs are released into the environment and placed on the market "as or in products," in both cases to be applied "[i]n accordance with the precautionary principle." Once more, the need to assuage GM-skeptical member governments and MEPs—each of whom enjoyed veto power in the legislative process—had led to a ratcheting-up of EU regulatory requirements for GMOs.[13] More specifically, under the directive's environmental release requirements, member-state and applicant obligations were enhanced to include a more extensive environmental risk assessment, further information concerning the conditions of the release, and monitoring and remedial plans. The directive instructed the member states to adapt their laws to comply with its requirements by October 17, 2002, at which time Directive 90/220 would be repealed.

6.1.3. Food and feed, labeling and traceability: Regulations 1829/2003 and 1830/2003

Although touted by the EP's rapporteur David Bowe as "the toughest laws on GMOs in the whole world,"[14] the adoption of Directive 2001/18 did not satisfy a core of member states (in particular, Austria, Denmark, France, Greece, Italy, and Luxembourg), which insisted on the continuation of the *de facto* moratorium and on the need to impose national safeguard bans in the absence of still more stringent EU regulations.[15] Seeking to break the logjam on approvals and secure the removal of existing national bans, the Commission therefore put forward proposals for still more revisions to the legislative framework. For this reason, only the directive's provisions governing the release of GMOs into the environment remain in effect,[16] while its provisions governing the marketing of GMOs used for commercial crops were replaced within a mere eighteen months by two new EU regulations regarding the labeling and traceability of GM foods and their use in food and feed, respectively.

Proposed by the Commission in 2001,[17] these new regulations were finally adopted in September 2003 and went into effect in April 2004, once again after drawn-out bargaining among the Commission, Council, and European Parliament. Both legislative instruments took the form of regulations, and not directives, placing authority predominantly in the hands of Community institutions. Regulation 1829/2003, regarding the authorization of GMOs in food and feed, replaced the provisions of Directive 2001/18 governing the authorization for marketing of GMOs as or in products,[18] and the labeling provisions of the Novel Foods Regulation. Regulation 1830/2003, in turn, created

Table 6.2 EU legislation governing GMOs and GM products as of November 2008

Step-by-Step Activities in the Production Process	Applicable EU Legislation
GMO research in laboratories	Contained Use Directive 90/219
GMO experimental releases (trials)	Directive 2001/18
GMO environmental releases for crops	Regulation 1829/2003 and Directive 98/95/EC (common seed catalogue)
Authorization of marketing of GM seeds (for environmental releases for crops)	Regulation 1829/2003 and Directive 98/95/EC
Authorization of marketing of GM food and feed	Regulation 1829/2003
Labeling of GM seed, food, and feed	Regulation 1829/2003
Traceability and labeling of GM products	Regulation 1830/2003

new rules on the traceability of GM products throughout the production and distribution process. Both regulations became effective on April 18, 2004. The revised EU regulatory scheme is summarized in Table 6.2.

Again reflecting the pressure from member governments and MEPs, the new regulations tightened EU requirements in a number of respects. Regulation 1829/2003 broadened the definition of legitimate objectives that may be pursued in determining whether to approve a GM food or feed variety. The range of objectives has been expanded to include not only the protection of the environment and of human life and health (as under the 2001 Deliberate Release Directive), but also "consumer interest in relation to genetically modified food or feed," with the coverage of "consumer interest" being left open for interpretation.[19] Moreover, in preparing its draft decision over the authorization of a GM variety, the Commission is to take into account not only EFSA's scientific opinion, but also "other legitimate factors relevant to the matter under consideration."[20] It is unclear whether these "other legitimate factors" will only include the identified "objectives" listed in article 1, including consumer protection, or whether the Commission can also take into account yet other objectives, such as ethical ones, especially given that the new regulation specifically provides for Commission consultation with the "European Group on Ethics in Science and New Technologies."[21] Not surprisingly, the term "other legitimate factors" reflects the language used in Codex's amended Manual of Procedure, for which EU member states were instrumental.

The regulation also broadened the scope of product coverage. The regulation's authorization and labeling requirements cover GM animal feed for the first time, in addition to food for human consumption. Animal feed has been a particularly important European market for US corn producers. In addition, the regulation covers food and feed that do not contain or consist of GMOs, but nonetheless are "derived, in whole or in part, from GMOs" or contain ingredients that are "derived, in whole or in part, from GMOs."[22] Thus, the

241

new regulation codifies the end of the EU's simplified procedure for "substantially equivalent" food products under the Novel Foods Regulation.[23]

One of the most controversial elements of the new regulation was the establishment of a set of thresholds for permitted traces of GM ingredients, provided their presence is "adventitious." Recognizing that it is practically impossible to ensure that any shipment is entirely free of GM material because of the way crops are threshed, stored, and transported, the Commission initially proposed a threshold of 1 per cent GM material, below which any crop would not have to be labeled as containing GMOs. The Commission's proposed threshold was contested, however, by environmental groups such as Greenpeace and the European Consumers' Organization (BEUC), by the European Parliament, and by several member governments in the Council, all of which called for lower thresholds. European biotech companies and the US government, by contrast, criticized the 1 per cent threshold as unrealistic, unnecessarily costly, and scientifically unjustified. These divisions were mirrored in the Council, where the United Kingdom favored the Commission's proposed 1 per cent threshold, while Austria at the other extreme favored thresholds as low as 0.1 per cent.[24]

The final regulation established three distinct, increasingly stringent thresholds. First, it provided that food products need not be labeled as containing GMOs if they contain material consisting of or produced from EU-approved GMOs "in a proportion no higher than 0.9 per cent of the food ingredients considered individually ... provided that this presence is adventitious or technically unavoidable." Second, the regulation established a second and stricter threshold of 0.5 per cent for GMOs not yet approved for environmental release in the EU, provided that they have received a favorable EU scientific risk assessment. Third, however, it established a three-year window after which no residues of such non-approved GMOs would be allowed in food or feed product unless new regulations are enacted (i.e., "zero tolerance").[25] This zero tolerance threshold went into effect in April 2007, and has caused severe strains with the US because of the difficulty (if not impossibility) of segregating grains for bulk shipments in international trade.

The regulation's authorization and labeling requirements nonetheless do not apply to food produced with the aid of GMOs, such as meat, milk, and eggs produced from animals fed with GM feed. This exception is crucial for the US and other grain exporters, as imported GM soy and corn are frequently used in animal feed in the EU, and the EU does not require labeling of the resulting meat. Similarly, cheese is not covered by the regulation's authorization and labeling requirements when a GM enzyme is used, to the extent that the enzyme is only a processing aid so that no GM source material is in the cheese. As Europe is a leader in the development of GM enzymes, and cheese is a major product in many European states for domestic consumption and export, this exemption suggests that biotech business interests have prevailed at least in some areas.[26] The European Parliament, and European NGOs such as Greenpeace, had wanted the labeling requirements to cover these products, but did not prevail.[27]

Regulation 1829/2003 also created a more centralized authorization procedure to regulate the placing of GM food and feed on the EU market. With a more centralized procedure, the Commission hopes to better manage countervailing member-state and US challenges to the EU's regulatory regime. Although the procedural scheme at the EU level is similar to that provided under Directives 90/220 and 2001/18, it became more centralized in two primary respects. First, EFSA, a centralized EU agency, oversees the application file and works in conjunction with member-state competent authorities and a Community reference laboratory to conduct risk assessments and product evaluations. Second, the regulation restricts the grounds on which member states may ban GMOs unilaterally as a "safeguard" measure. A member state may adopt "interim protective measures...where it is evident that products authorized...are likely to constitute a serious risk to human health, animal health, or the environment," provided that it first informs the Commission of the "emergency" situation and the Commission does not act. The Commission's original proposal, it should be noted, provided for no member-state safeguard powers at all, but the Parliament and Council succeeded in including this clause, although the Parliament preferred to grant even greater autonomy to member-state authorities.[28]

The application process still begins when an operator submits an application file to the competent authority from one of the member states. That member-state authority, however, now immediately provides the file to the EFSA, which in turn provides a copy to the other member states and the Commission, and makes a summary of the file publicly available. EFSA is to issue its opinion, based on risk assessments, within six months from its receipt of the file, subject to extensions if further information is needed. EFSA submits its opinion to the Commission, the member states, and the applicant, and after the deletion of any confidential information, makes it publicly available. The Commission is then to issue a draft decision, which may vary from EFSA's opinion. The Commission's draft decision is again provided to a regulatory committee, the Standing Committee on the Food Chain and Animal Health, consisting of member-state representatives. If the committee supports the Commission's proposed decision by a qualified majority vote, then the decision is approved. If the committee fails to approve it by a qualified majority, then the Commission must submit its proposal to the Council. Unless the Council, in turn, opposes the Commission's proposal by a *qualified majority vote*, the proposed decision "shall be adopted by the Commission," unless the Commission independently withdraws its proposal.[29] These voting rules represent a significant change compared to those applicable under Directive 90/220, under which the member states could only overturn a Commission decision by a unanimous vote.[30] Under the new rules, by contrast, a qualified majority within the Council is sufficient to overturn a draft Commission decision.[31] Any authorization of a GM variety is now limited to a term of ten years, although it is subject to renewal. For a brief summary of current authorization procedures, see Table 6.3.

Regulation 1830/2003, in turn, complemented the new authorization and labeling rules with a more centralized framework for tracing and labeling GM products. In contrast, Directive 2001/18 had left this responsibility to the member states. The new regulation requires the Commission to establish a system of unique identifiers for each GMO in order "to trace GMOs and products produced from GMOs at all stages of their placing on the market through the production and distribution chain." More specifically, the regulation requires producers to collect and retain for five years data regarding the GM content of foods and crops one step backward and one step forward in the distribution chain. These strict traceability requirements have been bitterly criticized by many US producers (whose commodity system does not require and is not designed for tracing GMOs through the distribution chain), as well as by some European producers. A representative from the American Soybean Association called the EU's traceability requirements a "fantasy" that would be "impossible" to meet.[32] The Commission, however, has justified it as vital to the EU labeling system as well as for any future recalls of GM foods or crops.[33]

Finally, the Commission faced pressure to propose new EU rules governing the coexistence of GM and non-GM crops, and liability for contamination of non-GM crops. Responding to concerns that GM seeds grown on a neighboring field could contaminate a conventional farmer's crop, making it impossible for that farmer to meet EU purity obligations, the Commission held a roundtable in April 2003 with a range of "stakeholders representing the farm-

Table 6.3 Authorization process for GM food and feed under regulation 1829/2003

1. An operator submits an application to the competent authority from one of the member states.
2. The member state provides the file to the new European Food Safety Authority (EFSA).
3. The EFSA provides a copy to the other member states and the Commission, and makes a summary of the file publicly available.
4. Within six months, the EFSA, based on risk assessments, submits its opinion to the Commission, the member states, and the applicant, and after the deletion of any confidential information, makes it publicly available.
5. The Commission is then to issue a draft decision, which may vary from EFSA's opinion, based on the regulatory committee consisting of member-state representatives.
6. The committee is to deliver its opinion on the Commission's proposed decision by a qualified majority. If the committee delivers no opinion or a negative opinion, the Commission must submit its proposal to the Council. If the Council does not adopt (or indicate its opposition to) the Commission's proposal (but this time by a qualified majority vote, as opposed to a unanimous one), then the proposed decision "shall be adopted by the Commission."

ing sector, industry, NGOs, consumers and other players."[34] In July 2003, the Commission issued guidelines on coexistence but left rulemaking to the member states on the grounds of "subsidiarity"—the EU principle that regulation should be left to the national level where practicable.[35] Likewise, both of the regulations enacted in 2003 provided that "the member states shall lay down the rules on penalties applicable to infringements."[36] Despite these efforts, a number of European interest groups and national governments have continued to call for binding legislation establishing rules for the coexistence of GM and GM-free crops, together with a system of legal liability to deal with instances of contamination, and have resisted a resumption of approvals until such steps are taken.

6.1.4. The end of the moratorium—but not of controversy

With the "completion" of the EU's legislative framework for agricultural biotechnology, the Commission at long last sought to enforce existing legislation, resume approvals of new GM varieties after the six-year moratorium, and bring an end to the national bans on varieties that had been accepted by the EFSA as safe. Nevertheless, the implementation of the EU's new rules faced continuing political contestation, resulting in a pattern of formal voting and deadlock within both the regulatory committee and the full Council. The "completion" of the EU's regulatory structure, whatever its other merits, has not changed the contentious nature of GMO regulation in the EU.

6.1.4.1. THE RESUMPTION OF APPROVALS

Most significantly for our purposes in this chapter, after a six-year lapse, in the midst of the WTO legal proceeding (examined in Chapter 5), the EU resumed approvals of new GM varieties in May 2004. By that time, the Commission had received twenty-two notifications for approvals of GM varieties—eleven involving import processing only, and eleven for cultivation.[37] With the completion of the new regulatory framework, the Commission moved to resume approvals of new GM varieties. In November 2003, it proposed to approve the importation and marketing (but not the cultivation) of a variety of GM maize (Bt-11 sweet corn), for which EFSA had delivered a favorable opinion. It was the first time that the Commission had initiated a GM approval since 1998. The regulatory committee, however, again refused to approve the Commission's proposal, so the matter was referred to the Council, which was given until the end of April to act.[38] On April 26, 2004, a divided Agriculture Council failed to reach agreement on the Commission's proposal.[39] In the absence of a decision by the Council, the Commission was free to adopt the proposal the following month—the first new approval of a GM variety in nearly six years.[40]

Despite this apparent breakthrough, US officials and industry representatives noted that the Commission's decision had been taken over the objections of a bloc of implacably hostile member governments, and a chorus of condemnation among European environmentalists and consumer groups, with no guarantee that additional approvals were to follow or that EU risk managers would continue to be guided by the scientific risk assessments carried out by EFSA. As one US official put it, "The approval of a single product is not evidence that applications are moving routinely through the approval process in an objective, predictable manner based on science and EU law, rather than political factors."[41] Moreover, the Commission's approval applied only to importation and not cultivation, and was subject to the full range of EU regulations regarding traceability and labeling, with their attendant costs.

Subsequent approval procedures appeared to support this cautious interpretation of the "end" of the moratorium. One month following the approval of the Bt-11 sweet corn, a similar pattern emerged when the Environment Council of a newly enlarged EU of twenty-five states met to consider the Commission's recommendation to approve another Monsanto variety, the NK603 GM corn (for import and marketing, not cultivation). Here again, the Council was divided, with nine member states (including four of the new members) reportedly voting against, nine in favor, and seven abstaining.[42] Although the Commission approved NK603 in July 2004 in the absence of Council agreement, this case once more demonstrated the persistent divisions among the member governments on new approvals.[43]

Later that same year, in June 2004, a regulatory committee failed to agree on the Commission's proposed approval of GT73 GM rapeseed (canola), with six of the ten new member states (Cyprus, Estonia, Hungary, Malta, Lithuania, and Poland) joining six existing members (Austria, Denmark, Greece, Italy, Luxembourg, and the UK) in voting against the approval.[44] In December 2004, the Environment Council once again deadlocked, by a vote of six countries in favor and thirteen opposed, with six abstentions, once again leaving the final decision to the Commission.[45]

This pattern would be repeated again and again in the coming years. The Commission, after consulting the EFSA, proposed a series of draft decisions authorizing the importation and placing of various new GM crops on the market. In each and every case from May 2004 through November 2008, the relevant comitology committees deadlocked (i.e., failed to reach a qualified majority for or against approval), resulting in the submission of the draft decision to the Council of Ministers. Such deadlocks, it is worth noting, are extremely unusual within the EU's expert committees. In its report on the workings of comitology committees for 2005, for example, the Commission noted that out of 2,637 draft decisions submitted to the various EU expert committees that year, only eleven of those decisions (less than 0.5 per cent) were referred to the Council for a decision—and six of these eleven were draft decisions authorizing the placing of GM foods and crops on the market.[46] The pattern of deadlock, moreover, persisted in the Council of Ministers (meeting variously in its Environmental and Agriculture formations), which failed repeatedly to reach qualified majorities for or against the approval of one new GM variety after another, leaving the Commission in each case to author- ize the new variety unilaterally, to choruses of condemnation from member governments, members of the European Parliament, and environmental and consumer groups.[47]

These decisions, moreover, seemed to dispel initial expectations that the new member states—several of which were already engaged in small-scale cultivation of GM crops, often without adequate controls—might serve as a "Trojan horse" for the United States and the biotech industry.[48] Ensuring adequate testing facilities in the new member states remains a challenge for the EU post-accession, but it seems clear that the ambivalence toward agri- cultural biotechnology in "old" Europe is shared in the public opinion and governmental positions of the EU's new members as well. A 2003 survey of citizens of the EU's ten new members in relation to these countries' pending accession showed that sixty-eight percent held negative views toward GMOs, a result roughly similar to surveys of citizens of the existing members.[49] One of the new member states, Hungary, proceeded to adopt some of the strong- est anti-GMO policies in the Union, issuing a safeguard ban on Monsanto's genetically modified YieldGuard corn (MON810) and adopting a strict coex- istence law that would require growers of GM crops to establish 400-meter buffer zones between GM and conventional crops and procure the written consent of neighboring farmers.[50] Another new member, Poland, ran afoul of the Commission in October 2006 when its parliament adopted a strict law for- bidding the cultivation and marketing of all GMOs, leading the Commission to initiate infringement proceedings against Poland for violation of EU law.[51] More generally, they indicate the European Commission's awkward position in proceeding with new approvals amid consistent member-state opposition.

6.1.4.2. THE COMMISSION'S STRATEGY

In May 2005, the new Commission of President José Manuel Barroso held an "orientation debate" on GMOs, examining past Commission policy and lay- ing down guidelines for future Commission action to implement the EU's legis- lative framework for GM foods and crops.[52] In preparation for this meeting, an interservice group of commissioners prepared an internal communication to the College of Commissioners, which is remarkable for the candor with which it describes the state of affairs, and is worth citing at length. The com- munication begins by noting that the completion of the EU's strict regulatory framework had not succeeded in overcoming resistance to GMOs among the public or among the representatives of the member states. With regard to the latter, the commissioners noted the difficulty of resuming approvals and of overturning the member-state bans in the face of member-state opposition:

> In spite of the application of the new regulatory framework, so far it has proven impossible to obtain support of a majority of Member States when it comes to implementation, namely the adoption of decisions on specific products. However, the legislation itself is not contested.
>
> At the current time, only a few Member States tend to vote consistently in favour while several Member States tend to vote consistently against and many abstain. Other Member States' position varies; some of them consistently follow the advice of their own scientific bodies which sometimes diverge from the European Food Safety Authority (EFSA) assessments.
>
> Against this background, it will be difficult if not impossible to obtain a qualified majority either in favour or against the approval of the pending decisions in either the Regulatory Committee or the Council.
>
> In view of the above, it is expected that, in accordance with the legislation, **the Commission will** have to continue to take ultimate responsibility **for adoption of pending decisions for the placing on the market of new GMO products** at least for the immediate future.[53]

In light of this situation, the Commission laid out a plan of specific actions, including the continued submission to the Regulatory Committee and the Council of draft approvals for all new GM varieties "if there are no risks to human health and to the environment based on scientific information."[54]

In addition, the Commission indicated that it would pursue its challenge to the eight national bans that had persisted under the safeguard clause. Relying on the safeguard clauses of Directives 90/220 and 2001/18, five member states had adopted temporary bans on eight different GM varieties, which were claimed to be unsafe to put on the market.[55] The EFSA, however, had concluded that none of these bans was justified in scientific terms, and in November 2004 the Commission proposed to the Regulatory Committee that these bans be overturned, but the committee once again deadlocked, returning the question to the Commission. Following its March 22nd meeting, the Commission resolved to forward the eight draft decisions to the Council of Ministers, which it hoped would vote to overturn the bans.

Perhaps most strikingly, the Commission commented on the role of the member governments in the Regulatory Committee and the Council. "So far," it noted, **"every single one of the 13 Commission's proposals failed to get the required qualified majority**, even for those GMOs not intended for cultivation, but for import and processing only."[56] It continued with an explicit challenge to the member governments:

> Both the Commission and the Member States have a role to play in implementing this legislation. *However, so far, some Member States have tended to avoid taking a position in the Regulatory Committee and in the Council. Member States should be called upon to participate effectively in the process with a view to reaching clear positions.*

In the current legal context, when submitting proposals following an inconclusive opinion of the Regulatory Committee, *the concerned Councils should be requested to hold a thorough debate* in order to avoid adoption by abstention and to openly discuss the reasons for their reluctance to support the authorization of specific products which the Commission considers to be in compliance with the EU regulatory framework.[57]

In this remarkable passage, the Commissioners, in effect, call upon the Council to engage in the type of deliberation that had been called for by scholars such as Habermas and Joerges, but had been strikingly absent within the Council or its committees in this issue area. The Commission meeting itself revealed some differences among the various Commissioners in their attitudes toward GM foods and crops, and the full Commission reportedly decided to delete in full the previously cited paragraph calling on the Council to have a thorough debate (the idea of deliberation met *realpolitik*). This action itself suggests a lack of faith in such a deliberative process in the politically polarized EU and international context. The meeting nonetheless backed the substance of the communication, including the decision to proceed with new approvals and with the proposed overturning of the national safeguard bans.[58]

Our informants confirm the Commission's frustration with the lack of reasoned deliberation in the Regulatory Committee and Council over GMO approvals. As one member-state representative described the comitology process for GMO approvals:

The Commission presents a text for a variety's approval and sends it to the member states within the time limit provided for the Regulation (generally 15–30 days in advance). The member-state representatives come to the meeting and there is a first *tour de table* in which remarks are usually very general because if you clearly say you are for or against authorization, then the Commission won't listen to your proposals for changes in the text, such as the addition of further conditions for an approved variety. Then a representative will push further and countries start declaring their positions. The coffee break becomes an important time when countries discuss their positions, including regarding textual revisions. Generally some countries have engaged positions. Other countries are less clear. Those countries that are less clear can have an advantage because the Commission is more likely to take account of their textual amendments in order to obtain their vote. Germany is an example of a country that is excellent in creating suspense. For Austria, however, the Commission knows it will always vote no so the Commission has no reason to accept Austria's amendments to its draft decision. Having a clear position weakens your position vis-à-vis the Commission.[59]

Once again, we see very little evidence here, or in other sources, of deliberative decision-making when it comes to approval of new GM varieties. While some member-state votes have varied by the GM variety under consideration (so that, for example, a country may vote yes to approve the sale of a GM cut flower but no for the approval of a corn variety), most member governments appear to vote consistently for or against approval of any GMOs (or in some cases, consistently abstain), and national representatives appear to have little flexibility to change their national position on the basis of information presented in comitology meetings or in the Council.

6.1.4.3. SETBACKS AND OPPOSITION

In practice, the Commission's new strategy met with a number of immediate setbacks. On March 22nd, 2005, the day of its GMO debate, the Commission was notified by the Swiss agribusiness firm Syngenta of an accidental release of the experimental Bt-10 corn, an estimated 1000 tons of which had been distributed in the US along with the well-established Bt-11 corn. Bt-11 corn had been approved in both the US and the EU, but Bt-10 had yet to be submitted for approval in either jurisdiction. Within the US, the Environmental Protection Agency and the Department of Agriculture investigated the accidental release and the properties of the experimental corn, eventually ruling that Bt-10 posed no threat to humans, plants, or animals. By contrast, the European Commission, under pressure from public opinion and facing the prospect that some of the unauthorized corn might have entered the EU among US imports, issued a statement "deploring" the import of an experimental and unauthorized GM crop into the EU.[60] In April, the Commission went further, requiring all imports of corn gluten feed and brewers' grain from the US to be certified as free from Bt-10.[61] A similar crisis would arise in the summer of 2006, when US authorities announced the discovery of adventitious traces of Bayer Cropscience's unauthorized Liberty Link Rice 601 in US commercial rice exports, leading the Commission to declare that all US shipments of long-grain rice would be tested for the presence of LL 601 rice.[62]

A more serious setback to the Commission's plans, however, occurred on June 22, 2005, when the Commission submitted to the Environment Council a series of eight draft decisions which would overturn the eight national bans on GM varieties that had been declared safe by the EFSA. In contrast with the pattern of deadlocks over the approval of specific GM varieties, on this issue the Council was able to summon lopsided majorities of twenty-two member states voting to *reject* the Commission proposals—the first qualified majority that the Council had summoned for or against *any* Commission proposal on GMOs—and thus upheld the continuation of the member-state bans.[63] Luxembourg Environment Minister Lucien Lux, who chaired the meeting during the Luxembourg Presidency of the Council, expressed his "great satisfaction" at the outcome, noting pointedly that, "We were able to give a clear message to the European Commission."[64] In essence, member states agreed to protect their unilateral powers so that they would not be pressed to give reasons for the safeguards in a legal proceeding before the European Court of Justice, as provided for in the Regulation. Following the vote, Environment Commissioner Stavros Dimas called on his colleagues in the Commission to discuss the "political significance" of the vote, and to consider how to proceed regarding both the pending approval of new varieties as well as the national safeguard bans.[65]

Opposition to GMOs, attacks on the Commission and EFSA, and calls for stricter rules reached a crescendo during the Austrian Presidency of the Council in the first half of 2006. Longstanding opponents of GMOs, the Austrians used their Presidency as a bully pulpit, scheduling a public debate on the issue at the March 2006 Environment Council and convening a special conference on GMOs in Vienna in April.[66] Prior to the March 9th meeting of the Environment Council, Austrian Agriculture and Environment Minister Josef Proell raised two general questions for discussion, first, calling into question the risk-assessment procedures of the EFSA, and second, raising the possibility of a shift in decision-making rules, allowing member states to block proposed authorizations by a simple majority instead of the qualified majority called for in existing legislation.[67] The second issue, that of changing the decision rules within the Council, attracted limited support from a few ministers[68] but was opposed by other member states and by Commission Environment Minister Stavros Dimas, who pointed out that the member states had adopted the current system by a large majority just a few years earlier. By contrast, a majority of member states indicated dissatisfaction with the risk-assessment procedure and especially with EFSA, which was accused by various ministers of being non-transparent, uncooperative, closed to input from EU member governments, and biased in favor of industry.[69] Many of these critiques were repeated two months later at the May 22nd Agriculture Council, which called on the Commission to propose binding EU rules on coexistence between GM and conventional crops, as well as new rules on organic labeling and on thresholds for the "adventitious presence" of GMOs in organic products.[70]

Seeking to mollify such concerns, Dimas and the Commission decided to focus on the risk assessment process and on the role of the EFSA.[71] Accordingly, the Commission proposed, after consultation with "Member States and stakeholders," a set of "practical improvements" to the regulatory processes in both the risk-assessment and decision-making phases. Undertaken within (and thus requiring no amendment of) the existing legislative framework, the Commission's proposed changes focused primarily on the member states' grievances about EFSA, "inviting" the agency "to liaise more fully with national scientific bodies"; "to provide more detailed justification, in its opinions on individual applications, for not accepting scientific objections raised by the national competent authorities"; and "to address more explicitly potential long-term effects and biodiversity issues in their risk assessments for the placing on the market of GMOs."[72] While the Commission's initiative was applauded by several member governments and by environmental groups, the biotech industry association EuropaBio criticized the decision, suggesting that, "It is not EFSA's job to listen to public opinion."[73] EFSA itself held a follow-up meeting in May with some sixty member-state experts to explore ways of strengthening cooperation between the Parma-based agency and its national counterparts, yet EFSA Director Dr. Harry Kuiper also insisted on the agency's political independence and on the importance of science-based decision-making free of political pressure.[74]

Meanwhile, on the subject of coexistence, the Commission released a report in March 2006, rejecting member-state calls for binding EU regulations on the coexistence of GM and conventional crops. While indicating its support for member states that wished to adopt such rules, Agriculture Commissioner Mariann Fischer Boel pointed to the diverse environmental conditions across the EU's member states and the lack of experience in growing EU crops in Europe, which remained virtually free of GM cultivation outside of Spain, as reasons why binding EU rules would be premature. Nevertheless, the Commission indicated that it would continue to monitor developments and compare best practice across the EU's member states, which were gradually adopting national regulations, and issue a progress report in 2008.[75]

6.1.4.4. THE WTO DECISION

In contrast with these internal developments, further external pressure on the EU to proceed with new approvals and eliminate the existing national bans came in September 2006, as we saw in Chapter 5, when the WTO panel issued its final ruling in the case brought by the US, Canada, and Argentina against the EU in respect of its regulation of GMOs. To recall, the complainants' claims were set forth in three parts, in which they respectively challenged the EU's "general moratorium," its "product-specific moratoria," and EU member-state marketing and import bans on GM varieties approved at the EU level. The panel found in favor of the complainants, but largely on procedural and not substantive grounds, holding that the EU engaged in "undue delay" in its approval process. Regarding the member-state safeguards, however, the panel found that all of them violated the EU's substantive obligations under the SPS Agreement because they were "not based on a risk assessment." The panel noted that the EU's "relevant scientific committees had evaluated the potential risks...and had provided a positive opinion," and had "also reviewed the arguments and the evidence submitted by the member state to justify the prohibition, and did not consider that such information called into question its earlier conclusions."

As we saw, the WTO panel decision empowers those in the EU who wish to facilitate the approval of GMOs, thereby potentially mitigating the conflict with the US. By finding that a moratorium constitutes "undue delay" in violation of WTO commitments, the decision has empowered the European Commission to push forward new approvals of GM varieties against the wishes of member states. In addition, the decision empowers the Commission and private parties to challenge member-state bans on varieties approved at the EU level, at least so the US hopes.

As we now show, however, events following the WTO decision demonstrate the severe limits for the US of using WTO dispute settlement to catalyze changes in the policymaking dynamic in either comitology committees or in the Council of Ministers. If we look at three key developments—the pattern of continued approvals for new varieties for importation and use as food and feed; the intense controversy over the possible approval of new varieties for cultivation; and the Commission's ongoing efforts to overturn the existing national bans—the story of GM regulation since the WTO decision remains as contentious as before.

6.1.4.5. NEW EU APPROVALS FOR FOOD AND FEED

Since the initiation of the WTO case and the resulting WTO decision, the EU approval process for GM foods and crops continues to produce approvals for new GM crops, for a total of seventeen crops approved for food and feed between May 2004 and November 2008 (see Table 6.4), suggesting that the pursuit of the WTO case was, on the whole, beneficial for the US and other traders of GM varieties. In each case, however, the decision had been taken by the Commission after both the Standing Committee on the Food Chain and Animal Health and the Council of Ministers have failed to reach a qualified majority either for or against the varieties in question.

In October 2006, for example, the Environment Council met to consider a series of Commission proposals that had once again deadlocked in the relevant regulatory committees. The Commission, on the basis of a positive EFSA risk assessment, put forward three proposals to approve three new varieties of GM oilseed rape, or canola (Brassica napus L., lines Ms8, Rf3, and the hybrid Ms8xRf3), but in each case the Council was unable to reach a qualified majority, once again leaving the final decision to the Commission.[76]

This same pattern continued, moreover, in 2007. On the one hand, the Commission approved seven new GM varieties during 2007—the fastest annual pace of approvals since the end of the moratorium—including three rapeseed (canola) varieties engineered by Bayer, three hybrid maize varieties from Pioneer, Dow AgroSciences, and Monsanto, and a GM sugar beet created by the alliance of KWS SAAT and Monsanto (see Table 6.4). Once again, however, each of these varieties went through the now-familiar routine of receiving a favorable risk assessment from EFSA and a proposed approval from the Commission, only to meet with deadlock in both the Standing Committee and the Council of Ministers, once again leaving the Commission to approve the variety unilaterally to a chorus of member-state and NGO condemnation. Take, for example, the case of the hybrid 1507xNK603 maize, a conventionally bred hybrid of the Herculex I and Roundup-Ready varieties, with "stacked" GM

Table 6.4 GM varieties approved by the EU, May 2004–November 2008

Crop/GM identifier	Company	Approved for Food and Feed	Approved for Cultivation	Date of EU Approval
Maize (Bt11)	Syngenta	Yes	No	19/05/2004
Maize (NK603)	Monsanto	Yes	No	18/10/2004 (feed) 03/03/2005 (food)
Oilseed rape GT73	Monsanto	Yes	No	31/08/2005
Maize (MON863)	Monsanto	Yes	No	10/08/2005 (feed) 13/01/2006 (food)
Maize (GA21)	Monsanto	Yes	No	24/10/2006 (feed) 28/03/2008 (food)
Maize (MON863 x MON810)	Monsanto	Feed only	No	16/01/2006
Maize DAS1507)	Pioneer and Dow AgroSciences	Yes	No	03/03/2006 (feed) 16/03/2006 (food)
Swede-rape (MS8)	Bayer	Yes	No	25/05/2007
Swede-rape (RF3)	Bayer	Yes	No	25/05/2007
Swede-rape (MS8xRF3)	Bayer	Yes	No	25/05/2007
Maize (DAS1507 x NK603)	Pioneer and Dow AgroSciences	Yes	No	24/10/2007
Maize (DAS59122)	Pioneer and Dow AgroSciences	Yes	No	24/10/2007
Maize (NK603 x MON810)	Monsanto	Yes	No	24/10/2007
Sugar beet (H7-1)	KWS SAAT and Monsanto	Yes	No	24/10/2007
Maize (MON863 x NK603)	Monsanto	Yes	No	18/04/2005
Cotton (LLCotton 25)	Bayer	Yes	No	29/10/2008
Soybean (A2704-12)	Bayer	Yes	No	08/09/2008

Sources: Community Register of Genetically Modified Food and Feed (http://ec.europa.eu/food/dyna/gm_register/index_en.cfm, consulted on November 15, 2008), and *GMO Products Authorised under Directive 2001/18/EC*, as of January 1, 2008 (http://ec.europa.eu/environment/biotechnology/authorised_prod_2.htm, consulted on November 15, 2008).

traits making it resistant both to the European corn borer and to Monsanto's Round-up herbicide. The developers of the new maize submitted an application to UK authorities in September 2004, and EFSA issued a positive scientific opinion in March 2006, after which the Commission submitted the draft approval to the member states. However, the Standing Committee on the Food

Chain and Animal Health was unable to reach a qualified majority either for or against in June 2007, and the October 2007 Agriculture Council similarly deadlocked, leaving the Commission to approve the variety a week later.[77]

In the face of this opposition, the Commission met on May 7, 2008 for another orientation discussion of the EU's policy toward GMOs. At the meeting, the Commission agreed that it would "continue to base its decisions on science as required by the legislation," noting its confidence in the scientific advice provided by EFSA. It also indicated that it would "proceed to finalize decisions on pending cases...and examine new cases accordingly." Nevertheless, the Commission decided to delay final decisions on seven pending applications (including three applications for cultivation discussed below and four for import and marketing) asking EFSA for more detailed scientific analyses about the potential environmental and other impacts of each of these varieties.[78] The Commission defended the delay as necessary to secure all necessary scientific information prior to approval of the various varieties, but industry group EuropaBio called the delay unacceptable, noting that more than forty products were awaiting EU approval, and US officials expressed similar concerns.[79]

6.1.4.6. APPROVALS FOR CULTIVATION—DIVIDING THE MEMBER STATES AND THE COMMISSION

Even more controversial than the aforementioned applications were a series of three applications for the *cultivation* of new GM crops in the EU. In the first of these, the German chemicals firm BASF put forward an application for the cultivation of a potato genetically modified to produce higher starch content in paper for industrial use. Although BASF's initial application was only for industrial use, and despite a favorable risk assessment by EFSA and a proposed approval by the Commission, the EU regulatory committee consulted in December 2006 and deadlocked on the decision, with 134 votes in favor of approval, 109 opposed, and 78 abstentions under the committee's weighted voting formula, leaving the decision in limbo.[80] Seven months later, in July 2007, the Council of Ministers also failed to reach a qualified majority on the cultivation of the "amflora" potato, which put the decision once again into the hands of the Commission.[81] After the meeting, Commission spokeswoman Barbara Helfferich reasserted the Commission's support for approving the new variety, telling the press that "the scientific evidence is irrefutable," and indicating the approval was likely in the "coming months."[82] Another EU official, speaking on the condition of anonymity, noted that, "These are elected officials and they have to face general unease at home about GMOs. They are passing the buck to the Commission, which is between a rock and a hard place on this issue."[83] Indeed, in contrast with previous approvals, the Commission did not move promptly to approve the BASF potato in the following months, raising charges that Environment Commissioner Stavros Dimas was "prevaricating," and prompting BASF to ask the Commission, in December, to take a decision as soon as possible and in time for the spring 2008 planting season.[84]

Also in 2007, the Commission received two other applications for cultivation, both for GM maize varieties—Syngenta's Bt11 and Pioneer/Dow's 1507—that had been previously approved for use, but not for cultivation, in the EU. Both varieties produce Bacillus thuringiensis, a substance that is toxic to the European corn borer worm, but which had also been implicated in possible damage to butterfly larvae in a laboratory study, although one that was later undercut by field studies.[85] The EFSA had issued positive opinions on both varieties in 2005, as had the FDA in the US, where both varieties were grown commercially. However, rather than adopt a draft opinion for consideration by the member states, the Commission asked EFSA in April 2006 to consider more explicitly the potential long-term effects of GMOs on the environment and called on the agency to review its decision on the two maize varieties in question.[86] In November 2007, Dimas (whose DG Environment was in charge of the file within the Commission) reportedly produced a draft decision for circulation within the Commission, calling for the rejection of both varieties. Calling the "potential damage to the environment irreversible," and citing scientific studies not considered by EFSA in its opinion, the draft concluded that "the level of risk generated by the cultivation of this product is unacceptable."[87] In a press conference, Commission spokeswoman Helfferich defended Dimas's draft decision. "Commissioner Dimas has the utmost faith in science," she said. "But there are times when diverging scientific views are on the table," she continued, noting that Dimas was acting in this case as a "risk manager."[88]

Dimas's draft decision, widely reported in the media, would represent the first time that a GM crop had been put up for approval but rejected by the Commission—even in the face of a positive opinion from EFSA—and it brought down both accolades from anti-GMO campaigners,[89] as well as intense criticism from outside and within the Commission. EuropaBio, the European industry lobby, condemned the decisions, which "would be setting a precedent for EU officials to reject products based on non-verified scientific data."[90] US officials responded in a similar fashion. "These products have been grown in the U.S. and other countries for years," said a spokesman for the USTR. "We are not aware of any other case where a product has been rejected after having been reviewed and determined safe" by EFSA.[91] More generally, the spokesman of the US Mission to the European Union expressed American frustration at the EU approval process: "The United States," he said, "has consistently stated that the EU continues to lack a predictable, workable process for approving these products in a way that reflects scientific rather than political factors."[92]

Dimas's draft decision was also intensely disputed within the Commission itself, where a number of commissioners and their respective services took a more favorable view than the Environment Commissioner, and were seeking to open up the approval process. Trade Commissioner Peter Mandelson, for example, had publicly called in the previous June for the EU to speed up its approval processing, arguing that, "If we fail to implement our own rules, or implement them inconsistently, we can—and probably will—be challenged."[93] On a related note, Agriculture Commissioner Mariann Fischer Boel and Health Commissioner Markos Kyprianou met with the members of the Agriculture Council on November 28th, where Fischer Boel called for a reconsideration of the EU's "zero-tolerance" policy toward the adventitious presence of non-EU-approved GMOs in shipments to the Union. While not explicitly proposing a relaxation of the rules, Fischer Boel reportedly noted the costs of a zero-tolerance policy for the EU, which relied heavily on imports of corn and soy for animal feed, prices of which had risen dramatically as a result of the zero-tolerance policy.[94] Not surprisingly, given the diversity of views within the Commission, Dimas's draft decisions proved controversial, with Commissioners Mandelson, Verheugen, and Fischer Boel reportedly opposing the decisions.[95]

The Commission took up these issues in an internal debate in May 2008 over GMO policy. On the issue of zero tolerance, the commissioners agreed to explore what they called a technical solution to the problem, asking the Commission services to put forward new guidelines that would allow small traces, within the testing limits of available technology, of unapproved GM varieties in EU imports. Subsequent reports, however, suggested that the new threshold levels might be as low as 0.1 per cent, far lower than the 0.9 per cent suggested by US exporters and by some European feed importers; and it remained unclear whether the Commission could, in any event, adopt such thresholds without returning to the Council of Ministers and the European Parliament for new legislation adopted by co-decision.[96] On the three pending applications for cultivation (the Amflora potato and the Bt11 and 1507 maize varieties), the Commission returned all three applications to EFSA for further evaluation, with a final decision not expected until 2009 at the earliest.[97]

This split within the Commission was mirrored in the continuing divisions among the member states, as well as increased skepticism about GMOs among some national governments. While some governments, such as those of the United Kingdom, remained generally favorable toward the promise of genetic engineering (notwithstanding opposing public opinion in that country), a number of member states—such as Austria, Italy, Luxembourg, and Greece—remained firm opponents of GM foods and crops, and several member states took increasingly *critical* positions toward GMOs in recent years. In Ireland, for example, the June 2007 elections brought the Irish Green Party into the Fianna Fáil-led government, and Green Party leader John Gormley, in his new position as Minister for the Environment, announced a radical shift in Ireland's position that month in the Council of Ministers, where he publicly backed Austria in its anti-GMO position and announced that "we want GM-free status for Ireland."[98]

Electoral turnover brought another stark change in France, where incoming President Nicolas Sarkozy indicated a fundamental reappraisal of GM foods and crops. While France had previously been ambivalent about the approval of GM crops, having joined in the earlier call for the moratorium and often abstaining from or voting against new approvals in Brussels, it had also seen increasing cultivation of GM maize crops, which had grown four-fold to some 22,000 hectares in 2007.[99] In October, at an "environmental summit" attended by Nobel Peace Prize-winner Al Gore, Sarkozy announced a raft of new environmental measures, including an immediate freeze on the cultivation of GM crops in France, and the creation of a vaguely defined GMO review body separate from the French Food Agency and EFSA.[100] The Commission responded quickly to the proposed freeze on the cultivation of GM varieties, noting that such a ban would violate European law, but this action did not deter France, which imposed the ban in early 2008.[101]

French eagerness to revisit and possibly strengthen EU regulation of GM foods and crops came more clearly into evidence in early 2008, as the country prepared to take up the rotating EU Presidency during the second half of the year. In February, France issued a paper to the other EU members proposing a review of the GMO approval process, including risk assessment (where it called for greater attention to environmental assessment as well as "socioeconomic" factors), scientific expertise (where it called for EFSA to pay greater heed to national risk regulation agencies), labeling thresholds for GM seeds (where it sought new legislation), and granting greater discretion to member states in authorizing the cultivation of GM seeds.[102] The Environment Council met in June and July to endorse the proposal, and France promptly announced, at the beginning of its presidency in July, the creation of a "group of friends of the presidency" to consider changes to the EU regulatory framework for GMOs. The group, France announced, would focus on two areas. First, it would address the approval process, where it called for greater transparency and greater consideration of long-term environmental effects and respect for national expertise. Second, the group would discuss "how the potential new effects—unknown at the time of authorization—would be taken into account," as well as how to deal with a "country who might wish to, for example, declare itself GMO-free."[103] Regardless of the outcome of the French Presidency group, these developments and others[104] demonstrate that the politics of GMOs remains deeply controversial within EU institutions and in EU public opinion, with considerable pressure to make EU regulations more, rather than less, restrictive.

6.1.4.7. CHALLENGING THE NATIONAL SAFEGUARD BANS

In addition to the slow pace of approvals for new GM varieties, the other major issue raised by the US before the WTO was the retention of the various national safeguard bans, which the EFSA, the Commission, and the WTO panel had all ruled were not supported by scientific risk assessments. Nevertheless, over-turning the member-state bans has proven to be one of the most difficult and politically sensitive areas in EU politics. As we have seen, the Commission ini-tially attempted in June 2005 to challenge all the member-state bans, only to have that decision rejected by an overwhelming majority in the Environment Council. The following year, the Commission put forward a more narrowly targeted set of two proposals to the December Environment Council, seeking to overturn the two Austrian bans on GM maize T25 and MON810.[105] Once again, a lopsided majority of twenty-one member states sided with Austria and voted against the Commission's proposal. The Council vote, taken over the objections of the UK, Netherlands, Czech Republic, and Sweden, was justified formally on various grounds, including the fact that the risk assessments for these products were undertaken under Directive 90/220 and should undergo "a procedure of re-approval and re-assessment in accordance with the new Directive" 2001/18/EC.[106]

A similar outcome followed when the Commission attempted to overturn the Hungarian ban on MON810 maize. The Hungarian government, for its part, actively lobbied its fellow governments, including the more GM-friendly new members Bulgaria and Romania, for support in retaining the ban.[107] Once again, at the Environment Council meeting in February 2007, a large majority of twenty-two member governments rallied around the right of member governments to retain national bans on specific GM varieties and not be challenged before the Court of Justice, with only Finland, the UK, the Netherlands, and Sweden supporting the Commission's proposal, and Romania (among others) abstaining.[108]

In light of its third overwhelming defeat on the issue of national bans, the Commission changed its tactics in October 2007, opting only to challenge Austria's ban on the use of the two aforementioned maize varieties (MON810 and T25) for sale as food and feed, while accepting, for the moment, Austria's continued ban on their cultivation. Arguing that "product safety is identical across Europe," unlike questions of cultivation and coexistence which var-ied by region and climate, the Commission put forward a proposal to over-turn Austria's ban on these two varieties for importation, processing, and use in food and feed, isolating this question from the environmentally sensi-tive question of cultivation.[109] This strategy paid off, to some extent, at the October 30th Environment Council, when the environment ministers failed to obtain a qualified majority against the Commission proposal.[110] Fifteen member states reportedly voted against the Commission's proposal, with only four voting in favor, but a larger group of eight member states abstained, denying the Austrians a qualified majority.[111] In principle, the failure to reach a qualified majority empowered the Commission to move ahead and at least partially overturn the Austrian ban, yet the contentious nature of the vote and the public protestations of many ministers put the Commission in

259

a politically difficult situation. In an unusual press conference, Portuguese Environment Minister Francisco Nunes Correia, who had chaired the meeting and whose country had abstained in the vote, described the discussions as "intense," adding that, "this is an uncomfortable position that we're all in." Many member states had supported Austria, he continued, either out of opposition to GMOs, or because of a belief that member states' positions should be respected. "The majority of member states are against the Commission's proposal, but the Commission's proposal will prevail against the will of one particular member state," he concluded, adding that, "this is something that has to give us pause for thought."[112]

Finally, after considerable delay, the Commission informed Austria in May 2008 that it was required to lift the ban on the marketing of the two varieties in question, and Austria complied later that month, formally repealing the two bans and informing the Commission of its actions.[113] The Austrian ban on cultivation of MON810, however, remained in place, with no evidence that the Commission intended to resume its failed challenge to the ban. Meanwhile, in January 2008, the French government indicated that, pursuant to its earlier informal decision to freeze cultivation of MON810 seeds, France would notify the EU of its intention to ban cultivation of MON810 seeds, under the safeguard clause provisions of EU law.[114] Far from being eliminated, the various national bans on GM foods and crops showed signs of spreading still further.

6.1.5. Reform without change?

In sum, notwithstanding the significant external pressures from the US and the WTO, there is little sign of change in the core principles of the EU regulatory system or in the underlying pattern of EU decision-making since the end of the moratorium in May 2004, and even less indication of convergence on the US model. To be sure, the EU's regulatory framework for the approval, marketing, tracing, labeling, and cultivation of GMOs has been substantially overhauled over the past years, and some of these reforms—such as the increased emphasis on scientific risk assessment by EFSA and the end of the moratorium on new approvals—appear to respond to foreign pressures, adopted in the hope of mitigating or forestalling WTO legal challenges. Indeed, the EU approved seventeen GM varieties between May 2004 and November 2008 for their consumption in feed or food, although these varieties are generally only used for animal feed.[115] Nevertheless, in contrast with the US, the EU continues to retain a system in which GMOs are regulated entirely by process rather than product standards, where strict regulations on traceability and labeling impose new constraints on GM foods and crops, and where "risk management" remains in the hands of political bodies which remain free to take decisions according to political, social, and economic as well as scientific risk assessment criteria. The pattern of voting in EU committees and in the

Council, moreover, reveals a distribution of preferences in which EU member governments are able to agree only upon the maintenance of national bans (against EFSA scientific advice) but not on the approval of new varieties, leaving the Commission to approve new varieties unilaterally in every instance since 2004.

Indeed, the review of developments through the middle of 2008 suggests caution in interpreting the new and ever more elaborate EU regulatory system, with its purported opportunities for deliberation and exchange of perspectives among multiple governmental representatives and affected stakeholders before different institutions and levels of government. For applicants seeking approval of new GM varieties, as well as for many non-European farmers and traders, the EU's increasingly complex, Byzantine system for authorizations and marketing of GM varieties, incorporating multiple governmental actors and non-government stakeholders, can appear to be a Potemkin village. That is, the *de jure* regulatory system with its many procedures for consultation, looks impressive on the surface, but in their view, appears to be largely a sham, as key member states work to block and otherwise restrict the importation, marketing, and cultivation of GM crops, regardless of scientific risk assessments. The result, in their view, is lots of costly show, possibly to meet formal WTO requirements, masking a deeply politicized process in which science and administrative procedures continue to take a back seat.

This combination of continued strong public and member-state opposition to GMOs, along with political institutions that provide veto power to relatively small coalitions of member states seeking to block any major changes, has produced a path-dependent pattern of change in which EU regulations have been altered at the margins, primarily in the direction of greater precaution and *away* from the US regulatory model. At the same time, however, there seems to be no majority support among EU member states for any major amendments to the regulatory framework, to either facilitate or block new approvals or to challenge the existing safeguard bans.[116] In the absence of some additional internal or external shock, the current EU regulatory system appears to be a stable—although politically contested and, to many on both sides of the debate, unsatisfactory—equilibrium. We call this *reform without change*.

6.2. Review of US regulatory developments: Change without reform

What, then, of the US? If the EU has not converged toward the American system of technocratic regulation by independent agencies, have US regulations converged on those of the EU? Or, in David Vogel's phrase,[117] has the US "traded-up" to EU standards in order to gain access for US farm products to the EU market? As Alasdair Young points out, the question of US "trading up" requires us to distinguish analytically among three inter-related phenomena: (1) *commercial adaptation*, which occurs when US firms or farmers voluntarily comply with EU standards (e.g., growing only EU-approved GM varieties) in order to gain access to the EU market; (2) *political mobilization*, which occurs when domestic US interest groups, spurred (at least in part) by events in Europe, mobilize for stricter GM regulations; and (3) *policy change*, when US authorities adopt stricter framework legislation or stricter implementing regulations.[118] The last type of change is the most demanding, since, in the GMO case, it would require not only the adoption of specific regulatory standards, but (potentially) a significant change in the national regulatory style and framework outlined in Chapter 2. In fact, a careful examination of recent US events provides some evidence of political mobilization, significant commercial adaptation, and some modest policy change. However, these policy changes largely reflect an incremental elaboration of the traditional US system rather than any regulatory overhaul in the direction of the EU's approach.

6.2.1. Commercial adaptation

With regard to commercial adaptation, Young finds some evidence of US farmers' and growers' associations taking decisions on which crops to plant based at least in part on the regulatory standards of the EU and other important markets such as Japan and Canada. The National Corn Growers' Association, for example, has established a "know before you grow/know where to go" program to advise farmers about the GM varieties accepted in various foreign

Table 6.5 Key events in US biotech regulation, 1999–2008 (November)

1999	FDA begins extensive public hearings into adequacy of US regulatory framework
1999	Scientific study links Bt-corn to damage to monarch butterfly larvae
2000	Starlink corn, an unapproved GM variety, found in US food supply
	FDA encourages biotech companies to submit safety data on GM foods prior to marketing
	Plant Protection Act revises guidelines for planting of GM crops
2001	FDA issues guidelines for voluntary labeling of GM foods
	Federal Plant Protection Act gives USDA authority to regulate GM crops as potential plant pests
2003	US launches WTO complaint over EU regulation of GMOs
	FDA decides not to regulate genetically modified "Glofish"
2004	USDA announces plans to strengthen biotech regulations
2004	Monsanto Corporation defers efforts to introduce GM wheat
2006	Unauthorized LL601 rice discovered in US shipments to Europe
2007	USDA releases environmental impact assessment of proposed regulatory reforms
	US federal court annuls APHIS approval of a GM alfalfa, which is subsequently re-regulated
2008	FDA releases risk assessment on meat and milk from cloned animals; no detailed regulations or approval on transgenic animals for food

markets, and Young points to evidence that US farmers have concentrated production in those varieties of GM corn and soybeans approved for marketing in the EU.[119]

Farmers' concerns about market reception have been particularly striking in relation to the controversy over the introduction of Monsanto's GM wheat. The new variety, a "Roundup-Ready" wheat resistant to Monsanto's Roundup herbicide, was submitted for regulatory approval by Monsanto in the US and Canada, only to encounter widespread concern among farmers that approval and adoption of GM wheat could imperil their markets in the EU and other countries. Several industry associations, including the National Association of Wheat Growers, called for a reliable system for segregating and tracing crops before GM wheat was introduced. In the words of the vice president of the North American Millers' Association, "Our customers are telling us that they have very serious concerns or are flat-out opposed to GM wheat....While this opposition may have nothing to do with science, the customer is always right."[120] Under these circumstances, Monsanto promised to wait for regulatory approval in both the US and Canada (the largest exporters of wheat) as well as Japan (one of the largest importers), before moving ahead with field trials. Finally, in May 2004, the company announced that it was "deferring all further efforts to introduce Roundup-Ready wheat."[121]

Moreover, US farmers and food processors respond to the demands of EU retailers that are influenced not by what EU regulation provides on its face, but rather by their concerns over real and potential European consumer and civil society pressures. US growers face considerable constraints imposed by processors and retailers operating in the EU market, in large part in reflection of anti-GMO campaigns.[122] To the extent that brand-name companies and retailers in Europe agree not to sell GM foods, formal regulatory approval will be of little solace to the agricultural biotech industry and farmers who use GM seeds. For example, Europe's largest supermarkets, such as Britain's Tesco, France's Carrefour, and Belgium's Delhaize, guaranteed that their private label foods are GM-free,[123] as did many of Europe's brand-name companies after being threatened with organized boycotts.[124] These market pressures constitute a form of private regulation, complementing public regulation.[125]

The EU's new traceability and labeling requirements for GM foods threaten potentially greater economic consequences. Even where a GM variety has been approved by the EU, European retailers may refuse all US-sourced grains and food products, even those derived from conventional plant varieties, where they fear that US grain shipments, or processed foods containing US grains may exceed the EU's low threshold requirements for adventitious traces of GMOs. There is evidence that retailers have done so. For example, a director at Kelloggs' Co. announced that Kelloggs' Pop Tarts sold in Europe are no longer made with US-origin ingredients, in order to avoid GMO content. An official at a US grain crusher with significant crushing presence in Europe, noted that pet food companies are also moving *en masse* to go non-GMO.[126] In fact, even US organic growers could face challenges selling their products in Europe. An organic maker of tortilla chips from Wisconsin had a large shipment destroyed because a Dutch importer "found a trace of genetically engineered corn" in the organic corn chips, allegedly because "wind probably blew corn pollen from a neighboring farm into [an organic grower's field]."[127] For these reasons, approved GM varieties are primarily used in the EU only for animal feed, as the resulting meat does not need to be labeled as being raised on GM feed.

A similar set of concerns arose in the summer of 2006, following the discovery of trace amounts of Bayer Cropscience's Liberty Link 601 rice, an unapproved GM variety, in the commonly grown Clearfield rice. Following the discovery of the GM rice, and subsequent actions by the EU and Japan to order testing of all US rice shipments for the unapproved variety, US farmers in the affected states of Arkansas and Missouri filed suit in federal court against Bayer Cropscience for damages suffered as a result of a temporary ban on exports of US rice to Japan, as well as the costs of testing for shipments to the EU.[128] The USDA's Animal Plant and Health Inspection Service (APHIS) subsequently "deregulated" LL 601 in November 2006, noting that the protein expressed by the "rice" was identical to two other Bayer varieties already approved but not marketed.[129] Critics, however, such as the Washington, D.C.-based Center for Food Safety, criticized the decision, which it said sanctioned an "approval-by-contamination" policy designed strictly to avoid legal liability, while other critics pointed to the possibility of cross-breeding with the wild red rice weed, which could thus acquire resistance to pesticides.[130]

Notwithstanding APHIS's approval, the plaintiffs in the suit claimed damages from Bayer Cropscience, including the costs of testing for US export markets as well as a 10 per cent across-the-board decline in the market price of US rice following the announcement of the LL 601 contamination. APHIS, for its part, issued a subsequent warning in March 2007, advising farmers not to plant Clearfield 131 rice, which continued to test positive for trace amounts of GM rice,[131] while the state of Arkansas banned the growing of two suspect rice seeds (Cheniere and Clearfield 131) and required that all rice seed planted in 2007 be tested and found negative for the presence of Liberty Link rice.[132] The USA Rice Federation, an industry group, supported moves by regulators to eliminate traces of GM varieties from conventional rice, but expressed frustration with "the apparent lack of ability on the part of private companies and federal regulators to control research and maintain accountability of the resulting products," noting further that "there must be market acceptance and regulatory approval prior to the production of genetically engineered rice in the United States."[133] Regardless of the outcome of the ongoing lawsuit, farmers' resistance to potential commercialization of GM rice remains high, and

Bayer has announced no intention to commercialize any of its three APHIS-approved varieties in the future.[134]

In 2007, it was the turn of US corn growers to risk losing major export markets because of the approval of a GM variety for commercial release in the US, before it was approved in other countries, in particular Japan. In January 2007, APHIS announced that it had received a petition from Syngenta seeking deregulation of its rootworm-resistant corn known as MIR604, and invited public comments. APHIS received positive comments from farm trade associations such as the Iowa Farm Bureau Federation, the Missouri Corn Growers' Association, and the National Chicken Council, which stressed the damage done by rootworm to corn crops, and the positive economic and environmental impact that such an approval would have.[135] However, after APHIS approved deregulation of the product, the National Grain and Feed Association, and the North American Export Grain Association criticized Syngenta for commencing commercialization before the variety received regulatory approval in major US export markets, namely Japan.[136] In a joint press release issued two weeks after APHIS granted MIR604 nonregulated status, the two trade associations stated that "[g]iven the U.S. government's deregulation of Agrisure RW corn [MIR604], we have no reason to question its safety for food or feed, but we do have important reasons for opposing its commercialization at this time because of its marketability."[137] In the end, the problem was avoided in this case as Japan approved the commercial sale of the grain in August 2007.[138]

In response to these demands, BIO, the biotech industry trade association, created a Product Launch Stewardship Policy in May 2007, constituting a form of self-regulation in response to US farmers' and grain traders' concerns about the lack of key foreign market approvals.[139] The policy aims to respond to "asynchronous authorizations" in which a product is approved in the US, but not yet approved in important foreign markets, creating a risk that grain could be mixed in low quantities, so that foreign regulatory authorities could impose expensive testing requirements to ensure purity, or if unapproved varieties are detected, turn back an entire shipment at immense costs to traders, raising prices and making US grain less competitive. The policy provides that BIO "member companies should, prior to commercialization, meet applicable regulatory requirements in key countries identified in a market and trade assessment that have functioning regulatory systems and are likely to import the new biotechnology-derived plant products." Key markets "at a minimum shall include the United States, Canada, and Japan."[140]

BIO arguably adopted this self-regulatory policy in order both to satisfy the industry's customers and to reduce pressure on US authorities to change US regulatory policy and add a foreign market impact criteria that would condition US regulatory approvals in relation to foreign approvals (as Argentina has done: see Chapter 7). As a result, Monsanto again "agreed to delay the commercialization of its Roundup Ready2Yield soybean to 2009 to obtain necessary regulatory clearances world-wide, even though the GMO has already been approved for cultivation in the U.S. and Canada."[141] In other words, agricultural biotech regulatory processes in the EU, Japan, and other major markets have exerted pressure on US agricultural biotech companies to synchronize their applications and commercialization of GM varieties in line with foreign regulatory processes, and to create a *de facto* policy of assuring US growers and grain traders that they will first obtain requisite government approvals abroad.

The commercial prospects for new GM foods and crops in the US, therefore, remain unclear in this writing. On the one hand, US farmers have showed little inclination to abandon established GM varieties, with the total acreage devoted to GM crops continuing to rise, to include about 89 per cent of soybeans, 83 per cent of cotton, and 61 per cent of corn, and with a new potential prospect for an important crop—GM sugar beets.[142] On the other, GM production in the US has increasingly concentrated in four crops (soybeans, cotton, canola, and corn), while notification of new varieties and commercial acceptance of other GM crops (including wheat, rice, and various fruits and vegetables) has decreased from the rapid pace of the late 1990s.

6.2.2. Political mobilization

With regard to political mobilization, the evidence suggests that media coverage of the US/EU dispute has provided some opportunities for US consumer and environmental groups to mobilize, although it has had much less effect on US public opinion about GMOs. Prior to the onset of the dispute, US consumers had indicated virtually no opposition to—and indeed virtually no awareness of—the existence of GM foods and crops. Surveys showed that while approximately 60 per cent of processed foods consumed in the US contain GM seeds, only about 33 per cent of Americans even knew that GM foods were available in supermarkets, and 60 per cent claimed that they would not buy foods labeled to contain GM ingredients.[143] Following the outbreak of the transatlantic dispute, as well as the opportunity for US interest groups to mobilize and to interact with their European counterparts, consumer groups and environmental groups such as Ralph Nader's Public Citizen, the Sierra Club, Friends of the Earth, and Greenpeace USA, adopted highly critical positions toward GM foods and crops, pressing for stricter regulations and mandatory labeling of GMOs.[144]

Public opposition to GM foods and crops was fed as well by internal US developments. In the May 1999 issue of *Nature*, Cornell University researchers reported that laboratory tests had shown that the use of a GM Bt-corn variety could kill not only targeted pests, such as the corn borer, but also monarch butterfly larvae if the corn variety's pollen were to travel to nearby milkweed, the larvae's source of food.[145] In the monarch butterfly, the Bambi of the insect world, opponents of GM foods have found a potentially powerful rallying symbol, even though subsequent field trials showed that the risk to the butterfly was negligible.[146] The following year, in October 2000, the first major GMO scandal emerged when Starlink corn—a GM variety marketed by Aventis and approved for use in animal feed but not for human consumption—was found to have worked its way into the food supply, turning up in Taco Bell taco shells, which was alleged to cause allergic reactions among some consumers, although "the incident, caused no documented harm to human health."[147] The Starlink incident, which led to a major recall of the corn by Aventis costing the company hundreds of millions of dollars, also stoked the debate over GM foods and over the adequacy of US regulatory oversight.[148] In response to consumer concerns, some US retailers—including Whole Foods supermarkets, Gerber and Heinz baby foods, and the makers of Frito-Lay corn chips—followed their European counterparts in announcing that henceforth their products would be GMO-free.

Despite this mobilization, there is little evidence that US public opinion shares the deep distrust toward GMOs felt by European publics. Indeed, a 2006 poll conducted for the Pew Initiative on Food and Biotechnology found remarkably little change over the previous five years in US awareness of, or support for, GM foods and crops. The poll showed a consistently low level of public awareness of GM foods over time, with only 26 per cent of those polled in 2006 believing that they have ever eaten GM foods. Similarly, generalized support for GM foods has held steady at a relatively low level (26 per cent in 2001, rising to just 27 per cent in 2006), but generalized opposition has declined, from 58 per cent of those surveyed in 2001 to 46 per cent in 2006, and support for GM foods tends to increase when individuals are given additional information about them.[149] In general, therefore, we see no dramatic increase in the salience of GM foods to the American public, which continues to know relatively little about them, and opposition has declined, rather than risen, over time.

Nevertheless, the Pew poll also includes findings about trust in federal regulation that are potentially ominous: for example, 41 per cent of respondents in 2006 claimed that there was too little regulation of GMOs (up from 35% in 2001), and trust in the FDA has declined, dropping from 41 per cent of people who trusted the FDA "a great deal" in 2001 to just 29 per cent in 2006.[150] The poll also points to difficulties ahead in the regulation of cloned and GM animals, with 64 per cent of respondents in 2006 saying that they are uncomfortable (46 per cent "strongly uncomfortable") with animal cloning.[151] As Michael Fernandez, Executive Director of the Pew Initiative, summarized the results, "public opinion remains 'up for grabs' on GM foods," potentially subject to contingent events, positive or negative.[152]

6.2.3. Regulatory change

With regard to regulatory change, finally, there have been significant debates among US legislators and regulators about possible reforms of the US regulatory process in recent years. Within both the US Congress and in various state assemblies, legislators have introduced dozens of bills with a potential impact on the regulation of biotechnology, including Congressional bills sponsored by Representative Dennis Kucinich (D-Ohio) and Senator Barbara Boxer (D-California) that would introduce mandatory labeling for all GM foods. These legislative initiatives have so far failed, however, and the few modest measures that were adopted at the federal level have merely instructed the US executive to support biotechnology internationally.[153]

In the absence of any federal legislative changes or, for that matter, any federal legislation specifically dedicated to the regulation of genetically engineered products, the FDA and the USDA conducted hearings and studies to consider administrative changes to the existing regulatory system, including the possibility of introducing mandatory labeling or premarket approvals of new GM varieties. In 1999, for example, the FDA held public hearings on the regulation of GM foods and crops, during which it received more than fifty thousand written submissions regarding the agency's rules on the approval and labeling of GM foods. After several years of collecting and analyzing these submissions, however, the agency rejected arguments for mandatory labeling, which it continued to hold was not required under the Federal Food, Drug and Cosmetic Act, opting instead for the issuing of guidelines for marketers who wish to "voluntarily" label whether their product has been made with, or without, the use of bioengineering.[154] In these guidelines, the FDA advised marketers of non-GM products regarding legal restrictions on labeling. The FDA specifically found that "the term 'GMO free' may be misleading on most foods" because the term "bioengineered" is technically more accurate and the term "free" is misleading in light "of the potential for adventitious presence of bioengineered material." It also found that "a statement that a food was not bioengineered or does not contain bioengineered ingredients may be misleading if it implies that the labeled food is superior to foods that are not so labeled." Such FDA statements warn market operators that if they so label their products, they could be subject to legal challenge from the FDA or a private party, including a biotech industry organization.

As a possible result, US marketers tend not to label their products as being GM-free, but rather use labels indicating that a product is "organic" pursuant to regulations for the labeling of organic foods under the US National Organic Program, which became effective in 2001. After strong public opposition to the USDA's initial proposal not to exclude food derived from genetically engineered varieties, the program excluded them from those that can be labeled as "organic."[155]

In 2001, the FDA also announced that it was considering whether to require mandatory prior *notification* of new GM varieties, such that much of the information provided by companies would be made publicly available. As of November 2008, however, the FDA had not finalized its proposal, which it did not consider to be a priority, and which in any event falls far short of the EU requirement of prior *authorization* of new GM varieties. As FDA Deputy Commissioner Lester Crawford testified before a subcommittee of the House of Representatives in the summer of 2003, "since the current system is working so well and since there is no public health reason to impose the mandatory requirement, it is not a high priority for FDA to finalize this rule at this point."[156]

Pending such a change, the only significant addition to the FDA's regulatory framework for GM foods and crops in recent years was its June 2006 guidance statement for industry regarding "early food safety evaluation" of "non-pesticidal proteins intended for plant use."[157] Responding to an earlier report by the Office of Science and Technology Policy, the FDA guidance noted that, while the regulation of GM field tests fell under the USDA and APHIS, the increasing number and diversity of field tests for GM plants "could result in the inadvertent, intermittent, low-level presence in the food supply of proteins that have not been evaluated through the FDA's voluntary consultation process for foods derived from new plant varieties.... FDA is issuing this guidance to address this possibility." More specifically, it continued, "FDA recommends that sponsors and developers of new plant varieties intended for food use consult with FDA about their evaluation of the food safety of any new proteins produced in these plants prior to the stage of development where the new proteins might inadvertently enter the food supply."[158] In effect, the new FDA guidance extends the existing voluntary notification procedure, encouraging companies to undergo food-safety evaluation of new GM varieties at the trial stage, even where the company in question has not made a formal decision to place those varieties on the market.

Like the FDA, APHIS and its Biotechnology Regulatory Service (BRS) have undertaken an ongoing review and possible revision of its policies on GM plants, developing an environmental impact statement that will weigh the costs and benefits of a number of possible regulatory reforms. "Among other changes, APHIS BRS is considering: expansion of the scope of regulation to include the noxious weed authority given by the Plant Protection Act of 2000, development of a multi-tiered risk system for field trials, use of compliance agreements between APHIS and the producer, for commercial production of plants not intended for food and feed."[159] At this writing, APHIS was still concluding its internal review of the environmental impact statement, which will be made available for public comment before the service draws up draft regulations for further notice and comment. In general, however, the revisions under consideration represent incremental changes to existing regulations, designed primarily to fill holes in the current regulatory framework and to prepare APHIS for new regulatory challenges posed by the next generation of GM foods and crops. In line with official US policy, moreover, the USDA has stressed that the system will continue to be "based on sound science principles and mitigation of risks."[160] In the meantime, APHIS has undertaken a number of smaller initiatives, including a clarification of its policy on the low-level presence (LLP) of regulated varieties and its response to instances, such as the Liberty Link rice case, in which GM seeds and crops are found in shipments of conventional crops.[161]

APHIS's role in the regulation of GM crops and field trials has raised increasing controversy in recent years among an attentive public of farmers and GM-skeptical NGOs. We have already seen, for example, how NGOs such as the Center for Food Safety questioned APHIS's decision to deregulate LL 601 rice *after* it had been discovered circulating in shipments of conventional rice. In another significant case, in 2005 the Center for Food Safety, together with the Sierra Club and a group of organic farmers, brought suit in federal court against APHIS for its decision to "deregulate" Monsanto's Roundup-Ready alfalfa.[162] The plaintiffs claimed that the deregulation of Roundup-Ready alfalfa created a significant risk that pollination would result in gene transmission from genetically engineered alfalfa to organic and other non-genetically engineered varieties. Gene transmission would be particularly damaging for organic farmers and for those exporting to markets such as Japan, where the genetically engineered alfalfa was banned. The plaintiffs argued that in light of these and other environmental risks, APHIS had failed to take all steps required by law to assess the environmental risks posed by the GM alfalfa, namely preparation of a full environmental impact statement (EIS).[163] In February 2007, US District Court Judge Charles R. Breyer decided in favor of the plaintiffs, ruling that APHIS had indeed violated the National Environmental Protection Act by failing to undertake an EIS before deregulating the crop.[164] One month later Judge Breyer "vacated" APHIS's decision to deregulate Roundup-Ready alfalfa, and issued a preliminary injunction order, prohibiting any additional sales of Roundup-Ready alfalfa seed and banning any future planting of the crop after March 30, 2007.[165] APHIS itself followed suit the same day, announcing the return to regulated status for the crop, together with all the accompanying precautions, pending the court's issuance of permanent injunctive relief.[166] Judge Breyer issued a permanent injunction on May 3, 2007, essentially maintaining the status quo established by the temporary injunction order until APHIS completes the court-ordered EIS, which APHIS estimated would take two years.[167]

The impact of the decision on GM alfalfa remains to be seen, both for the variety in question and more broadly. Alfalfa is the fourth most widely grown crop in the US, with approximately 23 million acres in production, yet Monsanto's GM alfalfa was only in its second year of commercial use at the time of the decision, with approximately 200,000 acres in production, making it far less significant than other crops such as corn and soybeans. More broadly, the decision set a potentially important precedent, both in its finding that APHIS's reasoning was "arbitrary and capricious", and in its decision to annul APHIS's action. "It's a very significant decision," said Charles Benbrook, chief scientist of the nonprofit Organic Center, "the next step in the pushback by the federal court system for the grossly inadequate environmental review of genetically engineered crops."[168] It is, however, unclear whether this case will indeed be followed by other challenges to, and annulments of, regulatory actions on GM foods and crops by APHIS or other US regulatory agencies.[169]

6.2.4. US regulation and transgenic animals

Not surprisingly, in light of the public opinion data reported earlier, the greatest controversy over US regulation of GMOs has emerged on the issue of GM or "transgenic" animals, as well as with regard to the related issue of meat and milk from cloned animals. The issue of meat and milk from cloned animals is, technically speaking, distinct from that of GMOs, as a cloned animal is an exact replica of the animal that donated the cells from which it is grown, with no modification of the genetic material. More specifically, the cloning of animals such as cows, pigs, and goats involves the substitution of genetic material from a particularly prized animal into a recipient egg, with the aim of creating an exact replica.[170] The scientific and technical problems encountered by the FDA in regulating meat and milk from cloned animals, however, are similar to those it is likely to face in the regulation of GM animals.

Faced with requests from biotech companies, the FDA in December 2006 issued three documents on milk and meat from cloned animals—a draft risk assessment, a proposed risk management plan, and a draft guidance for industry—for public comment. The draft risk assessment concluded that, "meat and milk from clones are as safe as food we eat every day," in the words of Stephen F. Sundlof, Director of the FDA Center for Veterinary Medicine, and that, "Cloning poses no unique risks to animal health when compared to other assisted reproductive technologies currently in use in U.S. agriculture."[171] In light of this finding, the FDA's proposed risk management plan envisaged the eventual marketing of meat and milk derived from cloned animals, and saw no reason to require mandatory labeling of cloned meat and milk products, but asked producers of clones and livestock breeders to voluntarily refrain from introducing food products from cloned animals, pending public comment and a final decision.[172]

The FDA decision, despite its release on December 28, 2006, during the winter holidays, raised a firestorm of opposition, in part from environmental and food safety groups opposed in principle to the sale of GM foods,[173] but also from retailers and farming industry associations concerned about the impact of the decision on consumer confidence and hence on sales. For example, the International Dairy Foods Association, a trade group, conducted a poll among US women, 14 per cent of whom indicated that they would turn away from milk products if milk from clones were introduced into the food supply.[174] Fearing a consumer backlash, the association reportedly sought to delay FDA approval of milk from clones,[175] and Dean Foods, the largest US milk company and owner of Land O'Lakes and Horizon Organics, later indicated that it would not use milk from cloned animals in its products.[176] Similarly, the American Meat Institute, while agreeing that cloning of livestock was safe, reportedly urged the FDA to exercise caution in approving meat from clones "if most consumers are unwilling to accept the technology."[177] Another sign of opposition came from the US Congress, where Senator Barbara A. Mikulski (D-MA) introduced a bill that would require mandatory labeling of meat and milk from cloned animals.[178]

Notwithstanding such opposition, the FDA in January 2008 released its long-awaited, 968-page "final risk assessment" for meat from cloned animals, noting the tendency for newborn cloned animals to be larger and sicker than their conventional counterparts, but also indicating that the tests had demonstrated no significant differences between meat and milk from cloned animals and from their conventional counterparts, and concluding that, "Food from cattle, swine and goat clones is as safe to eat as that from their more conventional counterparts."[179] Nevertheless, the FDA noted that cloned animals were likely to be used as breeding stock rather than for meat for the foreseeable future, and the USDA asked farmers to voluntarily withhold cloned animals from the food supply, out of concern for the market reception of US meat abroad.[180] Not surprisingly, the FDA assessment was widely reported and criticized in the media, but the agency defended its assessment, with Stephen Sundlof, Director of the FDA Center for Food Safety and Applied Nutrition, arguing that food from cloned animals is "indistinguishable" from that of conventional animals. "It is beyond our imagination to even have a theory for why the food is unsafe," he said.[181] More generally, for our purposes here, the debate over the regulation of food from cloned animals reflects themes from the FDA's earlier regulation of GM foods, with the agency insisting on its mandate to take decisions based solely on scientific grounds, rather than on ethical or consumer concerns, despite considerable opposition in this case from NGOs, public opinion, and nervous producers and retailers.

Just as the cloning of livestock animals has emerged as a public policy issue in the decade since the first cloning of a sheep in 1996, the genetic modification of animals has rapidly emerged as a public policy issue—one which the US regulatory framework addresses at best indirectly. The science of genetic modification of animals is now well established, and biotech companies envision the creation of a wide range of transgenic animals, including pigs, cattle, fish, and poultry, engineered for particular characteristics including increased growth rates, improved disease resistance, leaner meat, improved nutritional quality, improved wool quality, and so on.[182] Despite these potential advantages, opponents of genetic modification of animals point to the risks of such activities, including the welfare of the animals in question, the safety of foods derived from GM animals, and the environmental impacts of releasing GM animals into the wild where they may interbreed with conventional species or even drive those species out of existence by virtue of their GM traits.[183]

Federal approval of such transgenic animals falls under the primary jurisdiction of the FDA, insofar as those animals are intended for use in food or pharmaceuticals, and the agency has undertaken to regulate GM animals as "new animal drugs" under the Federal Food, Drug and Cosmetic Act.[184] In contrast with the FDA approach to GM foods and crops, which generally do not require premarket notification or approval, the FDCA *does* require premarket approval for new animal drugs, setting a higher regulatory standard for GM animals. By early 2008, however, the FDA had not yet offered any specific guidelines to companies seeking regulatory approval, and critics of the agency have pointed to problems in relying on the "new animal drug" provisions to regulate transgenic animals.[185] Under existing provisions, for example, the FDA must observe confidentiality rules for all new animal drug applications; the agency cannot, therefore, disclose even the existence of an application for approval, making the new animal drug procedure far less transparent than that for GM plants.[186] Other critics have suggested that the FDA lacks the expertise to assess fully the environmental as well as the food-safety aspects of transgenic animals, although defenders of the agency note that the FDA is subject to the regulatory requirements of the National Environmental Protection Act and the Endangered Species Act, and works together with other federal agencies to conduct environmental assessments of transgenic animals.[187]

The difficulties of regulating transgenic animals have been demonstrated by the case of an ornamental aquarium fish, the so-called Glofish, genetically altered using DNA from sea coral to glow under fluorescent light. The developer of the Glofish, Yorktown Technologies, contacted the FDA to assess the agency's potential concerns, but the latter concluded, in December of 2003, that the fish would not be used as either food or pharmaceuticals, and indicated that it "finds no reason to regulate this particular fish."[188] Critics of the decision, however, faulted the FDA for failing to consider the environmental risks if the fish should escape into the wild, and for failing to conduct a full environmental impact assessment, while others suggested that the failure of *any* regulatory agency to conduct a thorough review of the Glofish pointed to the inadequacy of the existing regulatory framework to deal systematically with new and novel GM animals.[189]

In addition to the Glofish, several other transgenic animals are known to be under development, including a transgenic salmon, the so-called "Advanced Hybrid Salmon" modified by its creator, Aqua Bounty Technologies, to grow from egg to market size in one and a half years, significantly less than the two to three years for conventional varieties.[190] As with the Glofish, this GM salmon has raised concerns among critics, most notably regarding the possible environmental impacts if the fish, which is designed to be harvested in a closed aquaculture setting, should escape and either interbreed with or replace wild population of salmon. The FDA has reportedly asked Aqua Bounty for more information about the transgene and about possible effects should the fish escape into the wild, and the company has stated publicly that it will market only sterile and all female salmon to minimize any adverse environmental impacts.[191] A final FDA ruling, pending at this writing, is likely to provoke intense controversy when it is announced.

6.2.5. Change without reform

In sum, recent developments in the US reveal significant elements of continuity and path dependence, with the core elements of the US regulatory framework remaining largely unchanged despite extensive studies by the regulators themselves; yet we also find limited evidence of "trading up," most notably in the commercial adaptation of US farmers and biotech firms to regulations in the EU and other key export markets, and we have suggested that the introduction of cloned and transgenic animals has the potential to arouse a largely dormant public opinion and call into question the adequacy of the Coordinated Framework to deal with new and emerging technologies.

There are significant elements of continuity, we find, in terms of Young's three categories of commercial adaptation, political mobilization, and regulation. At the commercial level, we find a high level of acceptance by farmers of the well-established GM crops such as corn, canola, soybeans, and cotton, with GM plantings now exceeding conventional US plantings in all four areas. Politically, we find that the transatlantic dispute has not, despite the best efforts of some NGOs, dramatically increased the generally low awareness of GMOs in American public opinion, and there are signs that opposition to GM foods and crops has weakened over the course of the past five years. In terms of regulation, finally, the US regulation of GM foods and crops continues to be undertaken under the 1986 Coordinated Framework, with no new Congressional legislation and only marginal changes to FDA, APHIS, and EPA regulations. Largely as a result of the early regulatory decisions of the 1980s, the US remains characterized by a strong pro-biotech constituency, low awareness and weak opposition to GMOs in public opinion, and a regulatory framework that remains unchanged in its essentials and relatively resistant to fundamental change in the absence of exogenous shocks that could trigger new Congressional legislation.

Nonetheless, a careful review of the US case does reveal new developments at all three levels, which demonstrate limited "trading up" as well as the potential for further change in response to contingent events down the road. At the level of commercial adaptation, we have seen that, while established GM crops retain strong market acceptance, new GM crops including wheat and rice have met with strong resistance among US farmers worried about access to export markets with little or no tolerance for the adventitious presence of unauthorized GM crops in US exports. This development in turn raises the prospect of a two-tiered market in which GM varieties remain dominant for some crops while encountering sharp market resistance for others. It also underlines the *de facto* influence of the EU and other foreign regulators on the business decisions of US farmers and biotech firms. At the level of political mobilization, both public awareness and support for GMOs remain relatively low, but US public opinion could change in response to a future food safety or environmental crisis, or in response to emerging regulatory challenges such as cloned and transgenic animals. Finally, while US regulatory policy toward biotechnology is often characterized as regulation based on products and not on the process of genetic modification itself, in practice, US regulatory agencies have increasingly subjected GM crops to a greater regulatory burden than conventional varieties.[192] US agencies have done so under their existing statutory authority, in part on account of "the difficulty of fitting biotechnology products into pre-existing legal categories," and "in part due to the perceived public interest in affording GE [genetically-engineered] products greater scrutiny."[193] For example, starting in May 2000, the FDA began to encourage biotechnology companies voluntarily to submit safety data prior to marketing food, which has since become a common practice. As the Pew Initiative reports, "developers of genetically engineered crops routinely consult with the FDA on a voluntary basis because of the practical marketplace reality that buyers would penalize products that had not been through the FDA's consultation process."[194] The result is a *de facto* higher level of regulatory scrutiny. Similarly, the USDA requires either notifications or permits for field trials of genetically engineered plants, but not for their conventional counterparts that can pose just as great (or greater) risk.[195] Because of the difficulty of fitting GM crops into existing regulatory categories, in July 2001, the EPA devised a new category for plants genetically modified to have pesticidal characteristics.[196] These varieties fall neither under the category of "plants" nor "pesticides," but rather that of "plant-incorporated protectorants," or PIPs. The EPA has enacted regulations that require premarketing approval for PIPs, such as Bt corn, whereas conventional corn is not subject to EPA oversight.[197] Transgenic animals, finally, would be subject under existing law to pre-market authorization, and face a higher hurdle of public acceptance than the GM foods and crops marketed in the US to date.

The debate about the regulation of biotechnology in the US is certainly not over. Some proponents of GMOs maintain that US regulation has already gone beyond what "science" requires, while other GMO advocates find that increased regulatory oversight may be necessary to satisfy national and international market demands, and ultimately, to ensure public trust that biotechnology products are safe.[198] Many environmentalists and consumer groups continue to press the FDA, the Congress, and the courts for stricter regulation and labeling of GM foods.[199] Others, such as the Pew Initiative on Food and Biotechnology, have highlighted potential incremental reforms to the system to deal with the imminent introduction of new and potentially more risky genetically engineered products, such as GM plants intended for pharmaceutical and industrial use and GM animals.[200]

Reform of the current system is thus likely to remain on the US agenda, resulting in administrative and possibly legislative changes to the current system. Nonetheless, there is little sign that the US is preparing to move away fundamentally from its core practice of regulations that are adopted by specialized agencies, and that are to be based on the risks posed by individual products, to be determined by scientific assessments (although these are largely conducted by the companies and only reviewed by the agencies themselves), as opposed to the EU approach of premarket risk assessment, political approval, labeling and traceability requirements imposed on each GM product because of the "novel" nature of genetic engineering. As Prakash and Kollman conclude, "while market competition and NGO activism can create pressures on polities to converge, such pressures are always filtered through domestic institutions and political processes."[201] Anti-GMO activists have simply not been able to use events in the US to create significant public pressure for change. As a result, changes in US regulation of GM crops and foods are likely to continue to be piecemeal and relatively modest in comparison to the agricultural biotech regulatory requirements of the EU. While in the EU, we found there has been much reform with little or no fundamental change, in the US we find *some change without reform* of the regulatory framework.

6.3. Conclusions: Path-dependent change and (lack of) convergence

The period since 2000 has produced extraordinarily far-reaching reform and controversy in the EU's regulatory framework for biotechnology, while in the US we have witnessed some signs of wariness about GMOs among farmers and consumers, as well as discussions of reforming a regulatory system that remained virtually unchanged since the adoption of the Coordinated Framework for Biotechnology in 1986. Some elements of these changes can be interpreted as responses to external pressures, and as modest steps by each side toward convergence with the other side. In the EU, we have witnessed the increased importance of scientific risk assessments, particularly with the creation of the EFSA, and we have seen the tentative resumption of approvals of new GM varieties since 2004. Both of these steps can be attributed in large part to external pressure from past WTO jurisprudence and from the specter of a US legal complaint before the WTO, which eventually resulted in the panel decision in 2006. In the US, we find significant evidence of market adaptation among farmers, who have expressed reticence about adopting and growing GM crops that are unlikely to be accepted for export to the EU, and more limited evidence of increased political mobilization and of regulatory reforms by the FDA and the USDA. Neither regulatory system has remained static, and both show evidence of external influences in recent years.

Despite these changes, we have found at best limited evidence of meaningful convergence between the two systems, and considerable evidence of continuity and path dependence in each one. In 2000, the EU regulatory system was one which regulated GM foods and crops in terms of the process by which they were produced, rather than the characteristics of the products; the regulatory process was based in part on scientific risk assessments, but left significant room for political actors to intervene, including on the basis of "other legitimate factors" in their decisions; regulatory standards for GM foods and crops were strict relative to the US; and the implementation of the Union's regulatory framework had been paralyzed by a *de facto* moratorium on new approvals and by member-state bans on already approved varieties. These fundamental characteristics of the EU system remained essentially unchanged in 2008, and indeed the overhaul and ongoing tinkering with the Union's regulatory framework have introduced regulatory requirements far more strict than those in place during the late 1990s. This continuity, moreover, demonstrates several characteristics of path-dependent development. More specifically, the early adoption of a highly restrictive regulatory framework, together with the food-safety crises of the mid-1990s, discouraged farmers from planting GM crops, prompted retailers to resist selling GM foods, and led to a flight of biotech investment from Europe, all of which undermined subsequent political support for GM foods and crops. Furthermore, the EU's supermajoritarian legislative rules, requiring a qualified majority among disparate states in the Council of Ministers as well as a majority in the European Parliament, have created a huge institutional hurdle to the future reform of the EU regulatory framework.

Similarly, the US's regulatory system in 2000 was one in which relatively independent federal agencies regulated GM foods and crops, according to the characteristics of the product rather than the process of genetic modification, and in which standards were relatively lax, with no requirement for premarket authorization of new GM varieties and no mandatory rules for traceability or labeling of GM foods and crops. These essential characteristics of the US regulatory system also remained essentially unchanged after more than a decade of US/EU conflict. Here again, we see strong elements of path dependence in the US case, where the early adoption of a welcoming regulatory framework has contributed to the growth of a strong biotech industry and the widespread acceptance of GMOs among farmers and (to a lesser extent) public opinion. Here again, moreover, institutional rules privilege the status quo, in which GMOs continue to be regulated under the two-decades-old Coordinated Framework, which has changed only at the margins in the absence of significant new Congressional legislation.

Path dependence does not, of course, mean immobility or the impossibility of fundamental change in response to exogenous shocks. In the EU case, pressures from the WTO have led to some, albeit limited, regulatory changes and to the tentative resumption of approvals by the Commission in the face of considerable opposition. In the US, by contrast, market pressures from Europe and elsewhere have taken a toll on the market acceptance of new GM crops such as wheat and rice, demonstrating the effect of EU decisions within the US market and prompting regulatory reviews by the FDA and APHIS. Perhaps most importantly, regulatory systems on both sides remain vulnerable to contingent events sufficient to overcome the social and institutional barriers to regulatory change.[202] European acceptance of GM foods, for example, could potentially increase in the future if a new generation of GM varieties with greater consumer benefits becomes available, while US acceptance could potentially decline in response to possible food safety or environmental crises or in response to emerging technologies such as transgenic animals that raise new and novel ethical and environmental challenges. Nevertheless, it remains a striking fact that, notwithstanding nearly a decade of negotiation, deliberation, and dispute, the essential features of the two systems and the underlying frictions between them, have endured.

7

Conclusions: The Lessons of Transatlantic Conflict, Developing Countries and the Future of Agricultural Biotechnology

The regulation of agricultural biotechnology continues to be the subject of one of the most difficult and intractable disputes in the transatlantic relationship, one that has expanded to have global dimensions. The dispute illuminates the challenges posed *when national legal diversity meets economic interdependence.* It illustrates the particular challenge of managing the risks and benefits of technological innovations that have potential, but uncertain consequences in an economically interdependent world. The outcome of the conflict will affect the future of agriculture around the world and thus, what we eat and what we wear.

Our study of this conflict shows how global markets, and national and international law, politics, and institutions interact to shape global regulatory and commercial outcomes. On the one hand, we see the *limits* of transnational networks and supranational and international institutions for harmonizing, coordinating, or even accommodating conflicting national regulations when issues become politicized within states and when these institutions and networks are subject to public distrust and resulting legitimacy challenges. On the other hand, we see that international and foreign law and practices can have real effects on national regulations, regulatory practices, and commercial decisions. One can no longer understand national and EU regulation and regulatory impacts without examining them in global context. Regulation in any single jurisdiction has real and potential economic, health, and environmental consequences for people unrepresented in that jurisdiction's regulatory system.

In this concluding chapter, we first offer a set of conclusions and lessons from our analysis of the transatlantic dispute, revisiting our five core arguments, and drawing policy lessons from them for the management of the transatlantic (and now global) dispute, and for the future global regulation of GM foods and crops. In the second section, we step back from the transatlantic perspective which has dominated the dispute and thus our approach to it in this book, to assess the impact of the GMO conflict globally when transatlantic cooperation fails, and in particular for less developed countries (LDCs) that have served primarily as observers and pawns in the transatlantic struggle over the governance of GMOs. A third and final section concludes with a look at the uncertain global future for agricultural biotechnology.

7.1. Transatlantic conflict and the failure of cooperation: Five lessons

Several years ago, we presented an early draft of this book to a colleague, asking what she thought of the story of GMO regulation and of our arguments about it. Her first reaction was that our account was essentially a story of failure, particularly in contrast to much of the literature on international political economy and law.[1] Indeed, our account, with its emphasis on the intractability of the GMO dispute and the strikingly limited contribution of bilateral networks and multilateral regimes, contrasts sharply with most other books and articles about international cooperation, which are essentially optimistic accounts of how international cooperation is possible and can be facilitated by international regimes in an anarchic international system without any centralized authority. The general form of such studies falls into what we have called the Home Depot theory of international cooperation: "You can do it, we can help."

Our account of agricultural biotech regulation, by contrast, has emphasized the difficulties, the limits, and in many cases the outright failure of international cooperation in the regulation of GMOs. The transatlantic dispute over GMO regulation was not inevitable, we have argued, but it is real and the two sides are deeply entrenched in their respective positions. Bilateral attempts at joint deliberation on GMOs through policy networks have failed almost entirely, reflecting the fact that deliberation is a hothouse flower unlikely to flourish in the glare of intense politicization of GMO regulation. Multilateral regimes offer a promising forum for resolving disputes over agricultural biotechnology, but cooperation has been similarly hampered by distributive conflicts, as well as by the tensions within the regime complex for GMOs. Reflecting these failures, and the path dependence of policy frameworks on each side of the Atlantic, we see at best limited evidence that the US and EU approaches to regulating GMOs have converged in recent years. Finally, the WTO dispute-settlement system can have a positive influence on the conflict (or so we have argued), but WTO litigation alone will not resolve the dispute, which is likely to continue for years to come.

Our conclusion, however, is not a counsel of despair. While we find no perfect solution to the transatlantic (and increasingly global) GMO conflict, whether through transnational deliberative networks, multilateral regimes, or international litigation, we *do* believe that the dispute can be managed in such a way that governments protect their societies from risk while simultaneously respecting the views of others and avoiding a full-scale international trade war. Successful management of the conflict, however, must begin with a clear understanding of the origins and tenacity of the dispute, the interactions of domestic and international institutions and markets, and the limits of both cooperative and litigious approaches. A fuller version of the Home Depot theory, we suggest, would be, "You can do it, and we can help, but only if you understand the problem, bring the right tools and adjust your expectations about what international law and institutions can achieve in deeply politicized issue-areas."

In this spirit, we return here to the five inter-related arguments that we put forth in Chapter 1, recasting these not just as social-scientific arguments but also as lessons to be considered by policymakers in managing transatlantic regulatory disputes and in pursuing effective global governance of agricultural biotechnology. These five lessons, summarized briefly, are the following:

Lesson 1. The transatlantic GMO dispute does not represent a deep civilizational divide, but it is real, and deeply entrenched.

As we have seen, beginning in the late 1980s with the establishment of their respective frameworks for regulating agricultural biotechnology, the US and the EU have regulated GM foods and crops in starkly different ways. The transatlantic dispute has brought into confrontation not only different laws, but distinctive *systems* of risk regulation for GM foods and crops. On the US side, the system of regulation by federal regulatory bureaucracies based primarily on "scientific" risk assessments conducted by private actors and provided to public authorities, is not without its critics, but nevertheless has the strong support of most US farmers, industry and government officials, as well as a high degree of social acceptance, and thus legitimacy, in the US public opinion. On the EU side, European public and institutions have favored a much more cautious approach to GMOs, spurring the enactment of specialized legislation and regulatory procedures for GM crops and foods on account of the GM *process* itself. The food safety scandals of the 1990s increased European public anxieties over the adequacy and trustworthiness of European regulation of food and agriculture, spurring EU member governments to put in place ever stricter regulations for the premarket approval, traceability, and labeling of GM seed, food and feed products, independent of their individual safety characteristics. These stark differences between the US and EU systems and standards have been referred to as regulatory polarization, and rightly so. The differences are indeed real and large, and they have created real-world tensions and the possibility of transatlantic trade retaliatory measures between the US and the EU.

Nevertheless, we should resist the temptation, so common in media coverage of the dispute, to depict US/EU regulatory differences as the inevitable result of two profoundly different approaches to risk grounded in deep cultural, institutional, or interest-group differences. Neither the US nor the EU regulatory systems for GMOs were predetermined in any straightforward way by the interest-group, institutional, or cultural characteristics of either polity, and neither was inevitable. The specific regulatory frameworks for GMO regulation in the EU and the US have complex, multicausal explanations, relying at least in part on contingent events, and either regulatory system might have looked different if, for example, a Democratic administration has been in office in the US in the mid-1980s, or if the EU food-safety scandals of the 1990s had not overlapped with the first commercialization of GM varieties there.

Hence, while we do believe that there are generalizable lessons to be learned from the GMO dispute—lessons that "travel" beyond GMOs to other issue-areas—we do *not* wish to argue that the US is inevitably and everywhere more accepting of risk than the more precautionary EU. Comparative studies on risk regulation demonstrate that risk regulation varies substantially by issue-area *within* the US and the EU, and that the US is often more precautionary than the Union in regulating carcinogens such as tobacco, and nuclear power, among others. We do not, therefore, predict on the basis of the GMO case that the US will again be more accepting and the EU more precautionary in the regulation of emerging technologies, such as nanotechnology, which is likely to reflect the complex panoply of interests, institutions, culture, and contingent events specific to those new technologies.

What our analysis *does* suggest, however, is that regulation in both polities can be subject to politicization, positive feedbacks, path dependence and self-reinforcement, such that a given system of regulation (and, in particular, risk regulation), once established, will become extremely difficult to change, even in the face of strong international pressures. Our understanding of the domestic roots of regulatory polarization, and the path-dependent effects that make domestic regulations so difficult to change, is therefore a necessary prerequisite to any international efforts at either cooperation or litigation.

Lesson 2. Deliberative decision-making is a hothouse flower, which has seldom bloomed in the intense politicization of GMO regulation. Our expectations for it should be tempered accordingly.

Not only academic commentators, but the US, the EU, and other countries, have placed much stock in the establishment of cross-border networks of government officials and non-governmental organizations, which are urged and expected to communicate, learn from each other, and deliberate about the optimal regulatory policy on the basis of common understandings and shared principles. As we have seen, however, the most careful theoretical accounts of deliberative decision-making concede that genuine deliberation is most likely to occur only *under certain conditions*. More specifically, we have argued, deliberation is akin to a hothouse flower that flourishes only where negotiators share a "common lifeworld," are uncertain about the current state of knowledge or about their own preferences, engage each other repeatedly, and are free to change their views without fear of pressure or punishment from domestic constituencies.

The issue of agricultural biotechnology is indeed characterized by some scientific uncertainty, and efforts have been made to institutionalize transatlantic and global exchanges among government regulators and civil-society groups, leading some scholars to posit that it might be a promising arena for deliberative decision-making. Both our domestic and our international case-study evidence, however, suggest that for international regulators and civil-society actors, the conditions for deliberation are at best partially met in the area of GMO regulation: US and EU regulators and civil-society actors often do *not* share a common lifeworld; scientific uncertainty has not prevented the actors from identifying their own preferences and ascertaining the distributive consequences of possible cooperation; and regulatory officials within transgovernmental networks negotiate in an intensely politicized issue-area in which the ability to compromise, much less defer to the strength of the better argument, is limited by strong interest-group pressure on both sides. These conditions are far more conducive to "bargaining" than to "arguing," that is, to deliberation.

Within the EU itself, policymakers have attempted to facilitate harmonization, mutual recognition, and adaptation of member-state food and agricultural regulatory systems through creation of networks of scientists and regulators. In many areas of food policy, EU regulatory committees have brought together member-state regulatory officials to engage in forms of deliberation that have facilitated member states' ability to harmonize rules governing the food-safety risks and generally to make national regulatory systems more compatible.[2] As a result, food products produced in any one EU member state can typically be sold throughout the EU, thereby alleviating the discriminatory and other adverse consequences of national regulation on other European producers, and enhancing intra-European competition, with resulting consumer benefits.

Many EU authorities have hoped that similar procedures could help structure European political and social understandings of GM products. The European Food Safety Authority (EFSA), for example, brings together a community of specialized scientists from throughout the EU to conduct risk assessments of GM varieties. This Food Safety Authority could, through its exercise of "normative authority" grounded in technical expertise, help "structure individual and institutional choices on food safety within the EU."[3] Knowledge is not just about discrete information, but also is an active "process" of social exchange that organizes individual "belief" about the world.[4] Knowledge produced by EFSA, in this sense, could help structure conceptions of the risks and benefits of GM technology and of the products that the GM technology spawns. Yet EFSA risk assessments of GM varieties still appear to have had little impact on actual risk-management decisions in the EU.[5] Member-state politicians, responding to their constituencies, have instead applied national safeguard bans against GM varieties approved at the EU level, and attempted collectively to constrain the Commission from challenging them before the European Court of Justice. Private companies likewise are reluctant to engage in protracted litigation which could have little commercial effect. As a result, EU law has lain largely in abeyance.

Under these circumstances, efforts to bridge the gap between the US and the EU through bilateral regulatory cooperation—in the hopes that US and EU regulators will engage in joint deliberation and come to common understandings of their regulatory tasks—have borne little fruit. Where there have been transatlantic dialogues on biotechnology that have reached some consensus, they have been undertaken either by experts participating in their personal capacity, by like-minded activists, or (in the case of the 1999 Biotechnology Consultative Forum) by non-governmental representatives who operate under no constraint to represent their respective countries. Transatlantic networks of regulatory officials, in contrast, have generally gone nowhere and been largely abandoned.

Deliberative decision-making has been more in evidence within multilateral regimes such as the Codex Alimentarius Commission and the Organization for Economic Cooperation and Development, but even there successful deliberation has been limited largely to technical questions of risk assessment among scientific experts, and has given way to bargaining behavior when the topic turns to risk management and when discussions are clearly linked to trade-related issues. Deliberative decision-making, then, is worth pursuing, particularly among scientists and technocratic officials who can quietly establish a common basis for risk assessment of GM foods and crops—but we should not expect it to succeed, or to offer any solution to the dispute, when it comes to deeply politicized questions of risk management.

Lesson 3. Multilateral regimes can help states cooperate, but they are hampered by the dual challenges of distributive conflicts and regime complexes.

It is a core tenet of regime theory that multilateral regimes can help states to cooperate, by lowering transaction costs, monitoring compliance, providing dispute settlement, and possibly encouraging deliberative decision-making among states. To start, we find that the US and EU have common overlapping interests, not only to avoid a trade conflict, but also to create and comply with basic principles that limit the ability of any one party to restrict imports from the other without reasoned justification. To avoid conflicts, both sides also have incentives to agree on common principles and practices for assessing and managing risks, and to coordinate their policies.[6] Indeed, we find that cooperative efforts in agricultural biotechnology have generally been more fruitful (and in some cases more deliberative) within multilateral regimes such as the OECD and the Codex Alimentarius Commission, than have bilateral efforts between the US and the EU. Regimes such as these have indeed lowered the transaction costs of negotiation in a global context, created a multilateral forum for scientific exchange about risk assessment, and (in the case of the WTO) helped to clarify the meaning of the parties' international legal obligations through dispute settlement.

Despite these benefits, we have argued that cooperation within international regimes is hampered in practice by the inter-related challenges of distributive conflicts and regime complexes. With regard to the first of these challenges, we have argued that international coordination of GMO regulations is indeed characterized by a stark distributive conflict, in which both the US and the EU seek to promote global regulatory standards that effectively export their own domestic standards, and in the process shift the economic, environmental, social, and political burdens of adjustment to the other side. Once again, we do not assume that the EU's strict regulations are motivated by trade protectionism, and indeed the evidence presented in Chapters 2 and 6 suggests that genuine public concern or fear of GMOs is the primary motivation behind the EU's rather draconian regulatory regime. Nevertheless, even if both sides are "sincere" in their views and motivated by environmental safety and consumer and social welfare rather than narrow economic concerns, the choice of a global standard still has distributive consequences, as the "loser" in such a negotiation is likely to bear both the economic costs of adjusting production and regulatory practice to a new set of standards, *and* the political cost of changing the status quo in a deeply politicized issue-area and in the face of strong political opposition.

International cooperation on agricultural biotechnology, in this sense, resembles not so much a Prisoners' Dilemma game as a Battle of the Sexes game, in which the central question is not enforcement of an agreement, but agreement on the *terms* of cooperation, with different standards leading to different distributions of costs and benefits among the participants. International cooperation under such conditions, regime theorists have argued, is likely to take the form of hardball bargaining among states with clear perceptions of their own interests, and outcomes are likely to be influenced not (or not primarily) by the quality of one's arguments, but also and especially by the bargaining power of the participants. In the GMO case, the US and the EU—as the two largest economies and regulatory jurisdictions on earth—have each sought to export their respective regulatory approaches to various international regimes, bargaining from fixed positions and maneuvering constantly for advantage in setting the terms of global regulation.

Related to the importance of distributive conflict is the second challenge, namely the existence of a regime complex for agricultural biotechnology, including regimes for the environment (the Convention on Biodiversity and the Cartagena Biosafety Protocol), food safety (the Codex Alimentarius Commission), trade (the WTO), and others (the OECD and other multilateral regimes). Consistent with the literature on regime complexes, we have seen both sides forum-shop for the regimes most conducive to their own position, each attempting to assert the legal primacy of their own positions. Within each regime, the result has been, not a reasoned consensus, but a series of inelegant compromises that paper over rather than resolve most of the key issues. Among regimes, the result has been inconsistency and fragmentation of international law, with different and sometimes incompatible approaches taken by regimes for the environment, food safety, and trade, with no clear hierarchy among these various approaches.

In addition, we have found that the various soft-law and hard-law regimes within this regime complex have interacted in mutually constraining ways, resulting in a "hardening" of soft-law regimes such as the Codex, which has become less flexible and adaptable, and a "softening" of hard-law regimes such as the WTO, which has become less predictable as judicial decision-makers have sought to avoid applying the WTO's substantive law provisions in a clear manner. These outcomes, we have argued, should not be entirely unwelcome, in that they provide incentives for international officials to take note of the effects of their actions outside the narrow sectoral contexts of issue-specific regimes—a result that we consider to be normatively desirable as well as sociologically legitimate. But they underline the fact that practitioners cannot pursue, nor can scholars understand, international cooperation in biotechnology regulation without taking note of the presence of overlapping regimes, forum-shopping by member states, and the inconsistencies and interactions of international regimes which can no longer be examined in isolation from each other.

Our point here is not that bilateral or multilateral dialogue is fruitless, since even "bargaining" can result in useful exchanges of information and sometimes mutual accommodation. Any cooperative efforts, however, should begin with a frank acknowledgement that there is significant disagreement on the *terms* of cooperation, which can create winners and losers, and that bargaining power is likely to play a significant role in determining the outcomes. Similarly, we have shown that, while international regimes continue to be defined largely in terms of individual issue-areas, analysis of the international law and politics of issues like agricultural biotechnology must examine the ways in which governments forum-shop among international regimes, and the ways in which these regimes interact with each other and with the domestic politics of their members.

Lesson 4. International pressures—from markets and multilateral regimes—can have domestic effects, but these effects are limited by the path-dependent nature of domestic regulatory systems, which are unlikely to converge on any common denominator.

Just as the nature of domestic law and politics influences the working of international regimes, so we might expect that international regimes, as well as global market forces, can create pressures (or opportunities) for change in domestic law and commercial practice. International regimes such as the WTO, for example, can create pressures for regulatory change in the EU, by empowering domestic European actors (such as the Commission, individual companies and trade associations) with a preference for liberalizing GMO regulation and regulatory practice. Foreign market requirements, similarly, can create pressures for the US to "trade up" to a more precautionary regulatory stance, through catalyzing changes in public opinion, spurring market adaptation by farmers and private firms, and triggering regulatory responses. While we indeed find evidence of international regime and market pressures in Chapter 6, we nevertheless argue that the impact of these pressures has been significantly blunted by the path-dependent and self-reinforcing nature of domestic regulatory systems. Even in response to strong international pressures, domestic regulatory systems are likely to change only incrementally at the margins.

In the US case, both societal adaptation from below and change-resistant decision rules from above have lent considerable stability to the US regulatory framework. At the societal level, two decades of regulation under the Consolidated Framework have led farmers to embrace GM crops, particularly in well-established niches, such as corn, soybeans, and cotton, while encouraging the development of a thriving and politically significant biotech industrial sector, both of which have provided strong domestic support for the existing regulatory system. These actors are not, to be sure, immune from international market pressures, and we have seen how US farmers have generally shied away from growing new GM varieties such as rice and wheat, for fear of exclusion from the EU and other markets. We have even seen farm groups ask regulators to restrict the approval of commercial plantings of new GM corn varieties until they have also been approved in major export markets, although regulators have yet to do so. By contrast, however, US plantings of established GM crops continue to grow year-by-year, and we see little evidence of support for a move to a strongly precautionary, processed-based approach similar to that of the EU. At the level of national political institutions, US regulators like the Food and Drug Administration and the United States Department of Agriculture have demonstrated a willingness to revisit specific aspects of GMO regulation and have introduced some changes at the margins, but neither has indicated any preference for switching to a process-based system involving mandatory premarket approval, labeling, or traceability requirements. Such a major regulatory overhaul would likely require a legislative act of Congress, in which the US House and Senate, their respective committees, and the US President would each serve as veto players. Barring a major shock, such as a food-safety or environmental crisis, therefore, the likely outcome in the US is an adaptation by US farmers and biotech firms to foreign market demands, together with some marginal reforms by specialized US regulators—what we have called *change without reform*.

The EU, despite its significantly different regulatory framework for GM foods and crops, is subject to precisely the same path-dependent and self-reinforcing tendencies, with societal adaptation from below and change-resistant decision rules from above. At the societal level, the EU's increasingly strict regulatory framework, coupled with a politicized and sometimes paralyzed regulatory process, have discouraged all but a small group of farmers from planting, and food retailers from selling GM crops and foods, while at the same time discouraging the growth of the European agri-biotech sector. Hence, the potential coalition of pro-GMO farmers, retailers, and biotech sectors has progressively weakened over time, while consumer resistance to GMOs grew in response to food-safety crises and NGO activism. Societal pressure, therefore, has taken an increasingly precautionary direction over time, and this pressure has been reflected in the strong anti-GMO stance of many EU member governments. The EU's decision rules, moreover, are even more sharply change-resistant than those of the US. The EU legislative process provides multiple veto points and high majoritarian thresholds, requiring a proposal from the Commission as well as approval by the GM-skeptic European Parliament and a qualified majority of the EU's member governments in the Council. Given the strong anti-GMO bloc in the latter body, the relaxation of the EU's strict standards has proven impossible, and indeed the direction of change has been in the opposite direction, with the steady addition of new regulatory hurdles in recent years. Furthermore, and unlike the US where specialized federal regulators oversee the "deregulation" and approval of new GM varieties with relative independence (delegating considerable "self-regulation" to the private sector), the EU's regulatory process also empowers member governments as veto players, either collectively in the approval of new GM varieties, or individually in the application of national safeguard bans. For this reason, the pressures from WTO dispute resolution, and the resulting strengthening of the position of the Commission, has had limited effects in practice, facilitating the contentious approval of some individual GM varieties, but without changing the EU's precautionary approach or national safeguard bans of dubious legality—what we call *reform without change*. We thus see few signs of EU regulation converging on the US model.

Our point here is simple: international regimes and market forces can indeed generate pressures for domestic change by creating new incentives for domestic actors (as in US market adaptation) and by providing new political, legal, and knowledge resources to actors with a preference for change (as in the EU approval process, with a key technocratic role played by EFSA). But these pressures are filtered through decades of societal adaptation to existing regulations, which are themselves subject to change-resistant decision rules.

For these reasons, and again barring a major exogenous shock to either system, we should not place our hopes in the gradual convergence of the two regimes whose central features remain deeply entrenched in law and politics.

Lesson 5. The World Trade Organization and its Dispute Settlement Body cannot definitively solve regulatory disputes such as those over GMOs, but they can help to manage conflict, clarify the obligations of all sides, and provide some opportunities and leverage in domestic political and judicial processes.

Against this backdrop, the US complaint before the WTO dispute-settlement system presented significant risks of backlash against agricultural biotechnology and the WTO itself. A substantive WTO legal decision against the EU could further polarize public opinion about the technology. It could be simply rejected by the EU, its member states, and their constituents. It could spur further mass protests against the organization. In contrast, a WTO legal decision that provided full discretion to the current EU system could further undermine US views on the efficacy of the WTO and of WTO law.

Yet the WTO case also held the promise of some clarification and mutual accommodation that nearly a decade of bilateral and multilateral negotiation had failed to produce. The legal setting of the WTO dispute-settlement system has indeed forced both sides in the dispute to argue their respective cases in the language of international law. Even though the WTO panel produced a ruling that shied away from making substantive findings about EU procedures, it still provided some clarity about the obligations of each side, and it has helped to stabilize this potentially explosive trade conflict. The WTO panel took a middle ground in which it avoided examining whether EU decisions met WTO substantive obligations, such as whether the specific EU-level measures in question were based on a scientific risk assessment. It rather found that the EU must not engage in "undue delay," but must make reasoned decisions in a timely manner in response to individual applications for the approval of GM varieties. It similarly returned the issue of member-state safeguards to the EU, implicitly indicating that they must either be challenged and discontinued as required by EU law, or validated on the basis of a new risk assessment.[7]

The European Commission and affected private actors continue to evaluate how to proceed, on a case-by-case basis, in light of the WTO panel report. While GM proponents find their position somewhat strengthened by the WTO decision, they still operate in an EU political system characterized by numerous veto players, as well as relatively high societal opposition to GMOs. By itself, therefore, the panel's decision is unlikely to bring about full EU compliance with the terms of WTO law, much less a resolution to the broader conflict or to related conflicts over issues like traceability and labeling. Nevertheless, the WTO dispute-settlement process has channeled a bitter dispute into a legal process, possibly forestalling a potential trade war. In the process, it has clarified to some extent the obligations of both sides and has provided tools for domestic actors in domestic political processes which, over time, could lead to greater mutual accommodation. In this sense, the significant gamble of bringing the WTO complaint has paid off, even in the short term.

As Keohane and Nye noted long ago, international institutions can play a role in fostering regular interactions among government policymakers, including through the formation of transnational and transgovernmental networks.[8] Keohane and Nye did not, however, assess the potential role of *international courts*, as international judicial processes were not active at that time. If the WTO's most powerful members allow it, the WTO judicial system can provide an important form of conflict management in a world characterized by jurisdictional diversity and economic interdependence, where regulatory decisions in one country can have real and potential consequences on others, whether because national regulation is too restrictive or too lax. The WTO legal system can, in particular, press countries to justify their regulatory decisions to those foreigners affected by them. It can do so, in part, by providing new political and legal opportunities to domestic actors with similar interests and holding similar ideas. States are not monolithic, but include actors who can use an international legal decision as a tool to attempt to drive changes within domestic state practice. It is often in this way that transnational legal processes most effectively work.[9]

Although our account is primarily a positive one, we have built from our empirical work to examine, in a structured manner, the tradeoffs among the alternative choices that WTO panelists can make in interpreting WTO agreements in light of their institutional implications and the legitimacy constraints that the WTO dispute-settlement process confronts. In doing so, we address the impact of different institutional choices on the participation of affected constituencies. For the foreseeable future, most risk management decisions will continue to be made primarily at the national level, or in the case of the EU, at a hybrid of the national and regional levels. In order to be normatively and socially legitimate, they should likely be made at these levels.[10] Yet, even though the WTO judicial process involves an unelected body operating at a remote level of social organization, it can play a positive role in helping to correct the parochialism of national and, in Europe's case, regional decision-making. It can press national decision-makers to justify their decisions in a transparent manner, knowing that their decision-making processes are subject to supranational judicial review. Decision-makers can thereby be pushed to take greater account of the impact of their decisions on outsiders, and in particular, to limit any adverse impacts on outsiders to the extent feasible while still meeting their regulatory objectives. In our view, when making national and regional regulatory decision-making more accountable to otherwise unrepresented constituencies, WTO judicial decision-making is normatively legitimate.

There are of course tensions between attempts by the WTO judicial process to make national decision-making more accountable to outsiders and the acceptance of WTO decisions by those same national decision-makers. That being said, WTO judicial decisions will also more likely be accepted by WTO members and their constituents over the longer run, if WTO panelists and the Appellate Body do not presume to make ultimate substantive determinations regarding risks which political bodies have been unable to resolve but rather take a procedural approach, as they largely have done. International institutions such as the WTO should tread softly in an area that implicitly raises concerns about "risk," and more generally about what the public want, where there is no evidence of discriminatory protectionism. That is why the WTO Appellate Body and panels have largely taken (and should continue to take) a more procedural approach in these cases.

Taken together, our five observations point to one final, overarching lesson: The transatlantic dispute over GM foods and crops is unlikely to be definitively resolved in the near future, whether by bilateral deliberation, multilateral negotiation, gradual convergence, or international litigation—yet it can and should be managed through a combination of these and other methods.[11] Bilateral consultations have thus far failed to live up to the high expectations of deliberative decision-making; nevertheless, we do find evidence of deliberation among scientific experts over risk assessment, and this effort to establish scientific (if not political) consensus is worth pursuing, possibly bilaterally but preferably within multilateral regimes like the OECD and Codex, which can have the additional advantage of including third parties who engage in trade of GM products. Multilateral regimes, while subject to distributive conflicts and to forum-shopping and inconsistency across regimes, also have a potentially positive role to play by lowering the transaction costs of negotiations, providing a common vocabulary for discussing GMO regulation, clarifying at least some of the mutual obligations of the parties,[12] and contributing to regulatory capacity building in less developed countries (discussed in the next section). Barring a major exogenous shock, we are less sanguine about the possibility of spontaneous convergence of US and EU regulations, although even here it is striking that market adaptation by US farmers to regulations in their core export markets may reduce system friction in practice, if not in law.[13] In that context, we also believe that the WTO, while treading relatively lightly in a deeply politicized area, can have—and has already had, in the 2006 panel decision—a positive impact on the conflict, by channeling the dispute into legal processes, clarifying the procedural obligations of both sides, and catalyzing pressure to increase the transparency of domestic regulatory processes. Once again, however, we believe that the WTO will attempt to avoid—and would indeed be wise to avoid if it can—deciding the substantive issues in disputes between the US and the EU over agricultural biotechnology, which will continue to be debated and bargained over in diverse international fora for years to come.

7.2. The implications of the transatlantic dispute for developing countries

Throughout this book, we have focused our analysis of GM food and crop regulation primarily on the disparate approaches taken by the US and the EU, and the resulting conflict between them. We have done so neither to simply point out the range of regulatory choices for which the US and EU largely reflect different poles, nor merely to underline the drama of the conflict that has played out both bilaterally and across a panoply of multilateral organizations. Rather, we have focused on the US and the EU primarily because the sheer size of their domestic markets, their vast entrepreneurial and technological resources (whether to create and evaluate new GM varieties or to detect and trace their "pollution" in minute quantities of bulk grain shipments), and their negotiating power in the global arena, make the US and Europe the leaders in global economic regulation and regulatory practice. Where the US and EU agree on a regulatory approach, they exercise considerable influence in shaping regulatory outcomes around the globe, with third countries often following the collective European and American lead. Where they disagree, however, multilateral regimes can find themselves at loggerheads, and third countries, especially those that wish to export to global markets, face considerable challenges. Whose approach should they follow, when the outcome of the transatlantic conflict remains uncertain, and both sides actively seek allies and deploy incentives and threats to obtain support? What opportunities arise for these countries, and what dilemmas do they face, when transatlantic cooperation fails?

7.2.1. Special challenges for LDCs

The challenges posed are particularly acute for less developed countries (LDCs), most of which have no domestic biotech industry, little or no public research capacity, and weak state capacity to regulate GM foods and crops, and yet have major economic and social welfare stakes in the outcome of the conflict. As Richard Stewart notes in an excellent survey of LDC regulations, "developing countries as a whole have more at stake in decisions regarding GMOs than industrialized countries. Both the potential economic, environmental and social benefits, and the potential risks from using GM crops are generally greater for developing countries."[14] Indeed, Stewart continues, developing countries have three basic characteristics that increase the stakes of GMO regulation. First, agriculture generally plays a larger role in the economies of

LDCs, both as a percentage of GDP and as a proportion of their international trade, thereby increasing the significance of GMO regulation for the economy as a whole. Second, "while the rich industrialized countries have large food surpluses stimulated by government subsidies, many developing countries face serious food security problems with large numbers of malnourished citizens and growing populations." In this context, the potential productivity gains from current and future GM crops loom particularly large for developing countries. Third, developing countries have far lower regulatory capacity to engage in either risk assessment or risk management of GM foods and crops, although traditionally "they also generally place a lower priority on EHS [environmental, health and safety] objectives than more affluent developed countries."[15]

Such considerations, however, have played relatively little role in the conflict between the US and Europe, both of which have treated third parties in general, and LDCs in particular, as little more than potential supporters or opponents in a conflict motivated primarily by domestic political and economic considerations. In this brief section, therefore, we widen our focus beyond the transatlantic dispute, focusing on the challenges of GMO regulation for LDCs, and noting the multiplicity of constraints facing them and the diversity of responses they have undertaken as they remain caught in the crossfire of an ongoing transatlantic dispute where transnational and international cooperation have failed.

7.2.2. Domestic and international influences on LDC regulatory choices

Throughout this book, we have argued that the regulatory decisions undertaken by the US and the EU cannot be attributed to any single factor, be it interests, institutions, ideas and culture, or contingent events, but represent a combination of these factors. Even a cursory survey of other countries—including advanced industrialized countries like Canada, Australia, New Zealand, and Japan, as well as the many developing countries—reveals a similarly broad range of factors that influence regulatory decisions. They include "the overall stage of a country's development, agro-ecological conditions, the character, organization, and productivity of the agricultural sector, the role of agricultural imports and exports, food security needs, public sector and private sector crop R&D capacities, intellectual property regimes, and EHS regulatory capacities."[16] Different domestic and international factors have interacted, complicating developing country governments' regulatory choices.

To start, agricultural interests are often important, but can cut in different directions. Some LDC farmers advocate the use of GM foods and crops, and where possible, have adopted them without government approval for reasons of efficiency and cost.[17] Others, in contrast, are concerned about the approval of GM crops for fear of potential exclusion from the markets of Europe, Japan, and other countries with restrictive regulations. Yet others grow local staples for which GM varieties have not been developed, whether because the private sector lacks the economic incentive to do so, or because national and international public agricultural research services have not invested in their development, in part because of the costs and uncertainties of regulatory approvals, including for trials. As Sakiko Fukuda-Parr states, agronomic factors and export markets appear to explain much of the variance in the approval and in the commercialization of GM crops in the developing world.[18] Soybeans were among the first GM crops to be developed and commercialized in developed country markets; for example, Brazil and Argentina, the second- and third-largest soy producers in the world, followed the US in adopting GM soybeans. Similarly, China, the US, and India are the three largest producers of cotton, and all three have moved into extensive commercialization of Bt cotton, as has South Africa.[19] By contrast, private industry has been slow in developing GM varieties commonly grown in other developing countries, such as cassava and millet, which explains in part the slow take-up of GM crops across most of the developing world.

Domestic and international market pressures have affected the bottom-up demands of local farmers for the approval of GM varieties in developing countries that are large agricultural producers. Large numbers of developing-country farmers in, for example, Brazil and India have adopted GM soy and cotton even when these varieties were banned by their respective governments, rendering regulation irrelevant.[20] They did so because many farmers felt that the GM varieties were more effective in combating pests, reducing costs, and increasing yields, and found that they would be at a competitive disadvantage if they did not adopt them.[21] In fact, they were able "to find and breed transgenic seeds underground" to the chagrin of intellectual property holders such as Monsanto.[22] In light of farmer demands, both Brazil and India have had to back down and legalize GM soy and GM cotton varieties, respectively. Agricultural practice in Brazil and India, and in particular by small-scale farmers, respectively, made the adoption of GM soy and cotton a "*fait accompli.*"[23]

In addition, non-governmental groups and social movements have actively attempted to halt research trials and regulatory approvals of GM varieties in developing countries. They have ripped up trial fields in India, and initiated court cases resulting in injunctions in Brazil.[24] Robert Paarlberg finds that NGOs "in Brazil and India took up the criticisms on GM crops they heard being made by their counterparts in the industrial world, and the approval processes in their countries slowed accordingly."[25] As a result, developing countries have tended to take a much more demanding regulatory stance on transgenic varieties than they have traditionally taken in health, safety and regulation, arguably slowing the adaptation and adoption of GM varieties in the developing world.[26]

Domestic institutions have also played a role. First, democratic regimes such as India and Brazil have been particularly subject to pressure from NGOs and from public opinion that may be skeptical of GMOs, while more authoritarian regimes like China have been freer to make policy in relative isolation from public opinion.[27] Second, while domestic governments in the developing world wish to pursue policies aimed at enhancing domestic food security and self-sufficiency, most face considerable capacity constraints in forming an indigenous biotech research sector, whether public, private or involving public–private partnerships. Only China has developed and commercialized its own GM varieties (for cotton) through its public agricultural research services, although Brazil and India have invested substantial resources in this area.[28] Not surprisingly, the price of GM cotton varieties in China is significantly less than in other LDCs because of the increased competition with varieties owned by multinational companies.[29] This factor, in turn, has helped to reduce prices on GM cotton varieties in India, as an Indian company has imported the Chinese variety under license, and Indian state governments brought a legal challenge against Monsanto under Indian competition law regarding the price differentials it charged in India compared to China, which resulted in a negotiated price reduction.[30] In addition, significant traffic in underground GM cotton varieties in India has helped to drive down their price.

LDC regulatory decisions are complicated and constrained by global politics which pose their own severe risks. In making their domestic regulatory choices, developing countries must take into account developments in the ongoing US/EU dispute, the economic and political pressure exerted on them by both sides, the resulting conflicts among and stalemates within most relevant international regimes, and the uncertainty about the future direction of biotech regulation and its impact on agricultural trade. The transatlantic conflict has, in effect, represented both an opportunity and a severe constraint on LDCs in formulating domestic biotech regulations. On the one hand, it can be argued that the divisions between the US and the EU and the resulting absence of a coherent global standard have given the LDCs some leeway in selecting their own domestic regulatory approaches, which have been highly diverse, running the gamut between the permissive US approach and the more precautionary EU stance. In this vein, Vogel et al. argue in their survey of GMO regulation:

> These variations reflect the absence of effective international governance of GMO foods and crops....National policy discretion and variations have been encouraged by the substantial differences between the regulatory policies of the EU and the US, the countries that have historically played a critical role in setting international trade and regulatory standards. In effect, countries are free to model their regulatory and IP [intellectual property] policies after either the EU or the US.[31]

The uncertainty generated by the US/EU dispute, however, while perhaps increasing the freedom of maneuver for LDCs compared to an alternative in which the US and EU adopt a unified policy, has also complicated governments' ability to make informed decisions about biotech regulatory policy. As Stewart states, "most developing countries have been trapped in the crossfire of conflict between the EU and US, which has also prevented international trade regulatory bodies from providing meaningful guidance on GM trade regulatory issues."[32] For example, the US has exerted pressure on less developed countries to accept US exports or food aid consisting of GM foods and crops, and has sought allies in international negotiations and co-complainants in its WTO litigation against the EU, which was joined by one developing country (Argentina), almost joined by another (Egypt), and indirectly supported by one as a third party (Chile). Regulatory policy on GMOs is a key issue for the US in negotiating bilateral free trade agreements, and the US Congress has even "tied funding to fight AIDS to acceptance of GMO policies."[33]

The EU, by contrast, has sought to gain support for its more precautionary approach in international negotiations such as the Cartagena Biosafety Protocol and through its various technical assistance programs, while the EU's strict regulations on approval, labeling, and traceability (including its zero tolerance policy for the adventitious presence of non-GM varieties) have discouraged the adoption of GM foods and crops by developing-country farmers fearful of losing access to the EU market.[34] The threat of exclusion from lucrative developed-country markets has arguably become the most important factor preventing developing-country agriculture from advancing with GMO technology.[35] Analysts of the situation in Africa, for example, find that "[n]egative perceptions [of agricultural biotech] are often based on local decision takers' concern for certain donors' positions rather than scientific analysis."[36] Even Argentina, the leading grower of GM varieties after the US (see Table 7.1), created a foreign market access review component to its domestic approval process, which resulted in a "mirror policy" in which Argentina would not approve a GM variety until it was approved in Argentina's major export markets—mainly the EU. This policy gave rise to a *de facto* moratorium on new approvals of GM varieties in Argentina from 2001–2004 in response to the EU's moratorium.[37] Similarly, a temporary EU ban on the import of soy from China because of the detection of adventitious GM content, "played a critical role" in causing China to go slow in considering approvals of GM soy and other varieties.[38]

The various multilateral regimes within the agricultural biotech regime complex have not sent consistent messages to LDCs. The Cartagena Protocol on Biosafety, for example, has encouraged a more precautionary approach to the regulation and shipment of GMOs, and its capacity-building program, implemented by UNEP, has had a similar effect in developing countries, promoting a "precautionary biosafety discourse" in those countries.[39] In particular, the Biosafety Protocol has increased the leverage of environmental ministries in developing countries (which generally take the lead in their implementation) vis-à-vis agricultural ones.[40] By promoting complex and costly biosafety regulation, by expressly allowing states to restrict shipments of LMOs, and by requiring exporters to certify whether shipments "may contain" LMOs, the Cartagena Protocol has arguably discouraged its regulatory approval and commercial adoption by developing countries.[41] In particular, it has created more significant challenges for states with less scientific and regulatory capacity. As Vogel et al. write, "regulatory barriers are of particular importance for smaller countries as they impinge on their ability to attract external R&D and capture spill-in benefits. These regulations can also discriminate against national public research institutions and national firms by limiting their ability to become actively involved in product development."[42] Similarly, Scoones concludes, "[t]he small players, perhaps those best able to produce crops adapted to local needs, will be squeezed out by the cost of regulatory compliance. The net losers will be the poorest farmers, the very people who the opposition groups argue are their constituency."[43] Such regulatory restrictions are legitimated by the Biosafety Protocol's provisions on precaution. As Ronald J. Herring states, costly "strict control and testing regimes raise costs of seed development beyond what is affordable by small firms, enhancing the power of deep-pocket corporations."[44] Ironically, costly and restrictive biotech restrictions help to "stifl[e] the development of any local biotechnology projects," and further ensure that biotech varieties are produced primarily by large multinational firms for developed-country farmers, and not for the poor in developing countries who could benefit from them most.[45]

By contrast, the 2006 WTO panel decision, although directed in the first instance at the EU, has significant implications for developing countries. As we have examined in Chapter 5, the panel decision interprets the SPS Agreement to require that GM regulations be based on a scientific risk assessment (despite the challenge that such assessment poses for many LDCs), while ruling out any potential reliance on the more precautionary language of the Cartagena Protocol to justify restrictive measures. The panel decision is unfavorable for developing countries that wish to restrict the authorization of GMOs. In addition, the panel's findings on "undue delay" in regulatory approvals also constrains the ability of developing countries to "wait and see" how the US/EU dispute turns out before adopting a regulatory posture of their own.[46] In contrast, however, the panel decision has also offered little solace for LDCs that wish to develop and grow GM varieties for their own policy reasons, and in particular for export markets, because the decision so far appears to have had limited effect on EU agricultural biotech policy.

7.2.3. The diversity of LDC regulatory responses

Not surprisingly, in light of these domestic and international pressures and in the absence of a generally accepted global standard, countries have adopted a wide range of regulatory stances toward agricultural biotechnology, running the gamut from those approaching the US model of acceptance and widespread commercialization to strict rejection of the new technology. Stewart usefully classifies the range of national biotech regulations in developing countries as falling into four clusters.[47] The first group, composed of some of the more advanced developing countries such as Argentina, Brazil, and South Africa, has embraced GM technology, with government regulators approving and farmers commercializing various GM crops (although Brazil has a more precautionary biotech approval process). Both Argentina and Brazil are major growers and exporters of GM soybeans, Argentina joined the US in bringing the 2003 WTO complaint against the EU, and South Africa has approved and grows GM corn and cotton commercially.[48]

A second group of larger developing states, including most notably China and India, have undertaken major public sector research programs in GM food technology, but have "hedged their bets" by largely limiting the cultivation of GM crops to cotton, which is not subject to the types of restrictions imposed on GM foods by the EU and other importers. China, driven largely by concerns about food security, invested heavily and early in GM technology and has reportedly developed over 100 varieties of GM farm produce. By 2006, however, it had only approved three GM crops for commercial cultivation, of which only cotton is grown on a large scale. GM cotton comprised around 70 per cent of total Chinese cotton production by 2005, and cotton is China's largest cash crop.[49] Faced with concerns about the commercial acceptance of new crops, as well as fears of inadequate Chinese regulatory capacity, the Chinese government has developed a precautionary stance toward commercialization of other GM varieties. For example, China approved the import of GM soybeans in 2005, but continued to approve cultivation of only conventional soybeans, which it exported to the EU, South Korea, and Japan. Similarly, China has withheld approval for commercial cultivation of a Bt rice variety that it developed, out of concern for environmental effects and market acceptance, after expending huge sums on it.[50] More generally, China has appeared to resist adopting GM crops developed by Western biotech firms until it has developed its own domestically engineered varieties, such as for GM soy.[51] India has taken a related approach, investing heavily in GM research and engaging in widespread cultivation of GM cotton, but with holding approval for commercial cultivation of GM food crops.[52] Unlike in China, however, most GM cotton production in India is of a variety commercialized by a joint venture between Monsanto and an Indian seed company.[53]

The third and largest group of LDCs includes the majority of developing countries, and almost all of the least developed countries, which do not have the capacity to engineer or regulate GM foods and crops, and have not approved GM crops for cultivation or, in most cases, for marketing. Nor, however, have most of these countries taken a strong anti-GMO stance *per se*. In effect, these countries, like those in the second group, have adopted a "wait and see" approach, either because of domestic stalemates or to get a better sense of the future direction of global GMO regulation before adopting an explicit domestic policy.[54] On the one hand, such wait and see approaches are potentially subject to challenge under WTO law as constituting "undue delay" in the unlikely event that the US or another WTO member state brought a legal challenge against them.[55] On the other hand, just as many of these countries were bypassed by the Green Revolution because they lacked the capacity to make use of its innovations, so they could be bypassed by the "gene revolution," in this case also because of the predominance of the private sector and the role of intellectual property rights in agricultural biotechnology.[56]

A fourth group of mostly poor African states have rejected GM foods and crops entirely as "an instrument of neo-colonial exploitation." They have banned the importation and cultivation of GM crops on many grounds, including nationalistic ones because they are patented by Western agricultural biotech corporations.[57] While this last group is relatively small, including such states as Zambia, Madagascar, and Robert Mugabe's Zimbabwe, developing-country wariness of relying on Western GM varieties, subjected to intellectual property controls, is more widespread.

7.2.4. General trends in LDC regulation and use of GM foods and crops

Looking across the developing world, we see two broad and somewhat contradictory trends in the regulation and commercial adoption of GM foods and crops, respectively. On the one hand, cross-national surveys of GMO regulation indicate that with a few exceptions such as the US, South Africa, and Argentina, the general trend in regulation has been toward greater stringency.[58] A growing number of countries have enacted new regulatory application, approval and monitoring requirements, as well as more restrictive rules for the importation and labeling of GM foods and crops, reflecting in part transnational capacity-building efforts catalyzed by the Cartagena Protocol. While few, if any, states have adopted the full panoply of approval, traceability, and labeling requirements of the EU, a growing number of states have shown greater reluctance to approve the importation and especially, the cultivation of GM crops, with countries like China, India, Australia, and Columbia largely limiting commercialization to GM cotton, rather than food crops.[59]

At the same time, however, studies on GMO cultivation and commercialization reveal continuing and impressive increases in the adoption and planting of GM crops, with the greatest growth rates in the developing world. Studies by the International Service for the Acquisition of Agri-Biotech Applications (ISAAA) find that cultivation of biotech crops increased by 13 per cent in 2006 and 12 per cent in 2007, making twelve consecutive years of such increases, with a total of 114.3 million hectares of GM crops being cultivated by around

12 million farmers in twenty-three countries worldwide in 2007.[60] Just as striking, for our purposes here, at least eleven less developed countries are now growing GM crops, with Argentina, Brazil, China, and India taking leading roles. GM cultivation in the developing world increased by 21 per cent in 2007, compared with a substantially lower but still impressive growth rate of 6 per cent in developed countries.[61] India's cultivation of GM cotton, for example, increased almost three-fold in 2007 (see Table 7.1). It has been estimated that over 180 countries have imported GM corn and around 145 imported GM soy.[62] Moreover, "over sixty developing countries have [agricultural biotech] research programs, almost all in public sector agricultural research institutions and universities."[63]

The ISAAA study thus concludes on an optimistic note for the future of GM foods and crops, predicting continued growth of GM cultivation in the developing world.[64] Indeed, it seems likely that larger developing countries such as Argentina, Brazil, China, India, and South Africa will continue to invest in GM technology, cultivating a select group of GM crops, and most

Table 7.1 Global area of biotech crops in 2007, by country

Rank	Country	Area (million hectares)	Biotech Crops
1	USA	57.7	Soybeans, maize, cotton, squash, papaya, alfalfa, potato, canola
2	Argentina	19.1	Soybeans, maize, cotton
3	Brazil	15.0	Soybeans, cotton
4	Canada	7.0	Canola, maize, soybeans
5	India	6.2	Cotton
6	China	3.8	Cotton, papaya, sweet pepper, tomato, petunia, poplar
7	Paraguay	2.6	Soybeans
8	South Africa	1.8	Maize, soybeans, cotton
9	Uruguay	0.5	Soybeans, maize
10	Philippines	0.3	Maize
11	Australia	0.1	Cotton, canola, carnation
12	Spain	0.1	Maize
13	Mexico	0.1	Cotton, soybeans
14	Colombia	<0.1	Cotton
15	Chile	<0.1	Maize, soybeans, canola
16	France	<0.1	Maize*
17	Honduras	<0.1	Maize
18	Czech Republic	<0.1	Maize
19	Portugal	<0.1	Maize
20	Germany	<0.1	Maize
21	Slovakia	<0.1	Maize
22	Romania	<0.1	Maize
23	Poland	<0.1	Maize

Source: James 2007.
* Note that France suspended authorization for cultivation of maize in 2008.

others will agree to import at least some GM products. Once again, however, one should not understate the potential obstacles to widespread regulatory liberalization and commercialization of GM crops in the developing world. Developing-country governments may refrain from approving new varieties out of concern for their agricultural exports, the environment, public health, and dependence on intellectual property owned by Western multinational companies. In particular, outside a few widely accepted crops, such as GM cotton, corn, and soy, farmers in developing countries may remain reluctant to adopt GM crops that might face exclusion from Europe, Japan, and other important markets. Furthermore, as Fukuda-Parr points out, the first generation of GM crops developed by Western companies includes a limited range of traits, and the crops are designed particularly for US farmers and not for farmers in the developing world, particularly those confronting tropical or drought conditions.[65] The future reception of GM foods and crops in the developing world, therefore, will depend in large part on whether second generation GM varieties include key crops for developing countries, such as rice, wheat, cassava, plantain, and millet, and key traits, such as drought resistance, disease resistance, higher yields, and greater nutritional content, designed to appeal to developing-country farmers, consumers, and governments.[66]

7.2.5. LDCs as participants in the debate over global GMO regulation

Given the diverse regulatory and commercial responses to agricultural biotechnology, we should not expect, and we do not see, a monolithic developing-country position emerging regarding the regulation of GM foods and crops. Rather, we see various LDCs adopting different positions on regulation at the global level, as a function of their different contexts. For this reason, there is no simple answer to the question of what will happen as an increasing number of developing countries join the global debate over GM foods and crops. Some of these countries, especially Argentina and Brazil, have already emerged as supporters of more permissive global regulation, others, such as China and India, have taken relatively cautious stances on the regulation of GM foods and crops domestically, but have not pressed for strict global rules, while yet others, in particular in Africa, have demanded more stringent rules in negotiations over the implementation of the Biosafety Protocol. Strongly precautionary global rules are a "double-edged sword" for developing countries, allowing them greater discretion to exclude imports of GM crops, but also providing cover for their key trading partners, such as the EU and Japan, to exclude their agricultural exports on precautionary grounds.[67] Perhaps for this reason, many developing countries have continued on the sidelines of the global regulatory debate over GMOs, which continues to be dominated by the conflict between the US and Europe.

7.3. The uncertain global prospects for agricultural biotechnology

What then can we say about the prospects for the global regulation and adoption of agricultural biotechnology? We offer two concluding observations on this point, one relating to current trends and the other to the possible effects of future contingent events.

First, if we extrapolate from current trends, it seems likely that agricultural biotechnology will be increasingly accepted over time, albeit within significant market and regulatory constraints for GM foods directly consumed by humans. As we have seen, the commercialization of GM foods and crops has grown at double-digit rates each year for the past decade, making for a more than sixty-fold increase since 1996. The ISAAA predicts further growth as more developing countries approve and cultivate GM crops, and as new GM varieties emerge that either "stack" multiple traits into a single crop, or feature new traits such as for higher yields and drought resistance.[68] We will likely see increasing adoption of GM cotton for clothing, and we may see adoption of GM biofuels as well.[69] GM crops should also continue to be adopted for animal feed, and in particular GM soy and corn.

Nevertheless, it is striking that around 99 per cent of global plantings of transgenic varieties have been so far restricted to four major crops (soybeans, maize, cotton, and canola) consisting of two traits (an insect-resistant trait and an herbicide tolerant trait), all of which were developed during the 1990s.[70] By contrast, other potential GM staples like wheat and rice have met with market and regulatory opposition, while the private biotech companies that dominate the industry have lacked the economic incentive to develop crops attractive to growers in the developing world.[71] Barring major new technological developments (see below), the likely trend is for increasing cultivation, importation, and market acceptance of a relatively narrow group of GM crops, with limited penetration of new GM foods for direct human consumption, at least in the near future.

Moving to regulatory trends, we have seen that domestic regulation of agricultural biotechnology in the developing world runs the gamut from US-style permissive frameworks to the more precautionary EU stance, with most states adopting intermediate positions. Moreover, the record of the past decade demonstrates little evidence of convergence in national regulations. We see some evidence of growing precaution and some increase in agricultural biotech regulatory capacity in the developing world, due in large part to the effects of the Cartagena Protocol, as well as market forces. Yet the number of states approving GM crops and foods for cultivation and importation continues to grow, while WTO jurisprudence creates pressure on countries not simply to reject GM products across the board. In the absence of a single global standard emerging from the stalemated and conflicted global regime complex for agricultural biotechnology, states are likely to continue pursuing a broad range of regulatory responses, reflecting their varying interests, institutions, and cultural proclivities within a global context.

Finally, however, we have seen throughout this book that contingency has played an important role in the development of biotech regulation in both the US and the EU. Extrapolating from trends in US environmental regulation in the 1970s, an early analyst of US biotech regulation might well have predicted that the US would adopt a highly precautionary regulatory system, yet the election of President Reagan and his administration's subsequent adoption of a Coordinated Framework for the Regulation of Biotechnology put the US on a path towards its current, relatively permissive system of agricultural biotech regulation. Similarly, a student of EU regulation in the early 1990s, when the EU touted its integration into a "single market" that would enhance its competitiveness against the US, could have imagined that the EU, led by a Commission eager to foster the development of a competitive biotech sector, would move quickly to approve the cultivation and importation of GM crops and foods, especially where they received consistently favorable EU scientific risk assessments. Such an analyst would have been unlikely to predict the food-safety crises of the mid-1990s, the emergence of a widespread anti-GMO movement in Europe, or the subsequent *de facto* moratorium, politicization and tightening of the EU regulatory framework.

In much the same way, the development of GM foods and crops, their regulation and their commercialization, are likely to be shaped, for good or ill, by future contingent events. A major food-safety or environmental crisis from GM foods and crops is, in our view, unlikely in light of over a decade of experience—yet such a crisis, if it occurred, would likely have a profound and negative impact on the commercial acceptance and future regulation of GM foods and crops around the world. Similarly, the impending introduction of transgenic animals, or the marketing of meat from cloned animals, has the potential to increase opposition to transgenic varieties among both European and American publics, which remain deeply skeptical of such innovations. Such a development could potentially swing US regulation toward a more precautionary, process-based approach, and with it the global regime complex governing the food-safety, environmental, and trade aspects of GM foods and crops.

At the other extreme, the development of revolutionary new GM crops also holds the potential of swinging public opinion, markets, and regulators in favor of agricultural biotechnology around the world. A rough consensus has formed in the development community that GM crops should be part of the "toolkit" for developing-country agriculture.[72] Many commentators go further, contending that there is an ethical obligation to pursue the development of GM crops in light of the over 800 million people who suffer from chronic hunger,[73] and the increasing stresses on crops in light of climate change and the relative scarcity of fresh water.[74] Rising food prices in 2008 also opened opportunities for increased acceptance of GM varieties for animal feed, including in Europe.[75] As we have noted repeatedly in this book, first generation crops have been geared largely to the needs of farmers rather than consumers (who saw little direct benefit), and have been adapted largely for the particular climates and soil conditions of developed countries rather than those of the developing world. The private biotech firms of the industrialized West, as well as the under-resourced public sector initiatives of the developing world, have been slow to develop second generation crops that provide distinctive benefits for consumers (such as enhanced nutritional content)[76] or hold the prospect of a genuine contribution to fighting global poverty (including through drought resistance, higher yields, and the development of GM crops common to tropical and subtropical climates).[77]

In our view, the future of agricultural biotechnology, and in particular, its contribution to human welfare, rests in large part on the question of whether future research will produce second generation crops that directly benefit consumers and the world's poor. Around "seventy per cent of the world's poor live in rural areas and about two-thirds of these rely primarily on agriculture for their livelihoods."[78] Independent organizations, from the FAO to the Nuffield Council on Bioethics, have explored how the technology can be used to benefit the world's poor, while noting the challenges that have arisen because the technology so far is held and being driven primarily by the private sector.[79] Because the private sector does not see a profit in food staples grown in the developing countries, it will not invest in them, and because the technology is often held by the private sector, the public sector has greater difficulty in developing these products than at the time of the Green Revolution.[80] National, foreign, and global regulatory requirements, as well as consumer concerns, further dissuade investments.[81] Some large developing countries, such as China, Brazil, India, and South Africa, have established public sector research programs to develop new GM varieties, but the resources of these programs (except in the case of China) pale by comparison with those of private sector firms such as Monsanto, Syngenta, and Bayer Cropscience.[82] Particularly noteworthy is that international agricultural research centers under the Consultative Group on International Agricultural Research (CGIAR) system have relatively limited funding for research and development of GM varieties.[83] From a normative perspective, therefore, many development analysts advocate regulatory and funding reforms that will provide greater incentives for the development of GM crops with the potential to reduce malnutrition and poverty in the developing world.

The debate over the future of agricultural biotechnology has been conducted in European, the US, and other regulatory agencies, in the agora of a varyingly mobilized public opinion, and in a range of multilateral forums such as the Codex Alimentarius Commission, the Convention on Biodiversity, and the WTO, each of which have haltingly attempted to deal with a bitter and intractable transatlantic and now global dispute. As a lawyer and a political scientist, respectively, we can provide no magic bullet, and we see no impending and final resolution of this conflict, although we have indicated some specific policy adaptations that could facilitate a better accommodation of persistent regulatory differences. To the extent that a resolution of the conflict is possible, it is most likely to be found neither in the courtroom nor at a negotiating table, but in the laboratory and in fields where the potential benefits of agricultural biotechnology might be realized and placed at the service of those who need it most.

Notes

Chapter 1

1. Hill and Battle 2000. European laws tend to use the term "genetically modified organisms," or foods or crops, while United States regulatory authorities tend to refer to "bioengineered," "transgenic" or "genetically engineered organisms," foods or crops. As Ronald J. Herring writes, "transgenic is more precise designation of organisms that result from rDNA technology: a biological category, rather than a political one," since all the crops are obtained through breeding, and in this way genetic modification takes place over time. Herring 2007*a*: 5. We use these terms interchangeably in reflection of the political and legal discourse. However, when we use the more common term "genetically modified" food, it should be clear that we are speaking of genetic engineering through the transfer of genes, and not conventional modification through the cross-breeding of plants.
2. Bren 2003.
3. APHIS 2007: 1. Cf. Brookes and Barfoot 2006: 6 (referring to 93 per cent of soy, 79 per cent of cotton, 57 per cent of corn, and 82 per cent of canola).
4. The terms EU and EC (European Community) are used interchangeably in this book. The Treaty on European Union in 1992, created three separate "pillars" of activities for the regional bloc: the traditional EC one, the Common Foreign and Security Policy, and Justice and Home Affairs. The first (EC) pillar covers the regulation of agricultural biotechnology. The term which encompasses all three pillars is the European Union (or EU), which is most frequently used by commentators.
5. Kahler 1995.
6. Pollack and Shaffer 2001: 5 (typologizing international relations theory and the practice of transatlantic governance in terms of three levels: intergovernmental, transgovernmental and transnational).
7. Home Depot is the largest home-improvement retailer in the US, with additional stores in Mexico and China. Its aforementioned marketing slogan—"You can do it, we can help"—was introduced in 2003. See Wikipedia contributors 2008.
8. Slaughter 2004.
9. See e.g., Vogel 2003.
10. See, however, our discussion of recent changes in US regulatory practice (examined in Chapter 6).
11. Pierson 2000, 2004; Büthe 2002.
12. Greif and Laitin 2004: 34.
13. Putnam 1988; Evans et al. 1993.
14. Keohane and Nye 1974.
15. Slaughter 2004.

16. Pollack and Shaffer 2001*a*; Shaffer 2002.
17. Habermas 1985, 1998.
18. Risse 2000.
19. Risse 2000: 19–20.
20. Krasner 1991, Morrow 1994, Fearon 1998. Note that "distributive conflict" here refers broadly to any dispute about the distribution of all costs and benefits from a potential agreement, and need not refer narrowly to trade-related issues or protectionism.
21. On regime complexes, see Raustiala and Victor 2004; on forum-shopping, see Jupille and Snidal 2006, and Braithwaite and Drahos 2000.
22. The exporting of domestic laws has emerged as an explicit aim of EU policymakers in recent years, as discussed in Chapter 4 and in Buck 2007*b*.
23. For an extended discussion of the spread of GM foods to the developing world, see Chapter 7.
24. In 2008, a divided and deadlocked Council, for the first time, did not oppose—but neither did it approve—the Commission's proposal to challenge Austria's ban on the consumption of two EU-approved GM maize varieties. See Chapter 6.
25. To the extent that the EU exerts normative influence on third parties, in excess of its market- and negotiating-power, such influence might be considered a manifestation of "soft power"—an intriguing claim, but one that we do not test in this book. See e.g., Nye 2005 on soft power, and Manners 2002 on the prospect of a "normative power Europe."
26. See e.g., the arguments in Vogel 1995; and Shaffer 2000.
27. Young 2003: 458.
28. See Pollack and Shaffer 2001; Petersmann and Pollack 2003; and Pollack and Shaffer 2005.
29. The average yearly growth for world agricultural exports since 1951 is 3.6 per cent (since 1980, it is 3.0 per cent). The average yearly growth for world agricultural production since 1951 is 2.5 per cent (since 1980, it is 2.2 per cent). See WTO 2005*a* (containing a table that lists yearly changes in world agricultural exports and world agricultural production since 1951).
30. In this sense, our analysis represents an explanatory or "disciplined-configurative" case study, in which existing theories are used to understand a particularly significant case. However, as we are particularly motivated by a desire to explain the puzzle of repeated failure of cooperation among longtime partners, the GMO dispute can also be viewed as a "deviant case," in which we seek to identify the various obstacles to cooperation, some of which have not previously been identified by international relations or legal theorists. In the process, we have also engaged in a "plausibility probe" in our theorizing, in Chapter 4, of the interaction of hard and soft law regimes where there is distributional conflict among powerful states. For good discussions of case study types, see Van Evera 1997: 22, n. 43; and George and Bennett 2004: 74–79.
31. On process-tracing, see George and Bennett 2004, Chapter 10.

Chapter 2

1. On the EU as a quasi-federal system, see Nicolaidis and Howse 2001; Kelemen 2003; and Kelemen and Nicolaidis 2007.

2. The term is Thomas Bernauer's (2003: 1–2).

3. Plant breeders throughout the world, and in particular in Europe, use radiation to create many varieties of grains, fruits, and vegetables. For example, Irish and British brewers use mutant varieties of barley bred through radiation to make premium beer and whiskey. See Broad 2007.

4. Pew Initiative 2004: 3. See also the comments of Hugh Grant, Chairman and Chief Executive Officer of Monsanto, who described the first decade of GM crop development as being about "weeds and bugs." "Yield," he continues, "is the holy grail of agricultural research.... The second decade is going to be about hundreds of genes that grow yield." Quoted in Cameron 2007.

5. "Bt" designates "a gene isolated from the soil bacterium, *Bacillus thuringiensis* (Bt) [which] encodes a pesticide and when this gene is inserted into the plant, the plant can then produce the Bt pesticidal substance." NBII Frequently Asked Questions, available at http://usbiotechreg.nbii.gov/faqall.asp, accessed on July 25 2007.

6. In July 2007, Monsanto obtained regulatory approval in the US and Canada for a new soybean it expected to replace Roundup-Ready soybeans in its product line. The new strain, Roundup-Ready2Yield, was also engineered to be resistant to the Roundup herbicide. Pending regulatory approval in key international markets, Monsanto targeted widespread commercial distribution of Roundup-Ready2Yield by 2010. Monsanto 2007.

7. Bernauer 2003: 7. See also Stewart 2007: 4 (noting the exception of mangos genetically engineered to be resistant to rust, which has resulted in a significant drop in the price of mangos).

8. Traits that enhance crop productivity and reduce farmer costs can be viewed as "input traits," and those that provide benefits to consumers in the form of enhanced nutrition, improved taste and shelf life can be viewed as "output traits." Pew Initiative 2007: 5.

9. See World Health Organization 2005: 6–8; Pew Initiative 2004a: 3 ("The next generation of GE crop varieties will likely include a wider range of desirable agronomic traits, including drought tolerance. Food crops may be modified with traits to improve freshness, taste, and nutrition").

10. See e.g., Sunstein 2005: 31–32 (pointing to Zambians turning down US genetically engineered corn, leaving "2.9 million people at risk of starvation," based on World Health Organization data).

11. FAO (2003–04: 76).

12. See e.g., Botkin 1996: 25–27 (maintaining that organic foods are "actually riskier to consume than food grown with synthetic chemicals").

13. Vogt 2005: 2.

14. See US Food and Drug Administration (FDA), *Spinach and E-coli Outbreak*, http://www.fda.gov/oc/opacom/hottopics/spinach.html (last visited Feb. 7, 2007, accessed on February 7 2007). The spinach implicated in the outbreak was traced back to Natural Selection Foods LLC of San Juan Bautista, California.

15. See e.g., Rosenthal 2007. Cf. Runge and Senauer 2007: 42.

16. Andrew Pollack 2006. Conversely, biotech varieties engineered for low-till farming could increase carbon sequestration in soils. Stewart 2007: 6.

17. See Dua et al. 2002; Sasson 2005; Acalde et al. 2006; and Cases and de Lorenzo 2005.

18. Biotechnology Industry Organization (BIO) 2007*b*: 41. To date, transgenic plants have been used to produce antibodies, antigens, hormones, enzymes, collagen, and other therapeutic proteins. Id.
19. See BIO 2007*b*: 45–64.
20. Bernauer 2003: 19.
21. See Jeffrey Smith 2007: 123–141; and Bakshi 2003.
22. See Gonzalez 2007: 604–5.
23. For a history of international regimes governing seeds, from the treatment of seeds as part of "the common heritage of mankind," to the recognition of some seeds as private property controlled through state-enforced patents, see Kloppenburg 1988. See also Raustiala and Victor 2004. Cf. Sunstein 2006: 178 (noting how "scientists have recently developed an open source technique for genetic engineering, one that is not controlled by the patent system"); and Karapinar and Temmerman 2007 (assessing "how IPRs can be domestically tailored within the existing international commitments so as to encourage the development of technologies that favour and are accessible to small-scale farmers in developing countries," through public–private partnerships).
24. We thank Jeff Dunoff for encouraging us to stress this point.
25. Pendleton 2004: 15–16. Cf. Herring 2007*a*: 8 ("the deep irony is that…gene use restriction technology would be the only certain means of preventing horizontal gene flow—the major environmental risk").
26. Pendleton 2004: 24–25. AstraZeneca, the world's second largest seed company behind Monsanto, has also publicly disavowed commercialization of TT. See Shand 2003.
27. Press Trust of India 2006.
28. Indo-Asian News Service 2006.
29. Indo-Asian News Service 2006.
30. Cullen 1999: A1 (citing Prince Charles). See also Thompson 2003: 14.
31. Martin 2003.
32. See Singer 1990.
33. A transgenic ornamental fish for aquarium has been commercialized, and genetically engineered salmon (containing an introduced growth hormone) could be commercialized soon. Genetically engineered animals could also be used to produce pharmaceuticals, as well as organs and tissues for transplants into humans. Pew Initiative 2004*a*: 101. See also NRC 2002*a*, 2004, and Chapter 6 below on the regulatory challenges of transgenic animals.
34. While the deliberate modification of human genes is a significant concern, it is one that we do not address in this book, which focuses on the regulation of GM foods and crops. For a critical analysis, see Sandel 2007.
35. Kahan, Slovic, Braman, and Gastil 2006: 1083–84.
36. An Open Letter to Mr. Jacques Diouf, Director General of FAO, "FAO Declares War on Farmers Not Hunger" (June 16, 2004), available at http://www.grain. org/front_files/fao-open-letter-june-2004-final-en.pdf.
37. See FAO Newsroom, "Biotechnology at Work," available at http://www.fao.org/ newsroom/en/focus/2004/41655/article_41669en.html, accessed on July 25, 2007 (noting disease-free bananas in Kenya, pearl millet in India, and Bt cotton in China; "Yields for insect-resistant cotton are about 20 percent higher than for conventional varieties, and pesticide use has been reduced by an estimated 78,000

tonnes—an amount equal to about one-quarter of the total quantity of chemical pesticides used in China.") See generally FAO 2004a. The first trials of a genetically modified crop variety developed in Africa, a variety of maize resistant to maize-streak virus, a disease carried by insects, may occur soon; see *Economist* 2007.

38. See Joint FAO/WHO Food Standards Programme, Codex Alimentarius Commission, Procedural Manual 43–44 (16th edn., 2006); Organization for Economic Cooperation and Development (OECD), Descriptions of Selected Key Generic Terms Used in Chemical Hazard/Risk Assessment 16–17, Ref. No. ENV/JM/MONO(2003)15 (Series on Testing and Assessment No. 44, 2003).

39. Wiener and Rogers 2002: 320, emphasis in original.

40. See Knight 1921 (risk refers to situations with knowable probabilities, while uncertainty refers to situations where only randomness applies, and a decision maker cannot assign probabilities to them).

41. Beck 1992: 171.

42. Sterlin and Mayer 2000: 39. See also Wynne 1992; Sunstein 2005: 60.

43. See e.g., Kahan, et al. 2006.

44. Weber 1947: 193; Weber 1958: 19.

45. Knight 1921. Knight contends that if risk were perfectly calculable in terms of probabilities, then in a world of perfect competition, there would be no profit.

46. In statistics, "Type I errors" refer to "the risk of taking an affirmative step which is a mistake," and "Type II errors" refer to "the risk of failing to take a step that would be good." Margolis 1996: 74.

47. See Beck 1992, 1999; and Giddens 1991.

48. Giddens 2000: 42. See also Giddens 1991: 123 ("the point . . . is that, in conditions of modernity, for lay actors as well as for experts in specific fields, thinking in terms of risk and risk assessment is a more or less ever-present exercise of a partly imponderable character"). See also Beck 1992: 34.

49. Beck 1992: 19.

50. Majone 2003a: 18–26.

51. Majone 2003a: 26. See also Breyer 1993; and Margolis 1996. For critiques of reliance on cost-benefit analysis, see e.g., Ackerman and Heinzerling 2004; and Kysar 2006.

52. Vogel 2001, 2003a.

53. Vogel 2003b: 557–58.

54. On risk regulation and EU governance, see e.g., Neyer 2000; Vos 2000; Joerges 2001b; Vogel 2001, 2003a, 2003b; Abels 2002; Chalmers 2003; Majone 2003b.

55. The literature on the precautionary principle in risk regulation has mushroomed. For a range of supportive and critical views, see e.g., Bodansky 1991; Cameron and Abouchar 1991; Commission of the European Communities 2000a; Wiener and Rogers 2002; Majone 2003b; and Sunstein 2005.

56. Such a backlash was symbolized by the Unfunded Mandates Reform Act of 1995 (UMRA), Pub. L. No. 104–4, 109 Stat. 48 (codified as amended in scattered sections of 2 U.S.C.). The expansion of federal environmental regulation of the 1970s and 1980s hit a road block when state and local governments objected to unfunded mandates that saddled them with the increasing costs of administration and compliance with the growing federal regulatory regime. See e.g., Lee 1994, Congressional Budget Office 1995. The fight against Congress's use of unfunded environmental mandates dovetailed nicely with the tenets of the conservative movement that

helped Republicans take control of both houses of Congress during the 1990s. See Steinzor 1996: 98–100. However, while it was politically noteworthy the UMRA contained little of substance to actually restrain Congress from passing unfunded mandates. See General Accounting Office 1998.

57. Wiener and Rogers 2002: 322–23.
58. The 1906 Act was enacted in response to reporting of food production scandals, the most important example of which was Upton Sinclair's *The Jungle* (1906). Merrill 1998: 737–38.
59. Act of June 30, 1948, ch. 758, 62 Stat. 1155; see also Percival 1995: 1155, 1160 ("During the 1970s alone, more than twenty major federal environmental laws were enacted or substantially amended, giving EPA and other federal agencies enormous regulatory responsibilities"). Once again, a scandal raised by a book, Rachel Carson's *Silent Spring* published in 1962, was a catalyst for federal legislation promoted by a growing environmental movement. Carson's book helped to raise awareness of environmental harm, particularly from the use of pesticides.
60. National Environmental Policy Act of 1969, 42 U.S.C. § 4321 et seq.
61. In addition, the Centers for Disease Control have primary responsibility for monitoring foodborne illnesses and conducting investigations into the causes of such outbreaks, in cooperation with the FDA, the USDA, and relevant state authorities.
62. Taylor 1997: 15–16; Stewart and Johanson 1999: 248–49; Hawthorne 2005: 233–252.
63. Taylor 1997: 016.
64. Taylor 1997: 16–18.
65. The FDA has approved the use of five growth-promoting hormones after concluding that there was no health risk to consumers of hormone-treated meat. Penner 1999.
66. Adelman and Barton 2002: 4–5. Pub. L. No. 80–104, 61 Stat. 163 (1947). Administration of FIFRA was originally the responsibility of the USDA but was transferred to the EPA upon the latter's creation in 1970, following which FIFRA was significantly revised in 1972, representing a shift in policy focus from the promotion of agriculture to the regulation of the health and environmental effects of pesticides. However, FIFRA does not regulate the level of use of approved pesticides.
67. Taylor 1997: 19.
68. This section draws in part from Cantley 1995; Patterson 2000; Pew Initiative 2004*a*; and Sheingate 2006.
69. See the excellent overview of this period of uncertainty in Sheingate 2006.
70. See EPA 1984, cited in Sheingate 2006: 249. See also Prakash and Kollman 2003: 624.
71. Sheingate 2006: 248 (citing interview with Jane Rissler, former science advisor at the Office of Toxic Substances, and now Union of Concerned Scientists, Nov. 17, 2003; as well as Stanley Abramson, Memorandum to John A. Todhunter, March 14, 1983: Status of Recombinant DNA and New Life Forms Under TSCA, EPA, Office of the General Counsel); and Hilts 1983.
72. US House of Representatives 1984, cited in Sheingate 2006: 247.
73. *Foundation on Economic Trends v. Heckler*, 587 F. Supp 753 (D.C. 1984); *Foundation on Economic Trends v. Heckler*, 756 F. 2d 143 (I.S. App. D.C. 1985), cited in Sheingate 2006: 248.
74. Sheingate 2006: 249, citing Office of Science and Technology Policy 1985.

75. Office of Science and Technology Policy 1986.
76. Robert Peterson (2004) has summarized US agency responsibilities as follows: the FDA asks, "is it safe to eat?" the USDA asks "is it safe to grow?" and the EPA asks, "is it safe for the environment?" In practice, however, there is overlap. The EPA, for example, also regulates tolerance levels for pesticide residues in food that can affect human health under the federal Food, Drug, and Cosmetics Act (21 USC § 346a(a)(1)), and the USDA undertakes environmental impact assessments (APHIS 2004).
77. See Pew Initiative 2004*a*: 3.
78. See Sheingate 2006: 258, providing graph of coded Congressional hearings, showing shift from negative focus on risks to positive emphasis of benefits.
79. Sheingate 2006: 253. See also Plein 1991.
80. See Cantley 1995: 572.
81. Sheingate, in comparing US treatment of biotech crops with its more precautionary approach toward stem cells and cloning, maintains, "decisions about biotechnology made during the critical juncture in the 1980s had self-reinforcing effects that, over time, contributed to the bifurcated policy domains observed today," Sheingate 2006: 245.
82. Pew Initiative 2004*a*: 35.
83. See 7 USC § 7701 et seq. The Federal Plant Protection Act consolidated and expanded former laws that relate to the regulation of plant pests and diseases, such as the Federal Plant Pest Act (formerly 7 USC §151 et seq.). APHIS 2005: 1.
84. Pew Initiative 2004*a*: 16, 35. For the applicable regulations regarding "deregulation" of GM varieties, see 7 CFR Part 340. The list of approved, withdrawn and pending petitions for deregulation is set forth at http://www.aphis.usda.gov/bbep/bp/pet-day.html.
85. Pew Initiative 2004*a*: 31–32. Confirmed in Shaffer telephone interview with USDA official. May 26, 2008.
86. See APHIS field test database at http://www.isb.vt.edu/cfdocs/isbtables2.cfm?t var=9.
87. Petition for Determination of Nonregulated Status, 7 C.F.R. 340.6(d)(3) (on six month time period); and Shaffer telephone interview with USDA official. May 26, 2008 (on practice).
88. See Information Systems for Biotechnology, at http://www.isb.vt.edu/; Petitions for Deregulation Status Table at http://www.isb.vt.edu/cfdocs/isbtables2.cfm?tvar=4, accessed March 30, 2008 (link to table provided by the APHIS website; table based on APHIS database at). The web site lists 27 petitions withdrawn, 12 pending, and 1 incomplete, as of March 28, 2008.
89. An Environmental Assessment is much more streamlined, pursuant to which the agency will determine if its proposed action requires an Environmental Impact Statement (EIS). If the agency finds that an EIS is not necessary, it will issue a Finding of No Significant Impact (FONSI). Shaffer telephone interview with USDA official. May 26, 2008.
90. APHIS, Biotechnology Regulatory Services, Programmatic Environmental Impact Statement (July 2007). The definition of "noxious weed" in the Plant Protection Act broadly includes "any plant or plant product that can directly or indirectly injure or cause damage to crops (including nursery stock or plant products), livestock, poultry, or other interests of agriculture, irrigation, navigation, the natural resources

of the United States, the public health, or the environment." 7 USC § 7702. See Chapter 6 concerning the judicial challenges.

91. Pew Initiative 2004a: 49. See also USDA 2004*a*.
92. See e.g., USDA 2004*b* (announcing $161,257,500 in allocations to 71 US trade organizations to promote US agricultural products overseas).
93. Three per cent of US farms accounted for 62 per cent of US farm sales; and 15 per cent of US farms account for 89 per cent of US farm sales in 2002. See USDA National Agricultural Statistics Service, Distribution of Farms by Economic Class, available at http://www.nass.usda.gov/census/census02/quickfacts/distribution. htm (visited Sept. 20, 2004). There is evidence that financial markets are betting on further consolidation of farm production in the United States and globally. See e.g., *Economist* 2004 (noting that the Dutch co-operative bank Rabobank is expanding in the United States, betting that "with the farming industry consolidating globally, farmers will want a global financial institution").
94. Newell 2003: 62. See also Bernauer and Meins 2003: 668; Prakash and Kollman 2003; and Ferrara 1998.
95. Prakash and Kollman 2003: 624.
96. Triplett 2004: 663.
97. The EPA has also applied the Toxic Substances Control Act (TSCA) to regulate genetically engineered microbes. To date, "the most widely used microbial pesticides are subspecies and strains of Bt." See EPA, "Regulating Biopesticides" web page, http://www.epa.gov/pesticides/biopesticides/, accessed on July 27, 2007. The regulations are set forth at 7 CFR Part 340. See also Pew Initiative 2004*a*: 44–49.
98. Pew Initiative 2004*a*: 37, 41, 65.
99. See 40 CFR 152.20(a)(4). For a complete list of PIPs registered with the EPA, see the EPA website at http://www.epa.gov/pesticides/biopesticides/reg_of_biotech/ eparegofbiotech.htm, accessed on July 27, 2007. See also EPA 2003; Young 2003.
100. Pew Initiative 2004*a*: 37. USDA, however, still applies its notification and petition procedures under the Plant Protection Act until a variety is deregulated.
101. Pew Initiative 2004*a*: 15, 28, 41 (noting that the EPA relies on state enforcement in this area). The applicable EPA regulations are set forth at 7 USC § 136d. APHIS can also impose such conditions on the cultivation of GM varieties, but typically those imposed by EPA are more stringent. Shaffer telephone interview with USDA official. May 26, 2008.
102. 21 USC § 346a.
103. Pew Initiative 2004*a*: 43. The EPA maintains a list of registered PIPs at http://www. epa.gov/pesticides/biopesticides/pips/pip_list.htm (noting 16 active registrations as of January 4, 2007).
104. On May 22, 1998 the EPA published its final rule in the Federal Register granting a permanent split tolerance exemption for Cry9C protein and Cry9C DNA residues (the GM protein or PIP in Starlink corn), allowing their use "only in corn used for feed; as well as in meat, poultry, milk, or eggs resulting from animals fed such feed" (63 Fed. Reg. 28252). The "split" exemption from the requirement of tolerance was based on unresolved questions about the human allergenicity of the protein (62 Fed. Reg. 63168).
105. FDA 1992; FDA 2001a. See also Pew Initiative 2004*a*: 70. FDA policymakers and scientists responsible for overseeing the safety of GM foods are located within the

Center for Food Safety and Applied Nutrition, the Office of Policy, Planning and Strategic Initiatives, and the Office of Premarket Approval.

106. See FDCA § 409; 21 USC § 348.

107. Pew Initiative 2004*a*: 14. Pre-market approval, however, is required by the EPA for products with bioengineered pesticidal characteristics, as noted above.

108. Pew Initiative 2004*a*: 14–15.

109. See FDCA § 402(a)(1); 21 USC § 342(a)(1).

110. FDA 2001*a* and *Alliance for Bio-Integrity v Shalala*, 116 F.Supp. 2d 166 (D.D.C. 2000) (deferring to FDA expertise in upholding FDA's policy regarding the lack of need to label GM foods). See also discussion in Cantley 1995: 566–73; Echols 1998: 538; and Stewart and Johanson 1999: 248–49.

111. See *International Dairy Foods Ass'n v Amestoy*, 92 F3d 67, 71–74 (2d Cir 1996). The court found, "The wrong done by the labeling law to the dairy manufacturers' constitutional right *not* to speak is a serious one that was not given proper weight by the district court.... We need not resolve this controversy at this point; even assuming that the compelled disclosure is purely commercial speech, appellants have amply demonstrated that the First Amendment is sufficiently implicated to cause irreparable harm.... It is not enough for appellants to show, as they have, that they were irreparably harmed by the statute; because the dairy manufacturers challenge government action taken in the public interest, they must also show a likelihood of success on the merits. We find that such success is likely.... Absent, however, some indication that this information bears on a reasonable concern for human health or safety or some other sufficiently substantial governmental concern, the manufacturers cannot be compelled to disclose it. Instead, those consumers interested in such information should exercise the power of their purses by buying products from manufacturers who voluntarily reveal it."

112. FDA 2001*a*.

113. See FDA 1994.

114. FDA 1994. FDA is the primary authority with oversight of misbranding.

115. See e.g., Miller 2000; Miller and Conko 2003.

116. For an example of alleged executive use of the USDA for political purposes, see Martin 2004.

117. See e.g., Glanz 2004; Revkin 2006; Wilson 2006.

118. For good discussions, see McCubbins and Schwartz 1987; Kiewiet and McCubbins, 1991; Epstein and O'Halloran 1999; Huber and Shipan 2002.

119. See e.g., Melnick 1983; McCubbins, Noll, and Weingast 1987; Shapiro 1988, 1997.

120. See e.g., Kagan 2001, 2004 on "adversarial legalism" and its impact on the US regulatory process.

121. See Lasker 2005. The Foundation is a conservative public interest group and thus favors a finding of preemption, but the paper clearly presents the risk to state legislation.

122. See e.g., the excellent survey in Taylor et al. 2004.

123. As we will see in Chapter 6, this factor, as well as market pressures, helps to explain why developers of GM products voluntarily notify the FDA of their new products, even though there is no requirement for them to do so.

124. Pew Initiative 2004*a*: 89. (" ... an administrative shift to an affirmative finding of safety approach would likely involve a significant change in the FDA's oversight

of GE [genetically engineered] foods and in the interaction between the FDA and the developers of such foods. This change would flow from the fact that the FDA would be taking a share of the responsibility for the safety of GE foods, rather than relying as heavily as it does today on the developers' safety determinations").

125. Moravcsik 1998: 86–158.
126. For good discussions of the harmonization process, and the internal market more generally, see e.g., Dashwood 1983; Young 2005.
127. Under the principle of "mutual recognition," foods recognized by one member state's authorization must be accepted by all other member states, provided that they meet certain "mandatory requirements" of consumer protection, health and safety. See *Rewe-Zentrale AG v Bundesmonopolverwaltung fur Branntwwin* (Cassis de Dijon), Case 120/78, ECR 649.
128. For general discussions of the EU as a political system, see Hix 2005; Pollack 2005.
129. Moravcsik 2001: 163–4.
130. Kelemen 2003: 185.
131. For good discussions of the EU as a federal system, see the essays in Nicolaidis and Howse 2001; Kelemen 2003; and Kelemen and Nicolaidis 2007.
132. Kreppel 2002: 5.
133. On the internal workings of the Commission, see e.g., Nugent 2001; and Stevens and Stevens 2001.
134. The definitive work on the Council is Hayes-Renshaw and Wallace 2006.
135. Tsebelis and Garrett 2000: 24. For good reviews of the EU's legislative process, including the respective roles of the Council of Ministers and the European Parliament and the co-decision procedure, see e.g., Kreppel 2001; Hix 2005: 72–110; and McElroy 2007.
136. The literature on the Commission as an executive agent of the member states is voluminous, as is that on the more arcane subject of comitology. For good overviews, see Pollack 2003; and Tallberg 2007.
137. In 1987, the Court of First Instance was created, with jurisdiction over particular types of cases including competition-policy and EU staff disputes, but the ECJ remains the most important institution in the EU's judicial branch to this day. For good overviews of the European judicial system, see e.g., Stone Sweet 2004; Arnull 2006; and Conant 2007.
138. Commission of the European Communities 1978.
139. See Cantley 1995: 519.
140. Council of Europe, *Thirty-Third Ordinary Session of the Parliamentary Assembly*, Recommendation 934 (1982), on Genetic Engineering, text adopted by the Assembly on January 26, 1982 (22nd sitting), reprinted in Council of Europe 2005: 12–14. The Council of Europe is a distinct institution from the European Union. It was founded in 1948 and is open to all European democracies which accept the rule of law and fundamental human rights. The Council, which had 47 members in April 2008, features less powerful supranational institutions than the EU, but has adopted a number of recommendations on biotechnology-related issues. See Council of Europe 2005.
141. Cantley 1995: 538–539.
142. Patterson 2000: 324–32; Cantley 1995: 534–35 and 543–47. The BSC was chaired by DG XII, while the BRIC was co-chaired by DG III (Industry) and DG XI (Environment).

143. Cantley 1995: 535, 543.
144. Cantley refers more bitingly to "the inconvenience of productivity increases for the managers (at Community and national levels) of a Common Agricultural Policy already wrestling with surpluses," Cantley 1995: 639.
145. Cantley 1995: 644–645
146. Mark Cantley comments, Feb. 5, 2007.
147. Commission of the European Communities 1986.
148. Commission of the European Communities 1986, quoted in Cantley 1995: 553.
149. See Bernauer and Meins, 2003: 655; Greenwood and Ronit 1995; and Patterson 2000. Cantley notes how industry's European Committee on Regulatory Aspects of Biotechnology had prepared a position paper (1995: 550), but European industry did not exercise the influence that US industry had. He points out that the leading European biotechnology companies did not form a lobbying organization that specifically focused on biotechnology until mid-1989 when it was too late, forming the Senior Advisory Group for Biotechnology. Id. at 561.
150. Commission of the European Communities 1988. See also European Parliament 1989, Cantley 1995, and Patterson 2000 for good discussions of existing national regulations.
151. US Government, "International Harmonization in the Biotechnology Field," July 7, 1989, quoted in Cantley 1995: 559; emphasis added.
152. Council Directive 90/220/EEC of April 23, 1990 on the deliberate release into the environment of genetically modified organisms, *Official Journal of the European Communities* L117 of 08/05/1990, pp. 15–27.
153. See the Supreme Court's decision in *INS v Chadha*. I.N.S. v. Chadha 462 U.S. 919 (1983).
154. Regulation (EC) No. 258/97 of the European and of the Council of January 27, 1997 Concerning Novel Foods and Novel Food Ingredients, *Official Journal* L 043, 14/02/1997, pp. 1–6. For excellent analyses of the regulation, see also Hunter 1999: 217–225; and Commission of the European Communities, "Novel Foods and Novel Food Ingredients," http://europa/eu.int/scadplus/leg/en/lvb/l21119.htm, accessed on April 11, 2004.
155. The regulation would not apply to food additives, flavorings, or extraction solvents, governed by other EU legislation (Article 2).
156. This simplified procedure was, however, eliminated in 2003, since which time foods containing or derived from GMOs have been authorized under the provisions of Regulation 1829/2003 and subject to the labeling and traceability provisions of Regulation 1830/2003. For details on the use of the Novel Foods Regulation to approve GM-derived foods and crops, see "Novel Foods and Novel Food Ingredients," http://europa.eu/scadplus/leg/en/lvb/l21119.htm, accessed on April 28, 2007. For the post-2003 regulations on the approval of GM and GM-derived foods, see Chapter 6.
157. Van Waarden 2006: 57.
158. In May of 1999, a second major scandal broke when it was learned that Belgian farm animals had been given dioxin-contaminated feed, resulting in the removal of Belgian chicken, eggs, pork, and beef from the entire EU market, and the subsequent fall of the Christian Democratic government of Jean-Luc Dehaene. Later and less significant scandals, including a scare regarding possible contamination

of Coca-Cola products in northern Europe and the French admission that it had discovered sewage sludge containing human and animal wastes in feed destined for pigs and chickens, only served to deepen Europeans' mistrust of public food safety authorities. See Barlow 1999.

159. Chalmers 2003: 534–538.
160. Vogel 2003*a*: 27.
161. See e.g., BBC News 2000.
162. See Vogel 2003*a*, citing Enriquez and Goldberg 2000.
163. Bernauer and Meins 2003: 646.
164. Slovic, 2000: 410.
165. Vogel 2003*a*: 10.
166. Ansell, Maxwell, and Sicurelli 2006.
167. See discussions in Sunstein 2006: 36–39; and Slovic 2000: 37–38. On heuristics, see Tversky and Kahneman 1982.
168. See Sunstein 2006: 6, 39; and Sunstein 2002.
169. Ansell, Maxwell, and Sicurelli 2006: 107.
170. See Henley 1999: A18. ("He [Mr. Bové] has had expressions of sympathy from all sides of the political spectrum, from the Greens to the far-right National Front, the Socialists to the Gaullists.")
171. Seifert 2006.
172. Ansell, Maxwell, and Sicurelli 2006.
173. Gaskell, Allum, and Stares 2003.
174. Bradley 1998.
175. Bradley 1998: 212. See overview of the procedure in Part C.
176. Sato 2006.
177. Commission of the European Communities 2004*b*.
178. Ansell, Maxwell, and Sicurelli 2006: 105.
179. Hill and Battle 2000: 7.
180. Vogel 2001: 10.
181. Vogel 2001: 12.
182. Vogel 2001: 11.
183. See Commission of the European Communities 2002*a*.
184. Council of the European Union 1999.
185. For the now-familiar classification of interests, ideas, and institutions as explanations for political outcomes, see Hall 1997.
186. See e.g., Bernauer 2003, which offers a relatively parsimonious explanation, stressing different interests in the two polities as well as the different federal and quasi-federal institutions in the US and EU, respectively. We find considerable merit in Bernauer's explanation, but consider that a full explanation requires consideration of both ideational or cultural differences and contingent events, which Bernauer (2003: 11) considers implicitly, but without naming them as causal variables.
187. The biotech industry was well represented in Europe in the 1980s and into the 1990s. Even in 2000, when the Commission prepared a profile of the leading agri-biotech firms, four of the seven were European, including AgrEvo in Germany, Novartis in Switzerland, Zeneca in the UK, and Rhone-Poulenc in France. See Commission of the European Union 2007b: Annex A.
188. Graff and Zilberman 2004.

189. The report continues, "US companies are market leading in agro-biotech but new competitors are emerging." Enterprise and Industry Directorate General 2007: 7. The document states that, with respect to biotechnology as a whole (including agricultural, marine, healthcare, industrial, etc.), the US biotech industry employs twice as many people as Europe's, spends three times more on R&D, raises twice as much through venture capital and equity, three times as much via debt financing, and generates twice as much total revenue. Id., at 12.

190. See Jackson and Anderson 2005: 208–209 (noting that from 1998–2002, the US had a 40 per cent share of global production of maize, and a 43 per cent share of soybeans, compared to an EU share of 6 per cent of maize and 1 per cent of soybeans).

191. See e.g., Rosenthal 2006a, 2006b (citing Mr. Lappas, of the Greek farmers' union. " 'If our market doesn't buy it, and insurers won't insure it, how can we grow it?' " The article further notes that "even mainstream supermarket giants like Migros will not stock them," and that "Europe's agricultural insurers will not cover farmers for liability should their genetically modified crops contaminate adjacent fields," through cross-breeding).

192. See Jackson and Anderson 2005: 208–10.

193. Id. See further discussion in Chapter 6 regarding the challenges of EU thresholds for the adventitious presence of non-GM varieties.

194. The European Association for Bioindustries calls this "white biotechnology," in contrast to "green biotech" and "healthcare biotech." See their web site at http://www.europabio.org/white_biotech.htm. European companies, it should be noted, are particularly competitive in the production of GM enzymes, with the Danish company Novozymes claiming a global market share of 46 per cent in the production of GM enzymes; see Novozymes 2006: 9. Not surprisingly, the US complains that "the EU rules do not require labeling of products like beer and cheese (major European agricultural exports) that are made using enzymes produced with biotechnology, while soy oil derived from GM soybeans would have to be labeled even if no GM protein could be detected." Pew Initiative on Food and Biotechnology 2003b: 12–13.

195. Sunstein 2006: 68. Sunstein points to different heuristics to show how the precautionary principle is "incoherent."

196. The "Delaney clause" provides "that no additive shall be deemed to be safe if it is found to induce cancer when ingested by man or animal or if it is found, after tests which are appropriate for the evaluation of the safety of food additives, to induce cancer in man or animal." 21 U.S.C. §348(c)(3)(A). The clause implies that there should be a zero tolerance.

197. Sheingate 2006: 265. Sheingate, for example notes, how "biomedical applications received more positive framing by the British press than agricultural applications," citing work by Bauer. He contrasts Congressional treatment of biomedical and agricultural applications, which he breaks down in terms of number of Congressional hearings, the focus on the technology's risks or benefits, and the characteristics of those testifying. Sheingate 2006: 256, fn. 65, citing Bauer 2002.

198. Echols 1998: 525–31; and more generally Echols 2001. Cultural differences also play a critical role in Sheila Jasanoff's masterly analysis of science, democracy, and public policy in the US; see Jasanoff 2005.

199. As one analyst writes, noting how material conditions interact with human perceptions, "material agricultural circumstances in the UK and the US differ quite
distinctly: most countryside in the UK is farmed, whereas most of it in the US is
not. This material difference is associated with an ideational difference, in that
British environmental groups have been concerned for a long time about the environmental impact of agriculture on wildlife, whereas the main wildlife groups in
the US are more often concerned with the protection of wilderness areas." See Toke
2004: 184.

200. Douglas and Wildavsky 1982; Douglas 1973; and Slovic 2000: xxiii (offering a
"psychometric paradigm [that] encompasses a theoretical framework that assumes
risk is subjectively defined by individuals who may be influenced by a wide array
of psychological, social, institutional and cultural factors"). See also the cultural
critique of Sunstein by Kahan et al. 2006: 1083–84.

201. Kahan et al. 2006: 1083; emphasis in original. Similarly, Slovic writes, "risk is
inherently subjective.... Risk does not exist 'out there', independent of our minds
and cultures waiting to be measured. Instead, human beings have invented the
concept of risk to help them to understand and cope with the dangers and uncertainties of life. Slovic 2000: xxxvi.

202. See e.g., Korthals 2004: 499 ("It is impossible to separate data and the message
from the broader value context").

203. Kahan et al., 2006: 1091, emphasis added.

204. Sato 2006.

205. Sato 2006.

206. The applicant was Ciba-Geigy, which became Novartis in 1996 when it merged
with Sandoz. The company's name changed again in 2000 to Syngenta when
Novartis merged with AstraZeneca's agribusiness divisions.

207. Baumgartner and Jones 1993: 60–61, citing Weart 1998 ("Opponents had won primarily by getting their vision of the issue accepted, and by altering the nature of
the decision-making process by expanding the range of participants involved").

208. For an overview of the situation in France, see Sato 2006; and Tiberghien and
Starrs 2004,

209. Sato 2006: 3.

210. Sato 2006: 16.

211. See Axelrod 1997: 148 (using an agent-based model). On processes of "social ampli-
fication" in the perception of risks, see Slovic 2000: 232–45; and Sunstein 2005:
89–106, referring to "social cascades" and "group polarization." Social cascades
result from the repetition, as in the media, of a risk with "high emotional valence."
Id., at 96. Group polarization occurs when individuals with similar experiences
reinforce a common perception.

212. Baumgartner and Jones 1993: 27, citing Plein 1991.

213. For good reviews of historical institutionalism as an approach, see Thelen 1999;
Pierson 2000, 2004; Pierson and Skocpol 2002; on applications to the study of the
EU, see Pollack 2004, 2007.

214. Pierson 2000, 2004.

215. Levi 1997: 28, quoted in Pierson 2004: 20.

216. For related discussions, see Baumgartner and Jones 1993: 37, who argue that
"The interactions of image and venue may produce a self-reinforcing system

characterized by positive feedback"; and Greif and Laitin 2004: 34, whose game-theoretic approach defines "*Self-reinforcing* institutions" as "those that change the quasi-parameters of the institution so as to make the institution, and the individual behaviors of the actors within it, more stable in the face of exogenous changes" (emphasis in original). To be sure, not all institutions and policies generate positive returns—indeed, some institutions may generate "negative feedbacks" and become "self-undermining" over time, for example by generating unexpected negative consequences and hence undermining their societal basis of support. The presence of increasing returns and positive feedbacks, therefore, should be thought of, not as a constant feature of institutions, but as a variable. By and large, however, the emphasis of the historical institutionalist literature, as in our discussion here, has been on institutions characterized by increasing returns, continuity, and path-dependent development over time. For good discussions of negative feedbacks and self-undermining institutions, see e.g., Greif and Laitin 2004; Streeck and Thelen, 2005; Hall and Thelen, 2006; Immergut, 2006; and Pollack 2007: 47–50.

217. In this sense, our analysis is informed by liberal theories of international relations and international law, as well as by Putnam's two-level games model, considered below. On liberalism as a theoretical approach, see e.g., Slaughter Burley 1993; Moravcsik 1997.

218. Putnam 1988; Evans et al. 1993.

219. Moravcsik 1993: 23.

220. Moravcsik 1993: 25.

221. Moravcsik 1993: 31.

222. Moravcsik 1993: 31.

223. See Pew Initiative on Food and Biotechnology 2003c; *International Trade Reporter* 2004*a* (citing a report of the International Service for the Acquisition of Agri-biotech Applications, "an industry funded effort to bolster biotechnology in developing countries").

224. In 2007, they had increased slightly to US$ 1 billion in exports. See USDA, Foreign Agricultural Service, *U.S. Trade Exports FATUS Commodity Aggregations*, http://www.fas.usda.gov/ustrade/ (accessed April 16, 2008); and Tiberghien: 13.

225. In 2007, they were still just under US$ 8 million. See USDA, Foreign Agricultural Service, *U.S. Trade Exports FATUS Commodity Aggregations*, http://www.fas.usda.gov/ustrade/ (accessed April 16, 2008); and Tiberghien 2006: 13.

Chapter 3

1. This situation approximates what we shall refer to in Chapter 4 as a "Battle of the Sexes" game, in which two or more actors agree on the importance of coordinating their activities, but disagree on the terms of that coordination.

2. Keohane and Nye 1974: 43. This section draws in part on our earlier discussion of transgovernmental relations in Pollack and Shaffer 2001*b*: 25–29.

3. Keohane and Nye 1974: 42.

4. Vogel 1997.

5. Slaughter 2004: 5. See also Slaughter 1997; Raustiala 2002.

6. See Haas 1992: 3. Haas defines an epistemic community as a "network of professionals with recognized expertise and competence in a particular domain and an

321

authoritative claim to policy-relevant knowledge within that domain or issue-area." From the organizational literature, see Wenger 1999 and Wenger et al. 2002.

7. See Canan and Reichman 2002, studying "the ozone regime as a social system of networks" (at 61), resulting in the "socialization" of an "ozone community" (153–160), noting how "networks provide flexibility, swift communication and diffusion" (187).

8. By horizontal networks, we refer to those constituted by state agencies, and by vertical ones, those linking supranational organizations with state agencies. Slaughter also distinguishes "horizontal networks among national government officials in their respective issue areas" from vertical networks that include a specialized "higher," "supranational" organization. She further sub-categorizes them in terms of networks focused on information exchange, enforcement and harmonization. Slaughter 2004: 19–22, 52–64.

9. Slaughter 2004: 8. On Coreper, see Lewis 2005, Hayes-Renshaw and Wallace 2006. On comitology, see e.g., Joerges and Neyer 1997*a*, *b*; Joerges 2001*a*; and Pollack 2003*b*: 114–145.

10. On the EU's transgovernmental networks of regulators, see generally Eberlein 2005; Eberlein and Grande 2005; Wallace 2005: 85–89; Coen and Thatcher 2007; Eberlein and Newman 2007; on the competition policy network and its evolution over time, see Wilks 2007.

11. On the OMC, see e.g., Hodson and Maher 2001; Scott and Trubek 2002; Jacobsson and Vifell 2003; Borrás and Jacobsson 2004; de la Porte and Nanz 2004; Trubek and Trubek 2005; Zeitlin and Pochet 2005; Rhodes and Citi 2007; and the discussion in the next section of this chapter.

12. Joerges 2001*a*.

13. Habermas 1985, 1998. For good discussions of deliberative democracy in domestic politics, see e.g., Bohman 1998; Elster 1998; and Ryfe 2005. For applications of Habermasian thinking to EU and international politics, see e.g., Risse 2000, 2001; Ericksen 2000; Checkel 2001; Joerges 2001*a*; Jacobsson and Vifell 2003; Magnette 2004; Mitzen 2005; Zeitlin and Pochet 2005; and Cohen and Sable 2005.

14. The literature on constructivism and sociological institutionalism is large and diverse, and our analysis here is necessarily selective. For good introductions and overviews of constructivist theory, see e.g., Adler 1998; Checkel 1998; Ruggie 1998; Wendt 1999; and Risse 2004. On sociological institutionalism, see e.g., Hall and Taylor 1996; Finnemore 1996.

15. Risse 2004: 161, references removed.

16. March and Olsen 1989: 160–62.

17. March and Olson 1989: 22.

18. March and Olsen 1989: 22.

19. March and Olsen 1989: 23.

20. March and Olsen 1989: 23.

21. For good discussions on the logic of appropriateness in international politics, see e.g., Finnemore 1993, 1996; Risse 2000: 5–7; and Checkel 2005: 810–812. For examples of applications in normative legal theory, see e.g., Koh 1998 (contending that international law is "internalized" within states through the interaction of the international and national levels, including the engagement of non-state actors operating within states, leading to changes in norms, beliefs, and identities within national societies and/or among their political representatives); and Goodman

and Jinks 2004 (calling for institutional prescription to be built from sociological insights, and in particular, those of the "world culture" theoretical and empirical projects of John Meyer and others).

22. Risse 2000: 1–2 ("I claim that processes of argumentation, deliberation, and persuasion constitute a distinct mode of social interaction to be distinguished from both strategic bargaining, the realm of rational choice, and rule-guided behavior, the realm of sociological institutionalism. Apart from utility-maximizing action on the one hand, and rule-guided behavior on the other, human actors engage in truth seeking, with the aim of reaching a mutual understanding based on a reasoned consensus").

23. Risse 2000: 7. Such deliberative processes, Risse argues, are distinct from the strategic use of information to manipulate other actors' beliefs and behavior. "Participants are not necessarily oriented toward their own success in communicative action.... In arguing mode, actors try to convince each other to change their causal or principled beliefs in order to reach a reasoned consensus about validity claims. And, in contrast to rhetorical behavior, they are themselves prepared to be persuaded. Successful arguing means that the 'better argument' carries the day, while one's (material) bargaining power becomes less evident." Risse 2000: 9.

24. Deliberation theory has some overlap with learning theory in the social sciences. Learning theorists similarly posit that national regulatory policy choices can change through transgovernmental interaction that generates new social knowledge to which they adapt their beliefs. See Dobbin et al. 2007.

25. Bohman 1998; Elster 1998; Cohen and Sable 2005; on deliberative supranationalism at the international level, see Joerges 2001a.

26. Risse 2000: 10, 19.

27. Risse 2000: 19.

28. Risse 2000: 19.

29. See e.g., Risse 2000: 11, 21–23; Johnstone 2003.

30. See e.g., the superb discussion in Naurin 2007, as well as Zürn 2000; Checkel 2001; Stasavage 2004; Steiner et al. 2004; Lewis 2008; and Niemann 2008.

31. See e.g., Naurin 2007; Lewis 2008; and Niemann 2008.

32. See e.g., de Búrca and Scott 2006 (on new governance); Bermann 2007 (on legal pluralism); Scott 2007 (on WTO SPS committee); Johnstone 2003 (on UN); and Slaughter 2004 (on transgovernmental networks).

33. Checkel 2001, Magnette 2004: 208.

34. The study of deliberation is far more advanced in the EU setting discussed here than in international politics more generally. Exceptional applications in non-EU settings include Johnstone 2003; Mitzen 2005.

35. Joerges 2001a; Eriksen 2000.

36. Joerges and Neyer 1997a, b.

37. Pollack 2003b: 114–145.

38. Zürn 2000; Weiler 1999.

39. Maurer 2003; Closa 2004.

40. Magnette 2004: 220. In any event, of course, the resulting Constitutional Treaty was later rejected in referenda in the Netherlands and France in 2005, and ultimately abandoned by the EU's member states in favor of a slightly more modest "Reform Treaty."

41. Optimistic accounts of the OMC include Hodson and Maher 2001; Scott and Trubek 2002; Borrás and Jacobsson 2004; Trubek and Trubek 2005; and Zeitlin and Pochet 2005.
42. Jacobsson and Vifell 2003: 21.
43. For skeptical views of deliberation within the OMC, see e.g., Jacobsson and Vifell 2003; Borrás and Jacobsson 2004; de la Porte and Nanz 2004; Rhodes 2005; Rhodes and Citi 2007.
44. Pollack and Shaffer 2001*b*: 14–17.
45. See e.g., Slaughter 2004: 47; Pollack and Shaffer 2001*b*, 2001*c*, 2005; Pollack 2005. The discussion in this section draws on the broader analyses in Pollack 2005: 1038–1045 and Pollack and Shaffer 2005: 5–6.
46. Pollack 2005.
47. See e.g., the studies in Bermann et al. 2001; Petersmann and Pollack 2003; Andrews et al. 2005.
48. Devuyst 2001; Kovacic 2005.
49. Shaffer 2003; Nicolaidis and Steffenson 2005.
50. Shaffer 2003; Asinari and Poullet 2005.
51. Posner 2005.
52. Peterson et al. 2005: 46–51.
53. White House 2008.
54. *Inside US Trade* 2008*i*. For details and key documents on the Transatlantic Economic Council, see the website of the EU Commission Delegation to the US, at: http://www.eurunion.org/partner/euusrelations/TEC.htm. The Transatlantic Economic Council, it should be noted, has not played a significant role in the ongoing GMO dispute from its creation through late 2008.
55. Pollack 2003: 101–102.
56. Pollack and Shaffer 2005: 8.
57. Posner 2005.
58. Pollack and Shaffer 2005: 7.
59. Murphy 2001: 339.
60. Risse 2000: 19, 33.
61. Sunstein 2005: 126 ("If the public demand for regulation is likely to be distorted by unjustified fear, a major role should be given to more insulated officials who are in a better position to judge whether the risks are real").
62. Sunstein 2005: 82 ("There is an unambiguous lesson for policy here: It might not be helpful to present people with a wide range of information, containing both assuring and less assuring accounts. The result of such presentations will be to scare people").
63. See Chapters 2 and 6, this volume.
64. Sunstein 2005: 25, emphasis in original.
65. Sunstein 2005: 128. See the views on this point of the European Court of Justice as well, in Case 207/83 Commission v United Kingdom [1985] ECR 1201, ¶ 17 (in the context of a UK requirement of a certificate of origin that enables consumers "to assert any prejudices which they may have against foreign products"). Cf. Dhar and Foltz 2005 (in the context of regulated voluntary standards, finding that "results show significant consumer benefits from organic milk and to a lesser extent from rBST-free milk," having "implications for present U.S. labeling standards." In their

study of revealed pricing preferences of consumers from aggregated supermarket data, they show that "consumers benefit both from the competition induced by labeled milk and by the benefits of an increased choice set").

66. Sunstein 2005: 226.

67. See e.g., Kahan et al. 2006: 1096 ("The cultural-evaluator model, in contrast, supports an approach to risk regulation that is much more consistent with participatory and deliberative visions of democracy").

68. Boyle 1985: 751. Compare Fischer 2000.

69. See e.g., Slovic 2000: 191 (laypeople's "basic conceptualization of risk is much richer than that of the experts and reflects legitimate concerns that are typically omitted from expert risk assessments").

70. Slovic 2000: 231. For example, Slovic has "observed a strong 'affiliation bias' indicating that toxicologists who work for industry see chemicals as more benign than do their counterparts in academia and government." Id.

71. Slovic 2000: 325.

72. Kahan et al. 2006: 1104.

73. Kahan et al. 2006: 1102.

74. Slovic 2000: xxxvi. For other critiques of granting preeminence to "scientific" knowledge because of the limitations of science, see e.g., Barnes, Bloor and Henry 1996; Jasanoff 1990. Cf. Sokal and Bricmont 1998.

75. Cf. Scott 2003 (noting nonetheless that EU decision-making remains relatively centralized).

76. See Federal Food, Drug, and Cosmetic Act, 21 U.S.C. § 383(c)(2) (1994).

77. Commission of the European Communities 2000*c*.

78. European Commission 2000.

79. European Commission Directorate General for Trade (DG Trade) 2000: 13.

80. Food and Drug Administration 1998.

81. Shaffer telephone interview, Jan. 29, 2008.

82. Id. Shaffer's prior analysis of the 1997 transatlantic MRA regarding pharmaceuticals and medical devices confirms this point. See Shaffer 2003 (noting the MRA's limits and the reasons it has been difficult to implement).

83. Shaffer interview with Mark Cantley, Feb. 7, 2007.

84. Shaffer interview with Canice Nolan, European delegation in Washington, April 24, 2006.

85. Shaffer interview with Mark Cantley, Feb. 5, 2007.

86. The TEP Biotech Group is known to have met three times in 1999 and twice in 2000, but the authors know of no subsequent meetings of this group. See DG Trade 2000: 13. Pollack interviews with US and EU Commission officials, Brussels, July 2002, and Washington, DC, March 2005.

87. *Official Journal of the European Communities*, C 96 E, 27.3.2001, p. 247.

88. See European Food Safety Authority, *How the European Food Safety Authority is Funded*, http://www.efsa.europa.eu/en/about_efsa/efsa_funding.html (last visited Feb. 5, 2007) (giving figures for the end of the financial perspectives period ending in 2013). This compares to 250 people, with a budget of 40 million euros, as cited in Buonanno 2006 (with figures based on earlier data).

89. Quoted in Yerkey 1999: 2025.

90. Some participants missed the key final meeting because of illness and were upset with the result. One even considered filing a minority opinion but was persuaded by the US chair not to do so. Shaffer interview with Cantley, Feb. 7, 2007, and with Derek Burke, a UK scientist who was a member of the Consultative Forum, Feb. 14, 2007. Burke suggests that the report represented a "least common denominator," as consensus with the NGO representatives could only be obtained if such issues as mandatory pre-approvals, mandatory labeling, a form of the precautionary principle and traceability were included.

91. Shaffer interview with former member of DG Science, Research and Technology, Feb. 7, 2007.

92. See EU–U.S. Biotechnology Consultative Forum (2000); and Commission of the European Communities 2000d.

93. The report held that, "[t]he fact that a biotechnology food is held to be substantially equivalent to a conventional food should not be taken to mean that it needs automatically less testing or less regulatory oversight than 'non-substantially' equivalent biotechnology foods." EU–U.S. Biotechnology Consultative Forum 2000: 10.

94. See US Department of State 2000.

95. Pollack interview with a former high-level official of the European Commission, Florence, June 2004.

96. TABD 2002.

97. TABD 2004; Corporate Europe Observatory 2001.

98. TACD 2004; see also TACD 2003.

99. Shaffer interview, April 10, 2007.

100. Shaffer interview with official of the Italian Ministry of the Environment, June 18, 2007.

101. Shaffer telephone interview with EFSA scientist in agricultural biotech field, June 7, 2007.

102. See Chapter 2, this volume.

103. The one exception to this rule concerned an incomplete application, although EFSA is under considerable political pressure not to give positive risk assessments; see Chapter 6.

104. Agricultural biotechnology also raises issues in multiple areas within science. As Cantley points out, "the scientific base of biotechnology is multi-disciplinary, drawing upon elements from biochemistry, microbiology, molecular biology, genetics, process engineering." Cantley 1995: 508.

105. Thus, although Cantley notes that "a permanent technical working group on biotechnology was set up between DG XI and staff of the US Environmental Protection Agency," the two agencies played different roles with different powers on each side of the Atlantic. See Cantley 1995: 645.

106. Nicolaidis and Steffenson 2005.

107. Slaughter 2004: 29.

108. Risse 2000: 11.

109. After the WTO panel decision against the EU in 2006, the two sides intensified discussions about agricultural biotechnology. At that stage, however, the negotiations were led largely by trade negotiators, and were geared primarily toward EU compliance with the WTO decision; see Chapter 5.

Chapter 4

1. For good reviews of the literature on international regimes, see Krasner 1983; Keohane 1984; Hasenclaver, Mayer and Rittberger 1997.
2. Raustiala and Victor 2004.
3. Saegusa 1999; Aritake 1999; Pruzin 2000.
4. Aritake 1999; Vogel et al. 2008: 27.
5. Hill and Battle 2000.
6. Shea 2000.
7. Strauss 2002.
8. Quoted in United States Mission to the European Union 2003. The African situation was also invoked in the US's WTO complaint; see e.g., *Bridges Weekly Trade News Digest* 2004. Similarly, Monsanto adopted the slogan "Food, Health, Hope" for its website and publications, suggesting "that biotechnology product lines are the best bet when it comes to meeting the needs and hopes of the world's poor." See Drahos 2004 (critiquing the self-interested merchandising of hope, noting "Monsanto is in this sense a merchant of hope"). For an academic assessment that is favorable to this prospect, see Evenson 2003.
9. Herring 2007.
10. See James 2007. For a more detailed discussion of GMO cultivation and regulation in the developing world, see Chapter 7, this volume.
11. Krasner 1983.
12. Krasner 1983; Keohane 1984; Hasenclaver, Mayer and Rittberger 1997.
13. On the role of international regimes in monitoring and enforcing compliance, see inter alia: Chayes and Chayes 1993; Mitchell 1994; Downs, Rocke and Barsoom 1996; Raustiala 2000; and Raustiala and Slaughter 2002.
14. Keohane (1984: Chapter 7) also addresses the usefulness of regimes for states where rationality is "bounded"—that is, where actors are limited in their ability to process available information.
15. In its simplest form, the classic PD story goes as follows. Two individuals are caught after a crime. As rational egoists, each individual cares only about their own welfare, being indifferent to that of their counterpart. A prosecutor separates them and tells each of them, "You may choose to confess or remain silent. If you confess and your accomplice remains silent I will drop all charges against you and use your testimony to ensure that your accomplice does serious time. Likewise, if your accomplice confesses while you remain silent, they will go free while you do the time. If you both confess I get two convictions, but I'll see to it that you both get early parole. If you both remain silent, I'll have to settle for token sentences on firearms possession charges. If you wish to confess, you must leave a note with the jailer before my return tomorrow morning." The dilemma for each prisoner is that each is always better off confessing ("defection") no matter what the other prisoner does—what game theorists refer to as a "dominant strategy." In such a situation, the equilibrium outcome is mutual defection, with both actors confessing and receiving jail sentences for their crimes. Yet if each confesses, each is worse off than if neither had confessed—that is, if they had been able to cooperate. The quote is taken from Kuhn 2003. For applications in law, see Baird et al. 1994.

16. In game-theoretic terms, a Nash equilibrium is "a set of strategies, one for each player, such that no player has incentive to unilaterally change her action." Shor 2005.

17. Concretely, if player 1 cooperates, player 2 can maximize his utility by defecting, and hence receiving his first-choice payoff (4 rather than 3). By contrast, if player 1 defects, player 2 is also best off defecting, and thus avoiding the "sucker's payoff" which will leave him worst off (2 rather than 1). In game-theoretic terms, defection is the "dominant strategy" for both players and double defection the equilibrium outcome.

18. For pioneering applications of the PD model to international cooperation, see e.g., Axelrod 1984; Keohane 1984; and Oye 1986.

19. For example, elected officials within a state may have an incentive to respond to well-organized protectionist groups and renege on the deal, such as by raising tariffs or non-tariff barriers on imported goods. If they do so without being caught or without incurring retaliation, then the political officials will satisfy both domestic exporting constituents (because of the foreign states' compliance with an agreement to open markets) and domestic protectionist constituents (through providing the protection). According to Bagwell and Staiger, "[t]he purpose of a trade agreement is to offer a means of escape from a terms-of-trade driven Prisoners' Dilemma" (2002: 3).

20. On the tragedy of the commons, and the role of institutions in allowing states to cooperate to preserve such commons, see e.g., Ostrom 1990; Young 1994.

21. Keohane 1984: 13.

22. Benedick 1998.

23. Risse 2000.

24. For the classic realist discussion of these dual impediments, see Grieco 1988; see also the essays in Baldwin 1993, and the excellent review in Hasenclaver et al. 1997: 113–135.

25. We focus here on classical and structural realists. See Waltz 1979; Greico 1988; Hasenclaver et al. 1997; Steinberg and Zasloff 2006 (classifying branches of realism).

26. Keohane 1984; Hasenclaver et al. 1997.

27. Waltz 1979: 105; Greico 1988: 487; Hasenclaver et al. 1997: 114–125.

28. Powell 1991; Snidal 1991; Baldwin 1993; Hasenclaver et al. 1997: 125–134.

29. See Chapter 2, which discusses the efforts by successive US administrations and by the European Commission to foster a cutting-edge biotech industry.

30. Once again, we are referring here, not to the problem of relative gains, but to the distribution of absolute gains from cooperation among two or more states. For a range of views on the challenge of distributive conflict in international cooperation, see Krasner 1991; Morrow 1994; Fearon 1998; Gruber 2000; Koremenos et al. 2001; Mattli and Büthe 2003; and Drezner 2007.

31. Morrow 1994: 395.

32. Stein 1982.

33. Stein 1982; on constructed focal points, see Garrett and Weingast 1993.

34. Krasner 1991: 339.

35. Stein 1982: 314.

36. For good overviews of the situation-structural approach, see e.g., Snidal 1985; Oye 1986; Martin 1992; and Hasenclaver et al. 1997. For a critique of the approach, see Fearon 1998: 272–275. For a pioneering application to international law, see

Goldsmith and Posner 2005; and the critiques in Guzman 2006; and Hathaway and Lavinbuk 2006.

37. Krasner 1991: 340. In the case of global communications, for example, powerful states can impede the development of a regime where they can achieve their aims unilaterally (e.g., broadcasting and remote sensing), and they can use their superior bargaining power to dictate the terms of cooperation to weaker states (e.g., allocation of the electromagnetic spectrum, telecommunications). Krasner 1991: 343.

38. On a related note, Lloyd Gruber (2000) focuses on the "go-it-alone" power of states, which can engage in international cooperation on terms that not only disproportionately benefit themselves, but may also indeed leave other states absolutely worse off than in the absence of cooperation.

39. Fearon 1998: 270.

40. Fearon 1998: 270–271.

41. In sociology, John Meyer and his collaborators have purported to find evidence of the development of a "world society" involving growing organizational and technocratic convergence around the world, maintaining that science "constitute[s] the religion of the modern world, replacing in good measure the older 'religions'." Meyer et al. 1997; and Loya and Boli 1999.

42. Mattli and Büthe 2003: 10–11. Mattli and Büthe go on to advocate an "institutional complementarities" approach, in which long-standing domestic standardization institutions may fit more or less well with international standardization bodies, giving some countries an advantage in promoting standards at the international level. More specifically, they find that the more inclusive and hierarchical standardization systems in the EU give European firms and standards bodies an advantage over firms and standards bodies in the more market-based US system. Mattli and Büthe focus largely on private standards bodies and voluntary standards, and we do not elaborate on their approach in this book, which deals with binding regulations adopted by governmental actors on both sides of the Atlantic. Nevertheless, we are sympathetic to the view that some domestic standards institutions provide a better fit with international standardization than others, and indeed we suggest in the next section that the US and EU both "forum-shop" among international regimes that provide the best fit with their own institutional procedures and substantive interests. On international standard-setting, see also Abbott and Snidal 2001; Mattli and Büthe 2005. On the institutional advantages of the EU in international standard-setting, see also Posner 2005.

43. Krasner 1991: 349; see also Bach 2000: 6. In two-level games terms (see Putnam 1988 and Chapter 2, this volume), the win-sets of the two players in a Deadlock game would no longer overlap, making agreement impossible. Like Putnam, we treat two-level games analysis as heuristic and metaphorical, and make no attempt here to quantify or measure the win-sets of either party to the dispute. Moreover, the characterization of the game as Deadlock assumes that one player could, by refusing to cooperate, block the emergence of a global standard. The ability to veto, first of all, is not available for smaller states. Moreover, it is not clear that the US or the EU alone can, in all instances, block the adoption of a global standard in various multilateral regimes. We are grateful to Tim Büthe for comments on this point.

44. Thomas Bernauer and Thomas Sattler point to two factors in particular: (i) the all-or-nothing character of the trade barrier in question which cannot be reduced as part of a compromise (as can a tariff or subsidy); and (ii) the fact that the environmental, health, or safety regulation may reflect broader public support so that a narrow interest group cannot be easily bought off through some sort of side payment. Bernauer and Sattler 2006: 8–9.
45. Morrow 1994: 393.
46. Morrow 1994: 413.
47. Draft Commission document quoted in Buck 2007*a*: 7. The final version appeared as Commission of the European Union 2007*a*. See also the excellent discussion in Buck 2007*b*.
48. Krasner 1983: 1, emphasis added.
49. Raustiala and Victor 2004: 279. For important related work, see Helfer 2004; Alter and Meunier 2006.
50. For an excellent discussion of forum-shopping in international relations, see Jupille and Snidal 2006.
51. We thank Kal Raustiala for his comments on this point.
52. Benvenisti and Downs 2007; Drezner 2007.
53. See e.g., International Law Commission (Koskenniemi) 2006; and Berman 2007 (on legal pluralism).
54. International Law Commission 2006.
55. See e.g., Delmas-Marty 1998: 104 ("Le droit a l'horreur du multiple". Sa vocation c'est l'ordre unifié et hiérarchisé, unifié parce que hiérarchisé"); Dupuy 1999; Koskenniemi and Leinmo 2002; and Roberts 2004. On the WTO and public international law, see Marceau 2001; and Pauwelyn 2003*a*, 2003*b*.
56. International Law Commission (ILC) 2006: 9.
57. Compare, on the one hand, Charney 1996 and Koskenniemi and Leinmo 2002 (both contending that a positive development), and on the other, Benvenisti and Downs 2007 and Dupuy 1999 (contending that problematic).
58. Griffiths 1986.
59. See Macdonald 1998:76, 80; and Fisher-Lescano and Teubner 2004: 1004–07 (finding, from a legal pluralist perspective, that international legal fragmentation is "a reflection of a more fundamental, multi-dimensional fragmentation of global society itself" involving accelerating functional differentiation).
60. Snyder 2002: 94 (Snyder examines the interaction of private as well as public governance sites, finding that these sites "do not make up a legal system"). Cf. Pauwelyn 2003b: 578 (stressing "the unity of international law" while noting the diversity of sub-regimes).
61. Macdonald 1998: 77.
62. Simma 1985.
63. Abbott et al. 2000.
64. Abbott and Snidal 2000: 37.
65. Abbott and Snidal 2000: 38.
66. Abbott and Snidal 2000: 38.
67. See e.g., Hafner-Burton 2005 (on the weakness of the global soft-law human rights regime), Rhodes and Citi 2007 (on the purported failure of the EU's soft-law "Open Method of Coordination"), and more generally Sindico 2006: 846 ("Soft law, and

voluntary standards in particular, are a stage in the creation of international legal norms. It is as a pioneer of hard law that soft law finds its raison d'être in the normative challenge for sustainable global governance").

68. For good discussions on the purported strengths of soft law, see e.g., Lipson 1991: 500–501 (discussing the strengths of "informal agreements"); Raustiala and Victor 1998: 684–686; Abbott and Snidal 2000: 38–39; and Sindico 2006: 832. On the "depth" of cooperation, i.e., cooperative agreements that require a greater change in state behavior relative to the status quo, see Downs, Rocke and Barsoom 1996.

69. See e.g., Finnemore and Toope 2001 (taking a more sociological perspective, and critiquing Abbott et al's formal definition of legalization because it obscures how law and legal norms actually operate in practice). We also recognize that these formal definitions can obscure the relative roles of "hard" and "soft" law in sociological terms—that is, from the way law and norms operate in the world, which indeed is what interests us. Binding dispute settlement can be ignored or simply reflect existing power asymmetries, so that "hard" law may in fact not be so "hard" in practice. Similarly, softer forms of law can be much more transformative of state and constituent conduct, which should be the real measure of law's impact in the world. Despite these caveats, we believe that that hard/soft distinction does capture something important about the making and implementation of international law, and we find the distinction to be particularly useful for our analyses of how regimes interact.

70. Cf. Raustiala 2005: 586 ("There is no such thing as 'soft law'").

71. On the complementarity of hard and soft law, see e.g., Abbott and Snidal 2000; and Trubek, Cottrell and Nance 2006 ("[I]t is no surprise that the [European] Union has sought to draw on both hard and soft methods and processes and to marry them in a single system").

72. Shelton 2000.

73. Dunoff et al. 2006: 95.

74. Benvenisti and Downs 2007.

75. Within the EU, for example, the national ministry that represents a member state in Brussels can affect how that state votes regarding the approval of specific GM varieties. For example, as an Italian official told us, the Italian Ministry of Health is more favorable to GMOs, while the Italian Minister of the Environment is from the Green party and is strongly against them. When the dossier switched from the health to the environmental ministry, Italy's position was affected. Shaffer interview, June 18, 2007. The Italian representative made similar comments regarding votes by other EU member states, and in particular Portugal.

76. Shaffer interview, 19 April, 2007. A representative from the IPPC further noted that lawyers are more present at Biosafety Protocol meetings, unlike FAO and IPPC meetings. Shaffer interview, 24 April, 2007.

77. Shaffer interview, 24 April, 2007.

78. Shaffer 2003.

79. For example, the same US agricultural official complained how Greenpeace allegedly worked side-by-side with the representatives of the Seychelles in meetings of the Conference of the Parties to the Protocol. Shaffer interview with representative of the US Department of Agriculture, 19 April, 2007.

80. Shaffer interview with a second IPPC official, 24 April, 2007.

331

81. For brief discussions of the earlier roles played by many of these organizations, see Cantley 1995: 610, 618 (sections 6.1 & 6.3). For subsequent overviews, see Glowka 2003 and Josling et al. 2004.

82. See USDA 2005 (providing an overview of FAO work). In its 2002–7, the FAO identified two "priority areas of interdisciplinary action" addressing biotechnology, one on "Biotechnology in Agriculture, Fisheries and Forestry," and another on "Biosecurity for Agriculture and Food Production." Among many other initiatives, the FAO created a Biotechnology Consultative Forum in 2000, an e-mail based forum "with the aim of providing quality balanced information on biotechnology in developing countries and making a neutral platform available." In its 2006 report, it concludes, "[r]egarding GMOs, there was no evidence of the intensity and polarization of the debate declining." FAO 2006: 151.

83. UNEP created in 1995 "International Technical Guidelines for Safety in Biotechnology" that address risk management under uncertainty. More recently, UNEP has worked with the Global Environmental Facility (GEF) to provide funding and technical support to over 130 developing countries to develop National Biosafety Frameworks for the regulation of environmental risks from GMOs. See UNEP-GEF Biosafety Unit 2006a, 2006b. We are grateful to David Duthie, of the UNEP-GEF Biosafety Unit, for directing us to UNEP's activities in this area.

84. See International Standards Organization Technical Committee 2001 (noting that "[t]he most frequent demand submitted to ISO/TC 34 and its subcommittees, is to develop International Standards related to analytical and test methods").

85. For example, an UNIDO/UNEP/WHO/FAO working group on biosafety prepared a "Voluntary Code of Conduct for the Release of Organisms into the Environment."

86. CGIAR-supported research centers are conducting some work on GM varieties of interest to developing countries.

87. FAO 2004a. See also FAO 2004d. The report not surprisingly triggered a furious GMO reaction in a letter to the FAO's director general, entitled "FAO Declares War on Farmers, Not on Hunger." See Genetic Resources Action International 2004, contending that the adoption of agricultural biotech would exacerbate "the failures of the Green Revolution," resulting once more in "rural and urban impoverishment, and greater food insecurity." The full text of the letter and a list of its signatories is available at http://www.grain.org/front_files/fao-open-letter-june-2004-final-en.pdf. The NGO letter in turn triggered a counter-response signed by academic and other professional researchers defending the report as providing "the most comprehensive and up-to-date review of issues related to agricultural biotechnology and developing countries," and as "an important contribution to rationalizing the international debate on this topic." See International Consortium of Agricultural Biotechnology Research 2004.

88. Shaffer, telephone interview with US government official, January 28, 2008.

89. See OECD website, "Overview of the OECD," available at http:// www.oecd.org/document/18/0,2340,en_2649_201185_2068050_1_1_1_1,00.html#who.

90. Salzman 2005: 194.

91. Examples include the successful 1997 OECD Convention on Combating Bribery of Foreign Public Officials in International Business Transactions and the failed 1998 Multilateral Agreement on Investment.

92. Patterson and Joslin 2005: 191.

93. Salzman 2005: 217. On the importance of the "OECD model" for the EU's OMC, see Wallace 2005: 85.
94. See e.g., Cantley 1995: 613.
95. Telephone interview of Shaffer with ICGB chair, Feb. 12, 2007.
96. See Bull, Holt and Lilly 1982; and Cantley 1995: 613.
97. Cantley 1995: 614. The Blue Book, prepared by an *Ad Hoc* Group on Safety and Regulations in Biotechnology, is available at http://www.oecd.org/dataoecd/45/54/1943773.pdf.
98. Cantley 1995: 615.
99. See OECD 1993.
100. See OECD 2007.
101. See OECD 2005: 7–9.
102. See OECD 2000.
103. Id.
104. Id.
105. For early FAO/WHO work on this concept, see the FAO/WHO Expert Consultation on Biotechnology and Food Safety (1990). The OECD Green Book elaborated this concept, providing,
    "Three possible scenarios are envisaged as a result of a substantial equivalence evaluation:
    (a) When substantial equivalence has been established for an organism or food product, it is considered to be as safe as its conventional counterpart and no further safety evaluation is needed.
    (b) When substantial equivalence has been established apart from certain defined differences, further safety assessment should focus on these differences.
    (c) When substantial equivalence cannot be established, this does not necessarily mean that the food product is unsafe. Not all such products will require extensive safety testing....." Id. at 22.
106. See Codex Alimentarius Commission 2003*a*, referring in turn to a 2000 report from a joint FAO/WHO expert consultation addressing this concept, FAO/WHO 2000.
107. Id. at 6.
108. OECD 2006*a*.
109. Shaffer, telephone interview with member of the OECD secretariat, March 27, 2007.
110. Cantley comments on draft of 5 Feb. 2007.
111. Cantley 1995: 613.
112. Id., at 614
113. Comments of Mark Cantley, former member of Commission Directorate-General for Science, Research and Development and head of the OECD Biotechnology Unit in its Directorate for Science, Technology and Industry, by email of Feb. 1, 2007.
114. OECD Environment Directorate 2003.
115. Id. at 8.
116. Id. at 12.
117. Telephone interview of Shaffer with OECD secretariat member noting the US use of this "mantra." Feb. 12, 2007 (finding that there nonetheless clearly had been divisions with the US delegation as well).
118. See OECD 2007.

119. OECD 2006*b*.
120. OECD, Product Database, http://webdomino1.oecd.org/ehs/bioprod.nsf (last visited Jan. 9, 2008).
121. OECD, Biotechnology Statistics in OECD Member Countries: On-line Inventory, http://www.oecd.org/countrylist/0,2578,en_2649_37437_36428358_1_1_1_37437,00.html (last visited Jan. 9, 2008).
122. OECD 2004*a*: 6. "[BRCs] are fundamental to the harnessing and preservation of the world's biodiversity and genetic resources. They are part of the key infrastructure supporting biotechnology, bioprocessing and the development of new approaches in prevention, diagnosis and treatment of disease. They also have a vital role in ensuring the safe, regulated use of organisms that are known pathogens to humans, plants, or animals." Id.
123. Id. at 7.
124. Id. at 6. See also OECD 2004*b*, 2004*c*.
125. Shaffer interview, FAO official, 24 April, 2007.
126. Formally the EC is a member of the WTO, although the more encompassing term EU is typically used in the media. Individual EC members are also formally members of the WTO, but under internal EC rules (article 133 of the EEC Treaty, as amended), only the EC may speak for them on core trade-policy issues.
127. In its words, its aim is "the substantial reduction of tariffs and other barriers to trade and to the elimination of discriminatory treatment in international trade relations." Art. I, Agreement Establishing the WTO. The WTO offers a forum for ongoing negotiations, including through periodic trade negotiating "rounds," the current one being the Doha negotiating round which involves periodic meetings of trade ministers from around the world, such as the one in Hong Kong in December 2005.
128. There are approximately 630 persons in the secretariat, a large proportion of which are translators and support staff. See Shaffer 2005: 430 (citing WTO 2005*b*: 105).
129. Dunoff forthcoming (citing Tutwiler 1991).
130. World Trade Organization, Technical Information on Technical Barriers to Trade, http://www.wto.org/english/tratop_e/tbt_e/tbt_info_e.htm, accessed on June 20, 2006.
131. See Victor 2004.
132. Roberts 1998: 379.
133. The Ministerial Declaration launching the Uruguay Round called for "minimizing the adverse effects that sanitary and phytosanitary regulations and barriers can have on trade in agriculture, taking into account the relevant international agreements." See General Agreement on Tariffs and Trade (GATT) 1986: Part I(D). For an overview of the negotiations, see Croome 1995. Curiously NAFTA's language is based on an earlier more restrictive text, known as the "Dunkel draft," which was subsequently revised, in part in response to NGO complaints. See Steinberg 1997: 245.
134. Annex A of the Agreement provides that SPS "measures include all relevant laws, decrees, regulations, requirements and procedures, including inter alia, end product criteria; processes and production methods, testing, inspection, certificate and approval procedures; . . . and packaging and labeling requirements directly related to food safety."

135. SPS Agreement, art. 3.2.
136. Annex A of the SPS Agreement specifically identifies Codex, the Office International des Epizooties (the International Animal Health Organization, or OIE) and the International Plant Protection Convention (IPPC) as the relevant international standard-setting bodies.
137. SPS Agreement, Art. 3.2.
138. SPS Agreement, Art. 2.2.
139. Article 5.1
140. For good discussions, see Victor 2004; Covelli and Hohots 2003; Howse 2000: 2329.
141. On the notion that science is not purely objective and value-free, see Walker (1998); Wirth (1994: 857–859). See also Busch et al. 2004: 12 ("According to a growing body of social scientific research and expert panel reports, judgment enters into both risk assessment and risk management"). Compare Sykes 2002 with Sunstein 2003a: 294 (supporting the "centrality of science and expertise to the law of risk" and "sharply skeptical of populism") and Sunstein 2003b.
142. But compare Guruswamy (2002: 497) ("the risk assessment procedures under the SPS Agreement are fair and reasonable and are qualitatively superior to the pre-cautionary principle embodied in the Biosafety Protocol"). For a more radical perspective, taking from the social theory of Michel Foucault, see Andrée 2002: 184. Andrée views debates over GMOs in terms of "a biopolitical struggle between those scientists who would frame genetically modified organisms as a manageable risk and those who would have adopted a more precautionary framing," and finding that in practice, "the narrow focus of the risk assessment is not curtailed but entrenched by the renewed precautionary critique." Id.
143. Hudec 2003: 187. In a similar vein, Conrad writes, "It seems surprising, that of all the values listed in Article XX, the contracting parties chose that measures relating to the highest values, namely human health and life, should be viewed under the stricter standards of the SPS Agreement." See Conrad 2006: 29. See also Walker 1998; Howse, 2000; and Sykes 2002.
144. Wirth 1997: 334. See also Driesen 2001: 300 (finding that although "the WTO has not embraced laissez-faire government as an explicit goal, the WTO has taken a substantial step in that direction" by "creating burdens governments must meet in order to impose regulations"); Trebilcock and Soloway 2002: 557; Walker 1998; Howse 2000.
145. Hudec 2003: 188.
146. See Busch et al. 2004: 7.
147. Chang 2004: 771. Cf. Sykes 2002: 355 (concluding that, "Meaningful scientific evidence requirements fundamentally conflict with regulatory sovereignty. WTO law must then choose between an interpretation of scientific evidence requirements that essentially eviscerates them and defers to national judgments about 'science,' or an interpretation that gives them real bite at the expense of the capacity of national regulators to choose the level of risk that they will tolerate").
148. Howse 2000: 2330.
149. Id., at 2335.
150. Roberts and Unnevehr 2005: 480. See generally the important work of Joanne Scott on these issues in Scott 2007 (noting how the committee operates through information provision, peer review, and norm creation, as part of a "dynamic"

legal process, resulting in "a soft law elaboration of hard law obligations"). Id. at 70–72. WTO members also have invoked the good offices of the chair of the SPS Agreement three times. Two of the three referrals resulted in a settlement. Shaffer interview with member of the WTO secretariat, June 12, 2007.

151. Roberts and Unnevehr 2005: 485–86.

152. Id, at 493.

153. SPS Committee 1998.

154. See WT/DS3 and DS41 (Korea on fresh fruit inspection procedures); WT/DS5 (Korea, shelf-life requirements for frozen processed meets and other products); WT/DS21 (Australia, import restrictions on fresh, chilled or frozen salmon); WT/DS26 (EC, ban on imports of hormone-treated meat); and WT/DS76 (Japan, varietal testing requirements for fresh fruits).

155. The EU invoked the precautionary principle as "a general customary rule of international law." See Appellate Body Report, European Communities—Measures Concerning Meat and Meat Products, par. 16, WT/DS26/AB/R, WT/DS48/AB/R (January 16, 1998). Interestingly, the EU did not invoke the precautionary principle as set forth in Article 5.7 of the SPS Agreement, which provides for "provisional measures" adopted on precautionary grounds. See Mavroidis 2003: 235.

156. As for the consistency claim under Article 5.5, the AB created a three-part test and found that the EU had not acted inconsistently with the third component of the test because its treatment was not arbitrary or a disguised restriction on trade. AB Report, par. 236–246.

157. Guillaume Parmentier, quoted in Henley 1999.

158. Request for Consultations by the European Communities, United States—Continued Suspension of Obligations in the EC—Hormones Dispute, WT/DS320/1 (November 10, 2004); and Pruzin 2007b.

159. See the CBD website, at http://www.biodiv.org/world/parties.asp.

160. Article 19.3.

161. The Miami Group included Argentina, Australia, Canada, Chile, the US and Uruguay. Safrin 2002: 614.

162. Shaffer interview with former US negotiator, April 1, 2005.

163. Shaffer interview with former US negotiator, April 1, 2005. For an overview of how the Commission, led by DG Environment, became more proactive over time, with increased member-state support at the end of the 1990s, see Rhinard and Kaeding 2006.

164. Id.

165. Safrin 2002: 613–14.

166. For useful overviews of the Biosafety Protocol and its relation to the SPS Agreement, see Safrin 2002; Winham 2003.

167. *Inside US Trade* 2000. Cf Safrin 2002: 619 ("The fact that this language appears in the preamble does not diminish its effect of stating the parties' intention to preserve rights and obligations deriving from earlier agreements."). The 2005 UNESCO Convention on the Protection and Promotion of the Diversity of Cultural Expressions, which addresses sovereign rights over policies and measures concerning cultural goods and services, provides another example of strategic ambiguity regarding a treaty's legal relationship to WTO agreements. Article 20 of

such convention provides, on the one hand, that, "without subordinating [the] convention to any other treaty," the parties "shall foster mutual supportiveness" with other treaties and "take into account the relevant provisions" of the convention "when interpreting and applying...other treaties" and "when entering into other international obligations." At the same time, article XX provides that it does not "modif[y] rights and obligations...under any other treaties." The convention entered into effect on March 18, 2007.

168. The parties deferred creating a dispute-settlement system until a subsequent Conference of the Parties. See article 34 of the Protocol. Unless the parties otherwise reach agreement on this issue, as Stewart notes, "the [Convention on Biodiversity] dispute-settlement procedures, which apply to its protocols, essentially require parties to first seek solution though negotiation and then mediation." Stewart 2007: 44. In other words, there is no automatic reference of a dispute under the Biosafety Protocol to binding third party dispute settlement.

169. See the CBD website at http://www.biodiv.org/world/parties.asp.

170. United States Department of State 2004.

171. Burton 2004; *International Trade Reporter* 2004b.

172. For details of the second meeting, see the official CBD website at http://www.biodiv. org/biosafety/cop-mop/second-meeting.aspx?menu=mop2. See also International Institute for Sustainable Development 2005; International Centre for Trade and Sustainable Development 2005a; and *European Report* 2005d.

173. See CBD 2006; International Centre for Trade and Sustainable Development 2006; and *International Trade Reporter* 2006.

174. See International Institute for Sustainable Development 2006: 11.

175. See Kalaitzandonakes 2006 (involving samples of only 5 pounds per 25,000 metric ton cargo, and noting the severe impact on costs of the European Commission's draft proposal "for sampling and testing LMOs in bulk commodities"). See also Stewart 2007:33 (noting that GMO producers contend that costs could "increase up to 25% or more").

176. Article 26 of the Protocol provides that parties "may take into account, consistent with their international obligations, socio-economic considerations arising from the impact of [GMOs] on the conservation and sustainable use of biological diversity." The FAO has prepared a preliminary draft International Code of Conduct on Plant Biotechnology that addresses "possible adverse socio-economic effects of biotechnology," but though they have existed for years, they have yet to be adopted. See Glowka 2003: 36–37. In the US view, there is no reason to proceed further with them in view of the lack of consensus. Shaffer interview with USDA official, Rome, April 29, 2007.

177. Regarding liability, the EU has advocated a two-stage approach in which the parties would first negotiate a non-binding instrument, followed by a binding one. International Institute for Sustainable Development 2006: 9.

178. Shaffer interview, April 19, 2007. However, implementing legislation has been blocked in some developing countries where NGOs argue that such legislation could suggest that a country has the capacity to make decisions to authorize LMO varieties. They prefer a blanket ban. Shaffer interview with FAO official, April 20, 2007 (noting debates in Kenya). From a formal analytic perspective, the Biosafety Protocol is a form of "hard" law as its rules are binding on the parties to it, but it is

"softer" than the WTO regime along a hard–soft law continuum, since third-party dispute settlement is not central to its operation.

179. Shaffer interview, April 19, 2007.

180. See Mayr and Soto: 2005: 164. Mayr was Minister of the Environment of Colombia, and served as President of the Extraordinary Session of the CBD in 2000, which led to the Biosafety Protocol.

181. The International Institute for Sustainable Development 2006: 11.

182. Shaffer interview, USDA official, April 19, 2007.

183. Shaffer interview, April 19, 2007.

184. Even when Canada supported strongly by the EU suggested in the June 2006 SPS committee meeting that the CBD make a presentation to the SPS committee in 2007 on the issue of risks (in this case of invasive species of animal origin), the US and Brazil kept it off the agenda out of concern for "overloading" it. See SPS Committee 2006a. See also Scott 2007: 63 (regarding the link with US opposition to granting the Arab League observer status to the WTO General Counsel which triggered a response from Arab countries who have blocked acceptance of all new requests for WTO observer status).

185. The OIE is referred to in the SPS Agreement, although the organization uses both names.

186. See International Committee of the OIE 2005; 2006; and International Centre for Trade and Sustainable Development 2005b.

187. See Codex *Ad Hoc* Intergovernmental Task Force on Foods Derived from Modern Biotechnology 2006: par. 36.

188. Article 1 of the IPPC.

189. Shaffer interview with FAO official, April 25, 2007.

190. Shaffer interview with IPPC official, April 25, 2007. The IPPC has adopted 29 standards to date, with two being adopted at the governing body's session in March 2007.

191. Food and Agriculture Organization 2004c. This followed an IPPC working group report in 2001 calling for such a standard.

192. There was nonetheless significant exchange over the text, including regarding the US preference for the most positive sounding terminology "produced by modern biotechnology" in place of "genetically altered plants." In the end, the term "genetically altered plants" was retained and defined as "plants obtained through the use of modern biotechnology." See APHIS 2003.

193. Shaffer interview, IPPC official, April 24, 2007.

194. Shaffer interview, IPPC official, April 24, 2007.

195. Shaffer interview, IPPC official, April 24, 2007 ("I keep saying that if you don't like it then fix it, but don't create a new one").

196. International Plant Protection Convention, Commission on Phytosanitary Measures Business Plan, 2007–2011.

197. Shaffer interview, IPPC official, April 25, 2007. There was apparently one formal case prior to the 1997 amendments involving imports to Pakistan from the Seychelles of coconut. There appears, however, to be no written record of this process. Shaffer interview with official in FAO legal division, April 25, 2007.

198. Shaffer interview with IPPC official, April 24, 2007.

199. Id.

200. IPPC-SBDS 2006: par. 4.45.
201. Shaffer interviews, FAO April 2007.
202. See CAC website at http://www.codexalimentarius.net; and OECD 2001: 36 (noting low participation rates for low- and middle-income countries). For a critique of Codex in terms of the limited participation of developing countries, see Chimni 2006: 811–818.
203. Shaffer interview with Codex secretariat member, Feb. 13, 2007.
204. Email from FAO secretariat member to Shaffer, Feb. 22, 2007.
205. Codex *Ad Hoc* Intergovernmental Task Force on Foods Derived from Modern Biotechnology 2006 (also posing questions regarding "the direct introduction of nucleic acids into non-germline tissue of animals that will enter the food supply"). There have been seven joint FAO/WHO consultations regarding the safety of foods derived from agricultural biotechnology. The first, in 1990, sought to outline appropriate strategies for assessing the safety of applications of biotechnology to food production and processing. Subsequent consultations addressing narrower topics were held in 1996 (nutritional value of biotech foods), 2000 (GM foods of plant origin), 2001 (allergenicity of GM foods and foods derived from GM microorganisms) and 2003 (foods derived from genetically modified animals). The most recent consultation, held in early 2007, built on the recommendations and conclusions of the 2003 consultation regarding safety of foods derived from transgenic animals. The reports from each consultation, except for 1996, are available from the WHO website at http://www.who.int/foodsafety/biotech/meetings/en/.
206. See Codex Alimentarius Commission 2006*a*: 14–16 (Rules of Procedure, Rule XI; and 47–65 (Guidelines for Codex Committees and *Ad Hoc* Intergovernmental Task Forces).
207. For example, the Codex Committee on Food Labelling has created a Working Group on the Labelling of Foods/Food Ingredients Obtained through Certain Techniques of Genetic Modification/Genetic Engineering, which is jointly chaired by Argentina, Ghana, and Norway, and which held a meeting in Accra, Ghana in January 2008.
208. Boardman 1986: 57–58. The United States, for example, formed a "complex and regularised system of intra-governmental and government-industry concertation" for the purpose of the Codex system. Id., at 70–71.
209. Codex's membership has increased from around 45 in 1962 to 178 in 2007. Actual attendance at the Codex Commission annual meeting rose from 30 in 1962, to 77 in 1991, to 109 in 2006. See the Codex website and Post 2005.
210. Id., at 58, 72.
211. Victor 2004: 886.
212. Victor 2004: 886; Veggeland and Borgen 2005: 676. See also Victor 1997. Prior to Victor's analysis, David Kay published a 1976 study focusing on Codex standards for pesticides and David M. Leive published a 1976 study focusing on Codex commodity standards. See Kay 1976; Leive 1976. However, these figures overstate the overall impact of Codex since, in 1993 just before the WTO's creation, only 12 per cent of the Codex standards had been accepted by all of its members, and most acceptances were "with specific deviations" which permitted a country to apply its existing domestic standards. See Victor 2004: 891. Since 2005, Codex

has discontinued reporting on the adoption of Codex standards, at least in part because most standards reporting is now done through the SPS committee process discussed above.

213. Boardman 1986: 103.

214. Once countries adopted them, however, this "soft" law clearly had real legal effects.

215. Under an "accelerated procedure," three of the steps may be omitted upon agreement of two-thirds of the Commission, although the Commission attempts to work by consensus where possible. See Codex Alimentarius Commission 2006*a*: 19–29 (Procedures for the Elaboration of Codex Standards and Related Texts); and 31–33 (Guidelines on Cooperation between the Codex Alimentarius Commission and International Intergovernmental Organizations in the Elaboration of Standards and Related Texts).

216. Shaffer interview with a Codex secretariat member, Rome, Feb. 13, 2007 (although there have been votes on other matters, one example being the amendment of Codex rules in 2003 to permit a regional economic integration organization, such as the EC, to become a Codex member).

217. Veggeland and Borgen 2005: 684.

218. See e.g., Millstone and van Zwanenberg 2002. Put another way, in the words of an FDA official, Codex was a "backwater," as exciting as "watching the paint dry or the grass grow." Shaffer telephone interview Jan. 29, 2008.

219. Industry participates directly at the international level, as well as through national lobbying. A 2002 evaluation of Codex prepared for the FAO and WHO found that, "[o]f 151 INGOs with observer status in February 2002, around 71 per cent are industry bodies, 22 per cent professional and 8 per cent consumer/public interest." See FAO/WHO 2002: 19.

220. Victor 2004: 899; and Charnovitz 1997: 1786.

221. Quoted in Veggeland and Borgen 2005, at 683.

222. Id., at 689.

223. See Council Decision of 17, Nov. 2003 on the accession of the European Community to the Codex Alimentarius Commission, 2003 O.J. (L 309) 14. See also Poli 2004: 618.

224. Id., Annex III, Agreement between the Council and the Commission regarding preparation for Codex Alimentarius meetings and statements and exercise of voting rights.

225. Id. Annex II, par. 1. Single Declaration by the European Community on the exercise of competence according to Rule VI of the Rules of Procedure of the Codex Alimentarius Commission.

226. See Rule II.5 of the Rules of Procedure of the Codex Alimentarius Commission.

227. Michael Hansen of Consumers Union finds that the guidelines nonetheless can be viewed as a setback for the United States in that they indicate the need for risk assessments steps that go beyond those currently conducted by US authorities. Telephone interview with Gregory Shaffer, Jan. 25, 2008.

228. FAO/WHO 2002: par. 69. See e.g., the guidelines references to FAO/WHO 2001*a* and 2001*b*.

229. FAO/WHO 2002: par. 69 and Box 2.

230. In 2008, Codex was to vote on approval of a Proposed Draft Guideline for the Conduct of Food Safety Assessment of Foods Derived from Recombinant-DNA

Animals, and an annex to the Guidelines for the Conduct of Food Safety Assessment of Foods Derived from Recombinant-DNA Plants regarding genetic modification "for nutritional or health benefits."

231. Shaffer telephone interview with US government official, Jan. 29, 2008.

232. See Codex *Ad Hoc* Intergovernmental Task Force on Foods Derived from Modern Biotechnology 2006: par. 73. See also *Inside US Trade* 2007*b*.

233. The negotiation resulted in a draft Annex to the Guidelines for the Conduct of Food Safety Assessment of Foods Derived from Recombinant-DNA Plants, which would be submitted for approval by the Codex Commission in June 2008. Section 2 of the Annex covers safety assessments guidelines and Section 3 covers information sharing, including of a GM variety's "unique identifier" and "detection method protocols." See Codex Alimentarius Commission 2007; *Inside US Trade* 2007*g*. In addition, as part of the four-year mandate, Codex worked on a Proposed Draft Guideline for the Conduct of Food Safety Assessment of Foods Derived from Recombinant-DNA Animals, and an annex to the Guidelines for the Conduct of Food Safety Assessment of Foods Derived from Recombinant-DNA Plants regarding genetic modification "for nutritional or health benefits."

234. Quoted in Poli 2003: 134. See generally Poli 2004: 619–620; and Veggeland and Borgen 2005: 698.

235. See Codex Alimentarius Commission 2005: 102, par. 11.

236. See Codex Committee on General Principles 2006: 7–8, par. 58–77 (regarding Proposed Draft Working Principles for Risk Analysis). Confirmed in Shaffer interview with Codex secretariat member, Rome, Feb. 13, 2007.

237. Poli 2004: 624.

238. See Codex Alimentarius Commission 2005: 159–160. This contentious discussion was also linked to contention over the EU's ban on beef produced with meat-hormones. Shaffer interview with member of Codex secretariat, Rome, Feb. 13, 2007.

239. Codex Alimentarius Commission 2003: 3–4, par. 16.

240. The Codex Commission first agreed in 1991 that work on labeling of products derived from biotechnology would be undertaken. See Codex Committee on Food Labelling 1997: Appendix VI, par. 1.

241. See Poli 2004: 626–627.

242. See Codex Alimentarius Commission 2003b: 7–8, par. 52.

243. See International Centre for Trade and Sustainable Development 2005c.

244. Shaffer interview with Codex secretariat member, Rome, Feb. 13, 2006, and USDA representative, Rome, April 19, 2007 (noting that it had been agreed in the FAO executive council that work should be stopped on matters if there is no prospect of consensus after work over a long period of time). See also *Inside US Trade* 2008*e*. Codex Committee on Food Labelling 2006: 11–12, par. 89. A working group, however, continued to study the issue in 2007, in respect of a Draft Amendment to the General Standard for the Labelling of Prepackaged Foods: (Draft Recommendations for the Labelling of Foods Obtained through Certain Techniques of Genetic Modification/Genetic Engineering): Definitions; and Proposed Draft Guidelines for the Labelling of Foods and Food Ingredients Obtained through Certain Techniques of Genetic Modification/Genetic Engineering: Labelling Provisions." The working group, among other issues, was to address "[c]onsideration of the

rationale for Members' approach to the labeling of food and food ingredients obtained through certain techniques of genetic modification/genetic engineering." Id.

245. FAO/WHO 2002: par. 72.
246. See Poli 2003: 627–629.
247. Codex Alimentarius Commission 2003a: par. 21, fn. 9.
248. See Codex Alimentarius Commission 2006b: par. 9, fn. 7.
249. Poli 2003: 146–47.
250. Victor 2004: 933.
251. Veggeland and Borgen 2005: 698.
252. Victor 2004: 931.
253. Shaffer telephone interview, Jan. 25, 2008. Hansen finds, however, that the recording of minutes of Codex meetings can now be less transparent to the outside world, including for researchers. The minutes, he says, used to consist of transcripts, but they now are edited, leading to a risk of manipulation.
254. FAO/WHO 2002: par. 82.
255. Scott 2007: 68–69.
256. See Scott 2007: 68 (noting that this has led "in at least three cases to the adoption of new or revised standards," citing SPS Committee 2005: 3, par. 11, and SPS Committee 2001: 2, par. 8, and 3, par. 10.
257. See SPS Committee 2006b.
258. See Panel Report, European Communities—Measures Affecting the Approval and Marketing of Biotech Products, WT/DS291/R (Sept. 29, 2006), par. 7.3240 (referring to paragraph 25 of the Working Principles for Risk Analysis for Application in the Framework of the Codex Alimentarius)..
259. Huller and Maier 2006: 272–77. In this way, the WTO can be seen as "breathing new life into the Codex" and other standard-setting organizations. Shaffer interview with official of the WTO secretariat, June 12, 2007.
260. Quoted in Dunoff 2006: 16.
261. The observed distributive conflict between the US and the EU, we hasten to add, is not inherent to the subject of agricultural biotechnology, but reflects the prior regulatory polarization between the US and the EU, as well as the intense politicization of the issue and the entrenchment of the two regulatory systems over the past several decades. These factors could change over time, with the US becoming more precautionary and/or the EU becoming more interested in developing agricultural biotechnology within Europe. For example, if China, India or Brazil were to develop new biotech varieties that had not been tested in the US, then US regulators could be wary of relying on information from enterprises in those countries and more cautious in authorizing their use in agricultural production in the US or their sale to US consumers. (This point was raised by two members of the European Commission in separate interviews, Shaffer with Nolan in April 2006; and Pollack with Tony Van Der Haegen, March 9, 2005. Cf. International Trade Reporter 2007: 983, noting problems with Chinese food and consumer products and US politicians calling for an "import czar"). Similarly, were European consumer demand for GM foods to rise because new biotech varieties offer nutritional benefits or result in a significant reduction in price, and were European farmers to be placed at a competitive disadvantage as a result,

then European farmers would likely be mobilized to demand authorization for them to cultivate GM varieties, such that EU negotiating positions could correspondingly change.

These changes in US and/or EU attitudes, however, would be explained by factors exogenous to the regimes we discuss. The conflict would simply disappear because of the effect of these external factors, such as new technological developments or new foreign rivalry and accompanying regulatory concerns. *Exogenous* factors would, in that instance, have changed the nature of the game, from a distributive Battle of the Sexes to a pure coordination game. Deliberation theory, in contrast, provides an explanation for *endogenous* change, as US and EU officials engage in ongoing deliberation and learning, leading to the redefinition and convergence of national interests. For example, to the extent that conflicts like that over GMO regulation simply reflect differing scientific understandings, or different approaches to risks, international regimes can spur convergence if they can either (1) generate scientific knowledge beyond what any one party would otherwise generate on its own, whether because of economies of scale, the complementarity of expertise or otherwise; or (2) generate new knowledge that can change a party's approach to risk. While we do not rule out such an outcome categorically, we note once again the significant obstacles to genuine deliberation in the area of agricultural biotechnology, and the paucity of evidence of convergence between the US and EU positions—a point to which we return in Chapter 6. We thank Jeffrey Dunoff for his comments and expression of this point from which we have borrowed.

262. Benvenisti and Downs 2007.

263. For example, even before the WTO panel decision, an attorney-adviser at the Office of the Legal Adviser in the U.S. Department of State who helped to negotiate the "savings clause" in the Biosafety Protocol wrote, "I would anticipate, therefore, that with respect to parties to both the WTO Agreements and the Biosafety Protocol, the WTO Appellate Body would accord the requirements of the Protocol significant respect." Safrin 2002: 624 (although she limits her statement to cases between parties to both regimes, and she also maintains that it "would seem constitutionally unlikely to jettison WTO disciplines, such as the requirements under the SPS Agreement that sanitary and phytosanitary measures be based on scientific principles"). Id., at 625. As we will see, the panel found a way to avoid even having to get to the issue of whether the EU based its decisions on risk assessments.

264. For example, in the shrimp-turtle case, the Appellate Body found that the "term 'natural resources' in article XX(g) [of the GATT] is not 'static' in its content or reference but is rather 'by definition, evolutionary'." Appellate Body Report, United States—Import Prohibition of Certain Shrimp and Shrimp Products, par. 130, WT/DS58/AB/R (October 12, 1998). Similarly, under a GATT article XX(b) balancing test, the existence of the Protocol could be a factor weighing in favor of the EU. See e.g., Marceau 2001: 1098.

265. See also Nichols 1996: 464–465 ("If the World Trade Organization were to rule, for example, that the Convention on International Trade in Endangered Species or some other equally popular agreement violated the provisions of the trade agreements, popular acceptance of the World Trade Organization would probably decline").

Chapter 5

1. Alter 2003.
2. See DeYoung 2003. In the words of one administration official, "There is no point in testing Europeans on food while they are being tested on Iraq." Quoted in Becker 2003: 6.
3. Although the meat hormones dispute dated back to the 1980s and gave rise to earlier complaints under the GATT system, as well as bilateral and multilateral negotiations (including within the Codex Alimentarius Commission as noted in Chapter 4), the US was extremely quick in bringing a complaint before the new WTO dispute-settlement system after it was created.
4. See e.g., *Economist* 2003*b*; Rugaber 2003.
5. Jasanoff 2005: 109.
6. Congress granted the Bush administration trade promotion authority by one vote, subject to numerous conditions, including an expiration date of June 1, 2005, that was extended automatically until June 1, 2007, because under its terms, neither Congressional chamber adopted a resolution opposing extension. See *Trade Act of 2002*, Public Law Number 107–210, § 2103, 116 Statute 933 (August 6, 2002).
7. See the annual reports of the International Service for the Acquisition of Agri-biotech Applications (ISAAA), including James 2005, 2006.
8. Economist 2006: 66 (also noting that "On one estimate, GMOs made up more than half the world's soya crop by area, a quarter of its corn and over a tenth of its cotton").
9. See Table 7.1 in Chapter 7.
10. See Buck 2003.
11. Quoted in *Inside US Trade 2003*: 15.
12. See *Inside US Trade* 2004*a*. For a preliminary analysis of such a claim, see Scott 2007: 233–242.
13. See Checkagbiotech.org 2001 (quoting a USDA official and Neil Harl, an agricultural economist at Iowa State University). Similarly, it was estimated that Argentina would lose $1 billion a year in farm exports to the European Union. See Reuters 2003. These articles were forwarded to us from a representative of the American Soybean Association on April 14, 2004.
14. *Inside US Trade* 2004*a* (noting that the American Soybean Association is taking the lead in hiring private lawyers to prepare the background for such a WTO challenge).
15. Argentina's written submission was not made publicly available. The EU's public submission, however, responded to arguments made in the submissions of all three complainants. Canada and Argentina set forth their TBT claims either as "cumulative" or "alternative" claims. The US also made claims under GATT article III.4.
16. This substantive coverage is codified in Article 1.5 of the TBT Agreement which provides that "the provisions of this Agreement do not apply to sanitary and phytosanitary measures as defined in Annex A of the [SPS Agreement]."
17. See also Howse and Mavroidis 2000; Scott 2003.
18. The US pointed to twenty-eight product-specific moratoria. It claimed that in fourteen of them, the EU "has not put forth any risk assessments whatsoever." In the remaining fourteen, where the EU undertook risk assessments, the US stated that "the product-specific moratoria are not based on these assessments," as the

"scientific assessments...concluded that there was no evidence that these biotech products would pose a risk to human, animal or plant life or health or cause other damage." First Written Submission of the US, *EC-Biotech*, ¶¶ 47, 143, 145 (April 21, 2004).

19. In addition, Canada and Argentina noted the differential treatment of "biotech products that were approved for marketing prior to the imposition of the general moratorium, and novel non-biotech products such as those produced by conventional plant breeding techniques." Panel Report, *European Communities—Measures Affecting the Approval and Marketing of Biotech Products*, WT/DS291/R (Sept. 29, 2006) (hereinafter *EC-Biotech*), ¶ 7.1410.

20. Shaffer telephone interview with private US attorney following the case. June 5, 2007.

21. For EU concerns that the Bioterrorism Act does not respect US obligations under the SPS Agreement, including because its provisions have not been based on a risk assessment, see Commission of the European Communities 2002*b*; Yerkey 2004. The US notified the Bioterrorism Act to the WTO's SPS Committee, and WTO members have posed questions to the US in the Committee regarding the act's implementation.

22. Three of these measures were adopted by Austria (biotech corn products), two by France (oilseed rape), and one each by Germany (corn), Greece (oilseed rape), Italy (corn), and Luxembourg (corn).

23. First Written Submission of the US, *EC-Biotech*, ¶¶, at par. 170.

24. Id., at par. 174.

25. Emphasis added. Paragraph 3 of article 3 provides: "Members may introduce or maintain sanitary or phytosanitary measures which result in a higher level of sanitary or phytosanitary protection than would be achieved by measures based on the relevant international standards, guidelines or recommendations, if there is a scientific justification, or as a consequence of the level of sanitary or phytosanitary protection a Member determines to be appropriate in accordance with the relevant provisions of paragraphs 1 through 8 of Article 5."

26. See Appellate Body report, *European Communities—Import Restrictions on Meat and Meat Products (Hormones)* WTO Doc WT/DS26/AG/R, at par. 165. See discussion of this decision in Mavroidis 2003.

27. Scott 2004: 327, 333. But compare the *EC-Sardines* case in which the Appellate Body found the EU to be in violation of the TBT Agreement because the EU did not base its internal technical regulations on an international standard of the Codex Alimentarius Commission, and failed to demonstrate that this international standard would not be "effective" or "appropriate" in fulfilling the EU's "legitimate objectives" of ensuring "market transparency, consumer protection, and fair competition." *European Communities–Trade Description of Sardines*, Report of the Appellate Body, WT/DS231/AB/R (Oct. 23, 2002), paras. 259–291.

28. Victor 2004: 936.

29. The panel noted that "the residue level of hormones in some natural products (such as eggs and broccoli) is higher than the residue level of hormones administered for growth promotion and in treated meat." It found that the imposition of a "'no residue' level of protection against natural and synthetic hormones used for growth promotion," compared to an "'unlimited-residue' level of protection with regard to hormones occurring naturally in meat and other foods," constituted arbitrary and

unjustifiable discrimination. The quotations in this and the succeeding paragraph are taken from Appellate Body report, *European Communities—Import Restrictions on Meat and Meat Products (Hormones)* WT/DS26/AG/R, at paras. 210–246. However, in the *Australia-Salmon* case, the Appellate Body did find that Australia violated article 5.5 of the SPS Agreement when Australia imposed more restrictive measures for wild salmon than for other fish which posed at least as high if not higher degree of risk, such as herring used as bait and ornamental finfish. See *Australia–Measures Affecting Importation of Salmon*, Appellate Body Report, WT/DS18/AB/R (Oct. 20, 1998).

30. The panel noted that these two agents "are used for growth promotion in the pork meat sectors where the European Communities has no domestic surpluses and where international competitiveness is a higher priority," unlike the bovine sector where the EU has large surpluses and lacks international competitiveness. It held that this differential treatment again constituted arbitrary and unjustifiable discrimination.

31. Cho maintains that, in this decision, the AB rejected conventional laboratory science for a populist, "common-sense" based science. See Cho 2007: 23. Contrast, however, the Appellate Body's approach in the *Australia-Salmon* case, in which it compared Australia's "high" level of protection of uncooked ocean-caught Pacific salmon with its "definitely lower" standards for herring used as bait and live ornamental finfish. Appellate Body Report, par. 146. As Scott notes, the AB approach in *Salmon* "was predominantly, perhaps exclusively an objective one," while its approach in *Hormones* "seems somewhat more focused upon the subjective intent of the Member." Scott 2007: 154.

32. See Appellate Body report, *European Communities—Import Restrictions on Meat and Meat Products (Hormones)* WTO Doc WT/DS26/AG/R, at par. 253.

33. See Appellate Body report, *European Communities—Measures Affecting Asbestos and Asbestos-Containing Products*, WTO Doc WT/DS135/AB/R of March 12, 2001, especially at par. 122. The Appellate Body confirmed in *EC-Meat hormones* that members may enact measures so as to reduce a risk to zero where they have conducted an appropriate risk assessment, and provided that the risk is not merely a "theoretical" one.

34. See *Japan: Measures Affecting Agricultural Products, Report of the Appellate Body*, WT/DS76/AB/R (Feb. 22, 1999), at par. 84. Similarly, in the *EC-Meat hormones* case, the Appellate Body stated that: "The requirement that an SPS measure be 'based on' a risk assessment is a substantive requirement that there be a rational relationship between the measure and the risk assessment." *Hormones*, Report of the Appellate Body, at par. 163. The AB further maintained that "determination of the presence or absence of that relationship can only be done on a case-by-case basis, after account is taken of all considerations rationally bearing upon the issue of potential adverse health effects." Id.

35. See Scott 2007: 79; Guzman 2004 (arguing for procedural and not substantive review).

36. Appellate Body Report, *Australia—Salmon*, at ¶ 125.

37. Sykes 2002: 363–64 (asking what the EU actually could have done to show a risk, and concluding that "the Appellate Body's insistence [in *EC-Hormones*] that Europe point to highly particularized studies showing a risk from hormone residues in

meat likely presents an insurmountable hurdle. The effect is to make it impossible for national regulators to elect to eliminate low-level risks that are not susceptible to rigorous demonstration").

38. See Panel Report, *EC-Biotech*. The panel issued an interim decision to the parties on Feb. 7, 2006, which was leaked on the web. The final decision differed slightly from the interim decision. The panel found in the interim decision, for example, that the EU general moratorium had terminated, but withheld judgment on this issue in its final decision.

39. Panel Report, *EC-Biotech* par. 8.3.

40. Panel Report, *EC-Biotech,* par. 8.6.

41. This "decision not to decide" interestingly lay at the heart of the complainants' claims against the EU. Even the former EU Environmental Commissioner Margot Wallstrom had called the "moratorium" a "situation where we just simply decline to take a decision." Panel Report, *EC-Biotech*, par. 7.538. The quote, "decision not to decide" is included in the panel report from Canada's third written submission, paras. 202, 203 and 204 and Canada's replies to Panel question Nos. 172 & 179. Panel Report, *EC-Biotech*, par. 7.455, fn. 568.

42. General Agreement on Tariffs and Trade, Art. III, ¶ 4, Oct. 30, 1947, 61 Stat. A-11, T.I.A.S. 1700, 55 U.N.T.S. 194, provides: "The products of the territory of any contracting party imported into the territory of any other contracting party shall be accorded treatment no less favourable than that accorded to like products of national origin in respect of all laws, regulations and requirements affecting their internal sale, offering for sale, purchase, transportation, distribution or use."

43. Panel Report, *EC-Biotech*, ¶ 7.2514. See also the panel's rejection of Argentina's claim under the second clause of Annex C(1)(a), which provides that members shall ensure that "any procedure to check and ensure the fulfillment of sanitary or phytosanitary measures…are undertaken and completed…in no less favourable manner for imported products than for like products." The panel found that "it is not self-evident that the alleged less favourable manner of processing applications concerning the relevant imported biotech products (e.g., imported biotech maize) is explained by the foreign origin of these products rather than, for instance, a perceived difference between biotech products and novel non-biotech products in terms of the required care in their safety assessment, risk for the consumer, etc." Id. at ¶ 7.2411. In both cases, Argentina had failed to provide specific factual evidence and analysis in this respect. Id. at ¶¶ 7.2411, 7.2421, 7.2513 and 7.2157.

44. Agreement on Technical Barriers to Trade, Art. 2.2, Apr. 15, 1994, Marrakesh Agreement Establishing the World Trade Organization, Annex 1A.

45. Panel Report, *EC-Biotech*, at ¶ 7.150.

46. See Id. at ¶ 7.153;

47. See Id. at ¶¶ 7.162–7.170.

48. We calculate that the panel cited dictionaries fifty-nine times, involving the meaning of forty-two words.

49. See, e.g.,, Panel Report, *EC-Biotech*, ¶¶ 7.176–7.184.

50. See, e.g., Id. at ¶ 7.368. The EU cited concerns over "carbon and nitrogen recycling through changes in soil decomposition of organic material" as an important example. Id.

51. See, e.g., Panel Report, *EC-Biotech*, at ¶¶ 7.219 & 7.225–226 (the panel noted that footnote 4 of the SPS Agreement provides that the term "animal" includes "wild fauna," and that the term "plant" includes "wild fora"); ¶¶ 7.241–7.242 (the panel cited International Standard for Phytosanitary Measures (ISPM) no. 11, developed by the IPPC, as a relevant standard); and ¶¶ 7.300 and 7.314. The panel also stated, "to the extent that GMOs might cause damage to (as opposed to mere changes in) geochemical cycles, such that there would be damage to the environment other than damage to living organisms, we think such environmental damage could be considered as 'other damage' from the entry, establishment or spread of GMOs qua 'pests' within the meaning of Annex A(1)(d)." *Id.* at ¶ 3.374.

52. See Id. at ¶¶ 7.285–7.286; ¶¶ 7.343–7.344; ¶¶ 7.361–362; ¶¶ 7.379–380.

53. Id. at ¶¶ 7.2209–7.2218.

54. See Conrad 2006.

55. See Panel Report, *EC-Biotech*, at ¶¶ 7.2517 (re application of GATT III.4) and ¶¶ 7.2524 & 7.2528 (re application of the TBT Agreement).

56. Scott 2007: 17.

57. Pointing to a dictionary definition, the panel found that "the concept of a moratorium on approvals implies that the absence of approvals must be the consequence of a deliberate temporary suspension of approvals." Panel Report, *EC-Biotech*, at ¶ 7.534.

58. Id. at ¶ 7.474–7.483. Overall, the panel used the term "Group of Five" 401 times in the report.

59. This part of the panel's opinion reviewed the factual evidence regarding the approval process for each variety and alone comprised almost two hundred pages, complemented by a 54-page table attached as Annex B which summarized "the history of the individual approval procedures."

60. Panel Report, *EC-Biotech*, at ¶ 7.1382. The panel found that the words "the application" of requirements and procedures is not listed in Annex A, and thus such application is not included in the definition of the "nature" of an SPS measure. ¶ 7.1335 ("the application of such requirements and procedures would not, itself, meet the definition of an SPS measure"). See also ¶ 7.1697 ("while 'procedures, as such may according to the Annex A(1) definition constitute SPS measures, the application, or operation, of such procedures does not, itself, constitute an SPS measure within the meaning of Annex A(1)"). The panel stated that the moratoria constituted challengeable "measures" under the WTO agreements, but "all measures are not SPS measures." ¶¶ 7.1295 and 7.1333.

61. Id., at ¶ 7.1379. The panel noted that as the complainants did not challenge the underlying EU legislation, with its requirement of a pre-marketing approval, such legislation must be presumed to be WTO consistent. Since such approval by definition leads to a "provisional ban," "logic dictates that if the pre-marketing approval requirement must be presumed to be WTO-consistent, the same holds true for the provisional marketing ban.... The decision to delay final approval decisions merely had the effect of extending the duration of the provisional ban on the marketing of all non-approved biotech products." ¶ 7.1353 and 7.1357.

62. Panel Report, *EC-Biotech*, at ¶ 8.6.

63. It thus appears that the only EU acts reviewable under 5.1, in the panel's view, were "the pre-marketing approval requirement which results in a provisional marketing

ban" (i.e, the EU legislation itself) and any "final substantive approval decisions on individual applications." Id. at ¶¶ 7.1390–1391.

64. Article 8 provides that "Members shall observe the provisions of Annex C." The panel, however, found that the US failed to establish its claims under Annex C(1)(b) in respect of the moratoria.

65. The panel made similar findings regarding the claims against product-specific moratoria involving twenty-seven GM varieties, maintaining that the EU had engaged in "undue delay" in the approval process for twenty-four of them. The US initially listed forty-one applications in its request for the establishment of a panel, but in its first written submission only indicated twenty-five about which it was making claims. Canada identified four applications, two of which did not overlap with the US. Argentina indicated eleven applications, one of which did not overlap with the US, but was not examined by the panel because the applicant had withdrawn its application prior to the panel's establishment. Panel Report, *EC-Biotech*, ¶¶ 7.1638–7.1646. The reasons for the undue delay for different varieties included the "unjustifiably long" period of time for the Commission to convene a regulatory committee meeting or to forward a draft measure to the Council, and the "unjustifiably long" amount of time taken by the lead member state authority for its assessment of the application. See *id*. at ¶ 7.2391 (containing a chart indicating which varieties encountered undue delay in their approval). Of the twenty-four cases in which the panel found undue delay, three were on account of the Commission failing to call a meeting to approve the varieties, seven on account of the Commission failing to forward a draft decision to the regulatory committee, and fourteen on account of delay of the lead authority at the member-state level in respect of an application. In five of these latter cases, the lead member- state authority was Spain, in five cases it was the Netherlands, in two cases it was Belgium, and in two cases it was France. In the case of France, the government had initially approved the variety, but then changed its views and did not take action after the Commission approved the variety.

66. Panel Report, *EC-Biotech*, at ¶¶ 7.1517 & 7.1526.

67. In November 2003, the Commission proposed to approve the importation of a variety of GM maize (Bt-11 sweet corn), for which EFSA had delivered a favorable opinion. The EU regulatory committee again refused to approve the Commission's proposal so that the matter was referred to the Council, which was given until the end of April to act. On April 26th, a divided Agriculture Council failed to reach agreement on the Commission's proposal. In the absence of a decision by the Council, the Commission adopted its proposal. Commission Decision 2004/657/EC, 2004 O.J. (L 300) 48. Under the circumstances, Syngenta, the crop's manufacturer, indicated that it had no immediate intention of marketing Bt-11 sweet corn in Europe. See *European Report* 2004a and the discussion in Chapter 6, this volume.

68. The panel issued an interim decision to the parties on Feb. 7, 2006, which was leaked on the web. In the "interim decision," the panel held that the moratorium had ended and then added this footnote: "In view of its terms of reference, the Panel cannot, and does not, express a view on whether notwithstanding the approval of a biotech product which was subject to the general *de facto* moratorium in effect at the time of establishment of the Panel, an amended *de facto* moratorium continues to exist or whether a new general *de facto* moratorium has since been imposed." Interim Reports of the Panel, *EC-Biotech*, fn. 1,962, WT/DS291–293/INTERIM (Feb. 7,

2006), available at http://www.saveourseeds.org/downloads/WTO_conclusion_070206.pdf.

69. Panel Report, *EC-Biotech*, at ¶ 8.16.
70. Shaffer telephone interview with private US attorney following the case, June 5, 2007.
71. In the summer of 2007, the EU first considered approving a GM variety for cultivation, but no decision had been made as of late 2008. See Chapter 6, this volume.
72. See Panel Report, *EC-Biotech*, at ¶¶ 7.3412–7.3414. In one case, the panel stretched its analysis particularly far. In response to documentary evidence that one reason for the Austrian safeguard was the lack of an adequate EU labeling regime, the panel recalled its earlier finding that labeling regimes can have SPS and non-SPS objectives. The panel concluded that Austria's labeling objective "reflects a concern about risks to consumer health," and thus does not reflect a TBT-objective such as a consumer's right not be misled about the nature of the product. As a result, the panel avoided examining the Austrian safeguard measure under the TBT Agreement, which not only could have added hundreds of pages to its report, but also had institutional implications for the reasons we examined earlier. See, e.g., ¶¶ 7.2646–7.2651. The panel noted that the Austrian safeguard was enacted pursuant to the EU deliberate release directive which the panel found reflected SPS objectives. As it was, this section of the report comprised 200 pages.
73. For an excellent discussion of the panel's handling of the relation between Articles 2.2, 5.1, and 5.7 of the SPS agreement, see Broude 2007. Broude views Articles 5.1 and 5.7 as being applications of the general SPS Agreement obligation under article 2.2 to "two distinct situations—one, where there exists scientific evidence sufficient to establish an SPS measure on risk assessment; the second where scientific evidence is insufficient for such purpose." Broude 2007: 23. He finds the panel's discussion of a "qualified right" under article 5.7 unnecessarily confusing.
74. See Panel Report, *EC-Biotech*, at ¶¶ 7.3000 & 7.3004. By contrast, if article 5.7 was an exception, then the respondent should have the burden of proof to establish an affirmative defense.
75. See id. at ¶ 7.3006.
76. See id. at ¶ 7.3040. In total, the panel referred to the definition of a risk assessment elaborated by the Appellate Body in the *Australia—Salmon* case twenty-four times.
77. The panel recalled, in this respect, the Appellate Body's finding that "it is not sufficient that a risk assessment conclude that there is [only] a possibility" of the risk at issue. Id. at ¶ 7.3045. Commentators question the panel's factual findings. See, e.g., Scott 2007: 93, 108, 118 (concerning the panel's rejection of the Hoppichler study cited by Austria as a risk assessment); Gruszczynski 2008; and Perez 2007.
78. Panel Report, *EC-Biotech*, at ¶ 7.3217.
79. The four requirements that a respondent must meet in order for article 5.7 to apply are as follows: (i) the key threshold that "relevant scientific evidence [must be] insufficient;" (ii) the measure must be adopted "on the basis of available pertinent information;" (iii) the member invoking it must "seek to obtain the additional information necessary for a more objective assessment of risk;" and (iv) such member must "review the measure accordingly within a reasonable period of time." Panel Report, *EC–Biotech*, at ¶¶ 7.2929 & 7.3218 (citing *Japan–Agricultural Products*

II, at ¶ 89, and Appellate Body Report, *Japan–Measures Affecting the Importation of Apples*, ¶ 76, WT/DS245/AB/R (Nov. 26, 2003) [hereinafter *Japan–Apples*]).

80. Panel Report, *EC–Biotech*, at ¶ 7.3226.
81. Id. at ¶ 7.3238. The Appellate Body, however, subsequently overruled this particular legal position in the 2008 case *United States—Continued Suspension of Obligations in the EC–Hormones Dispute*. There the Appellate Body reversed "the Panel's finding that 'the determination of whether scientific evidence is sufficient to assess the existence and magnitude of a risk must be disconnected from the intended level of protection'." See Appellate Body Report, *United States—Continued Suspension of Obligations in the EC–Hormones Dispute*, WT/DS320/AB/R (Oct. 16, 2008), ¶¶ 684-686 and 736. We thus do not know how this would have affected the biotech decision had the panel applied such an analysis.
82. Id. at ¶ 7.3243.
83. Id. at ¶ 8.9. In addition, the panel was aided by earlier Appellate Body jurisprudence which found "that insufficiency of scientific evidence itself is not to be equated with scientific uncertainty." See discussion in Scott 2007: 116 (citing *Japan–Apples*, at ¶ 184).
84. See, e.g., Panel Report, *EC–Biotech*, at ¶ 7.3260
85. Panel Report, *EC–Biotech*, ¶ 7.3399. The panel, however, exercised "judicial economy" as regards Canada's and Argentina's claims under SPS articles 2.3, 5.5, and 5.6 and GATT article III.4, as well as all of the complainants claims under GATT article XI regarding the Greek safeguard, seeing "no need to examine and offer additional findings" on them. See id. at ¶¶ 7.3378, 7.3384, 7,3405, 7.3423 & 7.3429.
86. Id. at ¶¶ 7.3065.
87. Id.
88. See id. at ¶¶ 7.3244–7.3245.
89. See Annex K, Letter of the Panel to the Parties of May 8, 2006, WT/DS291/R/Add.9, WT/DS292/R/Add.9, WT/DS293/R/Add.9 (Sept. 29, 2006) (regarding its response to the breach of confidentiality).
90. See Commission of the European Union 2006*f* (in which the Commission "invite[s] EFSA to liaise more fully with national scientific bodies, with a view to resolving possible diverging scientific opinions with Member States" and it notes that "applicants and EFSA will also be asked to address more explicitly potential long-term effects and bio-diversity issues in their risk assessments for the placing on the market of GMOs"). We thank Sara Poli for pointing this out.
91. Panel Report, *EC–Biotech*, at ¶ 7.73–7.75 (Biosafety Protocol) and 7.76–7.89 (precautionary principle).
92. Panel Report, *EC–Biotech*, at ¶¶ 7.67–7.71.
93. Argentina and Canada have signed the Biosafety Protocol but not ratified it, while the US has not signed it. Argentina and Canada have signed and ratified the underlying Convention on Biodiversity, while the US has signed but not ratified it.
94. Panel Report, *EC–Biotech*, at ¶ 7.72.
95. Id. at ¶¶ 7.92–7.95.
96. Id. at ¶ 7.95.
97. Id. at ¶¶ 7.86–7.89.

98. Id. at ¶ 7.10. The professors were Lawrence Busch (Michigan State University), Robin Grove-White (Lancaster University), Sheila Jasanoff (Harvard University), David Winickoff (Harvard University), and Brian Wynne (Lancaster University); see Busch et al. 2004. The professors also wrote an article concerning the biotech case, and the role of judicial review of science in the WTO; see Winikoff et al. 2005.

99. Panel Report, *EC-Biotech*, at ¶ 7.11.

100. The term "judiocentric" is borrowed from Victoria Nourse, writing in respect of analogous questions concerning the analysis of questions of federalism and separation of powers under US constitutional law. See Nourse 2004: 835, 837, 856. ("I reject the judicocentric position that the separation of powers and federalism require recourse to descriptive texts or functions, and argue, instead, that our government *is*, in important structural senses, a set of popular relations....If we move a decision from Congress to the Court we have not only moved an activity, we have moved a decisionmaker (a decisionmaker whose incentives are governed by a particular relation to the people")).

101. See Komesar 1995 and Komesar 2002 (developing this approach in the US legal context); and Maduro 1998 (applying this approach to assess judicial choices in the context of the EU treaty provisions governing the EU's internal market). For earlier applications of this approach to WTO Appellate Body decisions, as in the *US–Shrimp-turtle* and *EC–GSP* cases, see Shaffer 2005*b*; and Shaffer and Apea 2005. For a powerful complementary approach assessing the role of judicialization in politics, see Stone Sweet 1999.

102. Dunoff 2006.

103. Petersmann 2000.

104. Walker 2003: 4 ("Constitutional pluralism...is a position which holds that states are no longer the sole locus of constitutional authority, but are now joined by other sites, or putative sites of constitutional authority, most prominently...and most relevantly...those situated at the supra-state level, and that the relationship between state and non-state sites is better viewed as heterarchical rather than hierarchical"); see also Walker 2001, 2002.

105. Trachtman 2006*a* (addressing different ways to approach the issue of WTO constitutionalism, including institutional ones), 2006*b*.

106. Jackson 1998.

107. Hudec and Farber 1994; Jackson 1991; McGinnis and Movsesian 2000.

108. In fact, Joel Trachtman, from his institutionalist constitutional perspective, explicitly notes this connection when he writes, "[t]he task of framers of constitutions, and of analysts, is to engage in comparative institutional analysis." Trachtman 2006*a*: 633.

109. On legal pluralism, for excellent overviews, see Griffiths 1986, and Berman 2007.

110. See Joerges 2006*a*; Joerges 2007; and Joerges and Neyer 2003. Joerges contends that his vision of conflicts of law in the WTO serves a constitutional function, but a very different one than a hierarchical version in which WTO law trumps. See also Fisher-Lescano and Teubner 2004 (calling for a conflicts approach involving mediation between sectoral regimes and in which transnational substantive norms are created). This approach, although it has some overlap in its concerns, should be distinguished from the literature on regime complexes and legal frag-

mentation, discussed in Chapter 4, which focuses on potential conflicts of law among international regimes as opposed to national ones.

111. Joerges 2006*b*.

112. Joerges and Neyer 2003: 224.

113. As Joerges convincingly argues, "[y]et, a meta-norm, referring to scientific knowledge as peacemaker, is not that innocent—actors involved know this quite well. Three reasons might suffice to illustrate this point: first, science typically provides no clear answers to questions posed by politicians and lawyers; second, it cannot resolve ethical and normative controversies about numerous technologies; third, consumer *angst* might be so significant that neither policy-makers nor the economy dare to ignore it, although scientific experts might assess a risk as tolerable or even marginal." Joerges 2006*b*: 11.

114. Kingsbury et al. 2005.

115. Kingsbury et al. 2005: 17.

116. The authors "define global administrative law as comprising the structures, procedures and normative standards for regulatory decision-making including transparency, participation, and review, and the rule-governed mechanisms for implementing these standards." Kingsbury et al. 2005: 17

117. Krisch 2006: 266.

118. Keohane (2003), for example, has categorized accountability mechanisms into seven types, which he terms hierarchical, legal, market, reputational, fiscal, supervisory, and participatory. See also Grant and Keohane 2005.

119. See Broude 2007; Perez 2007.

120. See e.g., Howse and Regan 2000.

121. Dunoff 1999: 756 (proposing new procedural mechanisms whereby WTO dispute-settlement panels would avoid controversial trade-environment cases on standing, ripeness, political question and related grounds, thereby permitting domestic trade restrictions imposed on environmental grounds to remain unchallenged before the WTO).

122. Nichols 1996 (proposing the creation of "an exception that would allow certain laws or actions to exist if they violate the rules of the World Trade Organization," provided that "the impediment to trade must be incidental," and the measure must be "undertaken for the purpose of reflecting an underlying societal value").

123. See e.g., Hamilton 1961: 107 ("Upon the principle that a man is more attached to his family than to his neighborhood, to his neighborhood than to the community at large, the people of each State would be apt to feel a stronger byass [sic] towards their local governments than towards the government of the Union").

124. See e.g., Tullock 1969 and Williamson 1967.

125. Olson 1965; McGinnis and Movsiean 2000.

126. See Appellate Body Report, *Japan–Agricultural Products II*, at ¶ 84; Appellate Body Report, *EC–Hormones*, at ¶ 163.

127. This comment of states having "the right to be irrational" was, in fact, made by a deliberative theorist at a conference attended by Shaffer in February 2007.

128. See e.g., FAO 2004*a*; Fukuda-Parr 2007*a*; Nuffield Council on Bioethics 2003; and Pinstrup-Andersen and Schioler 2000.

129. The 2003–04 FAO report notes, "some of these crops, especially insect-resistant, are yielding significant economic gains to small farmers as well as important social and environmental benefits through the changing use of agricultural chemicals." FAO 2004a: 6. It later continues, "[a]n expensive, unpredictable and opaque biosafety regime is even more restrictive for public research than private research, because public institutions have considerably less money to finance the research trials required to meet regulatory requirements." FAO 2004a: 88.

130. *Inside US Trade* 2007d. Moreover, unless there are strong penalties for illegally growing GM crops, EU farmers will have an incentive to gain an advantage against each other by illegally procuring them. In Brazil and India, farmers rebelled against restrictions on growing GM soy and cotton by procuring them illegally, which ultimately resulted in the regulatory approval of the use of GM soy in Brazil and GM cotton in India. See Herring 2007b; and Fukuda-Parr 2007d: 218.

131. To the extent that imported grains intended for consumption could escape into the environment, they would of course also raise environmental concerns, further complicating the analysis. The environmental risks, however, would be much reduced, especially in a highly regulated developed economy such as the EU, where farmers would be sanctioned for growing unapproved GM products.

132. See Bratton et al. 1996; Esty and Geradin 2000.

133. "GMO producers maintain that the [EU's] traceability requirements, coupled with the low labeling threshold for GMO content, will require complete segregation of GM and non-GM products throughout the production, transportation, processing and distribution chains, imposing major economic burdens (cost increase up to 25 per cent or more)." Stewart 2007: 33. To the extent that the labels were not private, but rather government-mandated or government-regulated, then this alternative would involve some degree of government intervention.

134. See e.g., Perez 2007 (criticizing the panel's application of article 5.7 of the SPS Agreement, and maintaining that there are always "different levels of insufficiency," and that "weights" or "thresholds" should be left to the "political domain," presumably at the member level, regardless of the effects on non-represented foreigners). See also Guzman 2004a and Sykes 2002.

135. Tinbergen 1965.

136. Keohane 1984: 93.

137. Guzman 2004b: 307–308.

138. Guzman 2004b: 309.

139. See e.g., Codex Alimentarius Commission 1995: § 3; Codex Alimentarius Commission 2003c: §§ 4 and 6; IPPC 2004: §§ 3.1 and 5.1.4; and OIE 2006: § 1.3.

140. See Marrakesh Agreement Establishing the World Trade Organization, Legal Instruments—Results of the Uruguay Round, arts. IX, X, XII, Apr. 14, 1994, 33 I.L.M. 1140 (1994). Under Article X, only a few provisions require a unanimous vote to be amended. From a technical perspective, most provisions can be amended by a two-thirds majority of the members, and will either take effect only with respect to those members or with respect to all members, depending on whether the provision alters the "rights and obligations" of the parties. See id. at art. X:1. In addition, WTO members may decide by a three-fourths majority that an amendment is of such importance that "any Member which has not accepted it within a

period specified by the Ministerial Conference...shall be free to withdraw from the WTO or remain a Member with the consent of the Ministerial Conference." Id. at art. X:3. For overviews, see Bhala and Kennedy 1998: § 4(f)(3); and Ehlermann and Ehring 2005.

141. See e.g., Sands and Klein 2001: 266 (noting "a trend towards a search for 'consensus' as opposed to reliance on the results of formal voting").

142. As Posner and Rief write, "At least one thing is clear about WTO interpretations and amendments: they are not designed to be taken regularly or readily. In fact, there has not been a single interpretation or amendment adopted since the WTO came into effect in 1995, and there were only six amendments (the last in 1965) in the previous forty-eight years of GATT." See Posner and Rief 2000: 504.

143. See CAC website at http://www.codexalimentarius.net; and OECD 2001 (noting low participation rates for low- and middle-income countries). For a critique of Codex in terms of the limited participation of developing countries, see Chimni 2006: 811–18 ("the overall participation of developing countries themselves is inadequate and ineffective," stating that "(1) developing countries most often do not participate in the meetings, given the inability to meet the travel and other expenses of participants; (2) members from developing countries have received little support from their governments; (3) developing countries have held few leadership positions in the primary committees; and (4) the complexities involved in 'tracking implementation requirements'.")

144. Stone Sweet and Mathews 2008.

145. Some may contend that judicial decision-makers are inevitably involved in some form of "balancing," including whether they wish to balance policy concerns in an explicit manner, as under this fourth institutional choice. Our interest lies in capturing the institutional implications, attributes, and deficiencies of this choice (as an ideal type) compared with the others.

146. See WTO Appellate Body, *Korea—Measures Affecting Imports of Fresh, Chilled and Frozen Beef* WT/DS161/AB/R & WT/DS169/AB/R ¶ 164 (Dec. 11, 2000). See also *Dominican Republic—Measures Affecting the Importation and Internal Sale of Cigarettes* WT/DS302/AB/R ¶ 70 (Apr. 25, 2005) (affirming the "weighing and balancing" of the judicial body of these factors). The WTO Appellate Body also took a balancing approach, in part, in the *US–Shrimp-turtle* case when it reversed much of the initial panel's decision. Rather than apply a generic analysis to all import bans based on foreign production and process methods, and thereby implicitly delegating decision-making to the market (under the second institutional alternative), the Appellate Body turned to the "facts making up" the "specific case," and sought to maintain "a balance...between the right of a Member to invoke an exception under Article XX and the duty of that same Member to respect the treaty rights of the other Members." WTO Appellate Body, ¶¶ 155–59.

147. See *Japan-Apples*, par. 8.198.

148. See Scott 2007: 110. Article 2.2 provides that "Members shall ensure that any [SPS] measure is applied only to the extent necessary, is based on scientific principles and is not maintained without sufficient scientific evidence, except as provided for in paragraph 7 of Article 5."

149. See Foster 2007 (suggesting that panels should rely more on article 5.6 than scientific assessments under article 5.1 to assess the legitimacy of member measures).
150. See Helwig 2007.
151. On the SPS Committee, see Scott 2007: 41–75. On the soft law dispute-settlement mechanism provided by the IPPC, see Chapter 4, this volume.
152. See Shaffer 2004 (concerning the *US–Shrimp-turtle* case); and Shaffer and Apea 2005 (concerning the *EC–GSP* case).
153. Under EU law, the member-state safeguards are only valid if adopted in an "emergency" in which it is "evident" that EU-authorized products "are likely to constitute a serious risk to harm human health, animal health or the environment." Commission Regulation 1829/2003, art. 34, 2003 O.J. (L 268) 1.
154. For example, were the complainants to challenge Switzerland's decision to apply a five-year moratorium on GM crop production, which resulted from a popular referendum in November 2005 that was supported by 56 per cent of Swiss voters and all 26 Swiss cantons, the panel's legitimacy challenges would have been much more stark. In contrast, had only one Swiss canton imposed a moratorium on GM varieties authorized by Swiss federal authorities based on Swiss risk assessments, and that Swiss canton's measures arguably violated Swiss law, a WTO panel's decision would be easier. The WTO decision would similarly provide leverage to public and private actors in the Swiss domestic law context to bring the canton into compliance. On the Swiss referendum, see Tiberghien 2007.
155. See Dabrowska 2006.
156. On the nature of international courts as agents of their member-state principals, and on the rational anticipation of state reactions by international organizations and courts, see e.g., Pollack 2003*b*: 59–60.
157. See Alter 2008.
158. See Bodansky 1999: 601–602 (speaking of sociological legitimacy as "popular legitimacy" and "normative legitimacy" as "whether a claim of legitimacy is well-founded—whether it is objective in some objective sense," and thus "whether it is worthy of support").
159. Laurence Helfer and Anne-Marie Slaughter thus define "effective adjudication in terms of a court's basic ability to compel or cajole compliance with its judgments." Helfer and Slaughter 1997: 278.
160. See e.g., Helfer and Slaughter 1997: 284 ("impartiality; principled decision-making; reasoned decision-making; continuity...; consistency of judicial decisions over time; respect for the role of political institutions at the federal, state and local levels; and provision of meaningful opportunity for litigants to be heard"); Bodansky 1999; and Franck 1990. Assessments of substantive legitimacy are often made in terms of broader theories of justice, morality or both, which is not our focus; we focus here on the role of the judicial process and challenges to its legitimacy. For a broader assessment of legitimacy and law that combines sociological and normative theory, see Habermas 1998. Habermas notes how law can be ineffective in terms of its effect on social practice (as in the market), as well as in terms of its failure to reflect civil society will-formation in the public sphere. See Habermas 1998: 386.
161. McDougal, Myres and Lasswell 1959: 10.

162. International Law Commission 2006: 21.
163. See Cho 2007.
164. See article 3.7 of the WTO Understanding on Rules and Procedures Governing the Settlement of Disputes, commonly known as the "Dispute Settlement Understanding." The Understanding provides further that, "Where a panel or the Appellate Body concludes that a measure is inconsistent with a covered agreement, it shall *recommend* that the Member concerned bring the measure into conformity with that agreement" (emphasis added). See Article 19.1 of the WTO Understanding on Rules and Procedures Governing the Settlement of Disputes.
165. Joerges and Neyer 2003: 224. See also Joerges 2007.
166. See Chapter 5. See also Bohanes 2002: 323–389.
167. See earlier discussion of Annex K, Letter of the Panel to the Parties of May 8, 2006, WT/DS291/R/Add.9, WT/DS292/R/Add.9, WT/DS293/R/Add.9 (Sept. 29, 2006).
168. The claim was filed in May 2003 and the Panel was formed on August 29, 2003, but not actually composed until March 4, 2004 (i.e., panelists actually designated by the director general because the parties could not agree on them). The procedure took 1,235 days between the Request for Consultations and the issuance of the Panel report. The report was finally adopted, without appeal, on November 21, 2006, 1,279 days after the initial request for consultations.
169. Panel Report, *EC–Biotech*, at ¶¶ 7.37–7.45.
170 On the uses and limits of counterfactual reasoning in international politics, see Tetlock and Belkin 1996.
171. See Hudec 1993: 452 (noting Danish Import Restrictions on Grains, L/3436).
172. The US and EU can, however, initiate cases of tit-for-tat litigation under the WTO, as arguably the EU did in the FSC case in which the WTO DSB authorized the EC to retaliate against the US in the amount of over $4 billion. See Hudec 2003.
173. Hudec 1993: 33–37, 246–49.
174. See Nonet and Selznick 2005: 58. In a related vein, Habermas writes how "[t]he transition from natural to positive law transforms these authorizations to use coercion…into authorizations to take legal action." Habermas 1998: 28.
175. *FDA Week* 2007*a*. The exception was the Austrian ban on MON810, which had been approved for consumption and cultivation in the EU. Commission efforts to resolve the dispute therefore focused largely on challenging these bans; see Chapter 6.
176. *FDA Week* 2007*a*; *Inside US Trade* 2007*a*; Shaffer telephone interview with private US attorney following the case, June 5, 2007.
177. Quoted in *FDA Week* 2007*b*.
178. For good discussions on the agreement on rice exports, see *FDA Week* 2007*b*; Bennett 2007.
179. See *Inside US Trade* 2008*f*.
180. See *Inside US Trade* 2008*a*, as well as Chapter 6 for a discussion of the Commission's actions against the member-state bans, and on Sarkozy's end-of-year announcement. Strikingly, the US did not respond to Sarkozy's announcement by breaking off talks or pressing for immediate sanctions, although a USTR official did point out that, "It is hard to overstate our disappointment" at the French move. Quoted in Ryan 2008.
181. *Inside US Trade* 2007*f*.

182. This had been a key concern of US negotiators throughout the talks. See *Inside US Trade* 2007*f*; and *Inside US Trade* 2008*f* (concerning search for a "technical solution" to the adventitious presence issue with which member states concur).
183. Quoted in Agence France-Presse 2008.
184. *Inside US Trade* 2008*a*.
185. Quoted in *Agence France-Presse* 2008. Dialogue continued as this book went to press, with the two sides continuing to meet regularly, and with the United States attempting to maintain the pressure on the EU to speed up new approvals and address the issue of low-level presence, without (yet) beginning the process of applying trade sanctions against the Union. See *Inside US Trade* 2008*j*, 2008*l*.
186. The US continued to refrain from imposing sanctions as of June 2008, although US officials and industry representatives indicated ongoing frustration with the EU for its slow approval process, including an extended delay in the approval of seven pending varieties announced by the Commission the previous month. Canada and Argentina announced similar decisions to extend the "reasonable period of time" allowed for EU compliance with the decision. See e.g., *Inside US Trade* 2008*g*. On developments in the EU since the WTO decision, see Chapter 6.
187. On the ability of international law and organizations to empower certain domestic constituencies, see e.g., Börzel and Risse 2007: 492–493.
188. James 2007. The crops, however, would likely be used predominantly for feed in light of ongoing consumer responses toward GM foods in Europe.
189. Commission of the European Communities 2002*c*, 2005*b*. Similarly, in a 1993 White Paper, the Commission remarked, "biotechnology has emerged as one of the most promising and crucial technologies for [the twenty-first] century." See Commission of the European Communities 1993: 100, fn. 1. In a 1996 report, the Commission expressed its concern over the low rates of approvals of GM products in Europe compared to the US; see Commission of the European Communities 1996.
190. Skogstad 2003: 336.
191. Commission 2004*d*. In Italy, commentators in the mainstream press criticize the environmental ministry's blocking of permits for field trial research on GM varieties on related grounds. See e.g., Veronesi 2007.
192. Quoted in Bounds 2007*c*.
193. The US certainly thought so. See Panel Report, *EC–Biotech*, par. 7.506 ("timing of the approval of Bt-11 maize (food) is no coincidence").
194. Panel Report, *EC–Biotech*, par. 7.1249.
195. See Judgment of the Court, Case C-236/01 (reference for a preliminary ruling from the Tribunale amministrativo regionale del Lazio): Monsanto Agricoltura Italia SpA and Others v Presidenza del Consiglio dei Ministri and Others (Sept. 9, 2003); and Battini 2007: 17–20. Similarly, the Court of Justice ruled in March 2000 that France could not ban the sale of GM crops that had been approved at the EU level without producing new information regarding health and environmental risks. The case was referred to the Court of Justice by a French court following a challenge by Greenpeace of France's initial approval of a GM maize variety. See Judgment of the Court, Case C-6/99 (reference for a preliminary ruling from the Conseil d'Etat): Association Greenpeace France and Others v French State, Ministere de

l'Agriculture et de la Peche and Others, In the presence of Novartis Seeds SA and another (March 21, 2000).

196. Scott 2007: 128, citing the Pfizer and Monsanto Agricoltora Italia court decisions.

197. *Eastbusiness* 2006, 2007.

198. *Inside US Trade* 2004*b*: 10.

199. See Sebastian 2007: 343–43.

200. Pruzin 2007; and *Inside US Trade* 2007*h*. Moreover, when the EU still failed to comply by the extended date (regarding, in particular, approvals of certain GM corn varieties and the safeguard ban imposed by Austria), the US formally notified the WTO that it would seek its right to retaliate, but moved slowly in order to give the EU more time. *Inside US Trade* 2008*c*.

201. Sek 2002. The EU later challenged the legality of the US tariffs before the WTO in light of new scientific risk assessments that the EU maintains justification of its ban under the SPS Agreement, but a WTO panel found that the EU still had not complied with WTO rules. See *Inside US Trade* 2008*d*. On October, 16, 2008, the Appellate Body reversed the panel findings regarding the EU's continued violations of Articles 5.1 and 5.7 because of the panel's application of an incorrect standard of review and allocation of burden of proof, among other grounds. The Appellate Body also found that it was unable to complete the analysis based on the factual record regarding the legality of the EU bans under the legal standards clarified by the Appellate Body in its decision. The Appellate Body nonetheless has signaled that the EU may have proper grounds to challenge the continuation of such US sanctions. See Appellate Body Report, *United States—Continued Suspension of Obligations in the EC-Hormones Dispute*, WT/DS320/AB/R (Oct. 16, 2008). The US sanctions, however, remain in effect.

202. We are grateful to Richard Morningstar for insightful comments on this point. Such noncompliance, we would add, may be normatively defensible as an "efficient breach" of WTO law, allowing the EU to compensate the US for damages rather than incur the extraordinarily high political costs of compliance. Critics of this view, however, are leery of allowing the EU to "buy its way out" of its legal obligations, and suggest that the EU would suffer reputational effects from extended noncompliance.

203. Shaffer interview with USDA official, April 19, 2007.

204. Each side may have feared that it could be worse off after an appeal. In addition, there is a sense that lawyers in the Commission wished to appeal in order to "clean up" certain legal aspects while diplomats felt that the decision was satisfactory. One can also read the Commission's non-appeal as another example of a decision not to decide. The deadline for appeal passed while the Commission had not reached a consensus as to what to do. Pruzin 2007a; Shaffer interview with Commission official, June 2, 2007. On the use of WTO dispute settlement for domestic political purposes, see e.g., Hudec 1996.

205. Shaffer interviews with USDA officials, April 10 and April 19, 2007.

206. See discussion in Chapter 7.

207. Howse 2000: 2330.

208. Similarly Scott, although she remains wary of the risk of "imposition of a methodological straightjacket operating in the name of false universalism," points to

how WTO law can "serve to open up decision-making, encouraging information generation and a healthy reflexivity." Scott 2007: 80.

209. See e.g., Case C 207/83 *Commission v United Kingdom* [1985] ECR 1201, ¶ 17 (in the context of a UK requirement of a certificate of origin that enables consumers "to assert any prejudices which they may have against foreign products"). Cf. Macmillan and Blakeney 2001: 114–115 (finding that "the issue of labeling of GM foods falls more properly within the ambit of the TBT Agreement," and that "at a time when the WTO is facing unprecedented, and increasingly well-organized opposition, the revelation that the WTO was antagonistic to the labeling of GMOs in food would be a publicity nightmare"); and Runge and Jackson 2000.

Chapter 6

1. Vogel 1995, Young 2003.
2. On the "second image reversed" tradition in international relations theory, see e.g., Gourevitch 1978, Keohane and Milner 1996. For a constructivist-oriented "transnational legal process theory" that contends that international law gives rise to domestic policy change through processes of "internalization," see Koh 1998. Koh theorizes the impact of international law in domestic systems through national internationalization of international legal norms in light of iterative processes involving interpretation over time.
3. Commission of the European Communities 2000a. In 2002, the Council and European Parliament adopted EC Regulation 178/2002 pursuant to which the new agency, named the European Food Safety Authority, was created. While member states debated and lobbied over its ultimate location, the EFSA was temporarily housed in Brussels. The European Council finally determined in December 2003 that its headquarters would be established in Parma, Italy. See European Food Safety Authority 2003.
4. Commission of the European Communities 2000b: 15, emphasis in original.
5. Commission of the European Communities 2000b: 15.
6. Commission of the European Communities 2000b: 15. The Commission observed that, although the precautionary principle is not defined in the EC Treaty (which only prescribes its use to protect the environment in article 174 EC), the European Court of Justice's case law had recognized the principle's application in other domains. See Scott 2003: 228.
7. Council of the European Union 2000. See also Vogel 2001: 28–29.
8. Kysar 2006: 41. See also Dana 2003.
9. See also Heyvaert 2006. Of course, the same holds true with the US's use of the term "science-based." Both of these terms—"precaution" and "science-based"—are used as forms of legitimation of decision-making.
10. For details, see Commission of the European Communities 1998.
11. For good accounts of the conciliation process, and the key issues separating the EP and Council delegations, see *European Report* (2000a, 2000b).
12. Directive 2001/18/EC of the European Parliament and of the Council, *Official Journal of the European Communities*, L 106 April 17 2001, pp. 1–38.

13. Young 2003.
14. Quoted in Evans-Pritchard 2001.
15. In a joint statement, France, Italy, Austria, Denmark, Greece, and Luxembourg "reaffirm[ed] their intention... of ensuring that the new authorizations for cultivating and marketing GMOs are suspended pending the adoption" of new provisions on traceability, labeling, and environmental liability. Quoted in Mann 2001, p. 8.
16. Part B of the directive covers the "deliberate release of GMOs for any other purpose than for placing on the market" that go beyond the "contained use" of GMOs (i.e., it addresses the testing of GMOs in pilot plots). Directive 90/219, as amended by the later Directive 98/81/EC, continues to govern the contained use of GMOs.
17. Commission of the European Communities, 2001*a*, 2001*b*.
18. Article 5.5 of Regulation 1829/2003 provides that articles 13–24, constituting Part C of Directive 2001/18, "shall not apply," but rather be replaced by the new, more centralized authorization procedures. In order to be marketed in the Union, GM seeds must also meet the standard requirements for all seed varieties to be placed in the Union's "common catalogue of agricultural plant species." See e.g., Council Directive 98/95/EC of December 14, 1998 (concerning requirements for such listings), as well as Directives 2002/53 and 2002/55.
19. Compare article 1 of Regulation 1829/2003 with article 1 of Directive 2001/18.
20. Article 7(1).
21. See article 33; and discussion in Scott 2004 (who notes the contrasting and limited legal basis of the regulation, which is based primarily on the Treaty's internal market provision, and does not expressly refer to the Treaty's environmental and consumer protection provisions).
22. See articles 2.10 and 3.1 (defining the scope of coverage).
23. Products from sixteen GMOs that were deemed "equivalent" to traditional foods, could be legally marketed under the exception set forth in the former regime under the Novel Foods Regulation. No other products were accepted for marketing under this regime. Commission of the European Communities 2004*a*: 4.
24. See *European Report*, 2002*a*, 2002*b*; Berg 2002.
25. See article 47 of Regulation 1829/2003.
26. Shaffer interview with Commission official, April 25, 2006. The Danish company Novozymes is the world leader in the production of GM enzymes, with a self-reported global market share of 46 per cent; see Novozymes 2006: 9.
27. Scott 2003: 15. The Commission later submitted a broader directive harmonizing member-state regulations on the use of industrially manufactured enzymes; this proposal, which would largely reproduce the existing rules on GM enzymes, was under consideration by the Council of Ministers and the EP at press time; see European Report 2007*b*.
28. See Commission of the European Communities 2001*a*; Scott 2003: 224.
29. See articles 7, 19 and 35 of Regulation 1829/2003 which in turn refer to the operation of the Standing Committee on the Food Chain and Animal Health, pursuant to article 5 of Council Decision 1999/468/EC of June 28, 1999 laying down the procedures for the exercise of implementing powers conferred on the Commission (known as the "comitology" decision).
30. In 1997, for example, a large majority of member governments voted against the Commission's proposed approval of a Bt maize, but the Commission proposal was sup-

ported by a single member state, which prevented the required unanimous vote in the Council and allowed the Commission decision to go into effect; see Chapter 2.

31. This has indeed occurred on several instances in which a qualified majority of member states voted to oppose the Commission when it sought to overturn national bans on individual GM varieties; see below.

32. Shaffer telephone interview, April 14, 2004.

33. See e.g., the comments of Tony Van Der Haegen, minister counselor at the EU Delegation in Washington, D.C., who referred to the US grain handling system as "very efficient, but... totally incompatible with the traceability system." Quoted in Reuters 2001.

34. See Commission of the European Communities 2003: para 1.3.

35. Article 43 of Regulation 1829/2003 added an article 26a to Directive 2001/18 to provide that "Member States may take appropriate measures to avoid the unintended presence of GMOs in other products."

36. See article 45 of Regulation 1829/2003 and article 11 of Regulation 1830/2003 (noting that the penalties "must be effective, proportionate and dissuasive").

37. Commission of the European Communities 2004*b*.

38. Within the regulatory committee, Austria, Denmark, France, Greece, and Luxembourg voted against the proposal, while Germany, Belgium, and Italy abstained. See Bridges BioRes 2003.

39. In the Council, six states (Ireland, Italy, the Netherlands, Finland, Sweden, and the UK) voted in favor of the Commission proposal, six others (Austria, Denmark, France, Greece, Luxembourg, and Portugal) voted against, and three states (Belgium, Germany, and Spain) abstained. Commission 2004*c*: 4. See also Andrew Pollack 2004*a*.

40. Commission of the European Communities 2004e.

41. Quoted in Browne 2004: 18.

42. See *Associated Press* 2004*a*, Spiteri 2004.

43. See *Bridges BioRes* 2004 (noting that "The approval only applies for the use of corn as feed—not for cultivation—and imports will only be allowed once the maize also has been approved for food use").

44. For good accounts of the debate, see *European Report* 2004*b*, Beatty 2004.

45. Commission of the European Union 2005e.

46. See Commission of the European Union 2006*a*: 6, which states that "This high concentration of referrals in the policy field of genetically modified organisms (GMOs) can be explained by the divided views of the Member States on the standards to apply to the scientific evaluation of new products." For details on the relevant votes, see Commission 2006*b*.

47. For up-to-date lists of new GM products authorized under various EU legislative instruments, see the following: European Commission, "GMO Products Approved under Directive 90/220/EEC As of March 2001," http://ec.europa.eu/environment/biotechnology/authorised_prod_1.htm; European Commission, "GMO Products Authorised under Directive 2001/18/EC As of January 31, 2006," http://ec.europa.eu/environment/biotechnology/authorised_prod_2.htm; European Commission, "GMO Authorized for Feed Use in the European Union in Accordance with Directives 90/220/EEC and 2001/18/EC," http://ec.europa.eu/food/food/biotechnology/authori

sation/2001–18-ec_authorised_en.pdf; European Commission, "Genetically Modified (GM) Foods Authorised in the European Union under the Novel Food Regulation (EC) 258/97," http://ec.europa.eu/food/food/biotechnology/authorisation/258–97-ec_authorised_en.pdf; European Commission, "EUROPA > European Commission > DG Health and Consumer Protection > Overview > Food and Feed Safety," listing newly authorized products under Regulation 1829/2003, http://ec.europa.eu/food/dyna/gm_register/index_en.cfm, all accessed on Nov. 15, 2008.

48. On the new members as a "Trojan horse," see Brown 2004. For an academic perspective, see Inglis 2003.

49. *Economist* 2003a.

50. See *Eastbusiness.org* 2006; *Deutsche Presse-Agentur* 2006a; Hungarian News Agency 2006.

51. See *European Report* 2006b. In a separate case, Poland banned the cultivation of sixteen varieties of GM and around 700 varieties of conventional maize, which it argued, under Directive 2002/53/EC on the Common Catalogue of Varieties of Agricultural Plant Species, were unsuited to cultivation in Polish climatic conditions. By contrast with the previous GM bans, and with the unanimous support of the relevant regulatory committee, the Commission approved the Polish ban on 8 May 2006. See *European Report* 2006a.

52. The Prodi Commission had held a similar orientation debate in January 2004, focusing on the completion of the legislative framework, the issue of coexistence of GM and conventional crops, and the resumption of approvals. For a good discussion, see Delegation of the European Commission to the USA 2004.

53. Commission of the European Union 2005a, bold in original.

54. Commission of the European Union 2005a: 7.

55. Corn varieties T25 and MON810 were banned in Austria; Bt-176 maize was banned in Austria, Germany, and Luxembourg; oilseed rape MS1xRF1 was banned in France; oilseed rape Topas 19/2 was banned in France and Greece; and Hungary declared a ban on MON 810 in 2005. See "Invocation of Article 16 under Directive 90/220/EEC and Article 23 under Directive 2001/18/EC (Safeguard clause) as of 15 March 2005," http://ec.europa.eu/environment/biotechnology/safeguard_clauses.htm, accessed on January 8 2008.

56. Commission of the European Union 2005a: 6, bold in original.

57. Commission of the European Union 2005a: 7, bold in original, italics added for emphasis.

58. *European Report* 2005c.

59. These are notes taken by Shaffer in an interview with a member-state representative, June 18 2007.

60. European Report 2005a.

61. Commission of the European Union 2005c. The certification requirement was later dropped, in January 2007, after EU experts had concluded that the spread of the grain had been contained and the measures were no longer required. See *Agence France-Presse* 2007a.

62. Commission of the European Union 2006c. LL 601 rice has been genetically engineered to resist Bayer's Liberty weedkiller, similar to Monsanto's Roundup-Ready GM varieties. For a good overview, see Bounds 2006a. The EU ended its mandatory testing require-

ment in January 2008, although EU member states can individually require the tests. See *Inside US Trade* 2008*b*.

63. According to official press releases and newspaper accounts, the British delegation voted in favor of the Commission's eight proposals to overturn the member-state bans, with Sweden and Finland abstaining and with the Czech Republic and Portugal each voting to overturn a single ban (on T25 maize and MON810 maize, respectively). See Council of the European Union 2005; Commission of the European Union 2005*d*; and *European Report* 2005*b*.

64. Quoted in Mahoney 2005.

65. Quoted in Reuters 2005.

66. Despite these public debates and attacks on the Commission and EFSA within the Council, no votes were taken in the Council on the approval of new GM varieties under the Austrian Presidency.

67. See *Austria Today* 2006*a*; *European Report* 2006*c*.

68. See e.g., the comments of German Environment Minister Sigmar Gabriel, who argued that the current rules put the ministers in an "unacceptable position." Quoted in *Associated Press* 2006*a*.

69. Numerous EU ministers called for changes to the risk assessment procedure, and the Hungarian environment minister, Mikos Persanyi, made a number of specific recommendations along those lines. For additional discussions of the March 9th Environment Council debate, see *European Report* 2006*d*; Johnson 2006.

70. See e.g., *European Report* 2006*e*; Brand 2006*a*; and Pesticide & Toxic Chemical News 2006. For a good discussion of the Council debate on organic foods, see *European Report* 2006*f*. The European Parliament, for its part, voted in March 2007 to tighten the proposed regulations on organic foods, reducing the allowable "adventitious presence" of GMOs in organic foods from the Commission's proposed 0.9 per cent to 0.1 per cent; see Meade 2007.

71. See, e.g., *European Report* 2006*g*, which reported that Dimas "…believes that it would be preferable to improve the credibility of the system by amending and clarifying the EFSA's operating rules rather than by meddling with committee procedure per se."

72. For the complete list of proposed changes, see Commission of the European Union 2006d.

73. EuropaBio representative Simon Barber, quoted in Spongenberg 2006.

74. For good discussions of the May 15 meeting, see Rosenthal 2006*a*; *European Report* 2006*h*.

75. Cultivation of GM products, as noted, was limited in 2005 largely to the planting of EU-approved GM corn varieties in Spain, where 12 per cent of the total corn crop was reported to be from GM seed; see Deutsche Presse-Agentur 2006b. There is also evidence, however, that cultivation of GM corn has spread beyond Spain to EU countries such as the Czech Republic, Portugal, Germany, and especially to France, where farmers have increasingly adopted GM corn, only to be confronted by anti-GM activists who have "outed" and protested against farmers planting GM crops; for a good discussion, see Miller 2006. The Commission's decision not to seek binding legislation on coexistence was condemned by several member governments and by environmental groups such as Friends of the Earth–Europe, whose representative characterized the Commission's policy as "first contaminate and then legislate";

quoted in Waterfield 2006. The Commission's approach was generally welcomed, however, at the Austrian Presidency's April 2006 conference, "Freedom of Choice," which focused on the issue of coexistence; see *BBC Monitoring International Reports* 2007.

76. Council of the European Union 2006*b*: 8. The Commission approved all three varieties for import and marketing, but not for cultivation, in March 2007; see European Report 2007*a*.

77. See *PR Newswire* 2007, and *International Herald Tribune* 2007.

78. Quoted in O'Donnell 2008: 3. For other accounts of the 7th May meeting, see *Inside US Trade* 2008*h*, Kanter 2008, and *EurActiv.com* 2008*a*. The Commission reportedly asked EFSA to analyze further scientific evidence on the environment and health effects of the Amflora potato and three Monsanto hybrid or "stacked" maize varieties; to review new scientific information on GMO maizes Bt11 and 1507, which Dimas had previously recommended for rejection; and to confirm that the scientific evidence is complete for GM rice LL62. See EurActive.com 2008a.

79. O'Donnell 2008, *Inside US Trade* 2008*g*.

80. Bounds 2006*b*.

81. See e.g., Rosenthal 2007*b*. Rosenthal notes the frustration of executives at BASF, which had established a $1.5 billion alliance with Monsanto and invested heavily in the amflora and other GM foods and crops. "You would think this approval would have been easy since this potato has no seeds, no wild relatives to cross with in Europe, and only industrial use," said one official. "But it didn't turn out that way." Quoted in Rosenthal 2007*b*: C3. See also Haxel 2007.

82. Quoted in Laitner 2007: 9.

83. Quoted in Rosenthal 2007*b*: C3.

84. See GMO Compass 2007 (on BASF's request) and Mortished 2008 (on charges of prevaricating). In late December 2007, the Commission reportedly submitted to the Council a second draft authorization for the amflora potato, this time for use in feed. Member states were reportedly split on this decision as well, with twelve member states opposed, ten in favor, and the rest abstaining; see *European Report* 2007*c*.

85. See Chapter 2.

86. *Inside US Trade* 2007*c*.

87. Quoted in Kanter 2007*a*: 1.

88. Quoted in Rosenthal 2007*c*: C5.

89. See e.g., Lean 2007: 18, which quotes Friends of the Earth campaigner Claire Oxborrow stating: "This could—and should—be the beginning of the end for GM crops in Europe." See also Brand 2007*a*.

90. Europabio spokeswoman Nathalie Moll, quoted in Kanter 2007*a*: 1. Later, Europabio questioned Dimas's use of the studies in question, suggesting that these studies had been misinterpreted and did not in fact call into question the safety of the varieties in question; see Kanter 2007*b*. For a discussion of the scientific arguments on both sides of the question, see Rosenthal 2007*c*. Other industry representatives speculated that Dimas, whose term was scheduled to end in 2009, might be seeking to avoid a positive ruling on these varieties, as on the amflora potato, for reasons related to his political career once he left the Commission; see *Inside US Trade* 2007*c*.

91. USTR spokesman Stephen Norton, quoted in Kanter 2007a: 1.
92. US Mission spokesman Rob Gianfranceschi, quoted in Kanter 2007a: 1.
93. Quoted in Bounds 2007c.
94. For good discussions, see Kanter 2007c; European Report 2007d; and Inside US Trade 2007d. "To postpone any new approvals will have dramatic consequences," Fischer Boel reportedly told the agriculture ministers, "and the result will be that production will move out of Europe and then we will have to import meat that is fed with GM products that are...not approved in Europe, but we will be eating it anyway because that is the only solution." Quoted in Brand 2007b. See also European Commission, DG for Agricultural and Rural Development ("there is an urgent need to take action in order to avoid negative implications for EU livestock production and agriculture overall"). Moving into 2008, there was growing attention to the zero-tolerance policy among European livestock farmers as well as within the European Parliament, where in April the Chair of the EP Agriculture Committee, Neil Parish (EDD-UK) called for a reconsideration of the policy. See European Report 2008a. Despite these indications, EU Health Commissioner Androulla Vassiliou told the press that, with the approval or pending approval of a number of disputed crops such as Monsanto's Roundup-Ready2Yield soybean, there would be "no need" for a change in the EU's zero tolerance policy. See Dalton 2008.
95. European Report 2007e.
96. European Report 2008c.
97. EurActiv.com 2008a, O'Donnell 2008. In late October, the European Food Safety Authority issued an opinion that the two maize varieties, Pioneer/Dow 1507 and Syngenta's Bt-11 were safe for cultivation within the EU. Despite this opinion, the Commission had not at press time (November 2008) brought to the member states a formal proposal to approve these two varieties for cultivation. Brand 2008, Casey 2008.
98. Quoted in Smyth 2007: 9.
99. Agence France-Presse 2007b.
100. Inside US Trade 2007e.
101. Associated Press 2007; Business Wire 2007; Inside US Trade 2008a.
102. See the excellent discussion in European Report 2008b.
103. French Secretary of State for Ecology Nathalie Kosciusko-Morizet, quoted in EurActiv. com 2008b. See also European Report 2008b, and République Française 2008. In addition to the French presidency group, Commission President Barroso had called in May 2008 for the creation of an ad hoc "sherpas group" of high-level officials, appointed by national prime ministers' offices, to consider the future of EU biotech regulation and report to the European Council; see EurActiv.com 2008b. The French Presidency group reported to an October 2008 meeting of the Environment Council, and focused on three primary issues: possible new guidelines for assessing environmental risks from GMOs; the controversial possible inclusion of socio-economic criteria in environmental risk assessments; and the possibility of allowing member states greater leeway to declare GM-free zones. The last possibility was reportedly floated by the Presidency as a potential "grand bargain," in which GM-skeptical states would accept GM approvals for import and marketing in return for greater controls over the more sensitive issue of cultivation. Whether such a

bargain could be struck remains unclear at press time (November 2008). See *Inside US Trade* 2008k.

104. See e.g., the critical comments of German Agriculture Minister Horst Seehofer, who called in November 2007 for a reconsideration of the entire EU approval process; Brand 2007*b*.

105. Commission of the European Union 2006*e*: 2. Austria's bans, according to the Commission, were the only ones affecting varieties being actively marketed in the EU, the other varieties since having been withdrawn from the EU market by their producers; see *Austria Today* 2006*b*.

106. Council of the European Union 2006*a*: 20. See also Brand 2006*b* and Bounds 2006*c*.

107. See e.g., *Eastbusiness.org* 2007 (quoting Hungarian State Secretary Kalman Kovacs saying that, "We hope that after the lobbying of recent weeks and knowing the development of EU countries' opinion it will be possible to preserve the ban"). EU officials widely expected that the Council would support Hungary, as it had supported national bans twice before in the previous two votes. "It could be that they (ministers) show solidarity (on the Hungarian GMO ban), the same way that they did with Austria," once Commission official was quoted as saying. "It's a very sensitive issue." Quoted in Smith 2007

108. Council of the European Union 2007*a*: 23–26. For good analyses see Bounds 2007*a*; Spongenberg 2007*b*.

109. European Report 2007*f*.

110. Council of the European Union, 2007*b*.

111. Voting in favor of the Commission proposal were the United Kingdom, the Netherlands, Sweden and Estonia. Opposed were France, Italy, Poland, Greece, Malta, Hungary, Austria, Ireland, Denmark, Lithuania, Luxembourg, Latvia, Slovakia, Cyprus and Germany. Abstaining were Romania, Spain, Portugal, the Czech Republic, Belgium, Bulgaria, Finland and Slovenia. See *European Report* 2007*g*, Goldirova 2007.

112. Austrian Environment Minister Josef Proell, for his part, was defiant, pointing to the fact that Austria's ban on cultivation had not been challenged: "It's an important point for environment and agriculture policy in Austria," he said, "We will remain free of gene-technology in cultivation." Quoted in Goldirova 2007.

113. *Inside US Trade* 2008*f*.

114. Reuters 2008. In late October 2008, the European Food Safety Authority issued a new opinion that MON810 does not pose a risk to animals or plants as the French government contended. The agency indicated that France had not identified "any new data...that would change previous risk assessments conducted on MON810," and that the French ban on cultivation was therefore unjustified. Quoted in Casey 2008. As of late November 2008, however, the Commission had yet to bring a challenge to the French ban.

115. Skogstad 2001; Scott 2003. Commission officials continued to put a brave face on the approval process in late 2008, pointing to the approval of two new varieties (one maize and one cotton variety, both by Bayer Cropscience; see Table 6.4) in September and October, and additional varieties such as Monsanto's Roundup-Ready2Yield soybean likely to be approved in early 2009, albeit over the continued objections of regulatory committees and the Council of Ministers. See e.g. *Inside*

U.S. Trade 2008j, Smith 2008. More broadly, the likely approval of the Roundup Ready2Yield soybean – over objections from member states in committee and in the Council – points to a central element of the Commission's recent approach, which is to retain the distinctive features of the EU's regulatory *system*, including its controversial zero-tolerance policy, but to secure approval for the most economically significant US varieties so that the impact of these systemic differences on transatlantic trade is minimized. Whether this pragmatic approach will facilitate a long-term management of the dispute remains to be seen.

116. Various anti-GMO groups nonetheless continue to press for further strengthening of the EU regulatory framework, including binding rules on coexistence and liability. In February 2007, moreover, Greenpeace presented the European Commission with a petition, signed by 1,000,000 Europeans, calling for new rules requiring labeling of meat, milk, and eggs produced from animals fed with GM crops. See Spongenberg 2007*a*.

117. Vogel 1995.

118. Young 2003: 458.

119. Young 2003: 468–70.

120. Jim Bair, quoted in Pew Initiative 2004*b*. See also Andrew Pollack 2004*b*.

121. Monsanto Company 2004.

122. Kettnaker 2001:224

123. Smith 2003.

124. Kettnaker 2001:224.

125. See Cutler, Haufler and Porter 1997 (on private authority in international relations).

126. See e.g., Reuters 1999.

127. "Altered Crops: Trouble in the Wind for Organic Foods," Aug. 11, 1999 (forwarded to us by email from a representative of the American Soybeans Association, April 14, 2004).

128. For good accounts of the suit see *Associated Press* 2006*b*, and Cole 2006, 2007*a*, 2007*b*.

129. Lee 2006.

130. For a good discussion, see Weiss 2006*b*.

131. APHIS 2007*b*. The rice in question was later discovered to be LL 604, another Bayer Cropscience variety approved for growing by APHIS, but not for sale or marketing. See also *Inside US Trade* 2007*a*, which notes the impact of the Liberty Link on US exports to Mexico, the largest export market for US rice, which announced a requirement that all US rice shipments be certified free of GMOs.

132. Cole 2007*a* (quoting Wayne Moery, chairman of the Arkansas Rice Research and Promotion Board: "If you want to sell rice to Europe, you've got to do something like this to prove to them that you've cleaned it up.")

133. USA Rice Federation 2007. See also the analysis in Russell 2007.

134. Weiss 2006*a*. ("In fact, many experts suspect that pressure from the food industry was a major reason why Bayer mysteriously dropped LL601 five years ago, without seeking USDA approval for it. The company has refused to answer questions about its biotech rice program, which produced two other varieties. The Agriculture Department has deemed these two safe for sale, but Bayer has opted not to market them.") Later studies revealed that US rice exports to Europe had indeed plunged by some 64 per cent in the first seven months of 2007 from a year earlier; *Bloomberg News* 2007. Bayer scored at least a temporary victory in October 2008, when US District Judge

Notes to pages 265–268

Catherine Perry refused to allow farmers' lawsuits against Bayer to be consolidated in a single, class-action lawsuit. Bayer remained potentially liable, however, in the face of some 1,200 individual claims for compensation. See Harris and Fisk 2008.

135. Comment from the Iowa Farm Bureau Federation, APHIS-2006-0157-2001; Comment from the Missouri Corn Growers' Association, APHIS-2006-0157-0019; and Comment from the National Chicken Council, APHIS-2006-0157-0043.

136. Producers of corn and corn-derived products had largely already lost the EU market because of EU restrictions on GM varieties. The NGFA's member companies operate grain, feed, processing, biofuel and export facilities that handle 70 per cent of the US grain and oilseed crop. The NAEGA consists of companies involved in the bulk grain and oilseed exporting industry. Its members are responsible for the vast majority of US grain exports. See Comment from the National Grain and Feed Association, APHIS-2006–0157–0045.1, March 12, 2007; Comment from the North American Export Grain Association, APHIS-2006–0157–0044.1, March 12, 2007 [NAEGA Comment]. The National Corn Growers' Association also criticized Syngenta's decision to proceed with commercial release of MIR604 before it received regulatory approval in Japan. See *Wallace's Farmer* 2007.

137. Press Release, *Grain Industry Urges Syngenta to Reconsider Plan to Commercialize Biotech Corn Seed Not Approved in Export Markets*, April 4, 2007. In their comments to APHIS, the two associations stated that if APHIS granted Syngenta's petition for nonregulated status, it should do so with certain conditions attached.

138. Shaffer telephone interview with Lisa Zannoni, Head, Global Biotechnology Regulatory Affairs of Syngenta, Jan. 28, 2008.

139. Biotechnology Industry Organization 2007*c*.

140. The EU was not listed, since BIO did not yet consider its regulatory system to be functioning, although the industry was in discussions with the EU on this issue. See *Inside US Trade* 2007*f*.

141. *Inside US Trade* 2007*b*.

142. APHIS 2007: 1. A sugar beet genetically engineered to withstand Monsanto's Roundup herbicide was to be marketed in the US in 2008, which would be used for processing sugar for candies, cereals, and other food products. If widely adopted, it would constitute a new breakthrough for GM foods. The commercialization decision could be easier because "only about 3 per cent of American sugar is exported ...compared with about half of wheat and rice." See Andrew Pollack 2007*b*.

143. Hill and Battle 2000.

144. Guy 2003.

145. Hill and Battle 2000: 95.

146. Pew Initiative 2002*a*; Pew Initiative 2004*a*: 15, 20; FAO 2004*a*: 71.

147. See e.g., Barnett 2002; Pew Initiative 2004*a*:3.

148. *Bridges Weekly Trade News Digest* 2000; Lueck 2000. See also the discussion in Busch et al. 2004: 34–35.

149. Mellman Group 2006: 2–4. See also Pew Initiative 2003*a*.

150. Mellman Group 2006: 5–7.

151. Mellman Group 2006: 9.

152. Pew Initiative 2006: 2.

153. For a detailed listing of biotech-related legislation introduced at the federal and state levels in the US, see the Pew Initiative on Food and Biotechnology's Legislative

Tracker, available on-line at: http://pewagbiotech.org/resources/factsheets/legisla-tion/index.php, accessed on April 17, 2008. See also Pew Initiative 2002. At the state level, Maryland enacted a law which limits the raising of aquacultured trans-genic fish to facilities which are not connected to any other water body and that are "constructed in a manner that assures that … genetically altered stocks are precluded from entering any other waters …" Md. Code Ann., Nat. Res. I § 4–11A-02 (2003). Similarly, but more strongly, California enacted a law which disallows the "artificial propagation, rearing, or stocking of transgenic freshwater and marine fishes, inver-tebrates, crustaceans, or mollusks" anywhere in the state, including in the Pacific. Cal Fish & Game Code § 15007 (Deering 2004). While many states have introduced bills to place moratoriums or bans on GM agricultural crops, none have passed so far. See Pew Initiative 2006. There has been action at the local level as well, however, with two California counties banning the production of all GM crops or animals. See Lucas 2004.

154. FDA 2001*b*. As the FDA reported, "Many of the comments expressed concern about possible long-term consequences from consuming bioengineered foods, but they did not contend that any of the bioengineered foods already on the market have adverse health effects. The comments were mainly expressions of concern about the unknown. The agency is still not aware of any data or other information that would form a basis for concluding that the fact that a food or its ingredients was produced using bioengineering is a material fact that must be disclosed under sections 403(a) and 201(n) of the Act. FDA is therefore reaffirming its decision not to require special labeling of all bioengineered foods."

155. See 7 C.F.R. sec. 205.301(f)(1) (2004), which provides that no product made using "excluded methods" can be labeled "organic" or "100 per cent organic." The defini-tion of "excluded methods" is as follows: "Excluded methods. A variety of methods used to genetically modify organisms or influence their growth and development by means that are not possible under natural conditions or processes and are not considered compatible with organic production. Such methods include cell fusion, microencapsulation and macroencapsulation, and recombinant DNA technology (including gene deletion, gene doubling, introducing a foreign gene, and chang-ing the positions of genes when achieved by recombinant DNA technology). Such methods do not include the use of traditional breeding, conjugation, fermentation, hybridization, in vitro fertilization, or tissue culture." The rule was published on Dec. 21, 2000 (65 Fed. Reg. 80548, 80638), and it became effective on Apr. 21, 2001.

156. US House of Representatives 2003: 16.

157. FDA 2006*a*. The guidance explicitly omits plant-incorporated protectants (PIPs) with pesticidal characteristics, which continue to be regulated by the EPA.

158. FDA 2006*a*, Section II.

159. APHIS website, "BRS News," http://www.aphis.usda.gov/biotechnology/current_initiatives.shtml, accessed on April 19 2007.

160. See Environmental Impact Statement; Introduction of Genetically Engineered Organisms, 69 Fed. Reg. 15,3271 (January 23, 2004) (to be codified at 7 C.F.R. pt. 340); and USDA 2004*c*.

161. APHIS 2007*a*.

162. As APHIS (2007a: 2) explains, "If a GE [genetically engineered] crop has gone through the regulatory process for USDA to determine that it can be safely com-

mercialized, it is commonly referred to as being a deregulated crop. This is necessary before it is sold and produced commercially. It allows the product to be moved and planted freely without the need for notification or permits. A developer may file a petition for deregulation only after a GE crop has been tested extensively and the developer can show that the product does not pose a plant pest risk."

163. "[A]n EIS must be prepared if 'substantial questions are raised as to whether a project may cause significant degradation of some human environmental factor.'" *Idaho Sporting Cong. v. Thomas*, 137 F.3d 1146, 1149 (9th Cir. 1998).

164. *Geertson Seed Farms, et al., v. Mike Johanns, Secretary of the United States Department of Agriculture, et al.*, Defendants, No. C 06–01075 CRB, United States District Court for the Northern District of California, 2007 U.S. Dist. LEXIS 14533, February 13, 2007.

165. The ruling did, however, allow farmers who had purchased their seed prior to March 12, 2007, to plant the seed and harvest existing crops. *Geertson Seed Farms et al., v. Mike Johanns, Secretary of the United States Department of Agriculture, et al.*, Defendants, No. C 06–01075 CRB, United States District Court for the Northern District of California, 2007 U.S. Dist. LEXIS 21491, March 12, 2007.

166. APHIS 2007*c*.

167. While the court declined to enjoin the harvest and sale of Roundup Ready alfalfa planted prior to March 30, 2007, it ruled that APHIS must issue an administrative order imposing conditions that will minimize the risk of gene transfer from existing Roundup-Ready alfalfa to organic and conventional varieties. *Geertson Seed Farms, et al., v. Mike Johanns, Secretary of the United States Department of Agriculture, et al.*, Defendants, No. C 06–01075 CRB, United States District Court for the Northern District of California, 2007 U.S. Dist. LEXIS 32701, May 3, 2007. APHIS announced the start of its EIS in early 2008; see Greenwire 2008.

168. Quoted in Lifsher 2007.

169. See e.g., the discussion of the "Glofish" case below, in which an FDA decision not to regulate a GM aquarium fish was upheld on several occasions by federal district courts.

170 . Given the high costs of producing cloned animals, on the order of $20,000 for a single cow, clones are not envisaged for use as meat, but rather for breeding purposes, to produce cattle or dairy cows with particularly desirable traits. For a good discussion, see FDA 2006*c*.

171. Quoted in FDA 2006*b*. This assessment, it should be noted, was limited to adult cattle, pigs, and goats, and did not extend to sheep, for which insufficient data was available to reach a conclusion.

172. FDA 2006*c*.

173. For example, Andrew Kimbrell, director of the Center for Food Safety, suggested that the FDA "has been trying to foist this bad science on us for several years," while Carol Tucker Foreman, director of the Food Policy Institute at the Consumer Federation of America, framed the issue as one of consumer choice: "I should have freedom not to spend my money and not to eat products that offend me.…This product, which causes great discomfort to a great number of people, goes on market with no labeling that enables me to make a choice." Quoted in Kaplan and Chong 2006.

174. Pollack and Martin 2006.

175. "There's a real trust in milk as a wholesome provider of core nutrition in your diet," said a representative of the association. "You don't want to fool around with that." Quoted in Weiss 2006c: A16.
176. A statement from the company indicated that, "Numerous surveys have shown that Americans are not interested in buying dairy products that contain milk from cloned cows." Quoted in *Grand Rapids Press* 2007: A8.
177. Pollack and Martin 2006.
178. Henderson 2007. Scientists have also weighed in on opposites sides of the issue: see e.g., the debate between David Schubert, a professor at the Salk Institute for Biological Studies, and Henry I. Miller, fellow of the Hoover Institution and former head of the FDA's Office of Biotechnology (1989–1993). See Miller 2007; Schubert 2007.
179. Food and Drug Administration 2008b, quoted in Weiss 2008b; Martin and Pollack 2008. The complete set of cloning-related documents from the FDA is available at its website, http://www.fda.gov/cvm/cloning.htm, accessed on November 19, 2008. By coincidence, the same month the EFSA released its own risk assessment of meat and milk from cloned animals, similarly concluding that meat and milk from healthy clones were "very unlikely" to pose risks to consumers. See European Food Safety Authority 2007; Bounds 2007b; Weiss 2008a. By contrast with the FDA's final assessment, however, the EFSA assessment was a nonbinding draft assessment for public consultation, and was followed by a second and more critical report from the European Group on Ethics in Science and New Technologies, which concluded: "Considering the current level of suffering and health problems of surrogate dams and animal clones, the EGE has doubts as to whether cloning animals for food supply is ethically justified. Whether this applies also to progeny is open to further scientific research. At present, the EGE does not see convincing arguments to justify the production of food from clones and their offspring." European Group on Ethics in Science and New Technologies 2008: abstract. A final EFSA opinion was scheduled for May 2008, and an official EU decision to approve or restrict the use of cloned animals appeared far off.
180. "We are very cognizant we have a global environment as it pertains to the movement of agricultural products," said Bruce I. Knight, Under Secretary of Agriculture for Marketing and Regulatory Programs. Quoted in Martin and Pollack 2008.
181. Quoted in Martin and Weiss 2008.
182. Biotechnology Industry Organization 2007a. For a good review of the regulatory issues raised by transgenic animals, see Pew Initiative 2004a.
183. Pew Initiative 2004a: 119–131.
184. "The Food, Drug and Cosmetic Act ('FDCA') provides that any new animal drug is considered unsafe prior to receiving FDA approval for its intended use. 21 U.S.C. §360b(a)(1)(A). To secure such approval, the FDCA requires the applicant to file a New Animal Drug Application ('NADA') that includes information demonstrating both the safety and the efficacy of the drug. Id. § 360b(d)(1)(A)." *A.L. Pharma, Inc. vs. Shalala*, 314 U.S. App. D.C. 152, 62 F.3d 1484, 1486 (D.C. Cir. 1995). See also Rodemeyer 2003; Pew Initiative 2004a: 79.
185. The FDA has been lobbied for clear and specific regulations on transgenic animals by both anti-GMO activists, who consider current regulations inadequate, and by industry representatives such as the Biotechnology Industry Organization, which

seeks legal certainty as well as consumer confidence in new transgenic animals. The FDA, for its part, has counseled patience, saying through a spokeswoman, "We want to get it out, but we also want to get it right." Quoted in Andrew Pollack 2007a: A14.

186. See Pew Initiative 2004a: 18; Andrew Pollack 2007a: A14; and Pew Initiative 2002c ("Companies realistically have to protect their [economic] interest in drugs," says Eric Hallerman, a GM fish researcher and professor at Virginia Polytechnic Institute and State University in Blacksburg, Va. "[But] what that means is that we do not know what is in the pipeline. There is a real transparency problem.")

187. See Pew Initiative 2002c (which notes that the FDA can supplement its expertise by consulting with the US Fish and Wildlife Service and the National Marine Fisheries Service); and Aqua Bounty Technologies 2006.

188. FDA 2003. The FDA statement reads, in its entirety, as follows: "Because tropical aquarium fish are not used for food purposes, they pose no threat to the food supply. There is no evidence that these genetically engineered zebra danio fish pose any more threat to the environment than their unmodified counterparts which have long been widely sold in the US. In the absence of a clear risk to the public health, the FDA finds no reason to regulate these particular fish."

189. Rodemeyer 2003; Pew Initiative 2004a.

190. Pew Initiative 2004a; Biotechnology Industry Organization 2007a.

191. Weintraub 2006.

192. See e.g., Miller and Conko 2003, 2004.

193. As the Pew Initiative 2004a report states, "in practice, agencies have developed a hybrid system that effectively treats biotechnology products differently." Pew Initiative 2004a: 10.

194. Pew Initiative 2004a: 11. See also FDA 2001a; and FDA 1997.

195. See Miller and Conko 2001:31. Former FDA official Henry Miller (2000: A27) claims that "USDA regulations have made experiments with gene-spliced plants ten- to twenty-fold more expensive than the very same field trials with virtually identical organisms crafted with older, less precise techniques."

196. See (40 CFR § 174.3). The overall regulations are contained in 40 CFR Parts 152–174.

197. As the Pew Initiative (2004a:13) notes regarding the differential treatment of different GM varieties, "foods derived from herbicide-tolerant crops [such as Roundup-Ready soybeans] have been considered to be as safe as comparable foods and therefore have gone to market without a mandatory premarket food safety approval (FDA 1992). In contrast, insect-resistant crops [such as Bt corn] is to be approved by the EPA as safe to eat under a mandatory premarket approval process for pesticide residues in food (40 CFR Part 1740)."

198. Compare McGarity and Hansen (2001) (re. insufficient regulation) and Miller and Conko (2003, 2004) (re. over-regulation). Miller and Conko note how the biotech industry itself has agreed to greater regulatory oversight, and claim that the real Frankenstein is the regulatory process itself.

199. See e.g., *Alliance for Bio-Integrity et al. v. Shalala*, 116 F. Supp.2d 166 (D.D.C. 2000).

200. The Pew report is particularly noteworthy in that, unlike many European critics of the US system, it does not call into question the fundamentals of the US system, but rather points to lacunae in the regulatory authority of the three core agencies under existing legislation, particularly with regard to bioengineered animals, and

puts forward a menu of possible reforms to clarify and strengthen the agencies' regulatory authority concluding with a proposal for a better coordinated "single door" approach. Pew Initiative 2004*a*: 141–147.

201. Prakash and Kollman 2003: 618.
202. See Baumgartner and Jones 1993: 37, 46 (assessing politics in terms of "punctuated equilibrium," noting the need for longitudinal studies because regulatory systems can be characterized by "long periods of no change or dramatic reversals in outcomes in relatively short periods of time," as issues are redefined).

Chapter 7

1. We are grateful to Emilie Hafner-Burton for this insight.
2. Joerges and Neyer 1997, 2003.
3. Chalmers 2003: 540.
4. Chalmers 2003: 543.
5. For an assessment of the constraints on EFSA compared to the European Agency for the Evaluation of Medicinal Products, see Krapohl 2004.
6. The results can be rules and principles for what has been termed negative integration (i.e., those that tell states what they cannot do), and rules and principles for positive integration (i.e., those that tell states what they must do).
7. Such validation would only protect the EU from future WTO challenge as of the date of the new risk assessment to the extent that the safeguards were deemed to be based on this new risk assessment. The initial safeguards would still be deemed to be inconsistent with WTO obligations as they were not based on a risk assessment. However, as the WTO does not provide for retrospective damages, this distinction would have little practical effect.
8. Keohane and Nye 1974.
9. The WTO panel decision, we have argued, also has implications for other regimes, such as the OECD, the Convention on Biodiversity, and Codex Alimentarius Commission, in the international regime complex for GM foods and crops. On the one hand, the WTO decision provides an impetus for regulatory work in those regimes; on the other hand, however, it has also had the effect of politicizing or "hardening' bargaining and thereby, bringing about paralysis within such soft-law regimes; see Chapter 4.
10. See Nicolaidis and Shaffer 2005.
11. For a related argument, see Pollack, 2003*a*, 2003*c*.
12. The Biosafety Protocol, for example, has provided a common definition of key terms like "living modified organisms" and "informed consent," and has begun to clarify international requirements for labeling of shipments containing LMOs. Similarly, Codex and the WTO have clarified what is required for risk assessments of GM foods and crops, and the OECD has provided guidance on the assessment of specific GM varieties through its consensus documents.
13. This is not to say that the US and the EU could not do more to manage system friction between them. Both scholars and practitioners, for example, have argued for the introduction of "trade impact assessments" of future regulations, to focus the

attention of regulators and legislators, reciprocally, on the effects of new legislation on each other's constituents, while others have called for greater transparency of rule-making, allowing all concerned actors (including foreign individuals or governments) some opportunity to provide input into the making of new regulations. Such domestic reforms would not eliminate friction between the two sides, but they would at least ensure that trade-related considerations, and the concerns of affected foreign constituencies would be heard in the regulatory process. See e.g., Pollack 2003c: 602.

14. Stewart 2007: 13. For other useful surveys of GMO regulation in less developed countries, see Paarlberg 2001; Nuffield Council on Bioethics 2003; FAO 2004a; Fukuda-Parr 2007a; and Vogel et al. 2007. James 2007 reports figures for the commercial cultivation of crops in various developing as well as developed countries; see below.

15. Stewart 2007: 13–14. See also Paarlberg 2001.

16. Stewart 2007: 13.

17. There have been significant informal markets for GM seeds in Argentina, Brazil, China, and India. See Herring 2007b (re India and Brazil); and Fukuda-Parr 2007d: 218.

18. Fukuda-Parr 2007e: 223–225.

19. Over 50 per cent of cotton grown in China and South Africa was GM cotton by 2006, and the percentage of GM cotton in India is rapidly expanding. Fukuda-Parr 2007d: 201. See also and Food and Agriculture Organization 2004b ("In China, 4 million small-scale farmers are currently growing insect-resistant Bt cotton on about 30 per cent of the country's total cotton area").

20. See Herring 2007a: 11; Nuffield Council on Bioethics 2003: 76. In fact, at one stage, the Gujurati provincial government simply ignored the demands of the central government to enforce federal rules because of the demands of local farmers. Roy et al. 2007: 160.

21. Herring 2007b: 130.

22. Herring 2007b: 140.

23. Roy et al. 2007: 160.

24. Herring 2007a; and Lopes and Sampaio 2005: 94–95.

25. Paarlberg 2001: 153. They have done so, however, with a relatively greater focus on intellectual property issues. See e.g., Shiva 2000; Shiva et al. 2000; and Herring 2006.

26. Paarlberg 2001.

27. See e.g., Fukuda-Parr 2007d: 217.

28. Fukuda-Parr 2007d: 204, 213, 215; Herring 2007a: 17.

29. Fukuda-Parr 2007d: 202–203; and FAO 2004a: 51.

30. See Fukuda-Parr 2007d: 213 (re license of Chinese varieties) and Mehta 2006 (recompetition case on pricing).

31. Vogel, Newell and Trigo 2008: 44.

32. Stewart 2007: 1.

33. Stewart 2007: 17; and Meijer and Stewart 2004: 253 (citing the United States Leadership Against HIV/AIDS, Tuberculosis, and Malaria Act, Public Law 108–25, 108th Congress, First Session, 27 May 2003).

34. See Meijer and Stewart 2004: 253 (comparing EU and US technical assistance programs).

35. Stewart 2007: 16.
36. See Nwalozie et al. 2007: 72.
37. Vogel, Newell and Trigo 2008: 17, 38, 39.
38. Vogel, Newell and Trigo 2008: 30, 32.
39. Gupta and Falkner 2006. See also Meijer and Stewart 2004: 259 ("Organizations such as UNEP and GEF push an agenda focusing on the risks of GMs, working with environment ministries"). GEF is the Global Environment Facility, which provides countries with funding to implement their obligations under the Biosafety Protocol. GEF-funded projects are implemented by UNEP, the UNDP and the World Bank.
40. This was confirmed by many of our interviewees. See also Vogel, Newell and Trigo 2008: 45; Fukuda-Parr 2007d: 217; Fukuda-Parr 2007e: 228; and Paarlberg 2001: 154.
41. See e.g., Stewart 2007: 22, 68 (noting asymmetric costs of segregation for developing countries to meet traceability and labeling requirements).
42. Vogel, Newell and Trigo 2008: 13. See also FAO 2004a: 88.
43. Scoones 2006: 337 (on India).
44. Herring 2007a: 17. Pray and Naseem (2007: 205) note that the cost for Monsanto "to obtain regulatory approval for Bt cotton in India…[was] more than the annual research budgets of most Indian seed companies."
45. Krent 2004: 65. See also Nuffield Council on Bioethics 2003: 38, 54; and Herring 2007a: 19. Moreover, there is evidence that multinational companies will not even sell existing GM products in countries that have not adopted biosafety and food safety regulations because of concerns over liability and over their reputation, should something go wrong.
46. See Stewart 2007: 84–86.
47. Stewart 2007: 17–18.
48. For good case studies on Argentina, Brazil, and South Africa, see Chudnovsky 2007; da Silveira and Borges 2007; and Gouse 2007, respectively.
49. See Huang et al. 2007: 131–133; Vogel et al. 2008: 29; and BBC Monitoring International Reports 2006.
50. Fields 2006; Huang et al. 2007: 132–133.
51. Cookson 2006 (stating that "GM soya will not be commercialized until China has its own varieties"). Despite this official caution, there is evidence that Chinese farmers have already begun to grow GM rice and other crops without official approval, with some GM rice turning up in Chinese rice shipments to the EU. For good accounts of Chinese biotech policy, see e.g., Huang et al. 2007; and Gupta and Falkner 2006.
52. For good reviews of the Indian case, see Ramaswami and Pray 2007, and the discussion in James 2007.
53. Ramaswami and Pray 2007: 161.
54. See e.g., Nwalozie et al. 2007: 72.
55. Stewart 2007: 18, 20; Fukuda-Parr 2007c: 27.
56. "The Green Revolution is the popular term for the development and spread of high-yielding staple foods in developing countries," which "was brought about almost exclusively through research undertaken by institutions in the public sector." Nuffield Council on Bioethics 2003: 6. See also World Bank 2007: 159–60 (noting "the limited green revolution in Sub-Saharan Africa").
57. Stewart 2007: 18.

58. See e.g., Fukuda-Parr 2007a; Vogel et al. 2007; Stewart 2007; and Paarlberg 2001.

59. See James 2007: xii, who counts fifty-two states which have explicitly approved GM foods and crops for importation—far more than the twenty-three countries that have proceeded with cultivation of GM crops. Cf Stewart 2007: 16 (citing a larger number importing GM products).

60. James 2006, 2007. Farmers growing on a small scale have rapidly adopted GM varieties, which appear to be scale-neutral, or even to be of greater benefit to small-scale farmers. See FAO 2004: 6, 49; Fukuda-Parr 2007b: 10; and Herring 2007a: 11

61. James 2007. Cf. Fukuda-Parr 2007c: 20 (also listing twelve LDCs).

62. Stewart 2007: 16 (citing FAO estimates). Cf James 2006 (citing a lower number).

63. Fukuda-Paar 2007b: 20.

64. James 2006, 2007.

65. See also the comments of Hugh Grant, chairman and chief executive of Monsanto, who describes the first decade of GM crop development as being about "weeds and bugs." By contrast, he predicts, "The second decade is going to be about hundreds of genes that grow yield." Quoted in Cameron 2007.

66. Fukuda-Parr 2007b: 6.

67. Stewart 2007: 68.

68. James 2006, 2007 (noting significant growth in growth of varieties with stacked traits).

69. See Pollack 2006 (noting how agricultural biotechnology can improve the chemical processes for more efficiently converting the crops into fuel).

70. See FAO 2004a: 5; and James 2006.

71. Fukuda-Parr 2007a; and Lipton 2007: 52 ("Below 0.05 per cent of transgenic area was planted with any main food staple," while noting that soy and maize are primarily for animal feed).

72. Herring 2007a: 7; FAO 2004a; Fukuda-Parr 2007a; World Bank 2007: 163; Nuffield Council on Bioethics 2003; and Pinstrup-Andersen and Schioler 2000.

73. See United Nations 2006: 5; Nuffield Council on Bioethics 2003: 5, 62 (citing World Bank 2003). Moreover, the WHO and FAO have found that over two billion people are afflicted with micronutrient malnutrition, and well over half of child mortality is associated with nutritional deficiencies. See FAO/WHO 2006: xviii; and FAO 2004: 9.

74. See Lipton 2007: 56; and Nuffield Council on Bioethics 2003: 42.

75. See Andrew Pollack 2008.

76. See Bouis 2007.

77. See Nuffield Council on Bioethics 2003: 35–42 (providing a number of case studies, including increasing yield in rice by dwarfing, improved micronutrients in rice, improved resistance to viruses in sweet potato, and improved resistance to diseases in bananas).

78. Nuffield Council on Bioethics 2003: 5 (citing World Bank 2003). See also World Bank 2007: xiii. ("Three out of every four poor people in developing countries live in rural areas, and most of them depend directly or indirectly on agriculture for their livelihoods").

79. See e.g., Nuffield Council on Bioethics 2003. The FAO notes that "there is clear evidence that the problems of the poor are being neglected. Barring a few initiatives

here and there, there are no major public- or private sector programmes to tackle the critical problems of the poor or targeting crops that they rely on." FAO 2004a: 4–5. Prabhu Pingali, the Director of the Agricultural and Development Economics Division of the FAO, notes how available peer-reviewed studies show that "the technology may in fact be strongly pro-poor." See Pingali 2007.

80. For example, "golden rice" that would produce enhanced levels of vitamin A reportedly required "licenses covering 70 patents belonging to 32 different owners," which "delayed progress in development about twelve months," although waivers of license fees were obtained. Nuffield Council on Bioethics 2003: 86. Vitamin A deficiency affects some 250 million children, "of whom some three million suffered xerophthamlmia, the primary cause of childhood blindness." Nuffield Council on Bioethics 2003: 37.

81. As the FAO notes, "[a]n expensive, unpredictable and opaque biosafety regulatory regime is even more restrictive for public research than private research because public institutes have considerably less money to finance the search trials required to meet regulatory requirements." FAO 2004a: 88. See also Nuffield Council on Bioethics 2003: 38, 54; World Bank 2007: 178–179; and Herring 2007b: 147–150 (focusing on opportunity costs). See, for example, the story of GM papaya, in Herring 2007a: 5 ("What eventually hurt small farmers who adopted transgenic papaya was not failure of the technology, nor intellectual property, but Pandoran logic: regulatory exclusion of transgenics from their major market, Japan. In a deeply ironic outcome, the beneficiary of Japan's restrictive policy was the transnational Dole, the world's largest producer and marketer of fresh fruits and vegetables", citing Gonsalves et al. 2007).

82. Fukuda-Parr 2007d: 213–214.

83. The CGIAR, or Consultative Group on International Agricultural Research, includes fifteen international agricultural research institutes, such as the International Rice Research Institute in the Philippines. See FAO 2004a: 28–33: 33; World Bank 2007: 178; and Fukuda-Parr 2007e: 231.

References

Abbott, Kenneth W., Robert O. Keohane, Andrew Moravcsik, Anne-Marie Slaughter, and Duncan Snidal. (2000). The Concept of Legalization, *International Organization*, Vol. 54, No. 3, pp. 17–35.

——and Duncan Snidal. (2000). Hard and Soft Law in International Governance, *International Organization*, Vol. 54, No. 3, pp. 421–56.

————(2001). International 'Standards' and International Governance, *Journal of European Public Policy*, Vol. 8, No. 3, pp. 345–70.

Abels, Gabriele. (2002),Experts, Citizens, and Eurocrats – Towards a Policy Shift in the Governance of Biopolitics in the EU, *European Integration online Papers* (EIoP), Vol. 6, No. 2002-019. Available at SSRN: http://ssrn.com/abstract=357501 or DOI: 10.2139/ssrn.357501, accessed on 24 November 2008.

Acalde, Miguel et al. (2006). Environmental Biocatalysis: From Remediation with Enzymes to Novel Green Processes, *Trends in Biotechnology*, Vol. 24, No. 6, pp. 281–7.

Ackerman, Frank and Lisa Heinzerling. (2004). *Priceless: On Knowing the Price of Everything and the Value of Nothing*, New York: New Press.

Adelman, David E. and John H. Barton. (2002). Environmental Regulation for Agriculture: Towards a Framework to Promote Sustainable Environmental Agriculture, *Stanford Environmental Law Journal*, Vol. 21, No. 1, pp. 3–43.

Adler, Emanuel. (1998). Seizing the Middle Ground: Constructivism in World Politics, *European Journal of International Relations*, Vol. 3, No. 3, pp. 291–318.

Agence France-Presse. (2007*a*). EU Drops Mandatory GM Tests on US Maize Gluten, *Agence France-Presse*, 16 January.

——(2007*b*). France Mulling Freeze on GM Crops: Report, *Agence France-Presse*, 20 September.

——(2008). US Holds fire on Sanctions Against EU in Biotech Food Dispute, *Agence France-Presse*, 14 January.

Alter, Karen J. (2003). Resolving or Exacerbating Disputes? The WTO's New Dispute Resolution System, *International Affairs*, Vol. 79, No. 4, pp. 783–800.

——(2008). Agents or Trustees? International Courts in their Political Context, *European Journal of International Relations*, Vol. 14, No. 1, pp. 33–63.

——and Sophie Meunier. (2006). Nested and Overlapping Regimes in the Transatlantic Banana Trade Dispute, *Journal of European Public Policy*, Vol. 13, No. 3, pp. 362–82.

————(2007). The Politics of International Regime Complexity, *Roberta Buffett Center for International and Comparative Studies Working Paper*, No. 3, http://ssrn.com/abstract = 996889, accessed on 7 July 2007.

Andrée, Peter. (2002). The Biopolitics of Genetically Modified Organisms in Canada, *Journal of Canadian Studies*, Vol. 37, No. 3, pp. 162–91.

Andrews, David, Mark A. Pollack, Gregory C. Shaffer, and Helen Wallace, eds. (2005). *The Future of Transatlantic Economic Relations: Continuity Amid Discord*, Florence: European University Institute, Robert Schuman Centre for Advanced Studies (available on-line at http://www.iue.it/RSCAS/e-texts/Future_Transat_EconRelations.pdf, accessed on 8 June 2007).

Ansell, Christopher, Rahsaan Maxwell and Daniela Sicurelli. (2006). Protesting Food: NGOs and Political Mobilization in Europe, in Christopher Ansell and David Vogel, eds., *What's the Beef? The Contested Governance of European Food Safety*, Cambridge, MA: MIT Press, pp. 97–122.

APHIS. (2003). International Standards for Phytosanitary Measures, Pest Risk Analysis for Living Modified Organisms, United States Comments, 10 October, http://www.aphis.usda.gov/ppq/pim/standards/LMOPRA-US.PDF, accessed on 7 July 2007.

——(2004). Environmental Impact Statement: Introduction of Genetically Engineered Organisms. *Federal Register* Vol. 69, 22 January p. 3271.

——(2005). APHIS Factsheet: Permitting Genetically Engineered Plants That Produce Pharmaceutical Compounds, July 2005, http://www.aphis.usda.gov/lpa/pubs/fsheet_faq_notice/fs_brspharmaceutical.pdf, accessed on 27 July 2007.

——(2007a). Low-Level Presence: Factsheet, March, http://www.aphis.usda.gov/publications/biotechnology/content/printable_version/fs_llppolicy3–2007.pdf, accessed on 29 May 2007.

——(2007b). APHIS Statement: Statement by Dr. Ron DeHaven Regarding APHIS Hold on Clearfield CL131 Long-Grain Rice Seed, 5 March.

——(2007c). Return to Regulated Status of Alfalfa Genetically Engineered for Tolerance to the Herbicide Glyphosate, *Federal Register*, Vol. 72, No. 56, Friday, 23 March, pp. 13735–6.

Aqua Bounty Technologies. (2006). Our Products: Frequently Asked Questions, http://www.aquabounty.com/faq.html, accessed on 1 May 2007.

Aritake, Toshio. (1999). Japan Moving on GMO Labeling Draft, Other Rules, as GMO-Free Market Surges, *BNA International Trade Reporter*, Vol. 16, p. 1741 (27 October).

Arnull, Anthony. (2006). *The European Union and Its Court of Justice*, New York: Oxford University Press.

Asinari, María, Verónica, Pérez, and Poullet, Yves. (2005). Privacy, Personal Data Protection, and the Safe Harbour Decision: From Euphoria to Policy, From Policy to Regulation? In Andrews et al., eds., *The Future of Transatlantic Economic Relations: Continuity Amid Discord*, Florence: European University Institute, Robert Schuman Centre for Advanced Studies, pp. 101–34.

Associated Press. (2004). EU Governments Deadlocked over Monsanto's Genetically Modified Corn Product, *Associated Press*, 28 June.

——(2006a). Majority of EU Governments Demand Changes to Biotech Crop Approval System, *Associated Press*, 9 March 2006.

——(2006b). 13 Biotech Rice Lawsuits Could Merge, *Associated Press*, 30 November.

——(2007). EU Warns French Biotech Crop Freeze Plan Could Infringe European Law, *Associated Press*, 26 October.

Austria Today. (2006a). Austria to Reopen EU GM Debate, *Austria Today*, 23 February.

——(2006b). EU May Order Austria to Lift Ban on GMO Maize Types, *Austria Today*, 5 September.

Axelrod, Robert. (1984). *The Evolution of Cooperation*, New York: Basic Books.

Axelrod, Robert. (1997). *The Complexity of Cooperation: Agent-Based Models of Competition and Collaboration*, Princeton, NJ: Princeton University Press.

Bach, David. (2000). International Cooperation and the Logic of Networks: Europe and the Global System for Mobile Communications (GSM), BRIE Working Paper 139, E-economy Project™, Berkeley, CA: University of California Press.

Baird, Douglas, Robert Gertner, and Randall Picker. (1994). *Game Theory and the Law*, Cambridge, MA: Harvard University Press.

Bakshi, Anita. (2003). Potential Adverse Health Effects of Genetically Modified Crops, *Journal of Toxicology and Environmental Health*, Vol. 6, No. 3, pp. 211–26.

Baldwin, David A. (ed.) (1993). *Neorealism and Neoliberalism: The Contemporary Debate*, New York: Columbia University Press.

Barnes, Barry, David Bloor, and John Henry. (1996). *Scientific Knowledge: A Sociological Analysis*, Chicago, IL: University of Chicago Press.

Barnett, Katherine. (2002). Food Fights: Canadian Regulators Are Under Pressure to Face the Uncertainties of Genetically Modified Food, *Canadian Business and Current Affairs*, Vol. 28, No. 1, pp. 28–33.

Barlow, Thomas. (1999). A Bug's Life Goes Global, *Financial Times*, 11 September, p. 2.

Battini, Stefano. (2007). *Amministrazioni Nazionali e Controversie Globali*, Milan: Dott. A Giuffre Editore.

Baumgartner, Frank and Bryan Jones. (1993). *Agendas and Instability in American Politics*, Chicago, IL: University of Chicago Press.

Bauer, Martin. (2002). Controversial Medical and Agro-Food Biotechnology: A Cultivating Analysis, *Public Understandings of Science*, Vol. 11, pp. 93–111.

BBC Monitoring International Reports. (2006). China Breeds 55 New Genetically Modified Cotton Varieties—Official, *BBC Monitoring International Reports*, 27 December, accessed at Lexis-Nexis Academic on 18 April 2007.

——(2007). Austrian EU Official Urging EU Legal Framework on GM Crops, *BBC Monitoring International Reports*, 7 April, accessed at Lexis-Nexis Academic on 5 May 2008.

BBC News. (2000). John Grummer, Beef Eater, *BBC News*, 11 October, available at http://news.bbc.co.uk/1/hi/uk/369625.stm, accessed 26 April, 2006.

Beatty, Andrew. (1992). *Risk Society: Towards a New Modernity*, London: Sage.

——(2004). Majority of New EU States Block GMO Approval, *euobserver.com*, 21 June, accessed on 22 June 2004.

Beck, Ulrich. (1999). *World Risk Society*, Cambridge: Polity Press.

Becker, Elizabeth. (2003). U.S. Delays Suing Europe over Ban on Modified Food, *New York Times*, 4 February, p. 6.

Benedick, Richard Elliot. (1998). *Ozone Diplomacy: New Directions in Safeguarding the Planet*, Cambridge, MA: Harvard University Press.

Bennet, David. (2007). U.S. Rice Leaps EU Testing Hurdle, Sets Up Easier Exports, *Western Farm Press Online*, 28 December.

Benvenisti, Eyal and George W. Downs. (2007). The Empire's New Clothes: Political Economy and the Fragmentation of International Law, *Stanford Law Review*, Vol. 60. Available at SSRN: http://ssrn.com/abstract = 976930, accessed on 3 January 2008.

Berg, Bettina. (2002) GMO Proposals Adopted in European Parliament, *euobserver.com*, 3 July.

Berman, Paul. (2007). Global Legal Pluralism, *Southern California Law Review*, Vol. 80, pp. 1155–237.

References

Bermann, George, Matthias Herdegen, and Peter Lindstreth, eds. (2001). *Transatlantic Regulatory Co-Operation: Legal Problems and Political Prospects*, New York: Oxford University Press.

Bernauer, Thomas. (2003). *Genes, Trade and Regulation: The Seeds of Conflict in Food Biotechnology*, Princeton, NJ: Princeton University Press.

——and Erika Meins. (2003). Technological Revolution Meets Policy and the Market: Explaining Cross-National Differences in Agricultural Biotechnology Regulation, *European Journal of Political Research*, Vol. 42, 643–83.

——and Thomas Sattler. (2006). Dispute-Escalation in the WTO: Are Conflicts over Environment, Health and Safety Regulation Riskier? Center for Comparative and International Studies Working Paper, Zurich, No. 21.

Bhala, Raj and Kevin Kennedy. (1998). *World Trade Law*, Charlottesville, VA: Lexis Law Publishing.

Biotechnology Industry Organization (BIO). (2007*a*). Transgenic Animals: Frequently Asked Questions, http://www.bio.org/animals/faq.asp, accessed on 19 April 2007.

——(2007*b*). *Guide to Biotechnology 2007*, available at http://bio.org/speeches/pubs/er/, accessed on 17 July 2007.

——(2007*c*). Bio, Product Launch Stewardship Policy, adopted 21 May 2007, available at http://www.bio.org/foodag/stewardship/20070521.asp (accessed on 28 Jan. 2008).

Bloomberg News. (2007). No U.S. Penalties for Tainted Rice, *The New York Times*, 6 October, p. C3.

Boardman, Robert. (1986). *Pesticides in World Agriculture: The Politics of International Regulation*, New York: St. Martin's Press.

Bodansky, Daniel. (1991). Scientific Uncertainty and the Precautionary Principle, *Environment*, Vol. 33, pp. 4–5, 43–4.

Bodansky, Daniel. (1999). The Legitimacy of International Governance: A Coming Challenge for International Environmental Law?, *American Journal of International Law*, Vol. 93, pp. 596–624.

Börzel, Tanja A. and Thomas Risse. (2007). Europeanization: The Domestic Impact of EU Politics, in Knud Erik Jorgensen, Mark A. Pollack, and Ben Rosamond, eds., *The Handbook of European Union Politics*, New York: Sage, pp. 483–504.

Bohanes, Jan. (2002.) Risk Regulation in WTO Law: A Procedure-Based Approach to the Precautionary Principle, *Columbia Journal of Transnational Law*, Vol. 40, pp. 323–89.

Bohman, James. (1998). Survey Article: The Coming of Age of Deliberative Democracy, *Journal of Political Philosophy*, Vol. 6, No. 4, pp. 400–25.

Borrás, Susana and Kerstin Jacobsson. (2004). The Open Method of Coordination and the New Governance Patterns in the EU, *Journal of European Public Policy*, Vol. 11, No. 2, pp. 185–208.

Botkin, Daniel. (1996). Adjusting Law to Nature's Discordant Harmonies, *Duke Environmental Law and Policy Forum*, Vol. 7, No. 4, pp. 25–38.

Bouis, Howarth E. (2007). The Potential of Genetically Modified Food Crops to Improve Human Nutrition in Developing Countries, *Journal of Development Studies*, Vol. 43, No. 1, pp. 79–96.

Bounds, Andrew. (2006*a*). EU States Order Tests on US Imports After Finding Illegal GMOs, *Financial Times*, 27 October, p. 7.

——(2006*b*). Andrew Bounds, EU Fails to Agree on Modified Crop, *Financial Times*, 5 December.

——(2006c). Austria Allowed to Keep its Ban on GM Corn, *Financial Times*, 19 December, p. 12.

——(2007a). EU Vote Reinforces Resistance to GMOs, *FT.com*, 20 February.

——(2007b). Cloned Meat Could Be Allowed in EU, *FT.com*, 9 March.

——(2007c). Mandelson urges EU to speed GM food approvals, *Financial Times*, 15 June.

Boyle, James. (1985). The Politics of Reason: Critical Legal Theory and Local Social Thought, *University of Pennsylvania Law Review*, Vol. 133, pp. 685–780.

Braithwaite, John and Peter Drahos. (2000). *Global Business Regulation*, New York: Cambridge University Press.

Bradley, Kieran St C. (1998). The GMO-Committee on Transgenic Maize: Alien Corn, or the Transgenic Procedural Maize, in M.P.C.M. Van Schendelen, ed., *EU Committees as Influential Policymakers*, Aldershot: Ashgate, pp. 207–22.

Brand, Constant. (2006a). EU Nations Divided over new Organic Food Labelling Rules, *Associated Press*, 22 May.

——(2006b). EU Environment Ministers Reject Appeal to Force Austria to Lift Bans on Biotech Crops, *The Associated Press*, 18 December.

——(2007a). Greenpeace Appeals to EU to Reject Use of 2 New Biotech Corn Products, *Associated Press*, 28 November.

——(2007b). Germany Calls for Review of How European Union Approves New Biotech, *Associated Press*, 26 November.

——(2008). EU Panel OK's 2 Genetically Modified Corn Products, *Associated Press*, 31 October.

Bratton, William, Joseph McCahery, Sol Picciotto, and Colin Scott, eds. (1996). *International Regulatory Competition and Coordination: Perspectives on Economic Regulation in Europe and the United States*, New York: Oxford University Press.

Bren, Linda. (2003). Genetic Engineering: The Future of Foods? *FDA Consumer Magazine*, Vol. 37, No. 6 (November–December), online at http://www.fda.gov/Fdac/features/2003/603_food.html, accessed on 15 October 2007.

Breyer, Stephen. (1993). *Breaking the Vicious Circle: Toward Effective Risk Regulation*, Cambridge, MA: Harvard University Press.

Bridges BioRes. (2003). European Food Committee Fails to End *de Facto* Biotech Moratorium, *Bridges BioRes*, Vol. 3, No. 22, 15 Dec., at http://www.ictsd.org/biores/03–12–15/story1.htm, accessed on 18 August 2005.

——(2004). EU Approves Another GM Import as WTO Dispute Drags On, *Bridges BioRes*, Vol. 4, No. 14, 23 July, at http://www.ictsd.org/biores/04-07-23/inbrief.htm#4, accessed on 12 June 2007.

Bridges Weekly Trade News Digest. (2000). Furor over Starlink Triggers Regulatory Rethink, *Bridges Weekly Trade News Digest*, Vol. 4, No. 45, 28 November.

——(2004). US Argues EC GMO Moratorium Hurts Developing Countries, *Bridges Weekly World Trade Digest*, Vol. 8, No. 15, 28 April.

Broad, William J. (2007). Useful Mutant, Bred with Radiation, *The New York Times*, 28 August, p. D1.

Brookes, Graham and Peter Barfoot. (2006). GM Crops: The First Ten Years Global Socio-Economic and Environmental Impacts (ISAAA Brief No. 36), available at http://www.isaaa.org/resources/publications/briefs/36/download/isaaa-brief-36–2006.pdf, accessed on 12 July 2007.

References

Broude, Tomer. (2007). Genetically Modified Rules: The Awkward Rule-Exception-Right Distinction in EC–Biotech, *World Trade Review*, Vol. 6, No. 2, pp. 215–31.

Brown, Paul. (2004). EU Races to Thwart Influx of GM Food from East, *The Guardian*, 14 February.

Browne, Anthony. (2004). Protests After Europe Ends GM Food Freeze, *The Times*, 20 May, p. 18.

Buck, Tobias. (2003). Blow to US as Egypt Pulls Out of Modified Crops Case, *Financial Times*, 29 May.

——(2007a). EU Wants Rest of the World to Adopt Its Rules, *Financial Times*, 19 February (London edition), p. 7.

——(2007b). Standard Bearer: How the European Union Exports Its Laws, *Financial Times*, 10 July (London edition), p. 13.

Bull, Alan T., Geoffrey Holt, and Malcolm D. Lilly. (1982). *Biotechnology: International Trends and Perspectives*, Paris: OECD.

Burton, John. (2004). International Conference Deals Blow to US on Labeling of Gene Modified Food, *Financial Times*, 28 February, p. 8.

Buonanno, Laurie, Sharon Zablotney, and Richard Keefer. (2001). Politics Versus Science in the Making of a New Regulatory Regime for Food in Europe, *European Integration online Papers*, Vol. 5, No. 12: http://eiop.or.at/eiop/texte/2001–012a.htm, accessed on 1 October 2007.

Buonanno, Laurie. (2003). The Creation of the European Food Safety Authority, in Christopher Ansell and David Vogel, eds., *What's the Beef? The Contested Governance of European Food Safety*, Cambridge, MA: MIT Press, pp. 259–278.

Busch, Lawrence et al. (2004). *Amicus Curiae Brief: Submitted to the Dispute Settlement Panel of the World Trade Organization in the Case of EC: Measures Affecting the Approval and Marketing of Biotech Products*, 30 April. Available at http://www.ecolomics-international.org/biosa_ec_biotech_amicus_academic2_ieppp_lancasteru_coord_0404.pdf, accessed on 17 July 2007.

Business Wire. (2007). BIO Statement: Biotechnology Industry Calls on French President to Reverse Illegal Ban on Biotech Crops, *Business Wire*, 26 October.

Büthe, Tim. (2002). Taking Temporality Seriously, *American Political Science Review*, Vol. 96, No. 3, pp. 481–93.

Cameron, Doug. (2007). Monsanto Lifted by Surge in Corn Planting, *FT.com*, 28 June, accessed on 10 August 2007.

Cameron, James and Juli Abouchar. (1991). The Precautionary Principle: A Fundamental Principle of Law and Policy for the Protection of the Global Environment, *Boston College International & Comparative Law Review*, Vol. 14, pp. 1–27.

Canan, Penelope and Nancy Reichman. (2002). *Ozone Connections: Expert Networks in Global Environmental Governance*, Sheffield: Greenleaf Publishing.

Cantley, Marc. (1995). The Regulation of Modern Biotechnology: A Historical and European Perspective: A Case Study in How Societies Cope with New Knowledge in the Last Quarter of the Twentieth Century, in H. J. Rehm and G. Reed (eds.) in cooperation with A. Pühler and P. Stadler, *Biotechnology, Volume 12: Legal, Economic and Ethical Dimensions*, Weinheim: VCH, pp. 506–681.

Cases, Ildefonso and Victor de Lorenzo. (2005). Genetically Modified Organisms for the Environment: Stories of Success and Failure and What We Have Learned from Them, *International Microbiology*, Vol. 8, No. 3, pp. 213–22.

Casey, Zoë. (2008). French Ban on GM Crop 'Unjustified', *European Voice*, 31 October, http://www.europeanvoice.com/article/2008/10/french-ban-on-gm-crop-unjustified-/62914.aspx?bPrint=1, accessed on 2 November 2008.

Chalmers, Damien. (2003). Food for Thought: Reconciling European Risks and Traditional Ways of Life, *Modern Law Review*, Vol. 66, No. 4, pp. 532–64.

Chang, Howard. (2004). Risk Regulation, Endogenous Public Concerns, and the Hormones Dispute: Nothing to Fear but Fear Itself? *Southern California Law Review*, Vol. 77, No. 4, pp. 743–76.

Charnovitz, Steve. (1997). The World Trade Organization, Meat Hormones, and Food Safety, *International Trade Reporter* (BNA), Vol. 14, pp. 1781–7.

Charney, Jonathan I. (1996). The Implications of Expanding International Dispute Settlement Systems: The 1982 Convention on the Law of the Sea, *American Journal of International Law*, Vol. 90, pp. 69–75.

Chayes, Abraham. and Antonia Handler Chayes. (1993). On Compliance, *International Organization*, Vol. 47, No. 2, pp. 175–205.

Checkagbiotech.org. (2001). Proposed EU Biotech Rules May Cost U.S. Producers $4 Billion, 19 September, www.checkbiotech.org, accessed on 21 April 2004.

Checkel, Jeffrey T. (1998). The Constructivist Turn in International Relations Theory, *World Politics*, Vol. 50, No. 2, pp. 324–48.

——(2001). Taking Deliberation Seriously, ARENA Working Papers, WP 01/14, http://www.arena.uio.no/publications/wp01_14.htm, accessed 11 June 2004.

——(2005). International Institutions and Socialization in Europe: Introduction and Framework, *International Organization*, Vol. 59, No. 4, pp. 801–26.

Chimni, B. S. (2006). Co-Option and Resistance: Two Faces of Global Administrative Law, *NYU Journal of International Law and Politics*, Vol. 37, pp. 799–827.

Cho, Sungjoon. (2007). Of the World Trade Court's Burden (March). Available at SSRN: http://ssrn.com/abstract=969437.

Chudnovsky, Daniel. (2007). Argentina: Adopting RR Soy, Economic Liberalization, Global Markets, and Socioeconomic Consequences, in Sakiko Fukuda-Parr, ed., *The Gene Revolution: GM Crops and Unequal Development*, London: Earthscan Publications, pp. 85–103.

Closa C. (2004). The Convention Method and the Transformation of EU Constitutional Politics, in E. O Eriksen, J. E. Fossum, and A. J. Menéndez, eds., *Developing a Constitution for Europe*, London: Routledge, pp. 183–206.

Codex *Ad Hoc* Intergovernmental Task Force on Foods Derived from Modern Biotechnology. (2006). *Report of the Sixth Session*, ALINORM 07/30/34.

Codex Alimentarius Commission. (1995). *Principles for Food Import and Export Inspection and Certification*, at § 3 (CAC/GL 20-1995).

——(2003a). *Principles for the Risk Analysis of Foods Derived From Modern Biotechnology*, CAC/GL 44-2003.

——(2003b). *Report of the Twenty-Sixth Session of the Codex Alimentarius Commission*, ALINORM 03/41.

——(2003c) *Principles for Food Import and Export Inspection and Certification*, at § 3 (CAC/GL 20-1995), and *Guidelines on the Judgment of Equivalence of Sanitary Measures Associated with Food Inspection and Certification Systems* (CAC/GL 53-2003).

——(2005). *Fifteenth Procedural Manual*, ftp://ftp.fao.org/codex/Publications/ProcManuals/Manual_15e.pdf, accessed on 15 July 2007.

——(2006a). *Sixteenth Procedural Manual*, ftp://ftp.fao.org/codex/Publications/Proc Manuals/Manual_16e.pdf, accessed on 15 July 2007.

——(2006b). *Principles for Traceability/Product Tracing as a Tool Within a Food Inspection and Certification System*, CAC/GL 60–2006.

——(2007). *Report of the Seventh Session of the Codex Ad Hoc Intergovernmental Task Force on Foods Derived from Biotechnology*, Chiba, Japan, 24–28 September.

Codex Committee on Food Labelling. (1997). *Report of the Twenty-Fifth Session of the Codex Committee on Food Labelling*, ALINORM 97/22A.

——(2006). *Report of the Thirty-Fourth Session of the Codex Committee on Food Labelling*, ALINORM 06/29/22.

Codex Committee on General Principles. (2006). *Report of the Twenty-Third Session of the Codex Committee on General Principles*, ALINORM 06/29/33.

Coen, David and Mark Thatcher. (2005).

————(2007). Beyond Delegation: The Rise of European Regulator Networks in Telecommunications and Securities, paper presented at the Biennial Conference of the European Union Studies Association, Montreal, CA, 17–19 May.

Cohen, Joshua and Charles Sabel. (2005). Global Democracy? *New York University Journal of International Law and Politics*, Vol. 37, No. 4, pp. 763–98.

Cole, Nancy. (2006). Farmers Sue Bayer, Riceland Affiliates; 3rd Suit Arises from Discovery of Unapproved Gene-Manipulated Grain in Supplies, *Arkansas Democrat-Gazette*, 30 August.

——(2007a). Arkansas Forges Path Toward Rice Fix, *Arkansas Democrat-Gazette*, 18 January.

——(2007b). Suit on Altered Rice Returns to State Court: Willman Case Ordered Back to Lonoke, *Arkansas Democrat-Gazette*, 11 October.

Commission of the European Communities. (1978). Proposal for a Council Directive Establishing Safety Measures Against the Conjectural Risks Associated with Recombinant DNA Work, *Official Journal of the European Communities*, C301/5–7 (1978).

——(1986). A Community Framework for the Regulation of Biotechnology, Communication from the Commission to the Council, COM(86)573 Final, Brussels, 4 November.

——(1988). Proposal for a Council Directive on the Deliberate Release to the Environment of Genetically Modified Organisms, COM(88)160 final—SYN 131 of 4 May.

——(1993). Growth, Competitiveness, Employment: The Challenges and Ways Forward into the 21st Century—White Paper, COM(1993)700 final of 5 December.

——(1996). Report on the Review of Directive 90/220/EEC in the Context of the Commission's Communication on Biotechnology and the White Paper, COM(1996)630 final of 10 December.

——(1998). Proposal for a European Parliament and Council Directive Amending Directive 90/220/EEC on the Deliberate Release into the Environment of Genetically Modified Organisms, COM(1998) 85 final of 23 February.

——(2000a). Commission Communication on the Precautionary Principle, COM (2000)1 final of 2 February.

——(2000b). White Paper on Food Safety, COM(1999)719 final of 12 January 2000.

——(2000c). Biotechnology in Bilateral and Multilateral Fora, mimeo, Brussels, 25 July 2000, available at http://trade-info.cec.eu.int/doclib/docs/2003/october/tradoc_111712.pdf, accessed on 19 June 2007.

——(2000*d*). Commentary on the Report by the Commission Services, http://europa. eu.int/comm/external_relations/us/biotech/ec_commentary.htm, accessed on 11 June 2004.

——(2001*a*). Proposal for a Regulation of the European Parliament and the Council on Genetically Modified Food and Feed, COM(2001)425 final of 25 July.

——(2001*b*). Proposal for a Regulation of the European Parliament and the Council Concerning Traceability and Labelling of Genetically Modified Organisms and Traceability of Food and Feed Products Produced from Genetically Modified Organisms and Amending Directive 2001/18/EC, COM(2001)182 final of 25 July.

——(2002*a*). Questions and Answers on the Regulation of GMOs in the EU, Commission Memo/02/160, 15 October.

——(2002*b*). Preliminary Comments from the European Commission on the USA Bioterrorism Act, 8 August, http://ec.europa.eu/food/international/trade/us_bio_act_ prel_com_en.pdf, accessed on 25 May 2005.

——(2002*c*). Communication from the Commission to the Council, the European Parliament, the Economic and Social Committee and the Committee of the Regions—Life Sciences and Biotechnology—A Strategy for Europe, COM(2002)27 final of 23 January.

——(2003). Recommendation of 23 July 2003 on Guidelines for the Development of National Strategies and Best Practices to Ensure the Coexistence of Genetically Modified Crops with Conventional and Organic Farming, *Official Journal of the European Communities* L 189, p. 0036–0047, 29/07/2003.

——(2004*a*). Question and Answers on the Regulation of GMOs in the EU, MEMO/04/16, 28 January.

——(2004*b*). State of Play on GMO Authorizations Under EU Law, MEMO/04/17, 28 January.

——(2004*c*). Commission Authorizes Import of Canned GM-Sweet Corn under New Strict Labeling Conditions—Consumers Can Choose, Press Release IP/04/663 of 19 May 2004, Brussels.

——(2004*d*). Communication to the Commission (from the President in association with Mrs Wallstrom, Mr Byrne, Mr Fishcler, Mr Lamy, Mr Likanen and Mr Busquin): For an Orientation Debate on Genetically Modified Organisms and Related Issues, 28 January, available at: http://www.eurunion.org/news/press/2004/20040010.htm (accessed 18 March, 2004).

——(2004*e*). Commission Decision of 19 May 2004 Authorizing the Placing on the Market of Sweet Corn from Genetically Modified Maize Line Bt11 as a Novel Food or Novel Food Ingredient Under Regulation (EC) No. 258/97 of the European Parliament and the Council, *Official Journal of the European Communities* L 300/48 of 25.9.2004. Accessed at http://ec.europa.eu/food/dyna/gm_register/gm_register2.cfm?gm_id = 29 on 4 February 2007.

Commission of the European Union. (2005*a*). Communication to the Commission (from the President in Association with Mrs Fischer Boel, Mr Dimas, Mr Kyprianou, Mr Mandelson, Mr Verheugen and Mr Potocnik). For an Orientation Debate on Genetically Modified Organisms, SEC (2005) 396/3 of 21 March.

——(2005*b*). Report from the Commission to the European Parliament, the Council, the Committee of the Regions and the European Economic and Social Committee. Life Sciences and Biotechnology—A Strategy for Europe. Third Progress Report and Future Orientations, COM(2005)286 final of 29 June.

—— (2005c). Bt10: Commission Requires Certification of US Exports to Stop Unauthorized GMO Entering the EU. RAPID press release, reference IP/05/437, 15 April.

—— (2005d). GMOs: Commission Reaction on Council Votes on Safeguards and GM Maize MON863, RAPID press release, reference IP/05/793, Brussels/Luxembourg, 24 June.

—— (2005e). Commission Decision 2005/635/EC of 31.08.05 Concerning the Placing on the Market, in Accordance with Directive 2001/18/EC of the European Parliament and of the Council, of an Oilseed Rape Product (*Brassica napus* L., GT73 line) Genetically Modified for Tolerance to the Herbicide Glyphosate, *Official Journal of the European Union*, L 228/11 of 3 September.

—— (2006a). Report from the Commission on the Working of Committees during 2005, COM(2006)446 final of 9 August.

—— (2006b). Commission Staff Working Document: Annex to the Report of the Commission on the Working of Committees During 2005, SEC(2006)1065 of 9 August.

—— (2006c), Commission Requires Certification of US Rice Exports to Stop Unauthorized GMO Entering the EU, IP/06/1120, Brussels, 23 August.

—— (2006d). Commission Proposes Practical Improvements to the Way the EU GMO Legislative Framework is Implemented, IP/06/498, Brussels, 12 April.

—— (2006e). Preparation European Council, 18 December 2006, MEMO/06/491, Brussels, 14 December.

—— (2006f). Report from the Commission to the Council and the European Parliament on the Implementation of EC No 1829/2003 of the European Parliament and of the Council on Genetically Modified Food and Feed, COM(2006)626 final, 25 October.

—— (2007a). Communication from the Commission to the Council, the European Parliament, the European Economic and Social Committee and the Committee of the Regions. A Single Market for Citizens: Interim Report to the 2007 Spring European Council, COM(2007)60 final, Brussels, 21 February.

—— (2007b). Economic Impacts of Genetically Modified Crops on the Agri-Food Sector, Working Document Rev. 2, Directorate-General for Agriculture, available at: http://ec.europa.eu/agriculture/publi/gmo/fullrep/cover.htm (last visited on 6 Dec. 2007).

Conant, Lisa. (2007). Judicial Politics, in Knud Erik Jørgensen, Mark A. Pollack and Ben Rosamond, eds., *The Handbook of European Union Politics*, New York: Sage, pp. 213–29.

Congressional Budget Office. (1995). *The Safe Drinking Water Act: A Case Study of an Unfunded Federal Mandate*, Washington, DC: Congressional Budget Office, September.

Conrad, Christine. (2006). PPMs, the EC-Biotech Dispute and the Applicability of the SPS Agreement: Are the Panel's Findings Build on Shaky Ground? Hebrew University of Jerusalem Research Paper No. 8–06 (August).

Convention on Biological Diversity. (2006). Report of the Third Meeting of the Conference of the Parties to the Convention on Biological Diversity Serving as the Meeting of the Parties to the Cartagena Protocol on Biosafety, UNEP/CBD/BS/COP-MOP/3/15, 8 May 2006.

Cookson, Clive. (2006). China Treads a Careful Path Towards Biotech Greatness, *Financial Times*, 1 February, p. A17.

Corporate Europe Observatory. (2001). TABD in Troubled Water: A CEO Issue Briefing, Global Policy Forum website, http://www.globalpolicy.org/socecon/tncs/2001/1008 tabd.htm, accessed on 13 June 2004.

Council of Europe. (2005). *Texts of the Council of Europe on Bioethical Matters, Volume II*, Directorate General I: Legal Affairs, Bioethics Department, CDBI/INF (2005) 2, Strasbourg (January).

Council of the European Union. (1999). 2194 the Council Meeting—Environment—Luxembourg, 24–25 June 1999, Press 203—Nr 9406/99.

——(2000). Council Resolution on the Precautionary Principle, Annex III to the Presidency Conclusions, Nice European Council Meeting, 7–9 December.

——(2005). Press Release: 2670th Council Meeting, Environment, Luxembourg, 24 June 2005, C/05/147, Luxembourg, 24 June, 10074/05 (Presse 147).

——(2006a). Press Release: 2773rd Council Meeting, Environment, Brussels, 18 December 2006, C/06/349, 16164/06 (Presse 249).

——(2006b). Press Release: 2757th Council Meeting, Environment, Luxembourg, 23 October 2006, C/06/287, 13989 (Presse 287).

——(2007a). Press Release: 2785th Council Meeting, Environment, Brussels, 20 February 2007, 6272/07 (Presse 25).

——(2007b). Communiqué de Presse, 2826ème session du Conseil, Environnement, Luxembourg, le 30 octobre 2007, 14178/07 (Presse 247), version provisoire.

Covelli, N. and V. Hohots. (2003). The Health Regulation of Biotech Foods Under the WTO Agreements, *Journal of International Economic Law*, Vol. 6, No. 4, pp. 773–95.

Croome, John. (1995). *Reshaping the World Trading System: A History of the Uruguay Round*, Geneva: WTO Publication Services.

Cullen, Kevin. (1999). Genetically Modified Food Fight Growing Unpalatable, *Boston Globe*, 3 August, p. A1.

Cutler, Claire, Virginia Haufler, and Tony Porter, eds. (1997). *Private Authority and International Affairs*, Albany, NY: State University of New York Press.

Da Silveira, José Maria F.J. and Izaias de Carvalho Borges. (2007). Brazil: Confronting the Challenges of Competition and Protecting Biodiversity, in Sakiko Fukuda-Parr, *The Gene Revolution: GM Crops and Unequal Development*, London: Earthscan Publications, pp. 104–29.

Dabrowska, Patrycja. (2006). *Hybrid Solutions for Hybrid Products? EU Governance of GMOs*, (unpublished Ph.D. thesis, European University Institute).

Dana, David. (2003). A Behavioral Economic Defense of the Precautionary Principle, *Northwestern University Law Review*, Vol. 97, No. 3, pp. 1315–45.

Dashwood, Alan. (1983). Hastening Slowly: The Community's Path Towards Harmonization, in Helen Wallace, William Wallace, and Carole Webb, eds., *Policy-Making in the European Community*, London: John Wiley & Sons, pp. 173–212.

Dalton, Matthew. (2008). EU Keeps 'Zero Tolerance' Policy on Biotech Imports for Now, 20 October, Dow Jones Newswires, on-line at http://greenbio.checkbiotech.org/news/eu_keeps_zero_tolerance_policy_biotech_imports_now, accessed on 24 November 2008.

De Búrca., Gráinne and Joanne Scott. (2006). *New Governance in the EU and the US*, Oxford: Hart Publishing.

Delegation of the European Commission to the USA. (2004). EU Commission Pushes for GMO 'Green Light', http://www.eurunion.org/news/press/2004/20040010.htm, accessed on 17 July 2005.

Delmas-Marty, M. (1998). *Trois défis pour un droit mondial*, Paris: Seuil.

de la Porte, Caroline and P. Nanz. (2004). OMC – A Deliberative-Democratic Mode of Governance? The Cases of Employment and Pensions, *Journal of European Public Policy,* Vol. 11, No. 2, pp. 267–88.

Deutsche Presse-Agentur. (2006a). Producers Blast Hungary's Stance on GM Crops, *Deutsche Presse-Agentur,* 29 November.

——(2006b). EU Will Postpone Laws on Genetically Modified Products: Commission, *Deutsche Presse-Agentur,* 10 March.

Deering. (2004). *California Fish & Game Code,* Newark, NJ: LexisNexis Matthew Bender.

Devuyst, Y. (2001). Transatlantic Competition Policy Cooperation, in Mark A. Pollack and Gregory C. Shaffer, eds., *Transatlantic Governance in the Global Economy,* Lanham, MD: Rowman & Littlefield, pp. 127–51.

DeYoung, Karen. (2003). Bush Proclaims Victory in Iraq; Work on Terror Is Ongoing, President Says, *Washington Post,* 2 May, p. A1.

Dhar, Tirtha and Jeremy Foltz. (2005). Milk by Any Other Name. . . . Consumer Benefits from Labeled Milk, *American Journal of Agricultural Economics,* Vol. 87, No. 1, pp. 214–28.

Dobbin, Frank, Beth Simmons, and Geoffrey Garrett. (2007). The Global Diffusion of Public Policies: Social Construction, Coercion, Competition, or Learning?, *Annual Review of Sociology,* Vol. 33, pp. 449–72

Douglas, Mary. (1973). *Natural Symbols: Explorations in Cosmology,* New York: Vintage Books.

——and Aaron Wildavsky. (1982). *Risk and Culture,* Berkeley, CA: University of California Press.

Downs, George W., David M. Rocke, and Peter N. Barsoom. (1996). Is the Good News About Compliance Good News About Cooperation? *International Organization,* Vol. 50, No. 3, pp. 379–406.

Drahos, Peter. (2004). Trading in Public Hope, *ANNALS of the American Academy of Political and Social Science,* Vol. 592, No. 1, pp. 18–38.

Drezner, Daniel W. (2007). *All Politics is Global: Explaining International Regulatory Regimes,* Princeton, NJ: Princeton University Press.

Driesen, David. (2001). What Is Free Trade? The Real Issue Lurking Behind the Trade and Environment Debate, *Virginia Journal of International Law,* Vol. 41, pp. 279–368.

Dua, M., A. Singh, N. Sethunathan, A. K. Johri (2002). Biotechnology and Bioremediation: Successes and Limitations, *Applied Microbiology and Biotechnology,* Vol. 59, Nos. 2–3, pp. 143–52.

Dunoff, Jeffrey. (1999). The Death of the Trade Regime, *European Journal of International Law,* Vol. 10, pp. 733–62.

——(2001). The WTO in Transition: Of Constituents, Competence and Coherence, *George Washington International Law Review,* Vol. 33, pp. 979–1013.

——(2006a). Constitutional Conceits: The WTO's 'Constitution', and the Discipline of International Law, *European Journal of International Law,* Vol. 17, pp. 647–75.

——(2006b). Lotus Eaters: Reflections on The Varietals Dispute, the SPS Agreement and WTO Dispute Resolution, in George Bermann & Petros Mavroidis, eds., *Trade and Human Health and Safety,* New York: Cambridge University Press, pp. 153–89.

——Steven Ratner, and David Wippman. (2006). *International Law: Norms, Actors, Process: A Problem-Oriented Approach,* 2nd edn., New York: Aspen.

Dupuy, Pierre-Marie. (1999). The Danger of Fragmentation or Unification of the International Legal System and the International Court of Justice, *NYU Journal of International Law and Politics*, Vol. 31, pp. 791–808.

Eastbusiness.org. (2006). Hungary Set to Impose Strict New GMO Crop Law, *Eastbusiness.org*, 27 November.

——(2007). Hungary Wants to Keep GMO Ban, *Eastbusiness.org*, 6 February.

Eberlein, Burkard. (2005). Regulation by Cooperation: The 'Third Way' in Making Rules for the Internal Energy Market, in Peter Cameron, ed., *Legal Aspects of EU Energy Regulation*, Oxford: Oxford University Press, pp. 59–88.

——and Edgar Grande. (2005). Beyond Delegation: Transnational Regulatory Regimes and the EU Regulatory State, *Journal of European Public Policy*, Vol. 12, No. 1, pp. 89–112.

————and Abraham Newman. (2007). Regulatory Networks: From Information Warehouses to Multi-Level Rulemakers? Paper Presented at the Biennial Conference of the European Union Studies Association, Montreal, CA, 17–19 May.

Echols, Marsha A. (1998). Food Safety Regulation in the European Union and the United States: Different Cultures, Different Laws, *Columbia Journal of European Law*, Vol. 4, pp. 525–43.

——(2001). *Food Safety and the WTO: The Interplay of Culture, Science and Technology*, Boston, MA: Kluwer.

Economist. (2003a). Genetically Modified Food, *The Economist*, 3 Apr., p. 5.

——(2003b). Invasion of the Transgenics, *The Economist*, 14 May (on-line edition).

——(2004). Agriculture and Banking: Against the Grain, *The Economist*, p. 68. 28 Aug.

——(2006). GMOs: Up from the Dead, *The Economist*, 6 May, p. 66.

——(2007). African Science: Local Heroes, *The Economist*, 13 February, pp. 75–6.

Ehlermann, Claus-Dieter and Lothar Ehring. (2005). Are WTO Decision-Making Procedures Adequate for Making, Revising, and Implementing Worldwide and 'Plurilateral' Rules? in Ernst-Ulrich Petersmann, ed., *Reforming the World Trading System: Legitimacy, Efficiency, and Democratic Governance*, New York: Oxford University Press, pp. 497–522.

Elster, Jon (ed.). (1998). *Deliberative Democracy*, New York: Cambridge University Press.

Enriquez, Juan and Ray A. Goldberg. (2000). Transforming Life, Transforming Business: The Life Science Revolution, Harvard Business Review, Mar.–Apr. 2000, published at http://www.hbsp.harvard.edu/products/hbr/marapr00/R00203.html.

Enterprise and Industry Directorate General, European Commission. (2007). *Competitiveness of the European Biotechnology Industry* (Working document, 12 July 2007), available at http://ec.europa.eu/enterprise/phabiocom/docs/biotech_analysis_competitiveness.pdf, accessed on 28 May 2008.

——(1984). Proposed Policy Concerning Certain Microbial Products, 49 FR 50880 (31 December).

EPA (Environmental Protection Agency). (2003). EPA's Regulation of Biotechnology for Use in Pest Management, http://www.epa.gov/pesticides/biopesticides/reg_of_biotech/eparegofbiotech.htm, accessed on 5 April 2004.

Epstein, David and Sharyn O'Halloran. (1999). *Delegating Powers: A Transaction Cost Politics Approach to Policy Making Under Separate Powers*, New York: Cambridge University Press.

Eriksen, Erik Oddvar. (2000). Deliberative Supranationalism in the EU, in Erik Oddvar Eriksen and John Erik Fossum eds., *Democracy in the European Union—Integration Through Deliberation?* London: Routledge, pp. 42–64.

Esty, Daniel and Damien Geradin, eds. (2000). *Regulatory Competition and Economic Integration*, New York: Oxford University Press.

EU–U.S. Biotechnology Consultative Forum. (2000). *Final Report*, http://europa. eu.int/comm/external_relations/us/biotech/report.pdf, accessed on 11 June 2004.

EurActiv.com. (2008*a*). Commission Hesitant to Approve More GM Crops, *EurActiv. com*, 8 May 2008, http://www.euractiv.com/en/environment/commission-hesitant-approve-gm-crops/article-172209, accessed on 14 July 2008.

——(2008*b*). France to Propose Concrete Solutions to EU's GMO Muddle, *EurActiv.com*, 8 July, http://www.euractiv.com/en/environment/france-propose-concrete-solutions-eu-gmo-muddle/article-174002, accessed on 14 July 2008.

European Commission. (2000). Biotechnology in Bilateral and Multilateral Fora, 25 July, http://trade.ec.europa.eu/doclib/docs/2003/october/tradoc_111449.pdf, accessed on 27 July 2007.

European Commission Directorate General for Trade [DG Trade]. (2000). Transatlantic Economic Partnership: Overview and Assessment, http://trade.ec.europa.eu/doclib/docs/2003/october/tradoc_111712.pdf, accessed on 11 July 2007.

European Council. (2000). Presidency Conclusions—Lisbon European Council, 23 and 24 March 2000, DOC/00/8 of 24/03/2000, http://europa.eu.int/ISPO/docs/services/docs/2000/jan-march/doc_00_8_en.html, accessed on 22 June 2007.

European Food Safety Authority [EFSA]. (2003). Press Release: EFSA Welcomes Decision on Permanent Location, 15 December, http://www.efsa.europa.eu/etc/medialib/efsa/press_room/press_release/32.Par.0001.File.dat/15–12–2003_en1.pdf, accessed on 10 April 2007.

——(2007). DRAFT Scientific Opinion on Food Safety, Animal Health and Welfare and Environmental Impact of Animals Derived from Cloning by Somatic Cell Nucleus Transfer (SCNT) and Their Offspring and Products Obtained from Those Animals, DRAFT Scientific Opinion of the Scientific Committee (Question No EFSA-Q-2007–092), endorsed for public consultation on 19 December 2007, http://www.efsa. europa.eu/EFSA/DocumentSet/sc_opinion_clon_public_consultation.pdf, accessed on 24 January 2008.

European Group on Ethics in Science and New Technologies. (2008). Ethical Aspects of Animal Cloning for Food Supply: Opinion No 23, 16 January 2008, http://ec.europa. eu/european_group_ethics/activities/docs/opinion23_en.pdf, accessed on 24 January 2008.

European Parliament. (1989). Report drawn up on behalf of the Committee on the Environment, Public Health, and Consumer Protection on the proposal from the Commission to the Council (COM(88)160 final—Doc. C 2–73/88) for a directive on the deliberate release to the environment of genetically modified organisms, PE Doc A 2–142/89. Rapporteur: Mr Gerhard Schmid.

European Report. (2000*a*). Genetic Engineering: Key Issues in GMO Conciliation Become Clearer, *European Report*, 1 November.

——European Report. (2000*b*). Biotechnology: Conciliation Talks Move Towards Agreement on GMOs, *European Report*, 13 December.

——(2002*a*). Biotechnology: MEPs Draw Inspiration from Hearing on Labeling and Tracing GM Food, *European Report*, 1 May.

——(2002b). Biotechnology: Member States Agree on GM-Labelling of Food, *European Report*, 4 December.

——(2004a). Biotechnology: Contrasting Reactions to Authorisation for Bt11 Transgenic Corn, *European Report*, 29 May.

——(2004b). Biotechnology: EU Member States Fail to Agree on GM-Rape GT-73, *European Report*, 19 June.

——(2005a). EU/US: Commission 'Deplores' Unauthorised Imports of Bt-10 Maize into EU, *European Report*, 2 April.

——(2005b). Genetic Engineering: Commission Faces Stinging About-Turn in Council on GMO Bans, *European Report*, 29 June.

——(2005c). Genetic Engineering: Commissioners Debate EU Policy and Say They Will Push Ahead with GM Approvals, *European Report*, 22 March.

——(2005d). Genetic Engineering: Agreement on GM Grain Trade Rules Blocked by Brazil and New Zealand, *European Report*, 8 June.

——(2006a). Genetic Engineering: Poland Allowed to Ban 16 Varieties of Unsuitable GM Maize, *European Report*, 9 May.

——(2006b). GMO: After Vienna, the Commission Turns Its Fire on Warsaw, *European Report*, 12 October.

——(2006c). Environment Council: Ministers Gear Up for Highly 'Political' Session on 9 March, *European Report*, 8 March.

——(2006d). Environment Council/Genetic Engineering: European Food Safety Agency under Fire over GMO Authorisations, *European Report*, 10 March.

——(2006e). Agriculture: Ministers Want Guidelines on Co-Existence of GM and Conventional Crops, *European Report*, 19 May.

——(2006f). Agriculture: Cool Reception for New Finnish Compromise on Organic Farming, *European Report*, 9 November.

——(2006g). Environment Council/Genetic Engineering: European Food Safety Agency Under Fire over GMO Authorisations, *European Report*, 10 March.

——(2006h). Agriculture: Ministers Want Guidelines on Co-existence of GM and Conventional Crops, *European Report*, 19 May.

——(2007a). Biotechnology: Green Light for Three Genetically Modified Oilseed Rapes, *European Report*, 28 March.

——(2007b). Consumers: EP Supports Proposal for a Regulation on Enzymes, *European Report*, 30 March.

——(2007c). Biotechnology: Four New GMO Products on the Council Table, *European Report*, 19 December.

——(2007d). Biotechnology: Ministers Equivocate on GMOs, *European Report*, 29 November.

——(2007e). Biotechnology: Dimas Proposes Banning Genetically Modified Corn, *European Report*, 29 October.

——(2007f). Biotechnology: Commission Tries to Force Vienna's Hand Again on GE Maize, *European Report*, 11 October.

——(2007g). GMOs: Commission is Thrown Back Against the Ropes, *European Report*, 31 October.

——(2008a). Biotechnology: MEPs Call into Question EU's Zero Tolerance Stance on GMOs, *European Report*, 25 April.

References

—— (2008*b*). Environment Council: Ministers Seek to Solve Impasse over GMOs, *European Report*, 9 June.

—— (2008*c*). Biotechnology: Consensus Reached on Revising GMO Regulation, *European Report*, 13 June.

Evans, Peter B., Harold K. Jacobson, and Robert D. Putnam, eds. (1993). *Double-Edged Diplomacy: International Bargaining and Domestic Politics*, Berkeley, CA: University of California Press.

Evans-Pritchard, Blake. (2001). Vote on GMO Legislation Today, *euobserver.com*, 14 February.

Evenson, Robert. (2003). *GMOs: Prospects for Increased Crop Productivity in Developing Countries*, Economic Growth Center, Yale University, Discussion Paper 878 (December 2003), available at http://www.econ.yale.edu/growth_pdf/cdp878.pdf, accessed on 24 April 2004.

Falkner, Robert, and Aarti Gupta. (2006). Implementing the Cartagena Protocol on Biosafety: Comparing Mexico, China and South Africa, *Global Environmental Politics*, Vol. 6, No. 4, pp. 23–55.

FDA Week. (2007*a*). U.S. Agrees to Discussion with EU to Normalize Biotechnology Trade, *FDA Week*, Vol. 13, No. 17, 27 April.

—— (2007*b*). U.S., EU Seek to Restore Rice Trade after GMO Contamination, *FDA Week*, Vol. 13, No. 42, 19 October.

Ferrara, J. (1998). Revolving Doors: Monsanto and the Regulators, *The Ecologist*, Vol. 28, No. 5, pp. 280–7.

Fearon, James. (1998). Bargaining, Enforcement, and International Cooperation, *International Organization*, Vol. 52, No. 2, pp. 269–305.

Fields, Robin. (2006). Bowlful of Worry, *Los Angeles Times*, 26 November, p. C1.

Finnemore, Martha. (1993). International Organizations as Teachers of Norms, *International Organization*, Vol. 47, No. 4, pp. 565–97.

—— (1996). Norms, Culture, and World Politics: Insights from Sociology's Institutionalism, *International Organization*, Vol. 50, No. 2, pp. 325–47.

—— and Stephen J. Toope. (2001). Alternatives to 'Legalization': Richer Views of Law and Politics, *International Organization*, Vol. 55, No. 3, pp. 743–58.

Fisher-Lescano, Andreas and Gunter Teubner. (2004). Regime Collisions: The Vain Search for Legal Unity in the Fragmentation of Global Law, *Michigan Journal of International Law*, Vol. 25, pp. 999–1046.

Fischer, Frank. (2000). *Citizens, Experts, and the Environment: The Politics of Local Knowledge*, Durham: Duke University Press.

Food and Drug Administration. (2003). *FDA statement regarding Glofish*. 9 December. Rockville, MD: FDA. http://www.fda.gov/bbs/topics/news/2003/new00994.html, accessed on 19 June 2008.

Food and Agriculture Organization. (FAO). (2004*a*). *The State of Food and Agriculture 2003–2004*.

—— (2004*b*). Biotechnology at Work, FAO Newsroom, 17 May, at http://www.fao.org/newsroom/en/focus/2004/41655/index.html, accessed on 7 May 2007.

—— (2004*c*). *Pest Risk Analysis for Quarantine Pests Including Analysis of Environmental Risks and Living Modified Organisms*, ISPM No. 11.

—— (2004*d*). *The State of Food Insecurity in the World*.

—— (2006). *Results from the FAO Biotechnology Forum*, FAO Research and Technology Paper, No. 11.

Food and Agriculture Organization/World Health Organization. FAO/WHO (1990). *Strategies for Assessing the Safety of Foods Produced by Biotechnology*, Report of a Joint FAO/WHO Consultation, Geneva, 5–10 November 1990.

——(2000). *Safety Aspects of Genetically Modified Foods of Plant Origin*, Report of a Joint FAO/WHO Expert Consultation, WHO/SDE/PHE/FOS/00.6, Geneva, 29 May–2 June 2000).

——(2001a). *Evaluation of Allergenicity of Genetically Modified Foods*, Report of a Joint FAO/WHO Expert Consultation, Rome, 22–25 January 2001.

——(2001b). *Safety Assessment of Foods Derived From Genetically Modified Microorganisms*, Report of a Joint FAO/WHO Expert Consultation, WHO/SDE/PHE/FOS/01.3, Geneva, 24–28 September 2001.

——(2002). *Report of the Evaluation of the Codex Alimentarius and Other FAO and WHO Food Standards Work*, 15 November 2002 [Codex Evaluation Report].

——(2003). *Safety Assessment of Foods Derived From Genetically Modified Animals*, Report of a Joint FAO/WHO Expert Consultation, Rome, 17–21 November 2003.

——(2006). *Guidelines on Food Fortification with Micronutrients*, Geneva: WHO.

Food and Drug Administration. (FDA) (1992). Statement of Policy: Foods Derived from New Plant Varieties. *Federal Register*, Vol. 57, pp. 22984–3001. 29 May.

——(1994). Interim Guidance on the Voluntary Labeling of Milk and Milk Products From Cows That Have Not Been Treated With Recombinant Bovine Somatotropin, *Federal Register,* Vol. 59, p. 6279, 10 February.

——(1997). Guidance on Consultation Procedures: Foods Derived from New Plant Varieties. Available at http://vm.cfsan.fda.gov/~lrd/consulpr.html.

——(1998). FDA/EC Bilateral Meeting: Summary, http://www.fda.gov/oia/bilat.htm, accessed on 12 June 2007.

——(2001a). Proposed Rule: Premarket Notice Concerning Bioengineered Foods, *Federal Register,* Vol. 66, p. 4706, 18 January.

——(2001b). Draft Guidance for Industry: Voluntary Labeling Indicating Whether Foods Have or Have Not Been Developed Using Bioengineering, *Federal Register*, Vol. 66, p. 4839, 18 January, http://www.cfsan.fda.gov/~dms/biolabgu.html, accessed on 5 April 2004.

——(2003) FDA Statement Regarding Glofish. December 9. Rockville, MD: FDA.

——(2006a). Guidance for Industry: Recommendations for the Early Food Safety Evaluation of New Non-pesticidal Proteins Produced by New Plant Varieties Intended for Food Use, CFSAN/Office of Food Additive Safety, Center for Veterinary Medicine, June, http://www.cfsan.fda.gov/guidance.html, accessed on 18 April 2007.

——(2006b). FDA Issues Draft Documents on the Safety of Animal Clones, *FDA News,* 28 December, http://www.fda.gov/bbs/topics/NEWS/2006/NEW01541.html, accessed on 24 November 2008.

——(2006c). *Animal Cloning, A Draft Risk Assessment,* 28 December, http://www.fda.gov/cvm/Documents/Cloning_Risk_Assessment.pdf, accessed on 24 November 2008.

——(2008). *Animal Cloning: A Risk Assessment—FINAL*, report available at http://www.fda.gov/cvm/CloneRiskAssessment_Final.htm, accessed on 19 January 2008.

Foster, Caroline. (2007). Genuine Fears: Accommodating the Right to Political Participation in Health-Related Decision-Making, Monash University and Sydney University Workshop on the World Trade Organisation and Human Rights Prato, Italy, on file.

References

Franck, Thomas. (1990). *The Power of Legitimacy among Nations*, New York: Oxford University Press.

Fukuda-Parr, Sakiko. (2007*a*). *The Gene Revolution: GM Crops and Unequal Development*, London: Earthscan Publications.

——(2007*b*). Introduction: Genetically Modified Crops and National Development Priorities, in Sakiko Fukuda-Parr, ed., *The Gene Revolution: GM Crops and Unequal Development*, London: Earthscan Publications, pp. 3–14.

——(2007*c*). Emergence and Global Spread of GM Crops: Explaining the Role of Institutional Change, in Sakiko Fukuda-Parr, ed., *The Gene Revolution: GM Crops and Unequal Development*, London: Earthscan Publications, pp. 15–35.

——(2007*d*). Institutional Changes in Argentina, Brazil, China, India and South Africa, in Sakiko Fukuda-Parr, ed., *The Gene Revolution: GM Crops and Unequal Development*, London: Earthscan Publications, pp. 199–221.

——(2007*e*). The Role of Government Policy: For Growth, Sustainability and Equity, in Sakiko Fukuda-Parr, ed., *The Gene Revolution: GM Crops and Unequal Development*, London: Earthscan Publications, pp. 222–37.

Garrett, Geoffrey, and Barry Weingast. (1993). Ideas, Interests, and Institutions: Constructing the European Community's Internal Market, in Judith Goldstein and Robert O. Keohane, eds., *Ideas and Foreign Policy,* Ithaca, NY: Cornell University Press, pp. 173–206.

Gaskell, George, Nick Allum, and Sally Stares. (2003). *Eurobarometer 58.0: A Report to the EC Directorate General for Research from the project 'Life Sciences in European Society'* QLG7-CT-1999–00286.

General Accounting Office. (1998). *Report to the Committee on Governmental Affairs: Unfunded Mandates: Reform Act Has Had Little Effect on Agencies' Rulemaking Actions*, February.

General Agreement on Tariffs and Trade (GATT). (1986). Ministerial Declaration on the Uruguay Round of Multilateral Trade Negotiations, 9 September.

Genetic Resources Action International (GRAIN). (2004). FAO Declares War on Farmers, Not on Hunger, GRAIN Press Release, 16 June.

George, Alexander L. and Andrew Bennett. (2004). *Case Studies and Theory Development in the Social Sciences*, Cambridge, MA: MIT Press.

Giddens, Anthony. (1991). *Modernity and Self-Identity: Self and Society in the Late Modern Age*, Stanford, CA: Stanford University Press.

——(2000) *Runaway World*, Routledge: New York.

Glanz, James. (2004). Scientists Say Administration Distorts Facts, *New York Times*, 19 February, p. A18.

Glowka, Lyle. (2003). *Law and Modern Biotechnology: Selected Issues of Relevance to Food and Agriculture*, FAO Legislative Study 78.

GMO Compass. (2007). BASF Expects EU Approval of Amflora within Weeks, *GMO Compass*, 14 December, http://www.gmo-compass.org/eng/news/310.docu.html, accessed on 15 January 2008.

Goldirova, Renata. (2007). Brussels Wins Final Say Over Controversial GMO Corn, *euobserver.com*, 30 October.

Gonsalves, C., D. R. Lee, and D. Gonsalves. (2007). The Adoption of Genetically Modified Papaya in Hawaii and Its Implications for Developing Countries, *Journal of Development Studies*, Vol. 43, No. 1, pp. 177–91.

Gonzalez, Carmen. (2007). Genetically Modified Organisms and Justice: The International Environmental Justice Implications of Biotechnology, *Georgetown International Environmental Law Review*, Vol. 19, No. 4, pp. 583–642.

Goodman, Ryan and Derek Jinks. (2004). How to Influence States: Socialization and International Human Rights Law, *Duke Law Journal*, Vol. 54, pp. 621–703.

Goldsmith, Jack and Eric Posner. (2005). *The Limits of International Law*, New York: Oxford University Press.

Gourevitch, Peter A. (1978). The Second Image Reversed: The International Sources of Domestic Politics, *International Organization*, Vol. 32, No. 4, pp. 881–912.

Gouse, Marnus. (2007). South Africa: Revealing the Potential and Obstacles, the Private Sector Model, and Reaching the Traditional Sector, in Sakiko Fukuda-Parr, ed., *The Gene Revolution: GM Crops and Unequal Development*, London: Earthscan Publications, pp. 175–96.

Graff, Gregory D. and David Zilberman. (2004). Explaining Europe's Resistance to Agricultural Biotechnology, *Agricultural and Resource Economics Update*, Vol. 7, No. 5. pp. 1–4.

Grand Rapids Press. (2007). Big Dairy Rejects Cloned Milk: Dean Foods Says Customers Don't Accept It, *Grand Rapids Press*, 23 February, p. A8.

Grant, Ruth W. and Robert O. Keohane. (2005). Accountability and Abuses of Power in World Politics, *American Political Science Review*, Vol. 99, No. 1, pp. 29–43.

Greenwire. (2008). Biotechnology: USDA Considering Enviro Effects of Genetically Modified Hay, *Greenwire*, 8 January.

Greenwood, Justin and Ronit Karsten. (1995). European Bioindustry, in Justin Greenwood, ed., *European Casebook on Business Alliances*, Hemel Hempstead: Prentice-Hall, pp. 75–88.

Greif, Avner and David D. Laitin. (2004). A Theory of Endogenous Institutional Change, *American Political Science Review*, Vol. 98, No. 4, pp. 633–52.

Grieco, Joseph. (1988). Anarchy and the Limits of Cooperation, *International Organization*, Vol. 42, No. 3, pp. 485–507.

Griffiths, John. (1986). What is Legal Pluralism? *Journal of Legal Pluralism and Unofficial Law*, Vol. 24, pp. 1–55.

Gruber, Lloyd. (2000). *Ruling the World*, Princeton, NJ: Princeton University Press.

Gruszczynski, Lukasz. (2008). The SPS Measures Adopted in Case of Insufficiency of Scientific Evidence—Where Do We Stand after EC-Biotech Products Case? in Julien Chaisse and Tiziano Balmelli, eds., *Essays on the Future of the World Trade Organization*, Geneva: Edis, forthcoming.

Guruswamy, Lakshman. (2002). Sustainable Agriculture: Do GMOs Imperil Biosafety? *Indiana Journal of Global Legal Studies*, Vol. 9, p. 461–500.

Guy, Sandra. (2003). Actor Asks Kraft to Take Biotech Out of Its U.S. Food, *Chicago Sun Times*, 23 April.

Guzman, Andrew. (2004). Food Fears: Health and Safety at the WTO, *Virginia Journal of International Law*, Vol. 45, No. 1, pp. 1–39.

——(2004a). Food Fears: Health and Safety at the WTO, *Virginia Journal of International Law*, Vol. 45, pp. 1–39.

——(2004b). Global Governance and the WTO, *Harvard International Law Journal*, Vol. 45, No. 2, pp. 303–51.

——(2006). The Promise of International Law, *Virginia Law Review*, Vol. 92, No. 3, pp. 533–64.

Haas, Peter. (1992). Introduction: Epistemic Communities and International Policy Coordination, *International Organization*, Vol. 46, No. 1, pp. 1–35.

Habermas Jürgen. (1985). *The Theory of Communicative Action*, Vols. 1 and 2. Boston, MA: Beacon Press.

——(1998). *Between Facts and Norms: Contributions to a Discourse Theory of Law and Democracy*, Cambridge: MIT Press.

Hafner-Burton, Emilie M. (2005). Trading Human Rights: How Preferential Trade Agreements Influence Government Repression, *International Organization*, Vol. 59, No. 3, pp. 593–629.

Hall, Peter A. (1997). The Role of Interests, Institutions and Ideas in the Comparative Political Economy of the Industrialized Nations, in Mark Irving Lichbach and Alan S. Zuckerman, eds., *Comparative Politics: Rationality, Culture, and Structure*, New York: Cambridge University Press, pp. 174–207.

——and Rosemary C.R. Taylor. (1996). Political Science and the Three New Institutionalisms, *Political Studies*, Vol. 44, December, pp. 936–57.

——and Kathleen Thelen. (2006). Varieties of Capitalism and Institutional Change, *APSA European Politics & Society: Newsletter of the European Politics and Society Section of the American Political Science Association*, Vol. 5, No. 1, pp. 1, 3–4, http://www.apsanet.org/~ep/newsletter.html, accessed on 1 May 2007.

Hamilton, Alexander. (1961). *The Federalist Papers*, Jacob Cooke, ed., Middletown, CT: Wesleyan University Press, No. 17, p. 107.

Harris, Andrew and Margaret Cronin Fisk. (2008). Bayer Avoided Class Actions, Faces 1,200 Rice Suits, Bloomberg.com, 22 October, http://www.bloomberg.com/apps/news?pid=20670001&refer=home&sid=aKFbx4hqs8xU, accessed on 24 November 2008.

Hasenclaver, Andreas, Peter Mayer, and Volker Rittberger. (1997). *Theories of International Regimes*, New York: Cambridge University Press.

Hathaway, Oona and Ariel N. Lavinbunk. (2006). Rationalism and Revisionism in International Law (Book Review), *Harvard Law Review*, Vol. 119, No. 5, pp. 1404–43.

Hawthorne, Fran. (2005). *Inside the FDA: The Business and Politics Behind the Drugs We Take and the Food We Eat*, Hoboken, NJ: Wiley.

Hayes-Renshaw, Fiona and Hellen Wallace. (2006). *The Council of Ministers*, 2nd edn., London: Palgrave.

Haxel, Stefanie. (2007). BASF Looks to Gene-Altered Potato for Profit, *International Herald Tribune*, 8 November, p. 17.

Helfer, Lawrence. (2004). Regime Shifting: The TRIPS Agreement and New Dynamics of Intellectual Property Lawmaking, *Yale Journal of International Law*, Vol. 29, pp. 1–83.

——and Anne-Marie Slaughter. (1997–1998). Toward a Theory of Effective Supranational Adjudication, *Yale Law Journal*, Vol. 107, No. 2, pp. 273–391.

Helwig, Alessia. (2007). Wither Science in WTO Dispute Settlement? June 2007, paper presented at workshop in Lecce, Italy, on file.

Henderson, Diedtra. (2007). Bill Calls for Labels on Food from Clones; Senator Says Shoppers Merit Right to Choose, *Boston Globe*, 27 January, p. A9.

Henley, Jon. (1999). McDonald's Campaign Spawns French Hero: Political Activist Turned French Peasant Has Fast Food on the Run, *The Guardian*, 11 September, p. 14.

Herring, Ronald J. (2006). Why Did 'Operation Cremate Monsanto' Fail? Science and Class in India's Great Terminator-Technology Hoax, *Critical Asian Studies*, Vol. 48, No. 4, pp. 467–93.

——(2007a). The Genomics Revolution and Development Studies: Science, Poverty and Politics, *Journal of Development Studies*, Vol. 43, No. 1, pp. 1–30.

——(2007b). Stealth Seeds: Bioproperty, Biosafety, Biopolitics, *Journal of Development Studies*, Vol. 43, No. 1, pp. 130–57.

Heyvaert, Veerle. (2006). Facing the Consequences of the Precautionary Principle in European Community Law, *European Law Review*, Vol. 31, No. 2, pp. 185–206.

Hill, Lowell B. and Sophia C. Battle. (2000). *Search for Solutions in the EU-US GMO Debate*, Proceedings of a Conference, *GMO Regulations: Food Safety or Health Barrier?*, held at the Wyndham Hotel in Chicago, Illinois, 22 and 23 October 1999, Department of Agricultural and Consumer Economics, Agricultural Experiment Station/Office of Research, College of Agricultural, Consumer and Environmental Sciences, University of Illinois as Urbana-Champaign, AE-4731.

Hilts, Philip. (1983). EPA Will Take over Regulation of the Gene Engineering Industry, *Washington Post*, 9 August, p. A13.

Hix, Simon. (2005). *The Political System of the European Union*, 2nd edn., London: Palgrave.

Hodson, Dermot and Imelda Maher. (2001). The Open Method of Coordination as a New Mode of Governance: The Case of Soft Economic Policy Co-ordination, *Journal of Common Market Studies*, Vol. 39, No. 4, pp. 719–46.

Howse, Robert. (2000). Democracy, Science and Free Trade; Risk Regulation on Trial at the World Trade Organization, *Michigan Law Review*, Vol. 98, No. 7, pp. 2329–57.

——and Petros Mavroidis. (2000). Europe's Evolving Regulatory Strategy for GMOs—The Issue of Consistency with WTO Law: Of Kine and Brine, *Fordham International Law Journal*, Vol. 24, No. 1/2, pp. 317–70.

——and Donald Regan. (2000). The Product/Process Distinction: An Illusory Basis for Disciplining 'Unilateralism' in Trade Policy, *European Journal of International Law*, Vol. 11, No. 2, pp. 249–89.

Huang, Jikun, Ruifa Hu, Scott Rozelle, and Carl Pray. (2007). China: Emerging Public Sector Model for GM Crop Development, in Sakiko Fukuda-Parr, ed., *The Gene Revolution: GM Crops and Unequal Development*, London: Earthscan Publications, pp. 130–55.

Huber, John D. and Charles R. Shipan. (2002). *Deliberate Discretion: The Institutional Foundations of Bureaucratic Autonomy*, New York: Cambridge University Press.

Hudec, Robert. (1993). *Enforcing International Trade Law: The Evolution of the Modern GATT Legal System*, Salem, NH: Butterworth Legal Publishers.

——(1996). International Economic Law: The Political Theatre Dimension, *University of Pennsylvania Journal of International Law*, Vol. 17, pp. 9–15.

——(2003a) Science and Post-Discriminatory WTO Law, *Boston College Int'l & Comp. L. Rev.*, Vol. 26, No. 2, pp. 185–95.

——(2003b). Industrial Subsidies: 'Tax Treatment' of 'Foreign Sales Corporations,' in Ernst-Ulrich Petersmann and Mark A. Pollack, eds., *Transatlantic Economic Disputes*, New York: Oxford University Press, pp. 175–205.

——and Daniel A. Farber. (1994). Free Trade and the Regulatory State: A GATT's-Eye View of the Dormant Commerce Clause, *Vanderbilt Law Review*, Vol. 47, pp. 1401–40.

Huller, Thorsten and Matthias Leonhard Maier. (2006). Fixing the Codex? Global Food-Safety Governance Under Review, in Christian Joerges and Ernst-Ulrich Petersmann, eds., *Constitutionalism, Multilevel Trade Governance and Social Regulation*, Oxford: Hart Publishing, pp. 267–300.

Hunter, Rod. (1999). European Regulation of Genetically Modified Organisms, in Julian Morris and Roger Bate, eds., *Fearing Food: Risk, Health and Environment*, Oxford: Butterworth Heineman, pp. 189–230.

Hungarian News Agency. (2006). GM Crop Companies Say New Regulations Unnecessarily Strict, *Hungarian News Agency*, 29 November.

Immergut, Ellen. (2006). From Constraints to Change, *APSA European Politics & Society: Newsletter of the European Politics and Society Section of the American Political Science Association*, Vol. 5, No. 2, pp. 4–6, accessed at: http://www.apsanet.org/~ep/news letter.html, on 8 January 2008.

Indo-Asian News Service. (2006). Monsanto to Reduce Bt-cotton Seeds Royalty, *Indo-Asian News Service*, 6 March 2006.

Inglis, Kirstyn. (2003). Implications of Enlargement for EC Agri-Food Law: The Accession Treaty, Its Transitional Arrangements, and Safeguard Clauses, December, *European Law Journal*, Vol. 10, No. 5, pp. 595–612.

Inside US Trade. (2000). GMO Protocol Offers Compromise on Crops, WTO Relationship, *Inside US Trade*, 14 February.

—— (2003). US' Zoellick Wants to File Case against EU on Biotech Issue, *Inside US Trade*, 10 January.

—— (2004a). ASA Takes Lead in Pushing for New WTO GMO Case Against EU, *Inside US Trade*, 12 March.

—— (2004b). European Member States Deadlocked on Two GM Applications, *Inside US Trade*, 2 July.

—— (2007a). U.S. Agrees to Discussion with EU to Normalize Biotechnology Trade, *Inside US Trade*, 20 April.

—— (2007b). Codex Task Force to Okay New Draft Guidelines for GMO Trace Elements, *Inside US Trade*, 24 August.

—— (2007c). EU Commissioner Recommends Rejecting GMO after Scientific Approval, *Inside US Trade*, 16 November.

—— (2007d). Commissioners Urge Reconsideration of Zero Tolerance GMO Policy, *Inside US Trade*, 30 November.

—— (2007e). French President Intends to Suspend GMO Cultivation Temporarily, *Inside US Trade*, 2 November.

—— (2007f). U.S. Ag Industry Could Consider Aligning GMO Approvals with EU, *Inside US Trade*, 20 July.

—— (2007g). Codex Task Force Sets New Guidelines on Low-Level GMO Traces, *Inside US Trade*, 26 October.

—— (2007h). U.S., EU Agree to Extend Deadline for Implementation of GMO Case, *Inside US Trade*, 23 November.

—— (2008a). France Moves Against GMO Cultivation in Advance of WTO Deadline, *Inside US Trade*, 4 January.

—— (2008b). EU to End Mandatory GMO Testing on Imports of U.S. Rice, *Inside US Trade*, 4 January.

—— (2008c). USTR Takes First Steps Toward Retaliation Against EU in GMO Dispute, *Inside US Trade*, 25 January.

——(2008*d*). Final Panel Report Upholds Sanctions in U.S.-EU WTO Hormone Case, *Inside US Trade*, 11 January.

——(2008*e*). U.S. Pushes for End to Work on International GMO Label Guidelines, 4 April.

——(2008*f*). Austria Lifts GMO Ban; EU to Examine Low Level Presence, *Inside US Trade*, 6 June.

——(2008*g*). U.S. Mulling Retaliation in EU GMO Case After Lack of Progress, *Inside US Trade*, 27 June.

——(2008*h*). Commission Debate leads to Mixed Results on GMO Access, *Inside US Trade*, 9 May.

——(2008*i*). TEC Agenda Suffers as Poultry Ban, Cosmetics Barrier Remain, *Inside US Trade*, 6 June.

——(2008*j*). U.S., Canada, Argentina Press EU to Approve GMOs More Rapidly, *Inside US Trade*, 24 October 2008.

—— (2008*k*). Environment Ministers Back New GMO Impact Rules, Fight Over Status, *Inside US Trade*, 24 October 2008.

——(2008*l*). Commission Presses for Continued GMO Dialogue, U.S. Non-Committal, *Inside US Trade*, 31 October 2008.

International Centre for Trade and Sustainable Development. (2005*a*). Divisions over Labeling Prove Insurmountable at Biosafety Meet, *Bridges Trade BioRes*, Vol. 5, No. 11, 10 June, accessed online at http://www.ictsd.org/biores/05–06–10/story1.htm on 17 July 2005.

——(2005*b*). Animal Health Organisation Looks at Trade, Biotech Standards, *Bridges Trade Bio-Res*, Vol. 5, No. 11, 10 June.

——(2005*c*). GMO Labelling Continues to Divide Codex, *Bridges Monthly*, Year 9, No. 5, May, p. 18.

——(2006*a*). Labelling Rules Adopted for 'Living Modified Organisms', *Bridges Trade Bio-Res*, Vol. 10, No. 2, March–April, p. 20.

——(2006*b*). DSU Review Picks Up after Hiatus, *Bridges Trade Bio-Res*, Vol. 10, No. 2, March–April, p. 9.

International Committee of the OIE. (2005). Resolution XXVIII, Applications of Genetic Engineering for Livestock and Biotechnology Products, 26 May 2005.

——(2006). Summary Report of the Fifth Meeting of the OIE Working Group on Animal Production Food Safety, 74th General Session, 21–26 May 2006.

International Consortium of Agricultural Biotechnology Research (ICABR). (2004). Open Letter to FAO Director General in Support of SOFA 2003/04—Biotechnology Report, http://www.economia.uniroma2.it/conferenze/icabr2004/open_letter/default.asp, accessed on 11 July 2007.

International Herald Tribune. (2007). EU Gives Ground on Some Modified Crops, *International Herald Tribune*, 25 October, p. 14.

International Institute for Sustainable Development. (2005). *Earth Negotiations Bulletin*, Vol. 9, No. 320, 6 June.

——(2006). Summary of the Third Meeting of the Parties to the Cartagena Protocol on Biosafety, *Earth Negotiations Bulletin*, Vol. 9, No. 351, 20 March.

International Law Commission. (2006). *Fragmentation of International Law: Difficulties Arising from the Diversification and Expansion of International Law*, para. 8, U.N. Doc. A/CN.4/L/682 (13 April 2006) (finalized by Martti Koskenniemi).

International Plant Protection Convention. IPPC (2004). *Guidelines for a Phytosanitary Import Regulatory System* (International Standards for Phytosanitary Measures (ISPM) No. 20, 2004).

International Plant Protection Convention, Subsidiary Body on Dispute Settlement. IPPC-SBDS (2006). *Dispute Settlement Manual*, https://www.ippc.int/servlet/CDSServlet?status = ND0xMzQxMiY2PWVuJjMzPSomMzc9a29z, accessed on 11 July 2007.

International Standards Organization Technical Committee. (2001). Business Plan ISO/TC 34 Food Products, 1 January 2001, http://isotc.iso.org/livelink/livelink/fetch/2000/2122/687806/ISO_TC_034__Food_products_.pdf?nodeid = 999731&vernum = 0, accessed on 20 June 2007.

International Trade Reporter. (2004*a*). Genetically Engineered Crops Up 15 Percent; China, South Africa Report Biggest Increases, *International Trade Reporter (BNA)*, Vol. 21, No. 4, p. 124, 22 January.

——(2004*b*). Delegates End Biosafety Talks with Rules on Documentation, Basis for Liability Regime, *International Trade Reporter (BNA)*, Vol. 21, No. 10, p. 388, 4 March.

——(2006). Cartagena Biosafety Protocol Parties Agree on Labeling for Living Modified Organisms, *International Trade Reporter* (BNA), Vol. 23, No. 12, p. 446, 23 March.

——(2007). Lawmakers Call for More Regulation of Hazardous Products Imported from China, *International Trade Reporter* (BNA), Vol. 24, No. 28, p. 983, 12 July.

Jackson, John. (1991). *U.S. Constitutional Law Principles and Foreign Trade Law and Policy*, Ann Arbor, MI: Institute of Public Policy Studies, University of Michigan.

Jackson, John H. (1998). *The World Trade Organization: Constitution and Jurisprudence*, New York: Routledge.

Jackson, Lee Ann and Kym Anderson. (2005). What's Behind GM Food Trade Disputes, *World Trade Review*, Vol. 4, No. 2, pp. 203–28.

Jacobsson, Kerstin and Asa Vifell. (2003). Integration by Deliberation? On the Role of Committees in the Open Method of Coordination, Paper Prepared for the Workshop on *The Forging of Deliberative Supranationalism in the EU*, European University Institute, Florence, 7–8 February 2003.

James, Clive (2005). ISAAA Brief 34, Executive Summary, Global Status of Commercialized Biotech/GM Crops: 2005, http://www.isaaa.org/Resources/Publications/briefs/34/executivesummary/default.html, accessed on 8 January 2008.

——(2006). ISAAA Brief 35, Executive Summary, Global Status of Commercialized Biotech/GM Crops: 2006, http://www.isaaa.org/Resources/Publications/briefs/35/executivesummary/default.html, accessed on 8 January 2008.

——(2007) ISAAA Brief No. 37, Global Status of Commercialized Biotech/GM Crops: 2007), Ithaca, NY: ISAAA.

Jasanoff, Sheila. (1990). *The Fifth Branch: Science Advisors as Policymakers*, Cambridge, MA: Harvard University Press.

——(2005). *Designs on Nature: Science and Democracy in Europe and the United States*, Princeton, NJ: Princeton University Press.

Joerges, Christian. (2001*a*). Deliberative Supranationalism—A Defence, *European Integration online Papers* (EIoP), Vol. 5, No. 8; http://eiop.or.at/eiop/texte/2001–008a.htm, accessed on 11 June 2004.

——(2001*b*). Law, Science and the Management of Risks to Health at the National, European and International Level—Stories on Baby Dummies, Mad Cows and Hormones in Beef, *Columbia Journal of European Law*, Vol. 7, No. 1, pp. 1–19.

——(2006a). Constitutionalism in Postnational Constellations: Contrasting Social Regulation in the EU and the WTO, in Christian Joerges and Ernst-Ulrich Petersmann, eds., *Constitutionalism, Multi-level Trade Governance and Social Regulation*, London: Hart Publishing, pp. 491–528.

——(2006b). Trade with Hazardous Products? The Emergence of Transnational Governance with Eroding State Government, EUI Working Paper Law No. 2006/05.

——(2007). Conflict of Laws as Constitutional Form: Reflections on International Trade Law and the Biotech Panel Report, RECON Online Working Paper No. 2007/03.

——and Jürgen Neyer. (1997a). Transforming Strategic Interaction into Deliberative Problem Solving: European Comitology in the Foodstuffs Sector, *Journal of European Public Policy*, Vol. 4, pp. 609–25.

————(1997b). From Intergovernmental Bargaining to Deliberative Political Processes: The Constitutionalisation of Comitology, *European Law Journal*, Vol. 3, pp. 273–99.

————(2003). Politics, Risk Management, World Trade Organization Governance and the Limits of Legalization, *Science and Public Policy*, Vol. 30, No. 3, pp. 219–25.

Johnson, Brian. (2006). EU Debates Changes to GM Approval Rules, *Eupolitix.com*, 9 March.

Johnstone, Ian. (2003). Security Council Deliberations: The Power of the Better Argument, *European Journal of International Law*, Vol. 14, No. 3, pp. 437–80.

Josling, Timothy, Donna Roberts, and David Orden. (2004). *Food Regulation and Trade: Toward a Safe and Open Global Food System*, Washington, DC: Institute for International Economics.

Jupille, Joseph and Duncan Snidal. (2006). The Choice of International Institutions: Cooperation, Alternatives and Strategies, unpublished paper, http://sobek.colorado.edu/~jupille/research/20060707-Jupille-Snidal.pdf, accessed on 15 January 2008.

Kagan, Robert A. (2001). *Adversarial Legalism: The American Way of Law*, Cambridge, MA: Harvard University Press.

——(2004). American Courts and the Policy Dialogue: The Role of Adversarial Legalism, in M.C. Miller and J. Barnes, eds., *Making Policy, Making Law: An Interbranch Perspective*, Washington, DC: Georgetown University Press, pp. 13–34.

Kahan, Dan, Paul Slovic, Donald Braman, and John Gastil. (2006). Fear of Democracy: A Cultural Evaluation of Sunstein on Risk, *Harvard Law Review*, Vol. 119, No. 4, pp. 1071–109.

Kahler, Miles. (1995). *Regional Futures and Transatlantic Economic Relations*, New York: Council on Foreign Relations Press.

Kalaitzandonakes, Nicholas. (2006). Cartagena Protocol: A New Trade Barrier, *Regulation*, Vol. 29, No. 2, pp. 18–25.

Kanter, James. (2007a). Europe Butterflies May Halt GMOs, *The International Herald Tribune*, 22 November, p. 1.

——(2007b). EU Claims on Gene-Altered Corn Disputed, *The International Herald Tribune*, 28 November, p. 11.

——(2007c). European Official Faults Ban on Genetically Altered Feed, *The New York Times*, 27 November, p. C4.

——(2008). EU Risks Ire with Delay on Genetically Modified Foods, *International Herald Tribune*, 8 May 2008.

Kaplan, Karen and Jia-Rui Chong. (2006). Milk from Cloning Called Safe; The Study Calls Labelling Unnecessary and Signals Likely FDA Approval; Many Consumers Leery, *Los Angeles Times*, 23 December, p. A1.

Karapinar, Baris and Michelangelo Temmerman. (2007). Benefiting from Biotechnology: Pro-poor IPRs and Public-Private Partnerships, *NCCR Trade Regulation Working Paper No. 2007/35*.

Kay, David. (1976). *The International Regulation of Pesticide Residues in Food: A Report to the National Science Foundation on the Application of International Regulatory Techniques to Scientific/Technical Problems*, St. Paul, MN: West Publishing Co.

Kelemen R. Daniel. (2003). The Structure and Dynamics of EU Federalism, *Comparative Political Studies*, No. 36, Nos. 1–2, pp. 184–208.

——and Kalypso Nicolaidis. (2007). Bringing Federalism Back In, in Knud Erik Jørgensen, Mark A. Pollack, and Ben Rosamond, eds., *The Handbook of European Union Politics*, New York: Sage, pp. 310–16.

Keohane, Robert O. (1984). *After Hegemony: Cooperation and Discord in the World Political Economy*, Princeton, NJ: Princeton University Press.

——(2003). Global Governance and Democratic Accountability, in David Held and Koenig-Archibuigi, eds., *Taming Globalization: Frontiers of Governance*, London: Polity, pp. 130–59.

——and Helen V. Milner, eds. (1996). *Internationalization and Domestic Politics*, Cambridge: Cambridge University Press.

——and Joseph S. Nye. (1974). Transgovernmental Relations and International Organizations, *World Politics*, Vol. 27, No. 1, pp. 39–62.

Kettnaker, Vera. (2001). The European Conflict Over Genetically Engineered Crops, 1995–1997, in Doug Imig and Sidney Tarrow, eds., *Contentious Europeans: Protest and Politics in Emerging Polity*, Lanham, MD: Rowman & Littlefield, pp. 205–32.

Kiewiet, D. Roderick and Matthew D. McCubbins. (1991). *The Logic of Delegation: Congressional Parties and the Appropriation Process*, Chicago, IL: University of Chicago Press.

Kingsbury, Benedict, Nico Krisch, Richard Stewart, and Jonathan Wiener. (2005). The Emergence of Global Administrative Law, *Law and Contemporary Problems*, Vol. 68, pp. 15–61.

Kloppenburg, Jack. (1988). *Seeds and Sovereignty: The Use and Control of Plant Genetic Resources*, Durham: Duke University Press.

Knight, Frank. (1921). *Risk, Uncertainty and Profit*, New York: Augustus M. Kelley.

Koh, Harold. (1998). The 1998 Frankel Lecture: Bringing International Law Home, Houston Law Review, Vol. 35, pp. 623–81.

Komesar, Neil. (1995). *Imperfect Alternatives: Choosing Institutions in Law, Economics and Public Policy*, Chicago, IL: University of Chicago Press.

——(2002). *Law's Limits: The Rule of Law and the Supply and Demand of Rights*, New York: Cambridge University Press.

Koremenos, Barbara, Charles Lipson, and Duncan Snidal. (2001). The Rational Design of International Institutions, *International Organization*, Vol. 55, No. 4, pp. 761–99.

Korthals, Michiel. (2004). Ethics of Differences in Risk Perception and Views on Food Safety, *Food Protection Trends*, Vol. 24, No. 7, pp. 498–503.

Koskenniemi, Martti and Päivi Leinmo. (2002). Fragmentation of International Law. Postmodern Anxieties? *Leiden Journal of International Law*, Vol. 15, No. 3, pp. 553–79.

Kovacic, W.K. (2005). Competition Policy Cooperation and the Pursuit of Better Practices, in David Andrews, Mark A. Pollack, Gregory C. Shaffer, and Helen Wallace, eds., *The Future of Transatlantic Economic Relations: Continuity Amid Discord*, Florence: European University Institute, Robert Schuman Centre for Advanced Studies, pp. 65–80; available online at http://www.iue.it/RSCAS/e-texts/Future_Transat_EconRelations.pdf, accessed on 8 June 2007.

Krapohl, Sebastian. (2004). Credible Commitments in Non-Independent Regulatory Agencies: A Comparative Analysis of the European Agencies for Pharmaceuticals and Foodstuffs, *European Law Journal*, Vol. 10, No. 5, pp. 518–38.

Krasner, Stephen D. (ed.). (1983). *International Regimes*, Ithaca, NY: Cornell University Press.

—— (1991). Global Communications and National Power: Life on the Pareto Frontier, *World Politics*, Vol. 43, No. 3, pp. 336–56.

Krent, Lawrence. (2004). What's the Holdup? Addressing Constraints to the Use of Plant Biotechnology in Developing Countries, *AgBioForum*, Vol. 7, pp. 63–9.

Kreppel, Amie. (2001). *The European Parliament and Supranational Party System: A Study in Institutional Development*, New York: Cambridge University Press.

—— (2002). The Environmental Determinants of Legislative Structure: A Comparison of the US House of Representatives and the European Parliament, Paper Prepared for the Conference on "Exporting Congress? The Influence of the U.S. Congress on World Legislatures," Jack D. Gordon Institute for Public Policy and Citizenship Studies, Florida International University, Miami, 6–7 December.

Krisch, Nico. (2006). The Pluralism of Global Administrative Law, *European Journal of International Law*, Vol. 17, No. 1, pp. 247–78.

Kuhn, Steven. (2003). Prisoner's Dilemma, in Edward N. Zalta, edn., *The Stanford Encyclopedia of Philosophy (Fall 2003 Edition)*, http://plato.stanford.edu/archives/fall2003/entries/prisoner-dilemma/, accessed on 8 January 2008.

Kysar, Douglas. (2006). It Might Have Been: Risk, Precaution, and Opportunity Costs, *Journal of Land Use and Environmental Law*, Vol. 22, No. 1, pp. 1–57.

Laitner, Sarah. (2007). Brussels Paves Way for First Cultivation of GM Crop, *Financial Times*, 17 July, p. 9.

Lasker, Eric. (2005). Federal Preemption and State Anti-'GM' Food Laws, *Washington Legal Foundation Legal Backgrounder*, Vol. 20, 2 December, p. 60.

Lean, Geoffrey. (2007). Safety Fears Prompt Europe to Consider First Ban on GM Crop, *The Independent on Sunday*, 25 November, p. 18.

Lee, Christopher. (2006). Genetically Modified Rice Wins USDA Approval; Grain Tainted U.S. Supply This Summer, *Washington Post*, 25 November, p. A3.

Lee, Martin R. (1994). *Environmental Protection and the Unfunded Mandates Debate*, Congressional Research Service, 94-739 ENR, 22 September.

Leive, David M. (1976). International Regulatory Regimes: Case Studies in Health, Meteorology and Food, 2 volumes, Lexington, KY: Lexington Books for the American Society of International Law.

Levi, Margaret. (1997). A Model, a Method, and a Map: Rational Choice in Comparative and Historical Analysis, in Mark I. Lichbach and Alan S. Zuckerman, eds., *Comparative Politics: Rationality, Culture, and Structure*, Cambridge: Cambridge University Press, pp. 19–41.

Lewis, Jeffrey (2005). The Janus Face of Brussels: Socialization and Everyday Decision Making in the European Union, *International Organization*, Vol. 59, No. 4, pp. 937–71.

References

—— (2008). Strategic Bargaining, Norms, and Deliberation: The Currencies of Negotiation in the Council of the European Union, in Daniel Naurin and Helen Wallace, eds., *Unveiling the Council of the EU*, New York: Palgrave, pp. 165–184.

Lifsher, Marc. (2007). Judge Halts Sale of Biotech Alfalfa Seeds: Activists Applaud the Preliminary Ruling as First Ban on Genetically Modified Crops, *Los Angeles Times*, 13 March, p. C1.

Lipson, Charles. (1991). Why Are Some Agreements Informal? *International Organization*, Vol. 45, No. 4, pp. 495–538.

Lipton, Michael. (2007). Plant Breeding and Poverty: Can Transgenic Seeds Replicate the 'Green Revolution' as a Source of Gains for the Poor? *Journal of Development Studies*, Vol. 43, No 1, pp. 31–62.

Lopes, Mauricio Antoñio and Maria José Amstalden Moraes Sampaio. (2005). Approaching Biotechnology: Experiences from Brazil and Argentina, in Ricardo Mélendez-Ortiz and Vicente Sánchez, eds., *Trading in Genes: Development Perspectives on Biotechnology, Trade and Sustainability*, London: Earthscan Publications, pp. 89–108.

Loya, Thomas and John Boli. (1999). Standardization in the World Polity: Technical Rationality over Power, in John Boli and George Thomas, eds., *Constructing World Culture: International Nongovernmental Organizations Since 1875*, Stanford, CA: Stanford University Press, pp. 169–97.

Lucas, Greg. (2004). Growing Genetically Altered Foods Banned, *San Francisco Chronicle*, 4 August, p. B3.

Lueck, Sarah. (2000). Corn-Recall Cost Could Reach into the Hundreds of Millions, *Wall Street Journal*, 3 November, p. A2.

MacMillan, Fiona and Michael Blakeney. (2001). Genetically Modified Organisms and the World Trade Organization, *Tulane Journal of Technology and Intellectual Property*, Vol. 3, pp. 93–116.

Maduro, Miguel Poiares. (1998). *We the Court: The European Court of Justice and the European Economic Constitution*, London: Hart Publishing.

Magnette, Paul. (2004). Deliberation or Bargaining? Coping with Constitutional Conflicts in the Convention on the Future of Europe, in E. O. Eriksen, J. E. Fossum, and A. J. Menéndez (eds.), *Developing a Constitution for Europe*, London: Routledge, pp. 207–25.

Mahoney, Honor. (2005). Member States Rebuff Commission in GM Vote, *euobserver. com*, 27 June.

Majone, Giandomenico. (2003a). Foundations of Risk Regulation: Science, Decision-Making, Policy Learning and Institutional Reform, in Giandomenico Majone, ed., *Risk Regulation in the European Union: Between Enlargement and Internationalization*, Florence: European University Institute, pp. 9–31.

—— (2003b). What Price Safety? The Precautionary Principle and Its Policy Implications, in Giandomenico Majone, ed., *Risk Regulation in the European Union: Between Enlargement and Internationalization*, Florence: European University Institute, pp. 33–53.

Mann, Michael. (2001). Six EU States Refuse to Lift Block on New Modified Crops, *Financial Times*, 16 February, p. 8.

Manners, Ian. (2002). Normative Power Europe: A Contradiction in Terms? *Journal of Common Market Studies*, Vol. 40, pp. 235–58.

Marceau, Gabrielle. (2001). Conflicts of Norms and Conflicts of Jurisdictions: The Relationship Between the WTO Agreement and MEAs and Other Treaties, *Journal of World Trade*, Vol. 35, No. 6, pp. 1081–131.

March, James G. and Johan P. Olsen. (1989). *Rediscovering Institutions: The Organizational Basis of Politics*, New York: Free Press.

Margolis, Howard. (1996). *Dealing with Risk: Why the Public and the Experts Disagree on Environmental Issues*, Chicago, IL: University of Chicago Press.

Martin, Andrew. (2003) One Fish, Two Fish, Genetically New Fish, *Chicago Tribune*, 12 November, p. A1.

——(2004). Mixing Milk and Politics: USDA Official Criticized for Linking Policy, Bush Votes, *Chicago Tribune*, 25 September, p. 1.

——and Andrew Pollack. (2008). F.D.A. Declares Cloned Animals Are Safe to Eat, *The New York Times*, 16 January, p. A1.

Martin, Lisa. (1992). *Coercive Cooperation*, Princeton, NJ: Princeton University Press.

Mattli, Walter and Tim Büthe. (2003). Setting International Standards, *World Politics*, Vol. 56, No. 1, pp. 1–42.

——and Tim Büthe. (2005). Global Private Governance: Lessons from a National Model of Setting Standards in Accounting, *Law and Contemporary Problems*, Vol. 68, Nos. 3–4, pp. 225–62.

Maurer Andreas. (2003). Less Bargaining—More Deliberation: The Convention Method for Enhancing EU Democracy, *Internationale Politik und Gesellschaft*, Vol. 1, pp. 167–90.

Mavroidis, Petros. (2003). Trade Disputes Concerning Health Policy Between the EC and the US, in Ernst-Ulrich Petersmann and Mark A. Pollack, eds., *Transatlantic Trade Disputes: The US, the EU, and the WTO*, New York: Oxford University Press, pp. 233–45.

Mayr, Juan and Adriana Soto. (2005). Balancing Biosafety and Trade: The Negotiating History of the Cartagena Protocol, in Ricardo Melendez-Ortiz and Vicente Sanchez, eds., *Trading in Genes: Development Perspectives on Biotechnology, Trade and Sustainability*, London: Earthscan, pp. 153–69.

McCubbins, Mathew, Roger Noll, and Barry Weingast. (1987). Administrative Procedures as Instruments of Political Control, *Journal of Law, Economics, and Organization*, Vol. 3, pp. 243–77.

——and Thomas Schwartz. (1987). Congressional Oversight Overlooked: Police Patrols Versus Fire Alarms, in Mathew McCubbins and Terry Sullivan, eds., *Congress: Structure and Policy*, New York: Cambridge University Press, pp. 426–40.

Madconald, Roderick A. (1998). Metaphors of Multiplicity: Civil Society, Regimes and Legal Pluralism, *Arizona Journal of International and Comparative Law*, Vol. 15, No. 1, pp. 69–91.

McDougal, Myres and Harold Laswell. (1959). The Identification and Appraisal of Diverse Systems of Public Order, *American Journal of International Law*, Vol. 53, No. 1, p. 10.

McElroy, Gail. (2007). Legislative Politics, in Knud Erik Jørgensen, Mark A. Pollack, and Ben Rosamond, eds., *The Handbook of European Union Politics*, New York: Sage, pp. 175–94.

McGarity, Thomas and Patricia Hansen. (2001). *Breeding Distrust: An Assessment and Recommendations for Improving the Regulation of Plant Derived Genetically Modified Foods*, Washington, DC: Consumer Federation of America Foundation.

References

McGinnis, John and Mark Movsiean. (2000). The World Trade Constitution: Reinforcing Democracy Through Trade, *Harvard Law Review*, Vol. 114, pp. 511–605.

Meade, Geoff. (2007). Euro Victory for Environmental Campaigners in GM Battle, *Press Association Newsfile*, 29 March.

Mehta, Pradeep S. (2006). Of Virus, Seeds, Patents, Competition, *Hindu Business Line*, 17 November.

Meijer, Ernestine and Richard Stewart. (2004). The GM Cold War: How Developing Countries Can Go from Being Dominos to Being Players, *Review of European Community and International Environmental Law*, Vol. 13, No. 3, pp. 247–62.

Mellman Group. (2006). Memorandum: Review of Public Opinion Research, 16 November.

Melnick, R. Shep. (1983). *Regulation and the Courts: The Case of the Clean Air Act*, Washington, DC: Brookings Institution.

Merrill, Richard A. (1998). The Importance and Challenges of 'Mutual Recognition, *Seton Hall Law Review*, Vol. 29, pp. 736–55.

Meyer, John W. et al. (1997). World Society and the Nation State, *The American Journal of Sociology*, Vol. 103, No. 1, pp. 144–81.

Miller, Henry I. (2000). Regulatory Gangs Maul Biotech, *The Wall Street Journal*, 18 May, p. A27.

——(2007). Food from Cloned Animals: These Products Are Long Overdue, *San Diego Union-Tribune*, 3 January, p. B7.

——and Gregory Conko. (2001). The Perils of Precaution: Why Regulators' Precautionary Principle is Doing more Harm than Good, *Policy Review*, No. 107, June–July, pp. 25–40.

————(2003). Children as Policy Pawns, *Washington Times*, 5 October, p. B4.

————(2004). *The Frankenfood Myth: How Protest and Politics Threaten the Biotech Revolution*, Westport, CT: Praeger Publishers.

Miller, John W. (2006). French Farmers, Activists Battle over Genetically Altered Corn, *Associated Press*, 12 October.

Millstone, Erik and Patrick van Zwanenberg. (2002). The Evolution of Food Safety Policy-Making Institutions in the UK, EU and Codex Alimentarius, *Social Policy and Administration*, Vol. 36, No. 6, pp. 593–609.

Mitchell, Ronald B. (1994). Regime Design Matters: International Oil Pollution and Treaty Compliance, *International Organization*, Vol. 48, No. 3, pp. 425–59.

Mitzen, Jennifer. (2005). Reading Habermas in Anarchy: Multilateral Diplomacy and Global Public Spheres, *American Political Science Review*, Vol. 99, No. 3, pp. 611–32.

Monsanto. (2007). Monsanto's Roundup RReady2Yield Soybean Completes Regulatory Process in U.S and Canada, Monsanto Press Release, 31 July 2007.

Monsanto Company. (2004). Monsanto to Realign Research Portfolio, Development of Roundup Ready Wheat Deferred, http://www.worc.org/pdfs/monsantodropsPR.pdf, accessed on 11 June.

Moravcsik, Andrew. (1993). Integrating International and Domestic Politics: A Theoretical Introduction, in Peter B. Evans, Harold K. Jacobson, and Robert D. Putnam, eds., *Double-Edged Diplomacy: International Bargaining and Domestic Politics*, Berkeley, CA: University of California Press, pp. 3–42.

——(1997). Taking Preferences Seriously: A Liberal Theory of International Politics, *International Organization*, Vol. 51, No. 4 (Autumn), pp. 513–54.

——(1998). *The Choice for Europe: Social Purpose and State Power from Messina to Maastricht*, Ithaca, NY: Cornell University Press.

——(2001). Federalism in the European Union: Rhetoric and Reality, in Kalypso Nicolaidis and Robert Howse, eds., *The Federal Vision: Legitimacy and Levels of Governance in the United States in the European Union*, New York: Oxford University Press, pp. 161–87.

Mortished, Carl. (2008). Frankenstein Foods Are Not Monsters, *TimesOnline*, 8 January, http://business.timesonline.co.uk/tol/business/columnists/article3155919.ece, accessed on 14 January 2008.

Morrow, James. (1994). The Forms of International Cooperation, *International Organization*, Vol. 48, No. 3, pp. 387–423.

Murphy, Sean. (2001). Biotechnology and International Law, *Harvard International Law Journal*, Vol. 42, No. 1, pp. 47–139.

National Resource Council. (2004). *Biological Confinement of Genetically Engineered Organisms*, Washington, DC: National Academy Press.

Naurin, Daniel. (2007). Safe Enough to Argue? Giving Reasons in the Council of the EU, ARENA Working Paper No. 11, accessed at: http://www.arena.uio.no on 15 January 2008.

Newell, Peter. (2003). Globalization and the Governance of Biotechnology, *Global Environmental Politics*, Vol. 3, No. 2, pp. 56–71.

Neyer, Jorgen. (2000). The Regulation of Risks and the Power of the People: Lessons from the BSE Crisis, European Integration online Papers (EIoP), Vol. 4, No. 6, http://eiop.or.at/eiop/texte/2000–006a.htm, accessed on 8 January 2008.

Nichols, Philip. (1996*a*). GATT Doctrine, *Virginia Journal of International Law*, Vol. 36, No. 2, pp. 379–466.

——(1996*b*). Trade Without Values, *Northwestern University Law Review*, Vol. 90, No. 2, pp. 658–719.

Nicolaidis, Kalypso and Robert Howse, eds. (2001). *The Federal Vision: Legitimacy and Levels of Governance in the United States and the European Union*, New York: Oxford University Press.

——and Gregory Shaffer. (2005). Transnational Mutual Recognition Regimes: Governance Without Global Government, *Law and Contemporary Problems*, Vol. 68, No. 3, pp. 267–322.

——and Rebecca Steffenson. (2005). Managed Mutual Recognition in the Transatlantic Marketplace, in Andrews David, Mark A. Pollack, Gregory C. Shaffer, and Helen Wallace, eds., *The Future of Transatlantic Economic Relations: Continuity Amid Discord*, Florence: European University Institute, Robert Schuman Centre for Advanced Studies, pp. 233–68; available online at http://www.iue.it/RSCAS/e-texts/Future_Transat_EconRelations.pdf, accessed on 8 June 2007.

Niemann, Arne. (2008). Deliberation and Bargaining in the Article 113 Committee and the 1996/97 IGC Representatives Group, in Daniel Naurin and Helen Wallace, eds., *Unveiling the Council of the EU*, New York: Palgrave, pp 121–143.

Nonet, Philippe and Philip Selznick. (2005). *Law and Society in Transition: Toward Responsive Law*, London: Transaction Publishers.

Nourse, Victoria. (2004). Toward a New Constitutional Anatomy, *Stanford Law Review*, Vol. 56, No. 4, pp. 835–900.

Novozymes. (2006). *The Novozymes Report 2006*, online at: http://report2006.novozymes.com/Material/Files/The + Novozymes + Report + 2006, accessed on 10 April 2007.

Nuffield Council on Bioethics. (2003). *The Use of Genetically Modified Crops in Developing Countries: A Follow-Up Discussion Paper*, available at: http://www.nuffieldbioethics.org/go/ourwork/gmcrops/publication_313.html, accessed on 8 January 2008.

Nugent, Neill. (2001). *The European Commission*, New York: St Martin's Press.

Nwalozie, Marcel, Paco Sereme, Harold Roy-Macauley, and Walter Alhassan. (2007). West and Central Africa: Strategizing Biotechnology for Food Security and Poverty Reduction, in Sakiko Fukuda-Parr, ed., *The Gene Revolution: GM Crops and Unequal Development*, London: Earthscan Publications, pp. 69–84.

Nye, Joseph. (2005). *Soft Power*, New York: Public Affairs.

O'Donnell, Peter. (2008). Commission GMO Debate Achieves No Resolution, *BioWorld International*, Vol. 20, No. 13, p. 3.

OECD. (1993). Safety Considerations for Biotechnology: Scale-Up of Crop Plants, Paris: Organization for Economic Cooperation and Development, http://www.oecd.org/dataoecd/26/26/1958527.pdf (accessed 22 June 2008).

——(2000). Report of the Task Force for the Safety of Novel Foods and Feeds, C(2000)86/ADD1, 17 May, http://www.olis.oecd.org/olis/2000doc.nsf/LinkTo/C(2000)86-ADD1 (accessed 22 June 2008).

——(2001). Non-tariff Measures on Agricultural and Food Products: The Policy Concerns of Emerging and Transition Economies, http://www.oecd.org/dataoecd/23/14/25498833.pdf (accessed 22 June 2008).

——(2004a). Biotechnology for Sustainable Growth and Development, Meeting of the OECD Committee for Scientific and Technological Policy at Ministerial Level, http://www.oecd.org/dataoecd/43/2/33784888.PDF (accessed 22 June 2008).

——(2004b). Guidance for the Operation of Biological Resource Centres (BRCs)—Part 1: General Requirements for All BRCs, OECD Global Forum on Knowledge Economy: Biotechnology, http://www.oecd.org/dataoecd/60/44/23547773.pdf (accessed 22 June 2008).

——(2004c). Guidance for the Operation of Biological Resource Centers (BRCs)—Certification and Quality Criteria for BRCs, OECD Global Forum on Knowledge Economy: Biotechnology, http://www.oecd.org/dataoecd/60/42/23547743.pdf (accessed 22 June 2008).

——(2005). An Introduction to the Biosafety Consensus Documents of OECD's Working Group For Harmonization in Biotechnology, Series on Harmonization of Regulatory Oversight in Biotechnology, No. 32, ENV/JM/MONO(2005)5, http://appli1.oecd.org/olis/2005doc.nsf/linkto/env-jm-mono(2005)5 (accessed 22 June 2008).

——(2006a). Biotechnology Update, Internal Co-ordination Group for Biotechnology, No. 17 (December), http://www.oecd.org/LongAbstract/0,3425,en_2649_37401_37769440_1_1_1_37401,00.html (accessed 22 June 2008).

——(2006b). Safety Assessment of Transgenic Organisms: OECD Consensus Documents, Vols. 1–2, http://www.oecd.org/document/15/0,2340,en_2649_34385_37336335_1_1_1_1,00.html (accessed 22 June 2008).

——(2007). Biotechnology Update, Internal Co-ordination Group for Biotechnology, No. 18 (September), http://www.oecd.org/dataoecd/3/44/39314743.pdf (accessed 22 June 2008).

OECD Environment Directorate. (2003). Output on the Questionnaire on National Approaches to Monitoring/Detection/Identification of Transgenic Products, ENV/

JM/MONO(2003)8 (27 May), http://www.oecd.org/LongAbstract/0,3425,en_2649_34385_2789664_119666_1_1_37401,00.html (accessed 22 June 2008).

Office of Science and Technology Policy. (1985). Establishment of the Biotechnology Science Coordinating Committee, 50 FR 4714, 14 November.

——(1986). Coordinated Framework for Regulation of Biotechnology, 51 FR 23302, 26 June.

OIE (World Organization for Animal Health). (2005). *Implementation of the Fourth Strategic Plan: Director-General's Programme of Work 2006–2008*, 74 SG 23, 1–15.

——(2006). *Terrestrial Animal Health Code*, 15th edn., http://www.oie.int/eng/normes/Mcode/en_sommaire.htm (accessed 22 June 2008).

Olson, Mancur. (1965). *The Logic of Collective Action: Public Goods and the Theory of Groups*, Cambridge, MA: Harvard University Press.

Ostrom, Elinor. (1990). *Governing the Commons: The Evolution of Institutions for Collective Action*, New York: Cambridge University Press.

Oye, Kenneth, ed. (1986). *Cooperation Under Anarchy*, Princeton, NJ: Princeton University Press.

Paarlberg, Robert L. (2001). *The Politics of Precaution: Genetically Modified Crops in Developing Countries*, Baltimore, MD: Johns Hopkins University Press.

Patterson, Lee Ann. (2000). Biotechnology, in Helen Wallace and William Wallace, eds., *Policy-Making in the European Union*, Oxford: Oxford University Press, pp. 317–43.

——and Tim Josling. (2005). Regulating Biotechnology: Comparing EU and US Approaches, in Andrew Jordan, edn., *Environmental Policy in the European Union: Actors, Institutions and Processes*, 2nd edn., London: Earthscan Publications, pp. 183–200.

Pauwelyn, Joost. (2003*a*). The Role of Public International Law in the WTO: How Far Can We Go? *American Journal of International Law*, Vol. 95, No. 3, pp. 535–78.

——(2003*b*). *Conflict of Norms in Public International Law: How WTO Law Relates to other Rules of International Law*, New York: Cambridge University Press.

Pendleton, Cullen N. (2004). The Peculiar Case of 'Terminator' Technology: Agricultural Biotechnology and Intellectual Property Protection at the Crossroads of the Third Green Revolution, *Biotechnology Law Report*, Vol. 23, No. 1, pp. 1–29.

Penner, Karen. (1999). Hormones and Meat: Food and Nutrition—the Link Between Agriculture and Health, National Food Safety Database, http://www.foodsafety.org/sf/sf083.htm (site under construction, accessed 22 June 2008).

Percival, Robert V. (1995). Environmental Federalism: Historical Roots and Contemporary Models, *Maryland Law Review*, Vol. 54, pp. 1141–82.

Perez, Oren. (2007). Anomalies at the Precautionary Kingdom: Reflections on the GMO Panel's Decision, *World Trade Review*, Vol. 6, No. 2, pp. 265–80.

Pesticide and Toxic Chemical News. (2006). EU Pushed to Lower 'Adventitious Presence for Organics', *Pesticide & Toxic Chemical News*, Vol. 34, No. 32, 29 May, p. 13.

Petersmann, Ernst-Ulrich. (2000). The WTO Constitution and Human Rights, *Journal of International Economic Law*, Vol. 3, pp. 19–25.

——and Mark A. Pollack, eds. (2003). *Transatlantic Economic Disputes: The US, the EU, and the WTO*, New York: Oxford University Press.

Peterson, John et al. (2005). *Review of the Framework for Relations between the European Union and the United States: An Independent Study*, Brussels: CEC.

411

References

Peterson, Robert. (2004). How Foods from Biotech Crops are Evaluated for Human Safety in the United States, University of Nebraska Agbiosafety, http://agbiosafety. unl.edu/food_safety.shtml, accessed 8 January 2008.

Pew Initiative on Food and Biotechnology. (2002a). Three Years Later: Genetically Engineered Corn and the Monarch Butterfly Controversy, http://pewagbiotech.org/ resources/issuebriefs/monarch.pdf, accessed on 8 January 2008.

——(2002b). 2001–2002 Legislative Activity Related to Agricultural Biotechnology, http://pewagbiotech.org/resources/factsheets/legislation/factsheet2002.php, accessed on 8 January 2008.

——(2002c). Biotech and the Deep Blue Sea, *AgBiotech Buzz: Transgenic Fish*, Vol. 2, Issue 3, 27 March.

——(2003a). News Release: Americans' Knowledge of Genetically Modified Foods Remains Low and Opinions on Safety Still Split, http://pewagbiotech.org/newsroom/ releases/091803.php3, accessed on 3 April 2004.

——(2003b). U.S. vs. EU: An Examination of the Trade Issues Surrounding Genetically Modified Foods, December.

——(2003c). Genetically Modified Crops in the United States, Pew Initiative on Food and Biotechnology Fact Sheet.

——(2004a). Issues in the Regulation of Genetically Engineered Plants and Animals, http://pewagbiotech.org/research/regulation/Regulation.pdf, accessed on 12 June 2004.

——(2004b). A Wheaty Issue: GM Wheat Enters the Regulatory Arena, http://pewag biotech.org/buzz/display.php3?StoryID = 96, accessed on 3 April 2004.

——(2006). Pew Initiative Finds Public Opinion about Genetically Modified Foods 'Up for Grabs' Ten Years after Introduction of Ag Biotech, Press Release, 6 December.

——(2007). Application of Biotechnology for Functional Foods. On-line at http://www. pewtrusts.org/uploadedFiles/wwwpewtrustsorg/Reports/Food_and_Biotechnology/ PIFB_Functional_Foods.pdf, accessed on 24 November 2008.

Pierson, Paul. (2000). Increasing Returns, Path Dependence, and the Study of Politics, *American Political Science Review* Vol. 94, No. 2, pp. 251–67.

——(2004). *Politics in Time: History, Institutions and Social Analysis*, Princeton: Princeton University Press.

——and Skocpol Theda. (2002). Historical Institutionalism in Contemporary Political Science, in Ira Katznelson & Helen V. Milner (eds.), *Political Science: State of the Discipline*, New York: W.W. Norton, pp. 693–721.

Pingali, Prabhu. (2007). Will the Gene Revolution Reach the Poor? Lessons from the Green Revolution, Mansholt Lecture, Wageningen University, 26 January 2007 (on file).

Pinstrup-Andersen, Per and Ebbe Schioler. (2000). *Seeds of Contention: World Hunger and the Global Controversy over GM Crops*, Baltimore, MD: Johns Hopkins University Press.

Plein, L. Christopher. (1991). Popularizing Biotechnology: The Influence of Issue Definition, *Science, Technology, and Human Values*, Vol. 16, No. 4, pp. 474–90.

Poli, Sara. (2003). Setting Out International Food Standards: Euro-American Conflicts within the Codex Alimentarius Commission, Giandomenico Majone, ed., *Risk Regulation in the European Union: Between Enlargement and Internationalization*, Florence: European University Institute, pp. 125–47.

412

——(2004). The European Community and the Adoption of International Food Standards within the Codex Alimentarius Commission, *European Law Journal*, Vol. 10, No. 5, pp. 613–30.

Pollack, Andrew. (2004*a*). Narrow Path for New Biotech Food Crops, *The New York Times*, 20 May, pp. C1, C3.

——(2004*b*). Monsanto Shelves Plan for Modified Wheat, *The New York Times*, 11 May 2004, p. C1.

——(2006). Redesigning Crops to Harvest Fuel, *The New York Times*, 8 September, p. D1.

——(2007*a*). Without U.S. Rules, Biotech Food Lacks Investors, 30 July, pp. A1, A14.

——(2007*b*). Round 2 for Biotech Beets: After Delay Over Safety Fears, Engineered Crop Will be Planted, *The New York Times*, 27 November, p. C1.

——(2008). In Lean Times, Biotech Grains Are Less Taboo, *The New York Times*, 21 April, p. A1.

——and Andrew Martin. (2006). F.D.A. Tentatively Declares Food from Cloned Animals to Be Safe, *New York Times*, 25 December, p. A1.

Pollack, Mark A. (2003*a*). The Political Economy of Transatlantic Trade Disputes, in Ernst-Ulrich Petersmann and Mark A. Pollack, eds., *Transatlantic Trade Disputes: The US, the EU, and the WTO*, New York: Oxford University Press, pp. 65–118.

——(2003*b*). *The Engines of Integration: Delegation, Agency and Agenda Setting in the European Union*, New York: Oxford University Press.

——(2003*c*). Managing System Friction: Regulatory Conflicts in Transatlantic Relations and the WTO, in Ernst-Ulrich Petersmann and Mark A. Pollack, eds., *Transatlantic Trade Disputes: The US, the EU, and the WTO*, New York: Oxford University Press, pp. 505–602.

——(2004). The New Institutionalisms and European Integration, in Antje Wiener and Thomas Diez, eds., *European Integration Theory*, New York: Oxford University Press, 2004, pp. 137–56.

——(2005). *JCMS* Annual Lecture: The New Transatlantic Agenda at Ten: Reflections on an Experiment in International Governance, *Journal of Common Market Studies*, Vol. 43, No. 5, December 2005, pp. 899–919.

——(2007). Rational Choice and EU Politics, in Knud Erik Jørgensen, Mark A. Pollack, and Ben Rosamond, eds. *The Handbook of European Union Politics*, New York: Sage, pp. 31–56.

——and Gregory C. Shaffer eds. (2001*a*). *Transatlantic Governance in the Global Economy*, Lanham, MD: Rowman & Littlefield Press.

————(2001*b*). Transatlantic Governance in Historical and Theoretical Perspective, in Mark A. Pollack and Gregory C. Shaffer, eds., *Transatlantic Governance in the Global Economy*, Lanham, MD: Rowman & Littlefield, pp. 3–42.

————(2001*c*). Who Governs? In Mark A. Pollack and Gregory C. Shaffer, eds., *Transatlantic Governance in the Global Economy*, Lanham, MD: Rowman & Littlefield, pp. 287–305.

————(2005). The Future of Transatlantic Economic Relations: Continuity Amid Discord, in David M. Andrews, Mark A. Pollack, Gregory C. Shaffer, and Helen Wallace, eds., *The Future of Transatlantic Relations: Continuity Amid Discord*, Florence: European University Institute, pp. 3–8.

————(2008). Risk Regulation, Genetically Modified Foods, and the Limits of Deliberation in the Council of Ministers, in Daniel Naurin and Helen Wallace, eds., *Unveiling the Council of the EU*, London: Palgrave, pp. 144–164.

References

Posner, Elliot. (2005). Market Power without a Single Market: The New Transatlantic Relations in Financial Services, in David M. Andrews, Mark A. Pollack, Gregory C. Shaffer, and Helen Wallace, eds., *The Future of Transatlantic Relations: Continuity Amid Discord*, Florence: European University Institute, pp. 233–68.

Posner, Theodore and Timothy Rief. (2000). Homage to a Bull Moose: Applying Lessons of History to Meet the Challenges of Globalization, *Fordham International Law Journal*, Vol. 24, No. 1–2, pp. 481–518.

Post, Diahanna. (2005). Standards and Regulatory Capitalism: The Diffusion of Food Safety Standards in Developing Countries, *Annals, AAPSS*, Vol. 598, No. 1, pp. 168–83.

Powell, Robert. (1991). The Problem of Absolute and Relative Gains in International Relations Theory, *American Political Science Review*, Vol. 85, pp. 1303–20.

PR Newswire. (2007). EU Approves Herculex® RW Corn for Food, Feed, Import and Processing, *PR Newswire*, 24 October.

Prakash, Seem and Kelly Kollman. (2003). Biopolitics in the EU and the U.S.: A Race to the Bottom or Convergence to the Top, *International Studies Quarterly*, Vol. 47, No. 4, pp. 617–741.

Pray, Carl and Anwar Naseem. (2007). Supplying Crop Biotechnology to the Poor: Opportunities and Constraints, *Journal of Development Studies*, Vol. 43, No 1, pp. 192–217.

Press Trust of India. (2006). AP Government Files Case Against Monsanto at MRTPC, *Press Trust of India*, 2 January.

Pruzin, Daniel. (2000). South Korea Notifies WTO of GMO Labeling Proposal, *International Trade Reporter*, Vol. 17, 20 January, p. 86.

——(2007a). EU Receives Deadline of Nov. 21 to Comply with WTO GMO Ruling, *International Trade Reporter*, Vol. 24, 5 July, p. 947.

——(2007b). WTO Ruling Said to Aid U.S., Canada in Beef-Hormone Disputes with Europe, *International Trade Reporter*, Vol. 24, 2 August, p. 1112.

Putnam, Robert D. (1988). Diplomacy and Domestic Politics: The Logic of Two-Level Games, *International Organization*, Vol. 42, No. 3, pp. 427–60.

Ramaswami, Bharat and Carl E. Pray. (2007). India: Confronting the Challenge—The Potential of Genetically Modified Crops for the Poor, in Sakiko Fukuda-Parr, *The Gene Revolution: GM Crops and Unequal Development*, London: Earthscan Publications, pp. 156–74.

Raustiala, Kal. (2000). Compliance and Effectiveness in International Regulatory Cooperation, *World Politics*, Vol. 49, No. 4, pp. 482–509.

——(2002). The Architecture of International Cooperation: Transgovernmental Networks and the Future of International Law, *Virginia Journal of International Law*, Vol. 43, No. 1, pp. 1–92.

——(2005). Form and Substance in International Agreements, *American Journal of International Law*, Vol. 99, No. 3, pp. 581–614.

——and Anne-Marie Slaughter. (2002). International Law, International Relations, and Compliance, in Walter Carlsnaes, Thomas Risse, and Beth A. Simmons, eds., *Handbook of International Relations*, New York: Sage, 2002, pp. 538–58.

——and David G. Victor. (1998). Conclusions, in David G. Victor, Kal Raustiala, and Eugene B. Skolnikoff, *The Implementation and Effectiveness of International Environmental Commitments: Theory and Practice*, Cambridge, MA: MIT Press, pp. 659–707.

————(2004). The Regime Complex for Plant Genetic Resources, *International Organization*, Vol. 58, No. 2, pp. 277–309.

République Française. (2008). Communiqué de Presse: Au Conseil informel Environnement, la Présidence française lance un groupe de travail pour renforcer l'évaluation des OGM à l'échelle européenne, Paris, 4 July 2008, http://www.developpement-durable.gouv.fr/IMG/pdf/CP_OGM_cle154a58.pdf, accessed on 5 July 2008.

Reuters. (1999). Leading European Dog Food Maker Eschews GM, *Reuters*, 9 September.

————(2001). US Grain System Said Incompatible with EU Rules, *Reuters*, 13 July.

————(2003). Argentina Sees $1 bn Loss a Year to EU GM Rule, *Reuters*, 20 December.

————(2005). Biotech Crop Policy in EU Gets Rethink after Rebuff, *Reuters*, 30 June.

————(2008). France Say Extend Ban on GMO Crop, *Reuters*, 11 January.

Revkin, Andrew C. (2006). Climate Expert Says NASA Tried to Silence Him, *The New York Times*, 29 January, p. A1.

Rhinard, Mark and Michael Kaeding. (2006). The International Bargaining Power of the European Union in Mixed Competence Negotiations: The Case of the 2000 Cartagena Protocol on Biosafety, *Journal of Common Market Studies*, Vol. 44, No. 5, pp. 1023–50.

Rhodes, Martin. (2005). Employment Policy: Between Efficacy and Experimentation, in Helen Wallace, William Wallace, and Mark A. Pollack, eds., *Policy-Making in the European Union*, 5th edn., New York: Oxford University Press, pp. 279–304.

————and Manuele Citi. (2007). New Modes of Governance in the European Union: A Critical Survey and Analysis, in Knud Erik Jørgensen, Mark A. Pollack, and Ben Rosamond, eds. *The Handbook of European Union Politics*, New York: Sage, pp. 463–82.

Risse-Kappen, Thomas. (2000). 'Let's Argue!' Communicative Action in World Politics, *International Organization*, Vol. 54, No. 1, pp. 1–39.

————(2001). Is Transnational Deliberation Possible in Europe? Paper Prepared for a Workshop on Ideas, Discourse and European Integration, European Union Center, Harvard University, 11–12 May.

————(2004). Social Constructivism and European Integration, in Antje Wiener and Thomas Diez, eds., *European Integration Theory*, New York: Oxford University Press, pp. 159–76.

Roberts, Donna. (1998). Preliminary Assessment of the Effects of the WTO Agreement on Sanitary and Phytosanitary Trade Regulations, *Journal of International Economic Law*, Vol. 1, No. 3, pp. 377–405.

————and Laurian Unnevehr. (2005). Resolving Trade Disputes Arising from Trends in Food Safety Regulation: The Role of the Multilateral Governance Framework, *World Trade Review*, Vol. 4, No. 3, pp. 469–97.

Roberts, Simon. (2004). After Government: On Representing Law without the State, *Modern Law Review*, Vol. 68, No. 1, pp. 1–24.

Rodemeyer, Michael. (2003). Technology Moves Faster than Regulators, *USA Today*, 29 December, pp. 13A.

Rosenthal, Elisabeth. (2006a). In EU, Front Lines in Global War Over Food, *The International Herald Tribune*, 25 May, p. 2.

————(2006b). Biotech Food Tears Rifts in Europe, *New York Times*, 6 June, p. C1.

————(2007a). European Farmers Sow Greener Future, *International Herald Tribune*, 29 May, p. 1.

——(2007*b*). A Genetically Modified Potato, Not for Eating, Is Stirring Some Opposition in Europe, *The New York Times*, 24 July, p. C3.

——(2007*c*). Both Sides Cite Science to Address Altered Corn, *The New York Times*, 26 December, pp. C1, C5.

Roy, Devparna, Ronald J. Herring, and Charles C. Geisler. (2007). Naturalising Transgenics: Official Seeds, Loose Seeds and Risk in the Decision Matrix of Gujarati Cotton Farmers, *Journal of Development Studies*, Vol. 43, No. 1, pp. 158–76.

Rugaber, Christopher S. (2003). Grassley, Baucus Pressure USTR Zoellick on WTO Case Against the EU, *BNA International Trade Reporter*, Vol. 20, No. 11, 13 March, p. 468.

Ruggie, John Gerard. (1998). What Makes the World Hang Together? Neo-utilitarianism and the Social Constructivist Challenge, *International Organization*, Vol. 52, No. 4, pp. 855–85.

Runge, C. Ford. and Lee Ann Jackson. (2000). Labelling, Trade and Genetically Modified Organisms: A Proposed Solution, *Journal of World Trade*, Vol. 34, No. 1, pp. 111–22.

——and Benjamin Senauer. (2007). How Biofuels Could Starve the Poor, *Foreign Affairs*, Vol. 86, No. 3, May–June, pp. 41–55.

Russell, Pam Radtke. (2007). Tough Row to Hoe: The Louisiana Rice Industry is Rebounding Nicely from a Difficult Year, but It's Not All Good News, *New Orleans Times-Picayune*, Money Section, p. 1.

Ryan, Missy. (2008). US Agrees to Give EU More Time on Biotech OK, *Reuters*, 15 January.

Ryfe, David M. (2005). Does Deliberative Democracy Work? *Annual Review of Political Science*, Vol. 8, pp. 49–71.

Saegusa, A. (1999). Japan Tightens Rules on GM Crops to Protect the Environment, *Nature*, June, p. 719.

Safrin, Sabrina. (2002). Treaties in Collision? The Biosafety Protocol and the World Trade Organization Agreements, *American Journal of International Law*, Vol. 96, No. 3, pp. 606–28.

Salzman, James. (2005). Decentralized Administrative Law in the Organization for Economic Cooperation and Development, *Law & Contemporary Problems*, Vol. 68, Nos. 3–4, pp. 189–224.

Sandel, Michael J. (2007). *The Case Against Perfection: Ethics in the Age of Genetic Engineering*, Cambridge, MA: Harvard University Press.

Sands, Philippe and Pierre Klein. (2001). *Bowett's Law of International Institutions*, 5th edn., London: Sweet & Maxwell.

Sasson, Albert. (2005). *UNU-IAS Report, Industrial and Environmental Biotechnology: Achievements, Prospects, and Perceptions 16*, Tokyo: Institute of Advanced Studies, United Nations University.

Sato, Kyoko. (2006). Politics and Meanings of Genetically Modified Food: The Case of Policy Change in France, Paper Presented at the Biennial Conference of Europeanists, Chicago, Illinois, 29 March–2 April 2006.

Schubert, David. (2007). Food from Cloned Animals: These Products Should Remain in the Lab, *San Diego Union-Tribune*, 3 January, p. B7.

Scoones, Ian. (2006). *Science, Agriculture and the Politics of Policy: The Case of Biotechnology Policy in India*, New Delhi: Orient Longman.

Scott, Joanne. (2003). European Regulation of GMOs and the WTO, *Columbia Journal of European Law*, Vol. 9, No. 2, pp. 213–39.

——(2004) International Trade and Environmental Governance: Relating Rules (and Standards) in the EU and the WTO, *European Journal of International Law*, Vol. 15, No. 2, pp. 307–54.

——(2007). *The WTO Agreement on Sanitary and Phytosanitary Standards*, New York: Oxford University Press.

——and David M. Trubek. (2002). Mind the Gap: Law and New Approaches to Governance in Europe, *European Law Journal*, Vol. 8, No. 1, pp. 1–18.

Sebastian, Thomas. (2007). World Trade Organization Remedies and the Assessment of Proportionality: Equivalence and Appropriateness, *Harvard International Law Journal*, Vol. 48 No. 2, pp. 337–82.

Sek, Lenore. (2002). Trade Retaliation: The 'Carousel' Approach, Congressional Research Service Report for Congress, CRS20715, Updated 5 March 2002, http://fpc.state.gov/documents/organization/23368.pdf, accessed on 11 July 2007.

Seifert, Franz. (2006). Synchronised National Publics as Functional Equivalent of an Integrated European Public: The Case of Biotechnology, *European Integration Online Papers*, Vol. 10, No. 8 (9 August).

Shaffer, Gregory C. (2000). Globalization and Social Protection: The Impact of Foreign and International Rules in the Ratcheting Up of U.S. Privacy Standards, *Yale Journal of International Law*, vol. 25, pp. 1–88.

——(2002). Managing U.S.-EU Trade Relations through Mutual Recognition and Safe Harbor Agreements: "New" and "Global" Approaches to Transatlantic Economic Governance?, *Columbia Journal of European Law*, vol. 9, pp. 29–77.

——(2003). Managing US–EU Trade Relations through Mutual Recognition and Safe Harbor Agreements: 'New' and 'Global' Approaches to Transatlantic Economic Governance? in Ernst-Ulrich Petersmann and Mark A. Pollack, eds., *Transatlantic Trade Disputes: The US, the EU, and the WTO*, New York: Oxford University Press, pp. 297–325.

——(2004). Power, Global Governance and the WTO: A Comparative Institutional Approach, in Michael Barnett and Robert Duvall, eds., *Power and Global Governance*, New York: Cambridge University Press, pp. 130–60.

——(2005*a*). The Role of the Director-General and Secretariat: A Comment on Chapter IX of the Sutherland Report, *World Trade Review*, Vol. 4, No. 3, pp. 429–38.

——(2005*b*). Power, Governance, and the WTO: A Comparative Institutional Approach, in Michael Barnett and Raymond Duvall, eds., *Power in Global Governance*, New York: Cambridge University Press, pp. 130–60.

——and Yvonne Apea. (2005). Institutional Choice in the GSP Case: Who Decides the Conditions for Trade Preferences: The Law and Politics of Rights, *Journal of World Trade*, Vol. 39, No. 5, pp. 977–1008.

Shand, Hope. (2003). *Terminator Technology—Five Years Later*, ETC Group Communique, Vol. 79, May–June, available at http://www.etcgroup.org/upload/publication/167/01/termcom03.pdf, accessed on 27 December 2007.

Shapiro, Martin. (1988). *Who Guards the Guardians?* Athens: University of Georgia Press.

——(1997). The Problem of Independent Agencies in the United States and the European Union, *Journal of European Public Policy*, Vol. 4, No. 2, pp. 276–91.

Shea, Griffin. (2000). Biotech Treaty Seeks to Stop Problems before They Start, *Agence France Presse*, 30 January.

Sheingate, Adam D. (2006). Promotion Versus Precaution: The Evolution of Biotechnology Policy in the United States, *British Journal of Political Science*, Vol. 36, No. 2, pp. 243–68.

Shelton, Dinah. (2000). *Commitment and Compliance: The Role of Non-Binding Norms in the International Legal System*, New York: Oxford University Press.

Shiva, Vandana. (2000). *Stolen Harvest: The Hijacking of the Global Seed Supply*, Cambridge, MA: South End Press.

——Afsar H. Jafri, Ashok Emani, and Manish Pande. (2000). *Seeds of Suicide: The Ecological and Human Costs of Globalisation of Agriculture*, New Delhi, Research Foundation for Science Technology and Ecology.

Shor, Mikhael. (2005). Nash Equilibrium, *Dictionary of Game Theory Terms*, Game Theory.net, http://www.gametheory.net/dictionary/ url_of_entry.html, accessed on 28 February 2008.

Simma, Bruno. (1985). Self-Contained Regimes, in *Netherlands Yearbook of International Law*, Vol. 16, pp. 111–36.

Sindico, Francesco. (2006). Soft Law and the Elusive Quest for Sustainable Global Governance, *Leiden Journal of International Law*, Vol. 19, No. 3, pp. 829–46.

Singer, Peter. (1990). *Animal Liberation*, 2nd edn., New York: New York Review of Books.

Skogstad, Grace. (2001). The WTO and Food Safety Regulatory Policy Innovation in the European Union, *Journal of Common Market Studies*, Vol. 39, No. 3, pp. 485–505.

——(2003). Legitimacy and/or Policy Effectiveness? Network Governance and GMO regulation in the European Union, *Journal of European Public Policy*, Vol. 10, No. 3, pp. 321–338.

Slaughter, Anne-Marie B. (1993). International Law and International Relations Theory: A Dual Agenda, *American Journal of International Law*, Vol. 87, pp. 205–39.

——(1997). The Real New World Order, *Foreign Affairs*, Vol. 76, No. 5, pp. 183–97.

——(2004). *A New World Order*, Princeton, NJ: Princeton University Press.

Slovic, Paul. (2000). *The Perception of Risk*, London: Earthscan Publications.

Smith, Jeffrey M. (2007). *Genetic Roulette: The Documented Health Risks of Genetically Engineered Foods*, Fairfield, IA: Yes! Books.

Smith, Jeremy. (2003). Europe's Supermarkets Still Wary about GMO Foods, *Reuters*, 24 October.

——(2007). EU to Debate Hungary GMO Ban, Flowers, and Potatoes, *Reuters*, 16 January.

——(2008). EU Approves Genetically Modified Soybean for Import, Reuters, 4 December.

Smyth, Jamie. (2007). Gormley Tells EU of State's Opposition to GM Foods, *The Irish Times*, 29 June, p. 9.

Snidal, Duncan. (1985) Coordination Versus Prisoner's Dilemma: Implications for International Cooperation and Regimes, *American Political Science Review*, Vol. 79, No. 4, pp. 923–42.

——(1991). International Cooperation Among Relative Gains Maximizers, *International Studies Quarterly*, Vol. 35, No. 4, pp. 387–402.

Snyder, Francis. (2002). Governing Globalization, in Michael Likovsky, ed., *Transnational Legal Processes: Globalisation and Power Disparities*, London: Butterworths, pp. 65–97.

Sokal, Alan D. and Jean Bricmont. (1998). *Fashionable Nonsense: Postmodern Intellectuals' Abuse of Science*, New York: Picador.

Spiteri, Sharon. (2004). Member States Split on GM Maize Approval, *euobserver.com*, 29 June.

Spongenberg, Helena. (2006). Commission Seeks More Transparent GMO Rules, *euobserver.com*, 12 April.

——(2007*a*). One Million Citizens Call for Labelling of GM Foods, *euobserver.com*, 5 February.

——(2007*b*). EU States Deal Blow to Brussels by Backing GMO Ban, *euobserver.com*, 21 February.

SPS Committee [World Trade Organization Committee on Sanitary and Phytosanitary Measures]. (1998). *Use of International Standards Under the SPS Agreement*, Submission by the United States, G/SPS/GEN/76, 4 June 1998.

——(2001). *Procedure to Monitor the Process of International Harmonization*, Third Annual Report, G/SPS/18, 19 September 2001.

——(2005). *Procedure to Monitor the Process of International Harmonization*, Seventh Annual Report, G/SPS/37, 19 July 2005.

——(2006*a*). *Summary of the Meeting of 27–28 June 2006: Note by the Secretary*, G/SPS/R/42, 25 September 2006, available at http://www.wto.org/english/tratop_e/sps_e/sps_e.htm (follow "Summary reports (minutes) > search" hyperlink; then follow "preview: html" hyperlink).

——(2006*b*). Sulfur Dioxide in Cinnamon, Letter from the SPS Chairperson and Response from the Codex Alimentarius Commission, G/SPS/GEN/716, 25 July 2006.

Stasavage, David. (2004). Open-Door or Closed-Door? Transparency in Domestic and International Bargaining, *International Organization*, Vol. 58, No. 4, pp. 667–704.

Stein, Arthur. (1982). Coordination and Collaboration: Regimes in an Anarchic World, *International Organization*, Vol. 36, No. 2, pp. 399–24.

Steinberg, Richard. (1997). Trade-Environment Negotiations in the EU, NAFTA, and WTO: Regional Trajectories of Rule Development, *American Journal of International Law*, Vol. 91, No. 2, pp. 231–67.

——and Jonathan Zasloff. (2006). Power and International Law, *American Journal of International Law*, Vol. 100, No. 1, pp. 64–87.

Steiner, Jürg, André Bächtiger, Markus Spörndli, and Marco Steenbergen. (2004). *Deliberative Politics in Action: Analyzing Parliamentary Discourse*, New York: Cambridge University Press.

Steinzor, Rena I. (1996). Unfunded Environmental Mandates and the 'New (New) Federalism': Devolution, Revolution, or Reform? *Minnesota Law Review*, Vol. 81, No. 1, pp. 97–228.

Sterling, Andrew and Sue Mayer. (2000). A Precautionary Approach by Technology Appraisal? A Multi-Criteria Mapping of Genetic Modification in UK Agriculture, *TA-Datenbenbank-Nachricthen*, Vol. 3, p. 9, Jg., October.

Stevens, Anne and Handley Stevens. 2001. *Brussels Bureaucrats? The Administration of the European Union*, New York: Palgrave.

Stewart, Terence P. and David S. Johanson. (1999). Policy in Flux: The European Union's Laws on Agricultural Biotechnology and Their Effects on International Trade, *Drake Journal of Agricultural Law*, Vol. 4, No. 1, pp. 243–95.

Stewart, Richard. (2007). The GMO Challenge to International Environmental Trade Regulation: Developing Country Perspectives (unpublished draft, on file).

Stone Sweet, Alec. (1999). Judicialization and the Construction of Governance, *Comparative Political Studies*, Vol. 31, pp. 147–84.

——(2004). *The Judicial Construction of Europe*, New York: Oxford University Press.

——and Jud Mathews. (2008). Proportionality Balancing and Global Constitutionalism, *Columbia Journal of Transnational Law*, Vol. 47, forthcoming.

Strauss, Stephen. (2002). Africans Must Wrestle with Food Dilemma, *Globe and Mail*, 23 August, p. A11.

Streeck, Wolfgang and Kathleen Thelen. (2005). Introduction: Institutional Change in Advanced Political Economies, in Wolfgang Streeck and Kathleen Thelen, eds., *Beyond Continuity: Institutional Change in Advanced Political Economies*, New York: Oxford University Press, pp. 1–39.

Sunstein, Cass. (2002). Probability Neglect: Emotions, Worst Cases, and Law, *Yale Law Journal*, Vol. 112, pp. 61–107.

——(2003*a*). *Risk and Reason: Safety, Law and the Environment*, New York: Cambridge University Press.

——(2003*b*). Beyond the Precautionary Principle, University of Chicago Public Law and Legal Theory Working Paper No. 38, available at http://www.law.uchicago.edu/academics/publiclaw/, accessed on 8 January 2008.

——(2005). *The Laws of Fear: Beyond the Precautionary Principle*, New York: Cambridge University Press.

——(2006). *Infotopia: How Many Minds Produce Knowledge*, Oxford: Oxford University Press.

Sykes, Alan. (2002). Domestic Regulation, Sovereignty, and Scientific Evidence Requirements: A Pessimistic View, *Chicago Journal of International Law*, Vol. 3, pp. 353–68.

TABD (Transatlantic Business Dialogue). (2002). *2002 TABD Chicago Conference Report*, http://128.121.145.19/tabd/ceoreports/2002%20Chicago%20CEO%20Report.pdf, accessed on 13 June 2004.

——(2004). Report to the US-EU Summit in Ireland, 26 June 2004—Establishing a Barrier-Free Transatlantic Market: Principles and Recommendations, http://128.121.145.19/tabd/media/TABDReportFINAL22AprilUS.pdf, accessed on 13 June 2004.

TACD (Transatlantic Consumer Dialogue). (2003). Transatlantic Consumer Groups Call Upon US Government to Drop WTO GMO Case at US–EU Summit in Washington, http://www.tacd.org/cgi-bin/db.cgi?page = view&config = admin/press.cfg&id = 29, accessed on 26 April 2004.

——(2004). Report: 6th Annual Meeting, http://www.tacd.org/events/meeting6/meeting_report.htm, accessed on 26 April 2004.

Tallberg, Jonas. (2007). Executive Politics, in Knud Erik Jørgensen, Mark A. Pollack, and Ben Rosamond, eds., *The Handbook of European Union Politics*, New York: Sage, pp. 195–212.

Taylor, Michael R. (1997). Preparing America's Food Safety System for the Twenty-First Century: Who is Responsible for What When it Comes to Meeting the Food Safety Challenges of the Consumer-Driven Global Economy? *Food and Drug Law Journal*, Vol. 52, No. 1, pp. 13–30.

——Jody S. Tick, and Diane M. Sherman. (2004). *Tending the Fields: State & Federal Roles in the Oversight of Genetically Modified Crops*, Pew Initiative on Food and Biotechnology, December, http://pewagbiotech.org/research/fields/report.pdf, accessed on 28 July 2007.

Tetlock, Philip E. and Aaron Belkin, eds. (1996). *Counterfactual Thought Experiments in World Politics*, Princeton, NJ: Princeton University Press.

Thelen, Kathleen. (1999). Historical Institutionalism in Comparative Politics, *Annual Review of Political Science*, Vol. 2, pp. 369–404.

Thompson, Paul. (2003). Crossing Species Boundaries is Even More Controversial than You Think, *American Journal of Bioethics*, Vol. 3, No. 3, pp. 14–15.

Tiberghien, Yves. (2007). Europe: Turning Against Agricultural Biotechnology in the Late 1990s, in Sakiko Fukuda-Parr, ed., *The Gene Revolution: GM Crops and Unequal Development*, London: Earthscan Publications, pp. 51–69.

——(2006). *The Battle for the Global Governance of Genetically Modified Organisms: The Roles of the European Union, Japan, Korea, and China in a Comparative Context*, Les Etudes du CERI, No. 124, Paris: Institut d'Etudes Politiques.

——and Sean Starrs. (2004). The EU as Global Trouble-Maker in Chief: A Political Analysis of EU Regulations and EU Global Leadership in the Field of Genetically Modified Organisms, Proceedings of the Council of European Studies 2004 meeting, Chicago, IL.

Tinbergen, Jan. (1965). *International Economic Integration*, New York: Elsevier.

Toke, Dave. 2004, *The Politics of GM Food: A Comparative Study of the UK, USA and EU*, New York: Routledge.

Trachtman, Joel. (2006a). The Constitutions of the WTO, *European Journal of International Law*, Vol. 17, No. 3, pp. 623–46.

——(2006b). Building the WTO Cathedral, *Stanford Journal of International Law*, Vol. 43, No. 1, pp. 127–67.

Trebilcock, Michael and Julie Soloway. (2002). International Trade Policy and Domestic Food Safety Regulation: The Case for Substantial Deference by the WTO Dispute Settlement Body Under the SPS Agreement, in Daniel Kennedy and James Southwick, eds., *The Political Economy of International Trade Law*, Cambridge: Cambridge University Press, pp. 537–74.

Triplett, William. (2004). Science and Politics, *Congressional Quarterly Researcher*, Vol. 14, No. 28, pp. 661–84.

Trubek, David, Patrick Cottrell, and Mark Nance. (2006). 'Soft Law', 'Hard Law' and European Integration: Toward a Theory of Hybridity, in Grainne De Burca and Joanne Scott, eds., *Law and New Governance in the EU and the US*, Oxford: Hart Publishing, pp. 65–96.

——Louise Trubek. (2005). Hard and Soft Law in the Construction of Social Europe: The Role of the Open Method of Coordination, *European Law Journal*, Vol. 11, No. 3, pp. 343–65.

Tsebelis, George and Geoffrey Garrett. (2000). Legislative Politics in the European Union, *European Union Politics*, Vol. 1, No. 1, pp. 9–36.

Tullock, Gordon. (1969). Federalism: Problems of Scale, *Public Choice*, Vol. 6, No. 1, 19–29.

Tutwiler, A. (1991). Food Safety, the Environment and Agricultural Trade: The Links, International Policy Council on Agricultural Trade Discussion Paper, June.

Tversky, Amos and Daniel Kahneman. (1982). Judgment Under Uncertainty: Heuristics and Biases, in Daniel Kahneman, Paul Slovic, and Amos Tversky, eds., *Judgment under Uncertainty: Heuristics and Biases* New York: Cambridge University Press, pp. 3–20.

UNEP-GEF Biosafety Unit. (2006a). *Building Biosafety Capacity: The Role of UNEP and the Biosafety Unit* (December), http://www.unep.ch/biosafety/development/devdocuments/ UNEPGEFBiosafety_BrochureDec2006.pdf, accessed on 20 June 2007.

——(2006*b*). A Comparative Analysis of Experiences and Lessons From the UNEP-GEF Biosafety Projects (December), http://www.unep.ch/biosafety/development/devdocuments/UNEPGEFBiosafety_comp_analysisDec2006.pdf, accessed on 20 June 2007.

United Nations. (2006). *Millennium Development Goals Report 2006*, New York: United Nations Publications.

United States Department of State. (2000). Fact Sheet: Report of U.S.-EU Biotechnology Consultative Forum, 19 December 2000, http://www.usembassy.it/file2000_12/alia/a0121901.htm, accessed on 11 June 2004.

——(2004). Fact Sheet: Cartagena Protocol on Biosafety, http://www.state.gov/g/oes/rls/fs/2004/28621.htm, accessed on 20 June 2007.

United States Mission to the European Union. (2003). Bush: Biotech Key to Fighting Global Hunger, Terrorism, 23 June 2003, http://www.useu.be/Categories/Biotech/June2303BushBiotech/html, accessed on 11 June 2004.

USDA (United States Department of Agriculture). (2004*a*). USDA Announces Funds to Promote US Food and Agricultural Products Overseas, USDA News Release 0234.04, 17 June.

——(2004*b*). United States Department of Agriculture, Foreign Agricultural Service. Bio Commodity Aggregations, 20 March. Available online at http://www.fas.usda.gov/ustrdscripts/USReport.exe/ accessed 20 March 2004).

——(2004*c*). USDA Announces First Steps To Update Biotechnology Regulations, USDA Release No. 0033.04, 22, Jan. 2004, http://www.usda.gov/Newsroom/0033.04.html.

——(2005). FAO and Biotechnology, *GAIN Report*, No. US5001, 9 December, http://www.fas.usda.gov/gainfiles/200601/146176552.doc.

US House of Representatives, Committee on Science and Technology, Subcommittee on Investigations and Oversight. (1984). The Environmental Implications of Genetic Engineering, US Government Printing Office: Washington, DC, Staff Report, Serial V.

US House of Representatives, Committee on Agriculture, Subcommittee on Conservation Credit, Rural Development and Research. (2003). Review of Biotechnology in Agriculture. Hearing. 17 June, 108th Congress, 1st Session, Washington DC.

USA Rice Federation. (2007). USA Rice Supports APHIS Action on Clearfield 131, but Questions Regulatory Practices, Press Release accessed at http://www.usarice.com on 19 April 2007.

Van Evera, Stephen. (1997). *Guide to Methods for Students of Political Science*, Ithaca, NY: Cornell University Press.

Van Waarden, Frans. (2006). Taste, Traditions, and Transactions: The Public and Private Regulation of Food, in Christopher Ansell and David Vogel, eds., *Why the Beef? The Contested Governance of European Food Safety*, Cambridge, MA: MIT Press, pp. 35–59.

Veggeland, Frode, and Svein Ole Borgen. 2005. Negotiating International Food Standards: The World Trade Organization's Impact on the *Codex Alimentarius* Commission, *Governance*, Vol. 18, No. 4, pp. 675–708.

Veronesi, Umberto. (2007) OGM: L'Illogico Blocco della Ricerca, *Corriere della Sera*, 15 July, p. 43.

Victor, David G. (1997). Effective Multilateral Regulation of Industrial Activity: Institutions for Policing and Adjusting Binding and Nonbinding Legal Commitments, unpublished Ph.D. dissertation, Massachusetts Institute of Technology, 184 (1997), available at: https://dspace.mit.edu/handle/1721.1/10038/ accessed on 2 April 2006).

——(2004). The Sanitary and Phytosanitary Agreement of the World Trade Organization: An Assessment After Five Years, *New York University Journal of International Law and Politics*, Vol. 32, No. 4, pp. 865–937.

Vogel, David. (1995). *Trading Up: Consumer and Environmental Protection in a Global Economy*, Cambridge, MA: Harvard University Press.

——(1997). *Barriers or Benefits? Regulation in Transatlantic Trade*, Washington, DC: Brookings Institution.

——(2001). Ships Passing in the Night: GMOs and the Contemporary Politics of Risk Regulation in Europe, *European University Institute Working Paper*, No. 2001/16, http://www.iue.it/RSCAS/WP-Texts/01_16.pdf, accessed on 14 May 2007.

——(2003*a*). The Politics of Risk Regulation in Europe and the US, *The Yearbook of European Environmental Law*, Vol. 3, New York: Oxford University Press, pp. 1–42.

——(2003*b*). The Hare and the Tortoise Revisited: The New Politics of Consumer and Environmental Regulation in Europe, *British Journal of Political Science*, Vol. 33, No. 4, pp. 557–80.

——Peter Newell and Eduardo Trigo. (2008). *A Framework for Policy-making on Trade, Agricultural Biotechnology and Sustainable Development*, Geneva: ICTSD (forthcoming).

Vogt, Donna U. (2005). Food Safety Issues in the 109th Congress, Congressional Research Service Report for Congress, updated on 16 June.

Vos, Ellen. (2000). EU Food Safety Regulation in the Aftermath of the BSE Crisis, *Journal of Consumer Policy*, Vol. 23, No. 3, pp. 227–55.

Walker, Neil. (2001). The EU and the WTO: Constitutionalism in a New Key, in Grainne de Burca and Joanne Scott, eds., *The EU and the WTO: Legal and Constitutional Issues*, London: Hart Publishing, pp. 31–58.

——(2002). The Idea of Constitutional Pluralism, EUI Working Paper Law No. 2002/1.

——(2003). *Late Sovereignty in the European Union*, in Neil Walker, ed., *Sovereignty in Transition*, London: Hart Publishing, pp. 3–32.

Walker, Vern. (1998). Keeping the WTO from Becoming the 'World Trans-Science Organization': Scientific Uncertainty, Science Policy, and Fact-Finding in the Growth Hormones Dispute, *Cornell International Law Journal*, Vol. 31, pp. 251–320.

Wallace, Helen. (2005). An Institutional Anatomy and Five Policy Modes, in Helen Wallace, William Wallace, and Mark A. Pollack, eds., *Policy-Making in the European Union*, New York: Oxford University Press, pp. 49–90.

Wallace's Farmer. (2007). Rootworm Biotech Trial Lacks Approval in Japan, NCGA Warns, *Wallace's Farmer*, 28 March.

Waltz, Kenneth N. (1979). *Theory of International Politics*, Reading, MA: Prentice-Hall.

Waterfield, Bruno. (2006). EU Ducks GM Rules, *eupolitix.com*, 10 March.

Wearth, Spencer. (1998). *Nuclear Fear: A History of Images*, New York: Cambridge University Press.

Weber, Max. (1947). *The Theory of Social and Economic Organization*, Talcott Parsons, ed., New York: The Free Press.

——(1958) (1904/05 original). *The Protestant Ethic and the Spirit of Capitalism*, New York: Charles Scribner Sons.

Weiler, Joseph. (1999). Epilogue: 'Comitology' as Revolution: Infranationalism, Constitutionalism and Democracy, in Christian Joerges and Ellen Vos, eds., *EU Committees: Social Regulation, Law and Politics*, Oxford: Hart Publishing, pp. 339–50.

References

Weintraub, Arlene. (2006). Online Extra: Salmon that Grow Up Fast, *Business Week Online*, 16 January, http://www.businessweek.com/magazine/content/06_03/b3967111.htm, accessed on 1 May 2007.

Wendt, Alexander. (1999). *Social Theory of International Politics*, New York: Cambridge University Press.

Wenger, Etienne. (1999). *Communities of Practice: Learning, Meaning and Identity*, New York: Cambridge University Press.

——Richard McDermott, and William M. Snyder. (2002). *Cultivating Communities of Practice*, Cambridge, MA: Harvard Business School Press.

Weiss, Rick. (2006a). Gene-Altered Profit-Killer: A Slight Taint of Biotech Rice Puts Farmers' Overseas Sales in Peril, *Washington Post*, 21 September, p. D1.

——(2006b). Biotech Rice Saga Yields Bushel of Questions for Feds: USDA Approval Shortcut Emerges as Issue, *Washington Post*, 6 November, p. A3.

——(2006c). FDA May Clear Cloned Food, but Public Has Little Appetite: Despite Safety Data, Americans Largely Find Idea Unappealing, *New York Times*, 25 December, p. A16.

——(2008a). Food from Clones Safe, E.U. Draft Says, *The Washington Post*, 12 January, p. A6.

——(2008b). FDA Says Clones are Safe for Food, *The Washington Post*, 15 January, p. A1.

White House. (2008). *Transatlantic Economic Council Report to the EU-U.S. Summit 2008*, White House Press Release, http://www.whitehouse.gov/news/releases/2008/06/200 80610-4.html, accessed on 14 July 2008.

Wiener, Jonathan B. and Michael D. Rogers. (2002). Comparing Precaution in the United States and Europe, *Journal of Risk Research*, Vol. 5, No. 4, pp. 317–49.

Wikipedia Contributors. (2008). The Home Depot, *Wikipedia, The Free Encyclopedia*, 27 February 2008, 21:56 UTC, http://en.wikipedia.org/w/index.php?title = The_ Home_Depot&oldid = 194511073, accessed 28 February 2008.

Wilks, Stephen. (2007). The European Competition Network: What Has Changed? Paper Presented at the Biennial Conference of the European Union Studies Association, Montreal, CA, 17–19 May.

Williamson, Oliver. (1967). Hierarchical Control and Optimum Firm Size, *Journal of Political Economy*, Vol. 75, No. 2, pp. 123–38.

Wilson, Janet. (2006). EPA Panel Advises Agency Chief to Think Again: Irate Scientists Say the Administrator Ignored or Misconstrued Their Recommendations in Proposed New Rules on Soot and Dust Pollution, *Los Angeles Times*, 4 February, Section: California Metro.

Winham, Gilbert. (2003). International Regime Conflict in Trade and Environment: The Biosafety Protocol and the WTO, *World Trade Review*, Vol. 2, No. 2, pp. 131–55.

Winickoff, David, Sheila Jasanoff, Lawrence Busch et al. (2005). Adjudicating the GM Food Wars: Science, Risk, and Democracy in World Trade Law, *Yale Journal of International Law*, Vol. 30, No. 1, pp. 81–123.

Wirth, David. (1994). The Role of Science in the Uruguay Round and NAFTA Trade Disciplines, *Cornell International Law Journal*, Vol. 27, pp. 817–59.

——(1997). International Trade Agreements: Vehicles for Regulatory Reform? *U. Chicago Legal Forum*, Vol. X, pp. 331–73.

World Bank. (2003). *World Development Report 2003*, Washington, DC: World Bank.

——(2007). *World Development Report 2008: Agriculture for Development*, Washington, DC: World Bank.

World Health Organization (WHO). (2005). *Modern Food Biotechnology, Human Health and Development: An Evidence-Based Study*, http://www.who.int/foodsafety/publications/biotech/biotech_en.pdf (accessed 22 June 2008).

World Trade Organization (WTO). (2005a). *International Trade Statistics 2005*, available at http://www.wto.org/English/res_e/statis_e/its2005_e/its2005_e.pdf, accessed on 25 May 2006.

——(2005b). *Annual Report 2005*, http://www.wto.org/english/res_e/booksp_e/anrep_e/anrep05_e.pdf (accessed 22 June 2008).

Wynne, Brian. (1992). Uncertainty and Environmental Learning: Reconceiving Science and Policy in the Preventive Paradigm, *Global Environmental Change*, Vol. 2, No. 2, pp. 111–27.

Yerkey, Gary G. (1999). US and EU Agree to Set Up Forum with NGOs to Review Biotech Dispute, *International Trade Reporter*, Vol. 16, p. 2025.

——(2004). Protectionist Pressures in U.S. Forcing Bush to Ignore WTO Obligations, EC Says, *International Trade Reporter*, Vol. 21, p. 19.

Young, Oran. (1994). International Environmental Governance, in Oran Young, ed., *International Governance: Protecting the Environment in a Stateless Society*, Ithaca, NY: Cornell University Press, pp. 12–32.

Young, Alasdair R. (2003). Political Transfer and 'Trading Up'? Transatlantic Trade in Genetically Modified Food and US Politics, *World Politics*, Vol. 55, No. 4, pp. 457–84.

——(2005). The Single Market, in Helen Wallace, William Wallace, and Mark A. Pollack, eds., *Policy-Making in the European Union*, New York: Oxford University Press, pp. 93–112.

Zeitlin, Jonathan and Philippe Pochet, with Lars Magnusson, eds. (2005). *The Open Method of Coordination in Action: The European Employment and Social Inclusion Strategies*, Brussels: P.I.E.-Peter Lang.

Zürn, M. (2000). Democratic Governance Beyond the Nation-State, in Michael Th. Greven and Louis Pauly, eds., *Democracy Beyond the State? The European Dilemma and the Emerging Global Order*, Lanham, MD: Rowman & Littlefield, pp. 91–114.

Subject Index

accountability 94, 207-209, 264, 290-91
ACP (African-Caribbean-Pacific)
  countries 163, 209, 214
Administrative Procedure Act (US 1946)
  49, 52
adventitious presence 156, 231-2, 242, 250,
  251, 263, 268, 275
  zero-tolerance policy toward 225,
  226, 257
advertising campaigns 67, 103, 206
Africa, *see* ACP; Egypt; Ethiopia; Madagascar;
  Mozambique; South Africa; Zambia;
  Zimbabwe
*Agence Française de Sécurité Sanitaire
  des Aliments* 74-5, 258
agribusiness 36, 37, 71
agricultural biotechnology 17-19, 28, 29,
  69, 76, 102, 107, 120, 125, 209
  attitudes toward 75
  Codex work on 164
  cooperative efforts in 285
  disputes over 280, 292
  distributive conflicts that
    characterize 5-6
  effective global governance of 281
  EU legislative framework for 245
  future of 8, 279-305
  OIE in 159
  practices (since 2000) 235
  role of intellectual property rights in 299
  transatlantic conflict over 143, 230
  transatlantic networks for 104-8, 281
agricultural biotechnology regulation 3-5,
  7, 9-10, 13, 15, 33-83, 85, 108,
  114, 130, 181, 199, 214-215,
  266, 279, 298
  developing countries and 29
  multilateral regimes and 136-74
  agricultural ministries 157, 159, 161
Agriculture Council (EU) 251, 255, 257
agriculture 37, 73, 279
Agrifood Biotech Group 105, 111
allergenicity 36, 50, 167, 170, 190, 267
American Meat Institute 272

American Soybean Association 183, 244
Amflora potato 255, 257
Amsterdam Treaty (1997) 56
anarchy 3, 121, 280
antibiotic resistance 36
anti-American sentiment 76
anti-globalization activists 65, 66, 75,
  76, 221
anti-GM activism 45, 70, 129, 181, 182,
  212, 236, 247, 256, 257, 263, 276,
  288, 298, 303
  spread from Europe to rest of
    world 116
anti-WTO movements/demonstrations 66,
  180-1
APHIS (USDA Animal Plant and Health
  Inspection Service) 47-8, 262, 264-5,
  269-71, 274, 278
Appellate Body (WTO) 22, 134, 184-7, 194,
  195, 198, 199, 209, 216, 218, 221-2,
  223, 228, 230, 291
approvals 1, 2, 9, 22, 29, 62, 68, 72, 74-75,
  79, 82, 116, 191, 230, 249, 253-8, 260-
  261, 265, 294, 296
  cultivation 255-8
  de facto moratoria on 20, 21, 24, 67, 72,
    182-3, 193, 277, 296
  delayed 272
  mandatory 27, 107, 108
  premarket 27, 50, 63, 108, 268, 273,
    281, 287
  regulatory 264, 265, 273, 294, 297
  resumption of 235, 245-7, 248, 277
  undue delay in process 187, 297
Argentina 83, 116, 170, 182, 187, 188, 190,
  226, 252, 265, 296, 299, 300, 301
  GM soy adopted 28-9, 294
  WTO complaint (2003) 2, 20,
    177, 183, 298
Asilomar conference (1975) 45, 139
Australia 116, 163, 170,181
Australia-Salmon case 186
Austria 67, 239, 240, 242, 246, 249, 251,
  257, 259, 260, 295